NANOTECHNOLOGY RESEARCH DEVELOPMENTS

NANOTECHNOLOGY RESEARCH DEVELOPMENTS

R. JIMÉNEZ-CONTRERAS
EDITOR

Nova Science Publishers, Inc.
New York

For permission to use material from this book please contact us:
Telephone 631-231-7269; Fax 631-231-8175
Web Site: http://www.novapublishers.com

Library of Congress Cataloging-in-Publication Data

Nanotechnology research developments / R. Jiménez-Contreras, Editor.
p. cm.
Includes index.
ISBN-13: 978-1-60021-899-6 (hardcover)
ISBN-10: 1-60021-899-7 (hardcover)
1. Nanotechnology. 2. Nanostructures. I. Jiménez-Contreras, R.
T174.7.N37524 2007
620'.5--dc22 2007029346

Published by Nova Science Publishers, Inc. ✢ New York

CONTENTS

PREFACE

Nanotechnology is a "catch-all" description of activities at the level of atoms and molecules that have applications in the real world. A nanometer is a billionth of a meter, about 1/80,000 of the diameter of a human hair, or 10 times the diameter of a hydrogen atom. Nanotechnology is now used in precision engineering, new materials development as well as in electronics; electromechanical systems as well as mainstream biomedical applications in areas such as gene therapy, drug delivery and novel drug discovery techniques. This book presents new and important breakthroughs in the field from around the world.

Chapter 1 - Nanotechnology quite certainly will become an important field in the science and technology of the next decades, allowing one to manipulate matter at the nanometer scale and to incorporate nanostructure and nanoprocesses into working technological innovation. Dendrimers are branched, synthetic polymers with layered architectures that show a fast growing field covering many areas of chemistry. Dendrimers are considered as a modern and an elegant class of branched macromolecules in the nanoscale dimension. They resemble treelike molecular architectures since they are built from repetitive monomers with branching point units that are radially connected around a template core. The hydrodynamic volume is dictated by the core, monomer units, branching points, dendrons, generation numbers, and peripheral functions. Dendrimers are discrete macromolecules with a high degree of molecular uniformity and monodispersity. Those features make for an easy characterization and analysis of their tailor-made properties. They are often macromolecules with few defects, whose dense outward (and/or inward) functional groups generate specific properties and functions such as a polyvalent ligand, a molecular receptor, molecular confinement, and a molecular translocator (vector). Multivalency as well as adhesive, amplification, recognition, additive, and cooperative effects have been observed. The expression "dendritic effects" was coined to illustrate those specific chemical behaviors. Dendrimers that used to be employed as standard, innovative stabilizers, media, ligand and supports have appeared to provided remarkable results with extremely high metal loading. By regulating dendrimer synthesis, it is possible to precisely manipulate both their molecular weight and chemical composition, thereby allowing predictable tuning of their biocompatibility and pharmacokinetics. Advances in our understanding of the role of molecular weight and architecture on the in vivo behavior of dendrimers, together with recent progress in the design of biodegradable chemistries, has enabled the application of these branched polymers as anti-viral drugs, tissue repair scaffolds, targeted carriers of chemotherapeutics and electrochemical sensors. Various types of dendrimers differing in composition and structure are mentioned, together with their practical use spanning from catalysis, transport vehicles to synthetic vaccines. This paper discusses the research advances development of functional and multifunctional dendrimeric and hyperbranched polymers with the objective of being applied in catalysis, biosensing, imaging, drug and gene delivery, nanoprinting, self-assembly, energy transfer, etc.

Chapter 2 - The perovskite nanoparticles such as BaTiO3, SrTiO3, BaZrO3 and their solid solutions can be prepared using a new method, DSS, near room temperature. Hydrogen interstitials are introduced into the lattice of perovskite nanoparticles synthesized by DSS method so that the nanoparticles take positive charges and adsorb the hydroxyl group with negative charges on the surface. Hydroxyl groups adsorbed on the surface of nanoparticles can be ionized freely, and the nanoparticles exhibit an alkalescent property. When this kind of perovskite nanopaticles disperses in a certain solution of inorganic salt such as Mg(NO3)2•6H2O, Zn(NO3)2•6H2O, Al(NO3)3•9H2O, Y(NO3)3•6H2O etc., the hydroxide related to these salts spontaneously deposit on the surface of the nanoparticles, as a result, a perovskite structure nanoscaled core materials are successfully self-wrapped by a non-ferroelectric hydroxide derived from a nitrate salt solution. After annealing at high temperature, core-shell structured perovskite nanoparticles with heterogeneous oxide shells can be synthesized. When the core-shell structured perovskite nanoparticles with heterogeneous oxide shells are composited with doped shell oxide materials, the pinning effect during sintering can be avoided, as a result, the sintering temperature are decreased and dense ceramics are obtained. The core-shell structures are analyzed by a conjunction of XRD, HRTEM and FE-SEM. The lattice match of shell and core in our ceramics of MZO wrapped BST nanoparticles probably follows the second match mode. The tunability and FOM values of the ceramics of MZO wrapped BST nanoparticles have been improved significantly compared with that of pure BST ceramics. The ceramics of MZO wrapped BST nanoparticles are promising materials for tunable microwave device applications and high energy density ceramic capacitors for their distinct dielectric loss improvement.

Chapter 3 - The universal models of nucleation thermodynamics and growth kinetics have been established for nanowire growth upon metal-catalyst-assisted thermal chemical vapor transport (CCVT) on the basis of vapor-liquid-solid (VLS) mechanism, with respect to the size-effects induced by the curvature of critical nuclei and nanowire in the nucleation and growth. Based on the proposed thermodynamic and kinetic analyses on nanoscale, the thermodynamic and kinetic size limit of the nanowire growth was addressed. It was found that the critical size of the nanowire nucleation is suggested to be the thermodynamic size limit, i.e., the theoretical minimum size of the nanowire grown by CCVT both thermodynamically and kinetically, and the size of the catalyst droplet could limit the growing size, i.e., the theoretical possible or maximum size of the nanowire to a certain extent in kinetically. For the issue of catalyst nanoparticles on nanowires tip or substrate, a nucleation thermodynamic criterion and a diffusion kinetic criterion were deduced for the selection growth of catalyst nanoparticles on nanowires tip or substrate upon CCVT. Importantly, theoretical predictions on the basis of the proposed models are in well agreement with the experimental data. Additionally, a growth kinetic approach was developed to theoretically elucidate the growth of one-dimensional nanostructures inside carbon nanotubes upon vapor deposition. Theoretical analysis showed that the growth rate of one-dimensional nanostructures would go much higher once nuclei formed inside carbon nanotubes, due to the effect of surface tension induced by the nano-sized curvature of carbon nanotubes. Theoretical results are in well agreement with the experimental data. The size-dependence of the shape evolution of nanowires such as SnO2 was pursued in theoretical. It was found that the thermodynamic driving force, the Gibbs free energy difference between two phases, is always responsible for the shape choice in the different growth stages such as nucleation, growth, and structural transition of 1D nanostructures, implying that the thermodynamic driving force is one of the physical origins causing the interesting morphological transformation in the growth of 1D nanostructures.

Chapter 4 - Because of the extremely high chemical reactivity and hence the considerably rigorous requirements for the equipments to preserve a high purity of the product, the preparation and characterization of nanostructured pure rare-earth metals are still a big challenge in nanoscience and nanotechnology. In this chapter, followed a thermodynamic description of the energy state and the thermal stability of rare-earth nanoparticles and nanocrystalline bulks, the practical preparation and characterization of nanoscale rare-earths are depicted in detail. Different from a number of existing techniques that have been developed to synthesize nanocrystalline bulk materials, a home-designed configuration of the "oxygen-free" in-situ synthesis system, where the inert-gas condensation is combined with the SPS technology into an entirely closed and vacuumized system, is mainly introduced. With the technically designed route of sequentially performed processes, i.e. amorphization of nanoparticles, nucleation and growth of the short-range ordered "clusters" inside the nanoparticle, and the complete nanocrystallization, the ultrafine nanocrystalline bulk can be obtained with the grain size much smaller than the initial nanoparticle size. Microstructures and properties of some typical members of the rare-earth family are presented ordinally. The prominently changed thermal, physical and mechanical properties imply the promising applications of the nanocrystalline rare-earth metals and alloys in the fields of functional and structural materials. Finally, although the chapter is focused on rare-earth nanoparticles and nanocrystalline bulks, the proposed thermodynamic models and preparation techniques are potentially applicable to a large variety of nano metal materials, hence may establish a new way for studies on microstructures and performances of metallic nanomaterials, especially of those highly reactive in the air.

Chapter 5 - Porous silicon (PS) films consisting of many pores and pillars are widely used to yield efficient visible photoluminescence (PL) and electroluminescence (EL) at room temperature. Such light-emission behaviors are primarily attributed to electron confinement in the nanocrystals that constitute the PS film. Also, the PS film shows a unique electrical property of negative difference conductance (NDC) for the carrier mobility difference in different sizes of pores and pillars.

The techniques of PS formation have been developed by many different methods. Among them, electrochemical anodization is the most commonly used. The anodization is performed in a hydrogen fluoride (HF) solution with an anodic current on the sample. However, it is very difficult to control the manufacturing parameters precisely in the manufacturing processes. Especially in nano-scale, the anisotropic etching or selective etching cannot be easily achieved in the wet-etching method. However, it is well known that nano-scaled PS (NPS) shows light emission more readily in the visible range than macro-PS does. Therefore, precise control of the NPS formation is very important.

In this chapter, the authors introduce three novel manufacturing technologies. In the first method, the Hall-effect is applied in the electrochemical anodization. By Lorentz force, the carriers can be well swept to the allocated areas. In the second method, a forward biased pn-junction is used for bottom-hole assistance. The NPS films with nano-scaled pores and high-aspect-ratio pillars can be formed. In the third method, the authors proposed a simple and mask-free method for the fabrication of patterned PS using a localized electric field. The electric field is applied by patterned electrodes (anode and cathode) which are placed horizontally underneath the sample. No masking-layer or related photo-lithography processes are needed in this method. Strong visible photoluminescence emissions in PS can be obtained by these novel methods.

In addition, the surface roughness of NPS leads to a high contact resistance to the metal layer of the applications on EL, and hence a waste of an electrical power. To enhance the EL efficiency and stabilize performance, a good metal contact to the light-emitting layer (NPS layer) is required. In the authors' study, a supercritical fluid (SCF) technique is explored to overcome the processing problem on metal contact. The SCF is a liquid or gas material used in a state above the critical temperature and critical pressure where gases and liquids can coexist. The use of a CO2 SCF and silver (Ag) nanoparticles is shown to improve the electrical contact between the metal and NPS. Therefore, the contact resistance can be largely reduced, and the power consumption of EL on the PS structure can be greatly reduced by the method.

The morphology, porosity, and photoluminescence of the NPS prepared by the proposed method are investigated. This novel technology will have much benefit in the applications of nano-electronic and opto-electronic devices.

Chapter 6 - Emission in mid-infrared range, 2 – 5□m, has very important applications in variety of areas including range-finding, laser surgery, remote trace-gas sensing, secure free-space communications, etc. Owing to reduction of the dimensionality carriers' confinement, quantum dots (QDs) based optoelectronic devices, e.g. semiconductor lasers, show novel properties when compared with their counterpart quantum well based devices.

In this article, self-assembly InAs quantum dots grown on InGaAs/InP matrix for mid-infrared emission using metal-organic vapor phase epitaxy is introduced. Effects of the growth conditions on the dot density and size uniformity of the quantum dots have been investigated in detail. A two-step growth method for growing high density and uniform quantum dots has been introduced. Improvement in QDs density and size uniformity in the two-step growth has been discussed in details. By changing the indium content of the InxGa1-xAs matrix layer of the InAs/InxGa1-xAs/InP quantum dot structure, emission wavelength of the InAs QDs structure has been extended to the longest >2.35 □m measured at 77K. Full-width at half-maximum (HWHM) of the photoluminescence emission spectrum of the InAs quantum dot structure was measured as narrow as 25.5 meV.

Chapter 7 - This chapter focuses on novel gas/chemical sensing systems that are based on nanotechnology. Recently, there has been an extensive interest in nanomaterials such as carbon nanotubes, nanowires, nanofibers, and nanoparticles. Their application on gas sensing systems has also been widely studied. Due to their unique micro/nano porous structures and ultra high surface-to-volume ratio, nanostructured materials have a high potential to enhance conventional gas sensors' performance, including the sensitivity, the operation temperature and the recovery time. In this chapter, various nanomaterials-based gas/chemical sensors are presented, including their structure, fabrication, and operation principles.

Chapter 8 - Over the past decade, liquid phase pulsed laser ablation (LP-PLA) has been used as an effective method for preparation of many nanostructured materials. This technique involves the interaction of a laser with a solid target and surrounding liquid. The laser causes the target surface to vaporise in the form of an energetic ablation plume, containing species such as atoms, ions, and clusters, travelling with high kinetic energy. These can collide and react with molecules of the surrounding liquid phase, producing new materials, often on the nanometre scale. Due the intensity of the laser and the nanosecond timescales, high temperature, high pressure, and high density conditions within the reaction volume are produced which provide a 'brute force' method of synthesising novel materials that have hitherto been inaccessible using milder, more conventional techniques. In this chapter, the authors discuss more recent progress in the synthesis of hierarchical, complex, and ordered crystal architectures by this novel technique. It is believed that the ability of nanostructures to self-assemble in this manner provide potential building blocks for microscale or nanoscale devices. The authors show that for C3N4, a Group IV-V covalent compound, a range of self-assembled arrays of nanoparticles, nanorods, nanoleaf, and mesoflowers can be produced, and this also suggests that the route described here can be applicable to the synthesis of nanostructures of other similar materials. Therefore, the authors expect LP-PLA may play a key role in advancing complex nanomaterials into many applications in optics, electronics, biology, medicine and energy/chemical conversion.

Chapter 9 - The bioinspired technique has developed into an intersection of materials science, soft chemistry technique, and biotechnology. In this chapter, the bioinspired strategies integrating biotemplate, biomineralization, and biomimesis are adopted to fabricate functional nanomaterials with hierarchical architectures and interrelated unique properties.

Some natural biomaterials, such as eggshell membrane, silk fiber, etc., are introduced into different liquid processes to construct a series of functional hierarchical nano-architectures (containing oxides, chalcogenides, noble metals, and/or their composites). Under the direction of the biotemplates or biosubstrates, the nanocrystallites units of about 2-10 nm are formed, developed, and assembled into hierarchical architectures from the nanoscale to the macroscale. The nucleation, growth, and assembly of nanocrystallite units depend on the collaboration of special proteins structures and functional biomacromolecules of relevant biomaterials as well as the processing conditions. Moreover, the biomaterials usually provide both the physical substrates and the inductive substances to direct some complicated reactions. In addition, some amino acids of the biomaterials function as the capping agents to achieve the organized growth and assembly of small-sized nanocrystallites. As-synthesized hierarchical nanoarchitectures take on outstanding chemical or physical properties and have valuable applications in photocatalysis, gas sensing, ductile ceramics, and semiconductor technology.

This method and the relevant ideas provide a novel and versatile avenue to synthesize a new family of functional nanomaterials by integrating material science, chemistry, and biotechnology.

Chapter 10 - Self-assembled monolayers (SAMs) present a 2D regular lattice, and the lateral force measured with an atomic force microscopy (AFM) scanning probe tip displays the typical stick-slip behavior. In this paper, the authors study the possible connection between an AFM induced damaged SAM surface and irregular behavior of the sticking phases.

In fact, the repeated passages of an AFM probe tip on an SAM surface can produce wear stripping off radicals from the carbon chain backbones, and irregularities of the SAM lattice take place. As a consequence, the frictional properties are changed because new extra space is made available to the final alkanethiol chain groups and the stick and slip phases are not time invariant and hysteresis between the forward and backward scanning direction is altered. The paper discusses if and how such friction irregularities can monitor the progressive modifications of the SAM worn surface.

Chapter 11 - Photovoltaic behavior in both organic thin films and organic/inorganic hybrid systems is an important subject not only in basic research but also in application fields such as organic solar cells, dyes sensitized solar cells etc. In order to improve the light collection efficiency, tuning film morphology is an efficient way to improve light harvest, then conversion efficiency. In this paper, some fabrication methods for different photoanodes with controlled morphology and their photovoltaic behaviors will be reviewed. While using electro-hydrodynamics (EHD) technique to prepare lotus leaf-like photoanodes will be discussed in detail to illustrate the effect of micro-/nano- composite structure on the enhancement of photo-electric conversion behavior in dye-sensitized solar cells (DSSCs). Further more, there will some discussion on the morphology control for the organic thin film system to increase photocurrent generation.

In: Nanotechnology Research Developments
Editor: R. Jiménez-Contreras, pp. 1-5

ISBN: 978-1-60021-899-6

Expert Commentary

Nanostructured Polymers: Developments and Trends, Highlights and Shadows

Maria Vittoria Russo and Ilaria Fratoddi
Department of Chemistry, University of Rome"
Sapienza", P.le A.Moro 5, 00185 Rome, Italy

The world of nano-materials burst out quite recently as a novelty in the scientific community, although macromolecules at the nanoscale have been built up by chemists and biologists and quantum effects have been studied by physicists since decades. Nowadays, since the growth of nanomaterial phases and structures can be controlled, applications in nanotechnology related to their chemical and physical properties are almost ready to be achieved. Dendrimers, fullerenes, carbon nanotubes and graphenes, metal and ceramic nanoparticles are all available for the development of advanced technological instruments. Atomic force and scanning probe microscopes, atomic-layer-deposition devices and nanolithography tools can manipulate matter in the atomic or molecular regime. These nanotools are able to control matter at the nano or atomic regime.

The first nanodevices on the market were quantum dot fluorescent biodetectors. MEMS devices are used as accelerometers in automotive airbags. Many other promising applications are pending such as nanoelectric memory devices, nanosensors, catalysts and drug delivery systems. Components for nanodevices envisage the use of semiconducting organic molecules, polymers and composites.

Among the variety of nano-materials, polymers are suitable for multipurpose technologies because their chemical structure can be varied to achieve functionalities that are improved by the nanosize morphology. Nanostructured polymers are easily synthesized in bulk amounts, while classical colloids or other nano-objects need the aid of ancillary techniques for the achievement of their nano shape. Moreover, unforeseen properties may be revealed merely because the polymers are formed in this peculiar structure. Indeed, nanoscience applied to polymers leads to a vision of unity of knowledge, involving integration of chemistry, physics, engineering and a bottom-up mastery of matter. Through two different mastering approaches, i.e. bottom-up and top-down, in principle any type of polymer can be prepared in a nanostructured fashion and some representative examples, which however are not exhaustive, will be reported in this commentary.

Polyamides, polystyrene and polymethylmethacrylate (S.Gross *et al.*, Eur. Polym. Journal (2007) 43, 673) are common multipurpose synthetic polymers used in a wide range of photonic crystals, gel opals, medical carriers, chromatographic industrial settings and consumer products such as spacers for display and biomedical media.

There have been many reports on the methods of preparation of their nanoscale particles, mainly concerning optimization of the emulsion polymerization conditions (bottom-up approach). Variations of temperature, reaction time, solvent/co-solvent ratio, addition of surfactants, water content as the dispersion medium, nature of the initiator, ionic strength of the suspension, functionalization of the monomer with hydrophilic/hydrophobic groups, allow the control of the shape, of the monodispersion and of the wideness of ordered domains for optical purposes. In general, the shape of nanoparticles for these materials is a sphere due to the isotropic surface energy. Otherwise, polymer nanofibers (with submicron scale diameters) can be formed with the electrospinning process. As a curiosity, some peculiar morphologies, such as raspberry-like nanospheres can be tuned for polymer composites obtained from a charge compensation process (Gouliang Li *et al*, J. Colloid and Interface Sci., (2006) 297, 705) and for magnetic polystyrene microspheres (Zhizhong Xu *et al*, Mat.Chem.Phys., (2007) in press).

Polymers, unlike other materials, are not considered suitable materials for hydrogen storage; however, when a intrinsic microporosity can be achieved by tailoring the chemical structure (for example, made from fused-ring subunits) so that large areas of free space are made available, hydrogen physisorption occurs rapidly and reversibly in amounts that compete with those of activated carbon (McKeown *et al.*, Angew.Chem.Int.Ed., (2006) 45,1804). A sector apart is represented by carbon nanotubes (CNT), which can be considered polymeric carbon and show a wide range of remarkable properties, such as high charge-carrier mobility and high mechanical strength associated with semiconductivity. The scientific literature on SWCNT and MWCNT is extremely ample and these materials are now finding successful commercialization, despite their cost.

The versatile properties of π-conjugated polymers, such as solubility, semiconductivity and easy processability, have led to the development and improvement of nanoscale organic transistors applied to chemical and biological sensors and light emitting diodes (A.Dodabalapur, Mat.Today (2006) 9(4), 24). The research on the synthesis and properties of polymeric semiconductors has been carried out thoroughly since the last two decades, because these materials are very promising and multipurpose for the achievement of flexible electronics. The fundamental figure of merit of charge transport is carrier mobility which is a prerequisite for the technological application of conjugated polymers. Despite the considerable amount of excellent experimental and theoretical work produced so far, there is still not a clear knowledge on what limits the charge transport in thin films of conjugated polymers. A basic picture could allow a universal view of the correlation between morphology, microstructure and electrical properties. In this contest, some indications come from the studies concerning the formation of crystallites as large as a few nanometers of a polythiophene with a particular chemical structure, i.e. PBTTT, which shows the best performance in semiconducting properties for this class of materials just because of its peculiar lamellar morphology (M. Heeney *et al.*, J.Am.Chem.Soc., (2005) 127, 1078; I. McCulloch *et al.,* Nat.Mater., (2006) 5, 328).

As a further example, among conducting polymers, polyaniline is also a carefully investigated material due to its excellent optical and electronic properties. Recently much research has been conducted on the nanostructure of polyaniline (PANI) because it exhibits enhanced performance in applications (sensors, catalysis) where a high surface contact area is needed between the nanostructures and its environment. To this purpose, polyaniline nanostructured particles i.e. rods, wires, fibres, spheres, belts, with unusual physical and chemical properties have been produced by various strategies including template synthesis, interfacial polymerization, self-assembly and stepwise electrochemical deposition. Although various methods for the synthesis of nanostructures have been proposed, yet there is lack of full understanding of the rationale which drives the nanostructure formation and most of the reports are based on phenomenological experiments. An attempt to explain the reasons about forming polyaniline nanofibers was suggested by applying a shaping theory based on hydrogen bond formation between polymer chains and intercalated alcohol molecules (J.Kan *et al.*, Eur. Polym. Journal, (2007) 43, 395). However, this theory cannot attain to every type of polymers.

A relationship between the supramolecular interactions between the individual π-conjugated molecules/polymers and the microscopic morphology appearing upon aggregation has been proposed recently. Approaches to control the self-assembling of conjugated systems in the solid-state have been developed with a joint experimental–theoretical approach, combining scanning probe microscopy characterization of thin deposits and force-field molecular modeling of supramolecular aggregates and adsorbates. With this methodology, a strong correlation is established between the polymer molecular structure, the degree of supramolecular order and the solid-state photoluminescence properties (for light-emitting applications), or the charge transport properties (for the design of field-effect transistors). (Ph.Leclerc *et al.*, Mat.Sci.Engin. R, (2007) in press).

It is expected that the use of nanosized polymers will further enhance the performance of these electronic devices and represents the challenge of the future, providing a real breakthrough in this field of science and technology. However, more effort in research has to be done for the assessment of a general pathway allowing the achievement of a desired nanostructure for any kind of polymeric material.

Block copolymers are able to self-assemble in ordered structures at the molecular scale with no phase separation, to form domain dimensions of 5-50 nm. Due to this peculiar property, these materials can be used for the fabrication of templates which give rise to patterns of periodic arrays of spots; these patterns based on block copolymers in some cases are being commercialized for the development of quantum dots, magnetic storage and flash memory devices, semiconductor capacitors, nanowires, photonic crystals and nanopores, besides the premises for integrating these materials into a conventional lithographic process by a bottom-up mode (M.P. Stoykovich and P.F.Nealey, Mat. Today (2006) 9(9), 20; *ibidem* M.Li and C.K.Ober, page 30).

An up to date and controversial topic in medicine has to deal with the growth and differentiation of stem cells which are able to repair damaged tissues and organs. The expectations of people for the resolution of health problems by the use of these cells are high, although accompanied by some ethic concerns. In the field of tissue engineering , of primary importance is the choice of a proper scaffold which enables the cell attachment and growth, mimicking the *in vivo* environment. In this contest, ideal scaffolds can be biocompatible nanostructured polymers e.g. natural proteins, polysaccharides and synthetic polymers. Some examples will give a glance on this topic. Synthetic hydrogels like polyethylene glycol, are water swollen polymeric networks, with micro/nano pores particularly suited for tissue engineering, because their surface can be modified to match with protein sequences of specific cell types. Peptide building blocks can self-assemble into nanofibers through hydrogen bonds and π-electron interactions, so that the hydrophobicity of the scaffolds can be driven by the types of amino acid used (N.D.Evans, E.Gentleman, J.M.Polak, Mat.Today, (2006) 9 (12), 26; *ibidem* J.R.Jones, page 34).

The electrospinning technique can be successfully used to form a three-dimensional hyaluronic acid nanofibrous scaffold which mimics the architecture of a natural extracellular matrix. NIH 3T3 fibroblasts were seeded on these fibronectin-adsorbed nanofibrous scaffolds showing that upon attachment the fibroblasts exhibit a dendritic morphology. Thus, potential applications of nanofibrous scaffolds in cell encapsulation and tissue regeneration is suggested also in this case (Yuan Ji *et al.*, Biomaterials (2006) 27(20) 3782).

Scaffolds based on a synthetic polymer can be prepared by a different way. A colloidal template-assisted method allows to obtain two-dimensional nanopore arrays, consisting of hydrophilic SiO2-like holes within hydrophobic poly(hydroxymethylsiloxane) (PHMS) surfaces. The nanostructured films are able to induce a general increase of micrometer width adhered cells, e.g. bacteria Pseudomonas aeruginosa, with respect to the unmodified hydrophobic surfaces and to endorse a spot grown of biofilm-like aggregates (C.Satriano *et al.*, Materials Science and Engineering C, (2006) 26, 942).

In general, the presence of nanoscale features increases cell attachment and proliferation. However, protein attachment and cell response to three dimensional nano-materials at the interface needs deeper insight and is still a great challenge for scientists.

Natural nanostructured polymers are also excellent carriers of catalyst such as enzymes. For example, spherical chitosan nano-particles prepared by the ionization gelation method are used for the immobilization of neutral lipase. This hybrid material exhibits improved catalytic performance in comparison with the free neutral lipase, i.e. remarkably improved stability properties to various parameters, such as temperature, reuse, and storage time. (Zhen-Xing Tang *et al.*, Materials Letters (2007) 61,37).

Among the important topics attaining at nano size, particularly appealing for technological applications are polymer-based nanocomposites. These materials are in some way apart from the main field of nano-polymers, since the nanostructure which provides new mechanical and physical properties to the final material, is a feature of the fillers. A recent review article reports the most investigated classes of nanocomposites with special attention to structure-property relationship (S.C.Tjong, Mat.Sci.Engin.R, (2007) 53, 73).

The reported spots on the world of nanostructured polymers may give an idea of the expectations arose from this field of science. In the ideal framework of structurally well defined nano-polymers there is still quandary about the number and complexity of methods for achieving architectures hierarchically organised in terms of structure and functions. At the state of art, too many parameters may affect a desired morphology and it is difficult to predict which one is the clue.

It must be mentioned that there is some concern in the society on the risks for health connected to the use of nanoparticles. At the moment, no evidence of potential hazard due to the interaction of nanoparticles with cells and organic tissues has been detected and nanosized polymers are probably safe. However, this topic cannot be ignored and needs to be investigated and submitted to regulation.

Looking to the future, there is no doubt that this amazing size discovered for new and known polymers, born from the fruitful activities in this research field still in its infancy, will open a land of promising applications in many areas: optics, electronics, energy, environment, biology, medicine, functional smart coatings, fuel and solar cells, catalysts, sensors, etc, that could not be all cited in this commentary, but are open to the fantasy and the curiosity of scientists.

In: Nanotechnology Research Developments
Editor: R. Jiménez-Contreras, pp. 7-93

ISBN: 978-1-60021-899-6

Chapter 1

ADVANCE AND APPLICATION PROSPECT OF DENDRIMERS

Bifeng Pan and Daxiang Cui*

Department of Bio-Nano Science and Technology,
National Dey Laboratory of Nano/Micro Fabrication Technology,
Key Laboratory for Thin Film and Microfabrication of Ministry of Education,
Institute of Micro and Nano Science and Technology,
Shanghai Jiao Tong University, 800 Dongchuan Road,
Shanghai 200030, China.

ABSTRACT

Nanotechnology quite certainly will become an important field in the science and technology of the next decades, allowing one to manipulate matter at the nanometer scale and to incorporate nanostructure and nanoprocesses into working technological innovation. Dendrimers are branched, synthetic polymers with layered architectures that show a fast growing field covering many areas of chemistry. Dendrimers are considered as a modern and an elegant class of branched macromolecules in the nanoscale dimension. They resemble treelike molecular architectures since they are built from repetitive monomers with branching point units that are radially connected around a template core. The hydrodynamic volume is dictated by the core, monomer units, branching points, dendrons, generation numbers, and peripheral functions. Dendrimers are discrete macromolecules with a high degree of molecular uniformity and monodispersity. Those features make for an easy characterization and analysis of their tailor-made properties. They are often macromolecules with few defects, whose dense outward (and/or inward) functional groups generate specific properties and functions such as a polyvalent ligand, a molecular receptor, molecular confinement, and a molecular translocator (vector). Multivalency as well as adhesive, amplification, recognition, additive, and cooperative effects have been observed. The expression "dendritic effects" was coined to illustrate those specific chemical behaviors. Dendrimers that used to be employed as standard, innovative stabilizers, media, ligand and supports have appeared to provided remarkable results with extremely high metal loading. By

* Tel and Fax: 0086-21-62933291 E-mail: dxcui@sjtu.edu.cn (D Cui) and panbifeng@tom.com (B Pan)

regulating dendrimer synthesis, it is possible to precisely manipulate both their molecular weight and chemical composition, thereby allowing predictable tuning of their biocompatibility and pharmacokinetics. Advances in our understanding of the role of molecular weight and architecture on the in vivo behavior of dendrimers, together with recent progress in the design of biodegradable chemistries, has enabled the application of these branched polymers as anti-viral drugs, tissue repair scaffolds, targeted carriers of chemotherapeutics and electrochemical sensors. Various types of dendrimers differing in composition and structure are mentioned, together with their practical use spanning from catalysis, transport vehicles to synthetic vaccines. This paper discusses the research advances development of functional and multifunctional dendrimeric and hyperbranched polymers with the objective of being applied in catalysis, biosensing, imaging, drug and gene delivery, nanoprinting, self-assembly, energy transfer, etc.

INTRODUCTION

During the past decade, there has been an explosive growth in studies of dendrimers, a new class of macromolecules that can be considered as monodisperse, precisely ordered polyfunctional polymers. Several applications of these relatively soluble compounds have been proposed. Their remarkable properties result from the presence of a large number of chain-end functionalities that provide highly controlled surfaces and interfaces, and from the presence of internal voids allowing, for example, encapsulation of guest molecules or molecular recognition at specific locations on the cascade superstructure.

1. SYNTHESIS ADVANCES OF DENDRIMER

A number of creative approaches to the synthesis of organic or inorganic "monodendrimers" have been reported. The reactivity of the core has been used to couple two or more dendrons leading, for example, to structures with two dissimilar halves. Coupling of cascade macromolecules through metal centers has also been achieved. Dendrons containing isophthalic acid units covalently attached to a rigid aromatic spacer have been shown to self-assemble through hydrogen bonding to form a hexameric dendritic aggregate whose stability varies in a generation- dependent manner. All these different routes to bis- or multidendrimers involve the association of dendrons by their core.

1.1. Control of Dendrimer Structure

Anne-Marie Caminade *et al.* [1] report two examples of the regioselective stepwise growth of six dendrimer units in the internal voids of a main dendrimer, illustrating that these cavities are accessible and that internal functional groups can be used for the synthesis of covalently linked multidendritic systems. The synthesis of the central dendrimer is based on the strategy using a divergent method. The formation of generation 1 involves three steps: (i) a condensation reaction between the flexible core, hexaaldehyde 1, and six equivalents of monomethylhydrazine; (ii) a "Mannich reaction" between the resulting product and six equivalents of the phosphane Ph2PCH2OH, giving an hexaaminophosphane; and (iii) a

"Staudinger reaction" between this latter compound and the azide N3P(S)(OC6H4CHO)2 to form the dendrimer 2 (generation 1) with six P5N–P(S) fragments in the framework..

Daniel Taton *et al.* [2] engineered polymer chains in star-like and dendrimerlike architectures using atom-transfer radical polymerization (ATRP) and anionic ring-opening polymerization (AROP) of ethylene oxide as synthetic tools. Daniel Taton *et al* first discussed the scope and limitations of ATRP when this method of controlled radical polymerization is applied to multifunctional initiators for the synthesis of star polymers by the core-first approach. The switch from ATRP to AROP of ethylene oxide and vice versa allows access to branched amphiphilic block copolymers exhibiting core–shell structures. Stress is also put on the methodologies for the selective branching of polymeric chain-ends, with a view to introducing ω-geminal functionalities from which further polymer branches can be grown. When linear polymer precursors are used, such a strategy leads to asymmetric and mikto-arm stars but the same can be applied to multi-arm stars so as to generate so-called 'dendrimer-like (co)polymers' that are dendrimers with true macromolecular generations.

George R. Newkome *et al* [3] presented an elongated structural design leading to more conical-shaped dendritic architectures by using a combination of 1→3, 1→(2+1), and 1→(2+1Me) C-branched monomers. Synthesis of the conifershaped macromolecule was achieved by reaction between isocyanate and amine in dry CH2Cl2. Aresultant extended focal adamantane-modified dendron was deprotected to generate the water-soluble product, which was subsequently complexed with β-cyclodextrin in D2O to create the desired tree-like product (Figure 1).

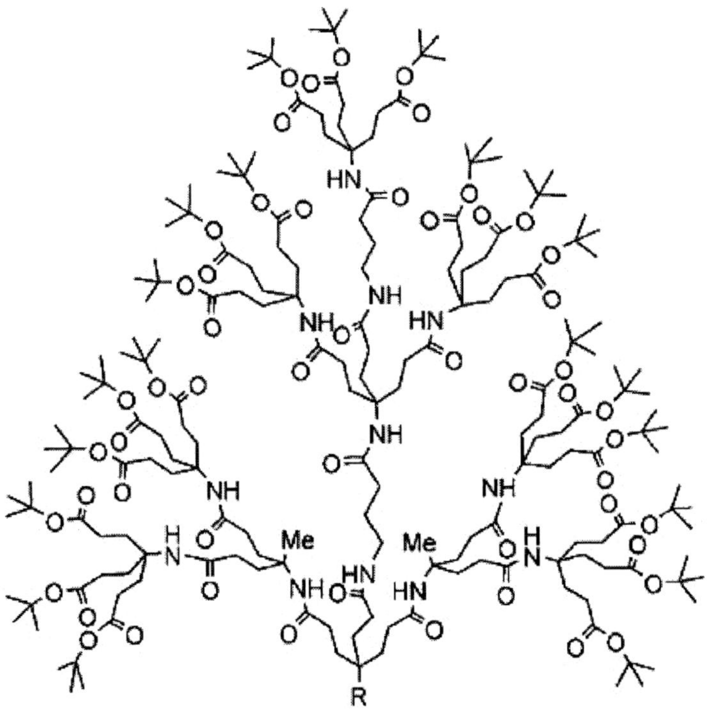

Figure 1. Elongated dendron reported by George R. Newkome et al.

Helmut Keul *et al.* [4] prepared monodisperse branched polyurethanes containing long alkyl chains by a new convergent synthesis. This synthesis comprises two steps, with hexamethylene diisocyanate uretdione as the starting molecule. The free isocyanate groups of this monomer are reacted with long-chain alkanols. These diurethane uretdiones are then reacted with amines or amino alcohols under ring opening and the formation of a biuret group. Branching points are not, as usual, part of the monomer but are formed during preparation.

A new reversible addition-fragmentation chain transfer (RAFT) agent, dendritic polyester with 16 dithiobenzoate terminal groups, was prepared by Chun-Yan Hong *et al* [5] for the RAFT polymerization of styrene (St) to produce star polystyrene (PSt) with a dendrimer core. It was found that this polymerization was of living characters, the molecular weight of the dendrimer-star polymers could be controlled and the polydispersities were narrow. The dendrimer-star block copolymers of St and methyl acrylate (MA) were also prepared by the successive RAFT polymerization using the dendrimer-star PSt as macro chain transfer agent (Figure 2).

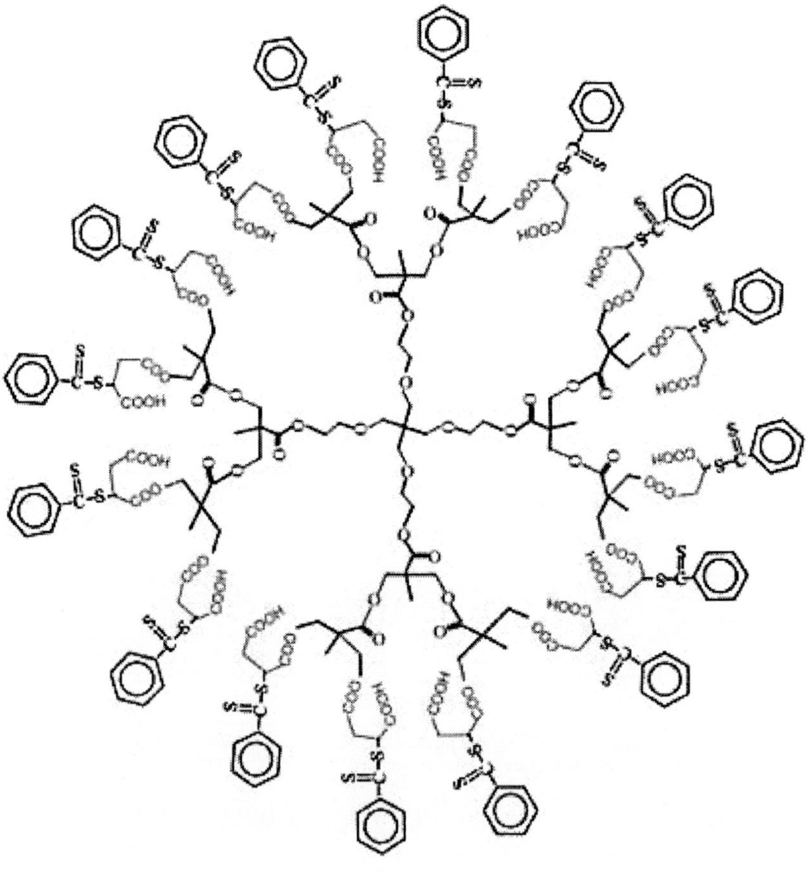

Figure 2. Dithiobenzoate-terminated dendrimer reported by Chun-Yan Hong et al.

Wenfang Shi *et al* [6] prepared a series of dendritic core-shell structural polymers with different shell densities based on dendritic polyester Boltorn H40 to form unimolecular micelles in chloroform. The encapsulation mechanism study using Congo red as a guest molecule by fluorescence and UV-vis methods showed that the interaction of Congo red with hydroxyl groups in the dendritic core-shell polymer led to encapsulation. Moreover, the results also indicated that the dendrimer with 43.4% hydroxyl groups endcapped by the long alkyl chains showed the best encapsulation capacity.

Fabrizia Negri *et al* [7] presented an atomistic molecular dynamics investigation of the structural time evolution of isolated polyphenylene dendrimers, carbon based dendrimers with a planar core formed by a 1,3,5 trisubstituted benzene ring. Fabrizia Negri *et al* presented a general classification of the conformations (core conformations) assumed by the three dendrimer branches with respect to the planar core.

1.2. Control of Surface Groups, Core, and Building Block

Chae Kyu Kim et al [8] described the synthesis and luminescence characterization of conjugated dendrimers with triazine peripheries and a distyrylanthracene core that are suitable for electroluminescence applications. The dendrimers consist of dendritic triazine wedges with high electron affinity, stilbene branches, and a distyrylanthracene core as an emitting moiety. The dendrimers have lowest unoccupied molecular orbital values of about -2.7 eV. Photoluminescence studies have indicated that a cascade energy transfer occurs from the triazine wedges to the stilbene bridges and finally to the distyrylanthracene core.

Cheol-Hee Ahn et al [9] reported the synthesis of dendrons and dendrimers with cationic amino groups at their periphery up to the third and second generation, respectively. The results obtained by 1H NMR spectroscopy and gel permeation chromatography analysis supported the formation of the targeted dendrons and dendrimers. The dendrons were grown via ester linkages, which endowed them with biodegradability in D2O at 37°C. The degradation rate depends upon the steric hindrance and reactivity caused by the bulkiness and compact structure of the dendrons. All of the synthesized dendrons were degraded within a month, while 60% of the ester groups in the sterically crowded dendrimers were degraded over the same time period.

Klaus Müllen et al [10] reported the systhesis first- and second-generation polyphenylene dendrimers starting from benzophenone as core. Variation of the size and the density of the dendrimer shell resulted in different isolation of the cores. To investigate the effect of shielding upon the reactivity of the core, chemical functionalizations as well as the alkali-metal reduction of the encapsulated benzophenone core were performed. The herein presented synthetic concept opens the way to spatially well-defined spherical nanoparticles bearing a single isolated function in the center.

Mackay B. Steffensen et al [11] reported a comprehensive and chronological account of dendrimers based on [1,3,5]-triazines. Mackay B. Steffensen et al discussed the synthetic strategies to install the triazine through cycloaddition, cyclotrimerization, and nucleophilic aromatic substitution of cyanuric chloride. Motivations and applications of these architectures are surveyed, including the preparation of supramolecular assemblies in the solution and solid states and their use in medicines, advanced materials, and separations when anchored to solid supports.

Well-defined, modular dendrimers enable processing techniques and electronic properties to be tuned independently by Weihong Zhu et al [12]. Moreover, the dendritic topology can isolate the core chromophore, thus reducing or eliminating strong intermolecular interactions.

Weihong Zhu et al presented the synthesis of three series of flexible, dendron-functionalized dendrimers as red-lightemitting materials by a convergent approach: (1) carbazole (CZ) or oxadiazole (OXZ) terminated imide-type dendrimers, (2) cascade energy-transferring imide-type dendrimers, and (3) CZ-terminated perylene bay-type dendrimers. They all consist of the luminescent core of perylenebis(dicarboximide)s with specific functional groups of CZ or OXZ at the periphery and are constructed from flexible Fréchet-type poly(aryl ether) dendrons (Figure 3).

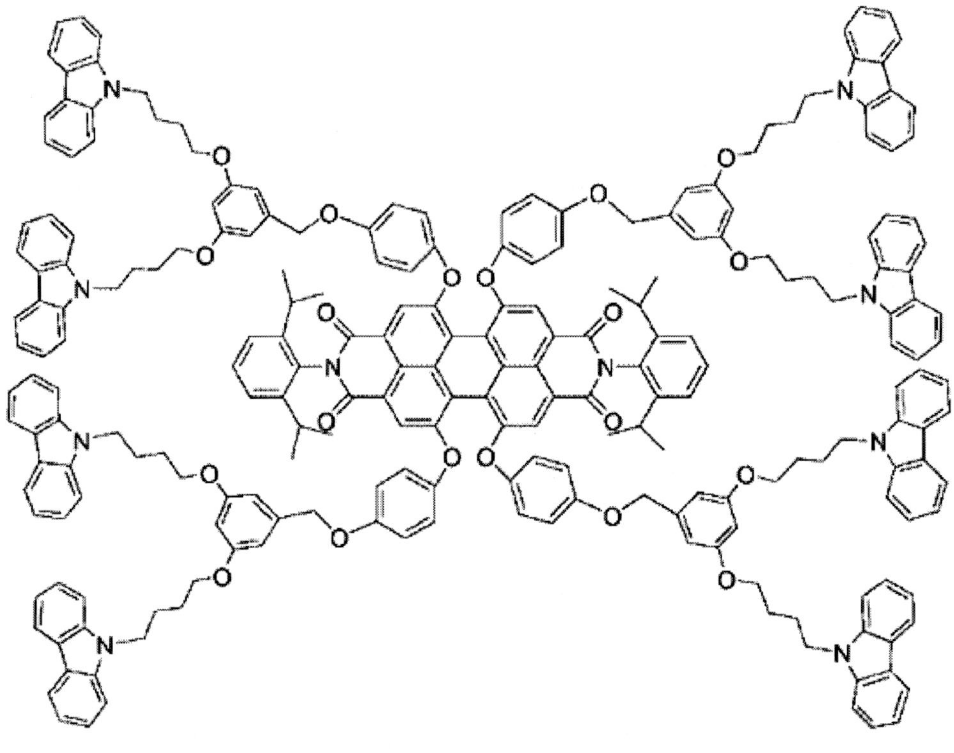

Figure 3. Carrier-transporting dendrimers with perylenebis(dicarboximide)s as a luminescent core.

Thermally labile dendrimers based on the reversible furan-maleimide Diels-Alder reaction are described by Dominic V. McGrath et al [13]. First through fourth generation benzyl aryl ether based dendrons that contained furan moieties at their focal point were allowed to react with bismaleimide central linker to provide the corresponding dendrimers.

As reported by Nicola Armaroli et al [14], multifullerene-terminated dendrimers Gn (n=1-5) were synthesized and structural, photophysical, and photoelectrochemical properties were studied for the fullerene dendrimers and their nanoclusters (Figure 4). The fullerene dendrimers formed clusters when toluene solutions of the fullerene dendrimers were injected into acetonitrile. Dynamic light scattering and atomic force and scanning electron microscopic measurements on these clusters revealed that the cluster size decreased with increasing the generation number of the dendrimers.

Cheol-Hee Ahn et al [15] reported the synthesis of aliphatic ester dendrimers based on 2,2-bis(hydroxymethyl)propionic acid (bis-MPA) growth units as a form of dendritic hybrids up to G4.0. Linear polystyrene supports in the dendritic hybrids were obtained through atom transfer radical polymerization at 130°C for 12 h using the CuCl/PMDETA catalyst/ligand system with 4-(chloromethyl)benzyl alcohol as the initiator.

Figure 4. Fullerene Dendrimer reported by Nicola Armaroli et al.

Two series of oligo (phenylene ethynylene)s (OPEs) with different dendrimer side groups have been designed and synthesized by Xiao-Ping Cao et al [16]. The molecules contain thiol groups at both ends to enable interconnection between nanoscale gapped metallic electrodes. The different dendrimer groups act as "shells", allowing tailoring to the nanoscopic environment surrounding the OPE "core". Meanwhile, the dendrimer shells also act as spacers for the precise control of the packing density and intermolecular interaction between the OPE cores.

Ye Tao et al [17] reported the synthesis and properties of a novel series of multi-triarylamine-substituted carbazole-based dendrimers G2-OTP(n)-G2(n=2-5) with an oligothiophene core. The length of the oligothiophene core varied from 2 to 5 thiophene units. It was found that both the absorption and fluorescence emission peaks red-shifted with increasing oligothiophene core length. As a result, the power conversion efficiencies of the as-fabricated solar cells using the composite of these dendrimers and PCBM as the photoactive layer increased with increasing oligothiophene core length.

As reported by Herbert Schumann et al [18], the catalytic hydrogermylation of $Si(CHdCH2)4$ with $Ph3GeH$ yielded $Si(CH2CH2GePh3)4$. In contrast, $Sn(CH=CH2)4$ and

Sn(CH2CH=CH2)4 reacted with Ph3GeH under the same conditions, forming Ph3Ge(CH2)2GePh3 and Ph3Ge(CH2)3GePh3, respectively, whereas the reaction of Sn(CH2CH2CH=CH2)4 and Ph3GeH afforded the expected tin-based metallodendrimer Sn[(CH2)4- GePh3]4. The reactions of Si[(CH2)2SnBr3]4 with the Grignard reagents RMgBr (R: CH=CH2, CH2CH=CH2, (CH2)2CH=CH2)) produced Si[(CH2)2Sn(CH=CH2)3]4, Si[(CH2)2Sn(CH2CH=CH2)3]4, and Si[(CH2)2Sn(CH2CH2CH=CH2)3]4.

Stéphane Campidelli et al [19] described the synthesis and characterization of a series of single-wall carbon nanotubes (SWNTs) functionalized with polyamidoamine dendrimers. Importantly, the dendrimers are linked directly to the SWNT surface using a divergent methodology. This approach allows the number of functional groups on the nanotubes to be increased without provoking significant damage to the conjugated π-system of the SWNTs (Figure 5).

Very high molecular weight (M_n up to 700 000) and narrowly polydispersed (PDI=1.1-1.4) hyperbranched polyglycerols (HPG) were synthesized by ring-opening multibranching polymerization of glycidol using dioxane as an emulsifying agent as reported by Donald E. Brooks et al [20]. Broader molecular weight distributions with low molecular weight fractions were obtained when diglyme was used as the emulsifying agent. The results show that these polymers are very compact and have spherical conformations in water with no indications of aggregate formation. Despite their self-similar structures, depending on the type of solvent used to synthesize them (diglyme vs dioxane), topologically restricted configurations are produced that result in completely different entanglement dynamics.

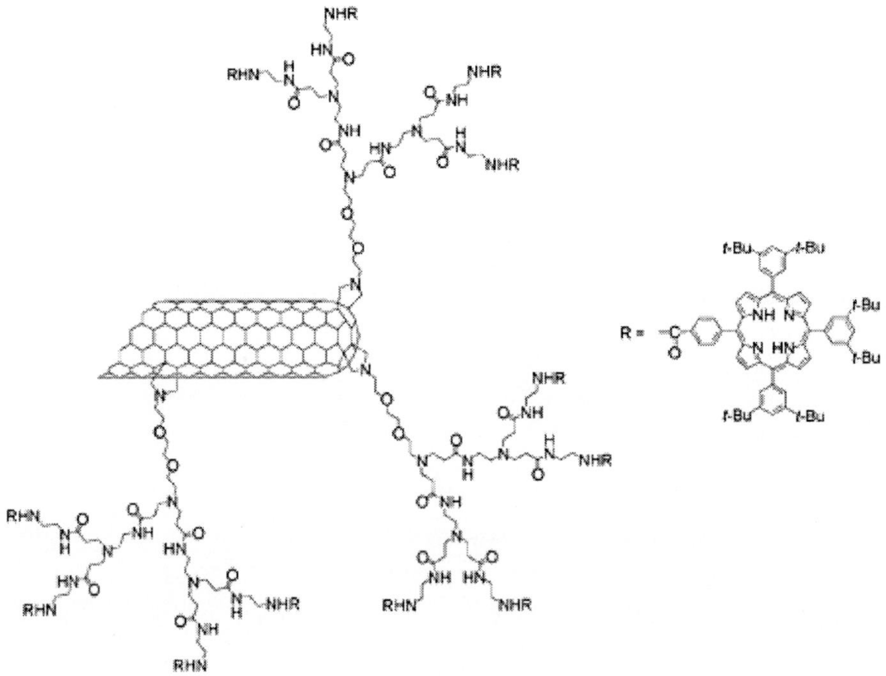

Figure 5. Dendrimer-functionalized single-wall carbon nanotubes reported by Stéphane Campidelli et al.

Two large π-conjugated dendrimers (G0 and G1) employing the stilbenoid moiety as the bridge unit have been developed by Yuguo Ma et al [21] through the Suzuki and the Horner-

Wadsworth-Emmons reactions. The molecular weight of G1 is 10 973 Da. Both G0 and G1 have good fluorescence quantum yields and exhibit similar absorption and emission behaviors in solutions and in thin films. They emit strong blue fluorescence in films under the irradiation of UV light.

As an application of a one-pot reaction involving Birch reduction and subsequent SN2 reaction in liquid ammonia, synthetic assembly of trisaccharidic moieties of globotriaosyl ceramide onto carbosilane dendrimers was accomplished using tris(3-bromopropyl)phenylsilane and tris[tris(3-bromopropyl)silylpropyl]phenylsilane as the core scaffolds as reported by Koji Matsuoka et al [22]. The common globotriaosyl derivative having benzylsulfide functionality at the terminal of the aglycon was efficiently prepared from D-galactose and D-lactose as starting materials. The glycosyl donor derived from galactose and the glycosyl acceptor derived from lactose were condensed in the presence of silver triflate as the best promoter to provide corresponding trisaccharide with newly formed α-1-4 linkages in 90% yield. Fully benzylated protection of the trisaccharide was deprotected under the Birch reduction condition followed by acetylation to give an acetate in which alkene was converted into benzyl sulfide by radical addition of α-toluenethiol in high yields. On the other hand, carbosilane dendrimers were prepared from appropriate chlorosilanes as starting materials by a combination of hydrosylation followed by alkenylation. The terminal C=C double bonds of the carbosilanes were converted into corresponding alcohols by means of the usual hydroboration reaction, and the alcohols underwent further chemical manipulation to give carbosilane dendrimers with peripheral bromine atoms.

Oleg Lukin et al [23] reported the synthesis of "designer" dendrimers and dendrons with sulfonimide units at every branching point (Figure 6). The synthesis is based on a series of (regio) selective functionalization reactions of amines and sulfonamides allowing precise control of the dendrimers' shape, the number of branches in each generation, and their peripheral decoration with functional groups. In principle, structurally different branches can be incorporated at any position within the dendrimer structure at will. Structurally perfect symmetrical and two-faced "Janus"-type dendrimers, as well as dendrimers and dendrons with intended interstices were synthesized on a preparative scale and fully characterized. Oligosulfonimide dendrons of various generations bearing an aryl bromide functional group at their focal points were attached to a p-phenylene core with the aid of Suzuki cross-coupling reactions resulting in dendrimers with photoactive terphenyl cores.

A novel family of diaminobutane-based poly(propyleneamine) ferrocenyl dendrimers with aza-crown ethers (Figure 7), DABdend-(Fc-A18C6)x (2-5: x=2, 4, 8, 16; Fc-A18C6) {[η5- C5H4CONH]Fe [η5-C5H4CONCH2 CH2(OCH2CH2)5]}), has been prepared by reaction of the noVel functionalized ferrocenyl derivative {[η5-C5H4COCl]Fe[η5-C5H4CONCH2CH2(OCH2- CH2)5]} as reported by Beatriz Alonso et al [24].

Mary J. Cloninger et al [25] functionalized dendrimers with mixtures of mannose, glucose, and galactose. Their association with concanavalin A was studied using precipitation and hemagglutination assays. With less idealized systems where the association was not optimized, mixtures of low- and high-affinity ligands caused smaller than the theoretically determined differences in binding activity, although linear binding trends were observed.

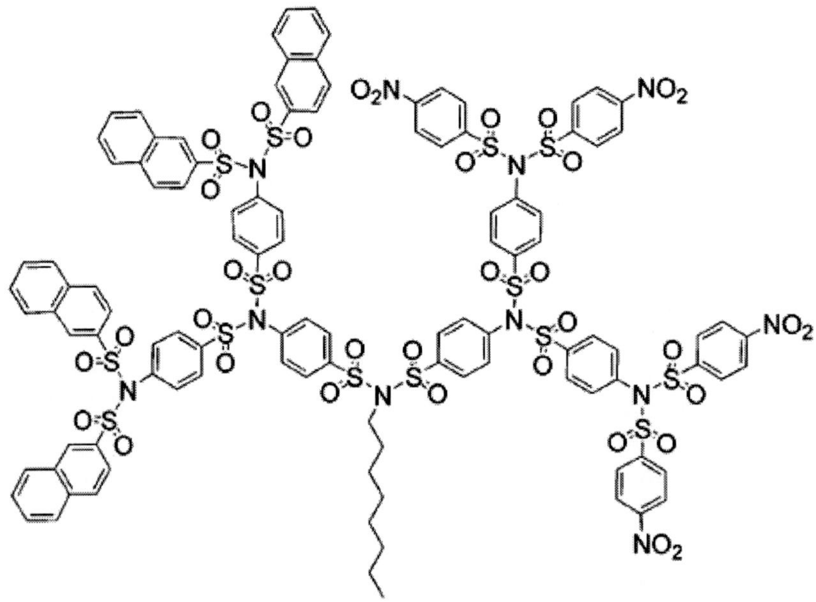

Figure 6. Branched Oligosulfonimides with Controllable Molecular Architectures.

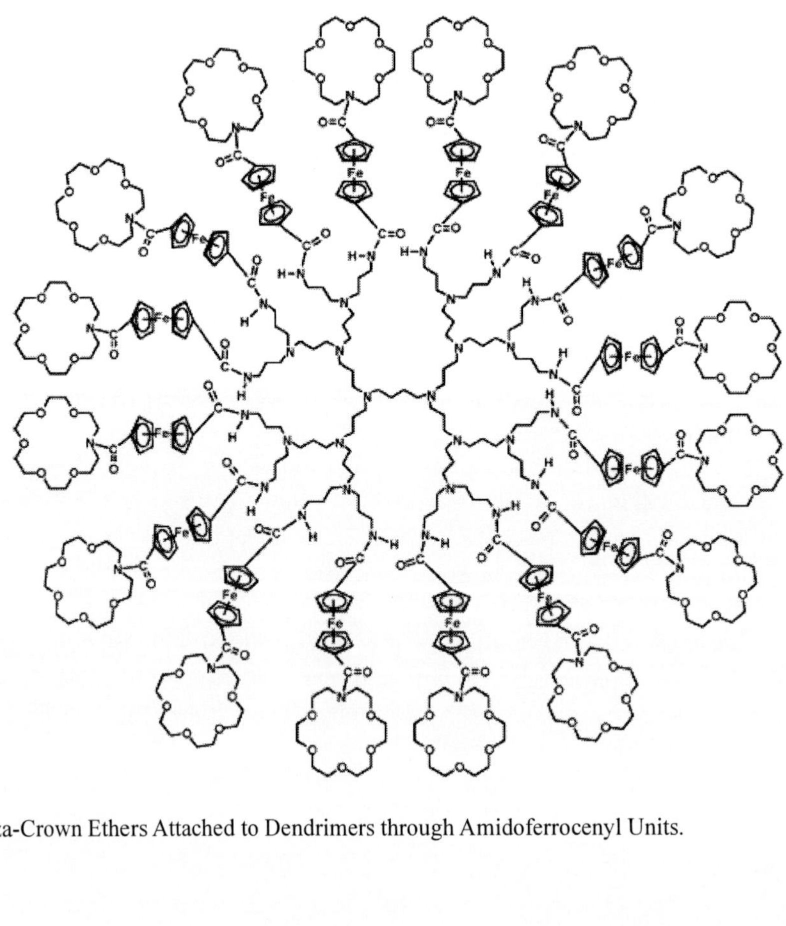

Figure 7. Aza-Crown Ethers Attached to Dendrimers through Amidoferrocenyl Units.

When systems were optimized so that high-affinity binding was achieved, then mixing low- and high-affinity ligands on the dendrimer's surface showed a predictable trend for lectin binding.

Jan H. van Maarseveen et al [26] described the convergent syntheses of three generations of carbosilane dendrimeric carbodiimides (Figure 8). The wedge-type building blocks were synthesized in a divergent way, starting from allyl chloride and a repetitive sequence of hydrosilylation with HSiCl3 and a Grignard reaction with allylmagnesium bromide. Hydrogenation of the terminal double bonds led to inert and stable wedges. The chloride substitutent at the focal point was transformed into several functional groups that eventually led to dendrimeric structures with a carbodiimide core.

Virgil Percec et al [27] reported the synthesis of methyl esters of 3-(4-hydroxyphenyl)propionic, 3-(3,4-dihydroxyphenyl)- propionic, 3-(3,5-dihydroxyphenyl)propionic, and 3-(3,4,5-trihydroxyphenyl)propionic acids and their use in a convergent iterative strategy to prepare up to four generations of three libraries, one of 3,4,5- and two of constitutional isomeric 3,4- and 3,5-substituted 3-phenylpropyl dendrons. Each library contains 3-[3,4,5- tris(dodecyl-1-oxy)phenyl]propyl-, 3-[3,4-bis(dodecyl-1-oxy)phenyl]propyl-, 3-{3,4-bis[3-(4-dodecyl-1-oxyphenyl) propyl-1-oxy]phenyl}propyl-, and 3-{3,4,5-tris[3-(4-dodecyl-1-oxyphenyl)propyl-1-oxy]phenyl}propyl ether first-generation dendrons on their periphery and -CO2CH3, -COOH, and -CH2OH groups at their apex. Regardless of their generation number and their periphery, internal, and apex structures, these dendrons self-assemble into supramolecular dendrimers that self-organize into all periodic and quasi-periodic assemblies encountered previously and in several unencountered with architecturally related benzyl etherbased supramolecular dendrimers. A variety of porous columnar lattices that were previously obtained only from dendritic dipeptides and hollow supramolecular spheres were also discovered from these building blocks.

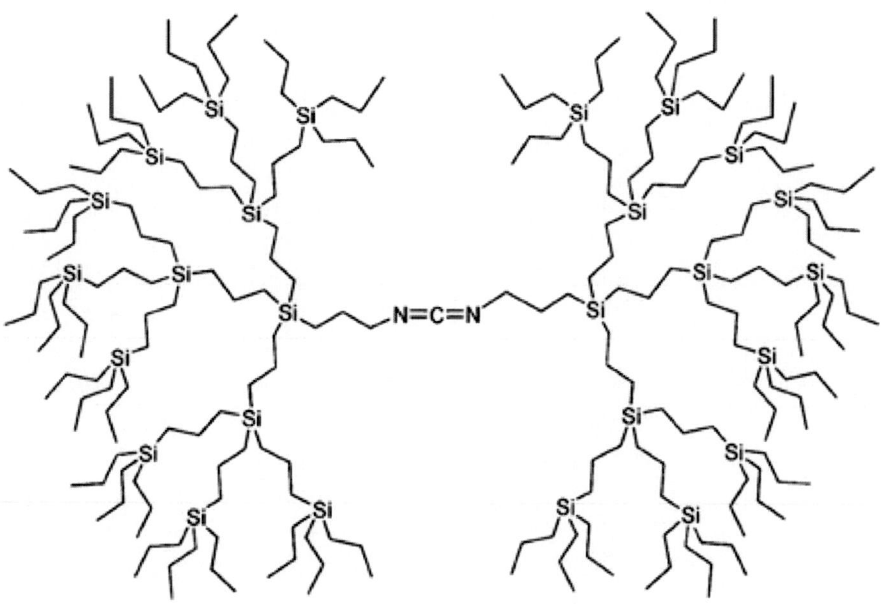

Figure 8. Dendrimeric Carbodiimide.

The more flexible and less compact 3-phenylpropyl ether repeat units are stable under acidic conditions, facilitate a simpler synthetic strategy, provide faster dynamics of self-assembly into higherorder supramolecular structures of larger dimensions, exhibit lower transition temperatures than the corresponding benzyl ether homologues, and demonstrate the generality of the self-assembly concept based on amphiphilic dendrons (Figure 9).

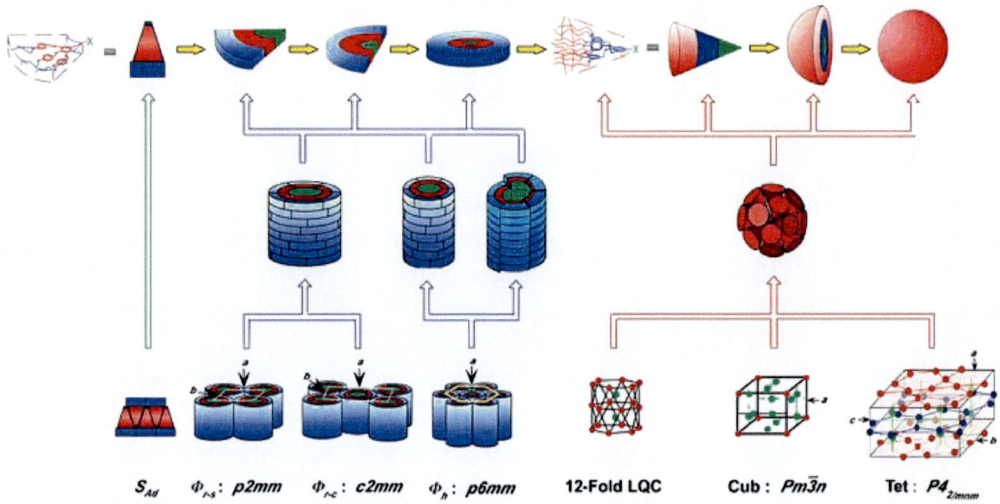

Figure 9. Retrostructural Analysis of 2D Smectic.

Vincenzo Balzani et al [28] have investigated the spectroscopic and electrochemical behavior of symmetric and unsymmetric first-, second-, and third-generation dendrimers comprising an electron-acceptor 4,4'- bipyridinium core (viologen type) and electron-donor 1,3-dimethyleneoxybenzene (Fréchet-type) dendrons. The quite strong fluorescence of the symmetrically and unsymmetrically disubstituted 1,3-dimethyleneoxybenzene units of the dendrons is completely quenched as a result of donor-acceptor interactions that are also evidenced by a low-energy tail in the absorption spectrum. In dichloromethane solution, the 4,4'- bipyridinium cores of the investigated dendrimers are hosted by a molecular tweezer comprising a naphthalene and four benzene components bridged by four methylene units. Host-guest formation causes the quenching of the tweezer fluorescence.

A new type of phenylazomethine dendrimer with a tetraphenylmethane core (Figure 10) was synthesized by a convergent method as reported by Kimihisa Yamamoto et al [29].

Anne-Marie Caminade et al [30] reported the synthesis of a new family of phosphorus dendrimers built from an octasubstituted metalfree phthalocyanine core up to generation 5. This core is used as a sensor and a probe for analyzing the properties of the internal structure and the influence of each structural part (core, branches, surface) upon the whole structure. UV-visible spectra show both a hyperchromic and bathochromic effect on the Q-bands with increasing generation, indicating that the chromophore is more isolated, and that the dendritic shell mimics a highly polar solvent. There is no evidence for aggregation, except for generation 0, showing again the isolation of the core (Figure 11).

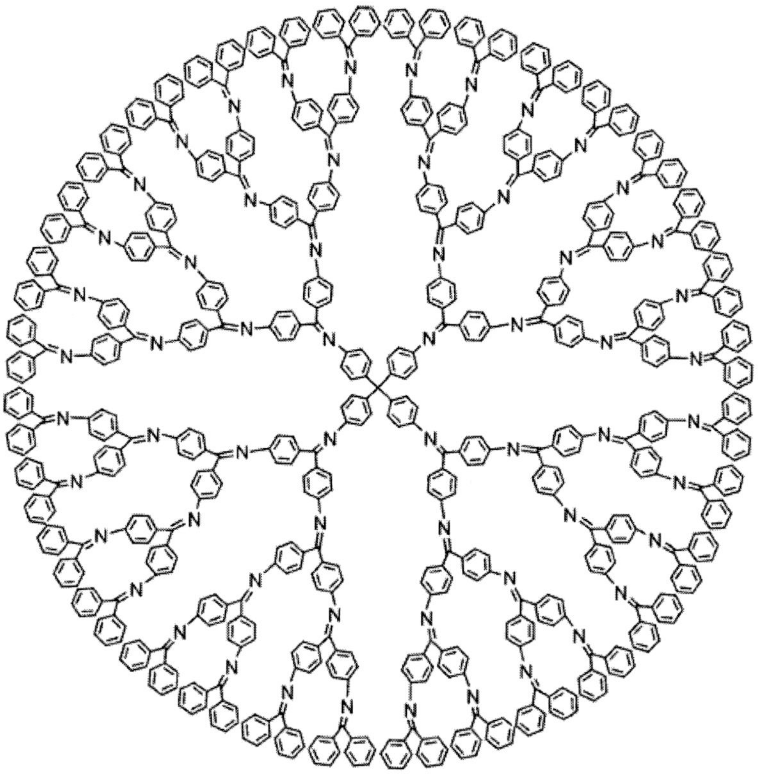

Figure 10. Phenylazomethine Dendrimer with a Tetraphenylmethane Core.

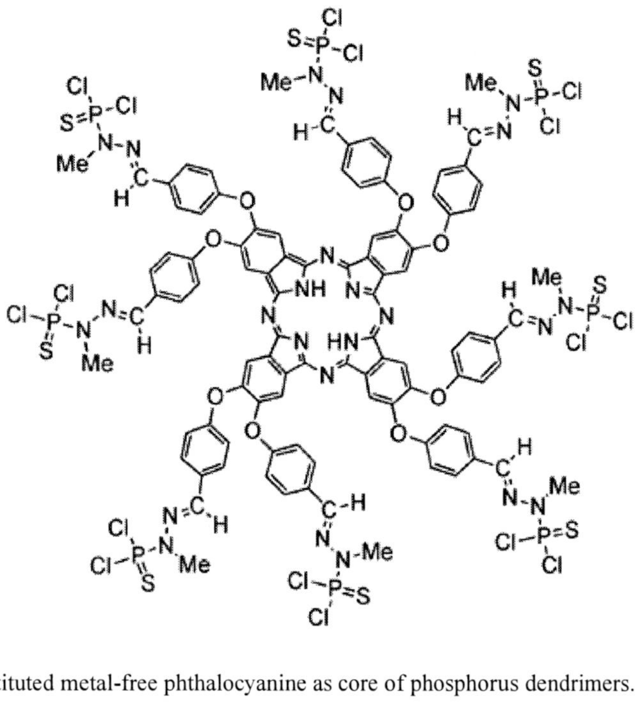

Figure 11. Octasubstituted metal-free phthalocyanine as core of phosphorus dendrimers.

Zinaida B. Shifrina et al [31] synthesized a new family of poly(phenylene-pyridyl) dendrimers (up to the fourth generation) with different periphery via a divergent route by using a cascade of Diels-Alder cycloadditions. Tetra(4-ethynylphen-1-yl)methane served as a core. A new A2B cyclopentadienone building block, namely, pyridine-containing cyclopentadienone with protected dienophile functions was prepared. By varying the building blocks at the last step of dendrimer construction, one could design molecules possessing either a phenyl- or a pyridyl-decorated outershell (Figure 12).

Natarajan Raju et al [32] described the synthesis of a new scaffold derived from iminodipropionic acid for the preparation of peptide dimers and tetramers.

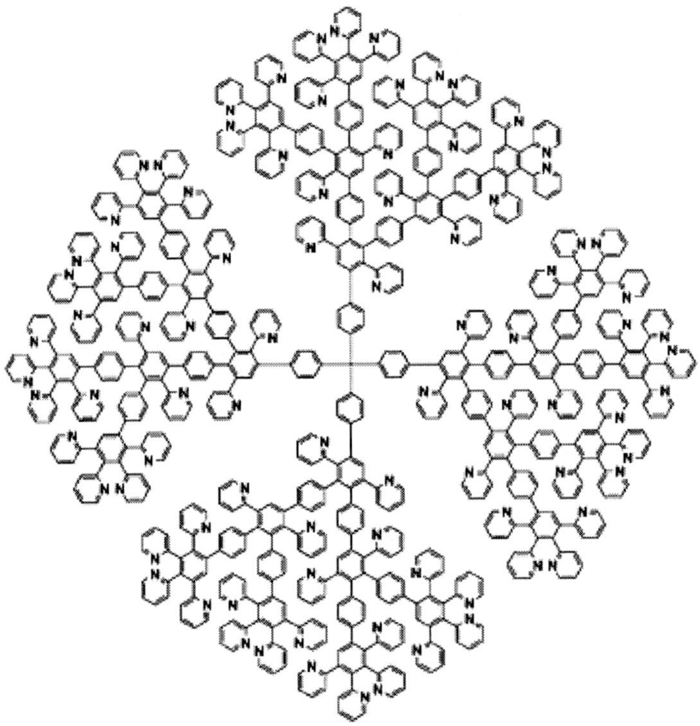

Figure 12. Poly(Phenylene-pyridyl) Dendrimers.

2. Characterization Advances of Dendrimer Physical-Chemical Properties

Johan Hofkens et al [33] used wide-field imaging of individual multichromophoric molecules and successive photobleaching to determine, accurately, the relative position of the chromophores in such systems. First, a polyphenylene dendrimer with well-defined geometry was used to establish the accuracy in localization that can be obtained by this methodology. For a signal-to-noise ratio of 20, interchromophoric distances could be measured with 4 nm accuracy. Next, the method was used to determine the end-to-end distribution of an end-capped polyfluorene polymer. From comparison between the experimental and simulated distributions, information on the conformation of the polymer could be deduced. It was found that the polymer has a nonlinear conformation. A conjugation length of sixmonomer units gave the best fit of the experimental data to the proposed model.

Shing Bor Chen et al [34] conducted brownian dynamics simulation to investigate the dynamics of dendrimer-polymer conjugates on the basis of a coarse-grained bead-spring model. The grafted polymers are found to slow the global dynamics of the dendrimer core. Because of the structural difference between the two moieties, the translational and rotational diffusivities of a conjugate show different scaling behaviors from those for a dendrimer, but in principle follow a Stokes-Einstein fashion, analogous to polymers of other types. When the grafted chains are long enough, they become brushlike, evidenced by the scaling behavior of the end-to-end distance as well as the chain-length dependence of the rotational relaxation time. For the range of chain length investigated, it is found that the brushlike grafted polymers still fall within the dilute regime. Thereby the interchain interactions are not strong enough to render a change in the scaling behavior of rotation relaxation in comparison with an isolated rodlike polymer.

Paul L. Burn et al [35] developed a simple convergent procedure for the formation of sterically encumbered phosphorescent dendrimers. The procedure is demonstrated with the preparation of a first-generation dendrimer composed of a fac-tris(2-phenylpyridyl)iridium(III) core and three dendrons (Figure 13). Each dendron is comprised of a branching phenyl unit with a further four phenyl groups attached. The lack of surface groups on the dendrons was found to reduce solubility and also reduced the level of control over the intermolecular interactions of the emissive and electroactive core in films. The 6-fold decrease in photoluminescence quantum yield in going from solution (69%) to the solid state (11%) showed that there were strong intermolecular interactions of the emissive cores in the solid state. Single-layer devices with the dendrimer blended with 4,4'-bis(N-carbazolyl)biphenyl showed an external quantum efficiency of 1.7% (5.4 cd/A) at 100 cd/m2 and 11.4 V, giving a power efficiency of 1.5 lm/W.

Ying Wang et al [36] reported nonlinear optical (NLO) and time-resolved fluorescence spectroscopic measurements of dendrimer-metal nanocomposites (DNCs). Third-order NLO effects were investigated by degenerate four-wave mixing (DFWM) measurements for DNCs and DNCs incorporated into thin polymeric films. The results show an enhancement of the third-order nonlinear susceptibility for the chromophore-functionalized dendrimer- metal nanocomposites. Investigations of the mechanism of the enhanced NLO effect suggested a strong contribution because of the metal's local field. These results show the potential usefulness of dendrimer-metal nanocomposites for nonlinear optical effects and biophotonic applications.

Natalia P. Yevlampieva et al [37] synthesized [60]fulleropyrrolidine-containing poly(benzyl ether) dendrimers via 1,3- dipolar cycloaddition of amino acid and aldehyde derivatives and [60]fullerene (Figure 14). Their dielectric and electrooptical (Kerr effect) solution properties were studied and compared with poly(propyleneimine) dendrimers, namely DAB dendr-(alkyl chain)n and DAB-dendr-(CN)n, and with known symmetrical and non-symmetrical poly(benzyl ether) dendrimers carrying benzyl groups ([Gx]-[C]-[Gx]) or benzyl and cyanobenzyl groups (CNn-[Gx]-[C]-[Gx+1]), respectively. It was shown that dendrimers display similar polar and electrooptical characteristics independent of their dendrimer generation. This is in contrast to the reference dendrimers for which a strong dependence of total dipole moment on dendrimer generation was observed.

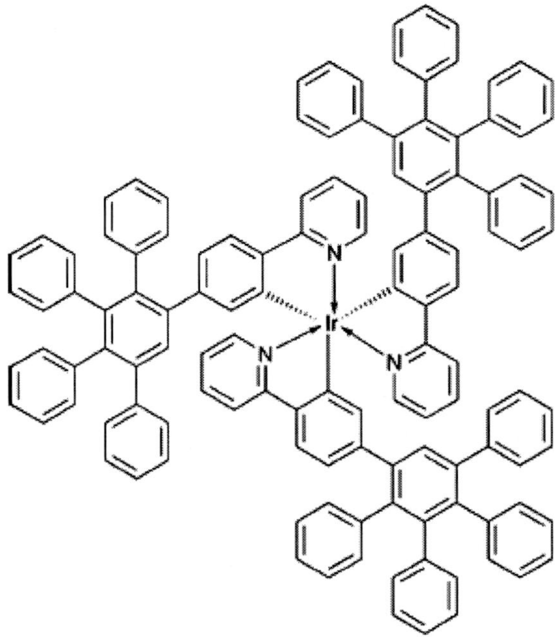

Figure 13. Phosphorescent dendrimer.

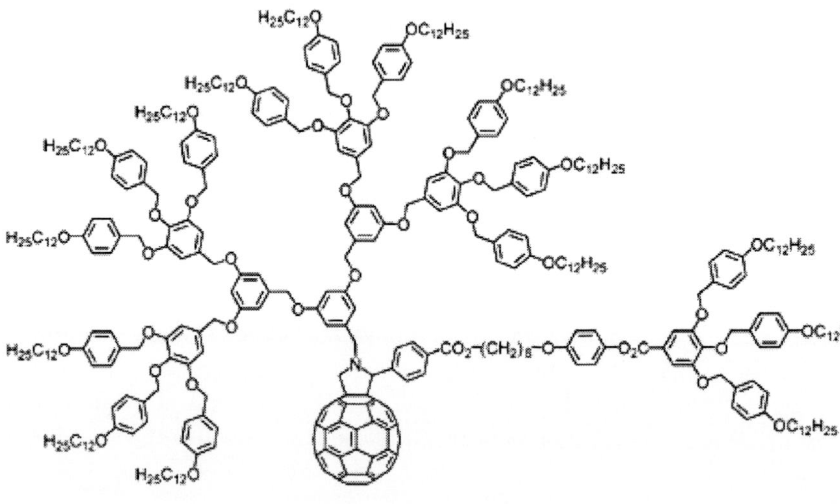

Figure 14. [60]Fullerene-containing poly(benzyl ether) dendrimers.

Robin L. McCarley et al [38] showed that generations 1-5 of poly(propyleneimine) dendrimers terminated with redox-labile, trimethyllocked quinone (TLQ) end groups released the TLQ units as lactone upon chemical or electrochemical redox action (Figure 15). Sodium dithionite was used as a chemical reducing agent to instantly convert the TLQ units to their deuteroquinone analogues (TLDQ), which were subsequently liberated from the dendrimers following zero-order kinetics in both DMSO-d_6-D2O (85:15, v/v, kobs =7 × 10–6 M s–1, t1/2 = 9 min) and THF-d_8-D2O (76:24, v/v, k*obs* = 4 × 10–6 M s–1, t1/2 = 15 min). Cyclic voltammetry of dendrimers in DMSO/0.1 M *n*-Bu4PF6 revealed that the reduction peak potential (Epc) values for the formation of TLQ2– were generally invariant (-1.85 V vs Fc/Fc+) with dendrimer generation.

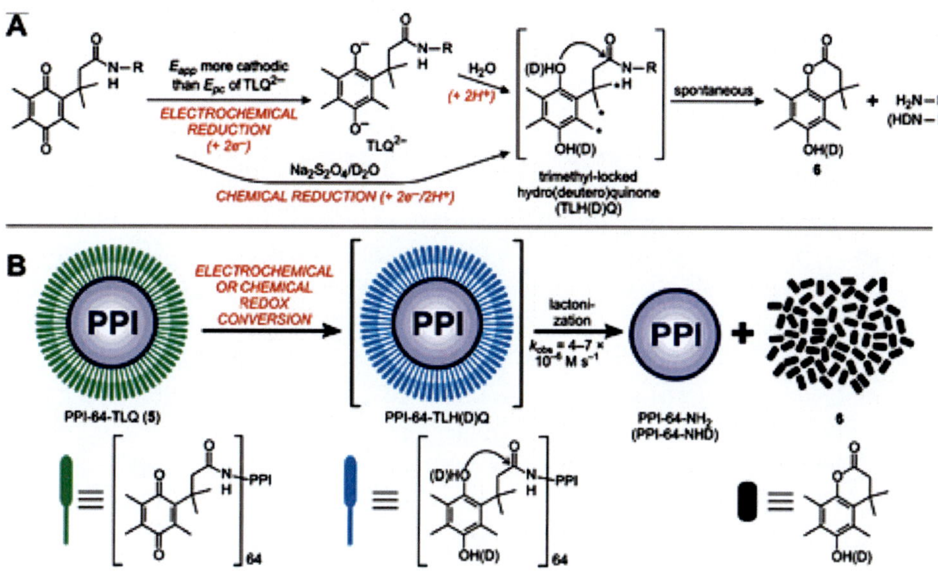

Figure 15. Chemically and electrochemically mediated release of dendrimer end groups.

Thus, bulk electrochemical reduction of 1-5 at an overpotential of -0.3 V vs the Epc of TLQ2– produced PPI dendrimers containing dianionic TLQ2– termini. Protonation of the electrochemically generated TLQ2– end-group units with H2O produced the transient and structurally labile hydroquinone (TLHQ) intermediates, which were liberated from the dendrimers as lactone 6 in a similar fashion as in the sodium dithionite-induced approach. In both the chemically and electrochemically mediated routes, 1H NMR spectroscopy verified that the redox-promoted disconnection of TLQ from the dendrimers was quantitative.

Jiri Horsky et al [39] reported the quantitative interpretation of MALDI-TOF mass spectra of imperfect carbosilane dendrimers. A general relationship is derived for the abundance of an imperfect dendrimer with a given number of missing constitutional repeating units in the two outmost layers. The relationship is used in the interpretation of the MALDI TOF mass spectrum of the second-generation carbosilane dendrimer prepared by the iterative divergent method. The model quantitatively describes the spectrum of the dendrimer and correctly predicts the MALDI TOF mass spectrum of its first-generation precursor. Thus, the use of well-resolved MALDI TOF mass spectra for assessing the purity of low-generation dendrimers with uniform end groups is substantiated for carbosilane dendrimers and to lesser extent for dendrimers in general.

Hak-Sung Kima et al [40] reported mass spectrometric analysis of affnity-captured proteins on a dendrimer-based immunosensing surface and the investigation of on-chip proteolytic digestion. The monolayer of fourth-generation poly(amidoamine) dendrimers was adopted to construct the immunoaffnity surface of an antibody layer. The antibody layer as a bait on the dendrimer monolayer was found to result in high binding capacity of antigenic proteins and a reliable detection. The affnity-captured protein at the immunosensing surface was subjected to direct on-chip tryptic digestion, and the resulting proteolytic peptides were analyzed by using matrix-assisted laser desorption/ionization time-of-flight mass spectrometry. The on-chip digestion method identiWed the affinity-captured bovine serum albumin (BSA), lysozyme, and ferritin at the level of around 100 fmol. The use of antibodies against the proteins involved in the metabolic pathway of L-threonine in Escherichia coli also led to discrimination of the respective target proteins from cell lysates.

Matthias Ballauff et al [41] reported the solution structure of stilbenoid dendrimers by a small-angle scattering study. The spatial structure of a stilbenoid dendrimer is investigated by small-angle X-ray scattering (SAXS) and small-angle neutron scattering (SANS) in dilute solution. All measurements are performed in toluene. The dendrimer consists of a stilbenoid scaffold with appended hexyloxy chains. SAXS is mainly sensitive to the dendrimer scaffold whereas SANS intensity, measured in fully deuterated toluene, derives from the solute molecules. The resulting SAXS and SANS intensities are analyzed by comparison with various models. It is found that the model of a circular disk gives the best description of the data. SAXS data demonstrate that the stilbenoid scaffold is flat as expected for benzene rings conjugated through vinylene units. SANS, on the other hand, shows that the hexyloxy chains are not confined in the plane defined by the core. The structure factor S(q), describing the interaction of the dendrimers at higher concentrations, is modeled quantitatively with the "polymer reference interaction site model" (PRISM) integral equation formalism for hard plates such as particles.

3. Advances in Nano Printing

Geoffrey A. Ozin et al [42] reported the synthesis of a new class of materials called as periodic mesoporous dendrisilicas. They are prepared from dendrimers with trialkoxysilyl groups at the outmost shell. Dendrimers of various cores and numbers of shells are used. The dendrisilica synthesis is based on acid- or base-catalyzed hydrolysis of the trialkoxysilyl groups and subsequent template-directed condensation of the dendrimers into an ordered template-dendrisilica nanocomposite. The template can be removed with organic solvents, producing a periodic mesoporous dendrisilica whose pore walls consist of interconnected dendrimer building blocks.

Byeongdu Lee et al [43] described the preparation of low- and ultralow-k (dielectric constant) nanoporous organosilicate dielectrics from blends of polymethylsilsesquioxane (PMSSQ) precursor with globular ethyl acrylate-terminated polypropylenimine dendrimers, which act as porogens (Figure 16). These dendrimers are found to mix well with the PMSSQ precursor and after their sacrificial thermal decompositions result in closed, spherical pores of <2.0 nm radius with a very narrow distribution even at high loading. This pore size and distribution are the smallest and the narrowest respectively ever achieved in porous spin-on dielectrics. The method therefore successfully delivers low- and ultralow-k PMSSQ dielectric films that should prove very useful in advanced integrated circuits.

Neal A. Rakow et al [44] described the dynamic imprinting of dendritic macromolecules with porphyrin templates to yield synthetic host molecules containing one binding site each (Figure 17). The process is based on our general strategy to prepare cored dendrimers, and involves covalent attachment of dendrons to a porphyrin core, cross-linking of the end-groups of the dendrons, and removal of the porphyrin template by hydrolysis.

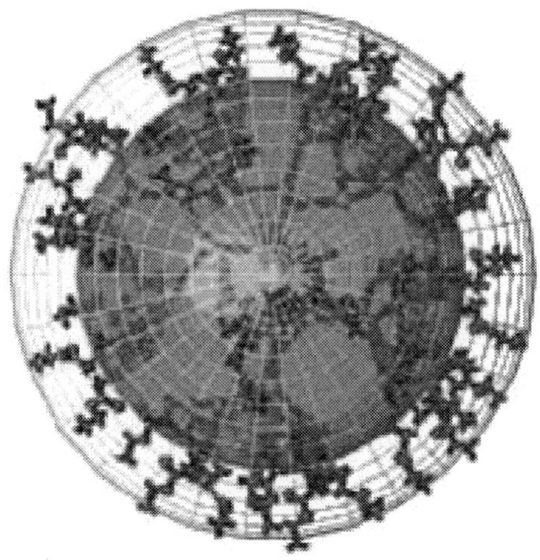

Figure 16. Globular ethyl acrylate-terminated polypropylenimine dendrimer.

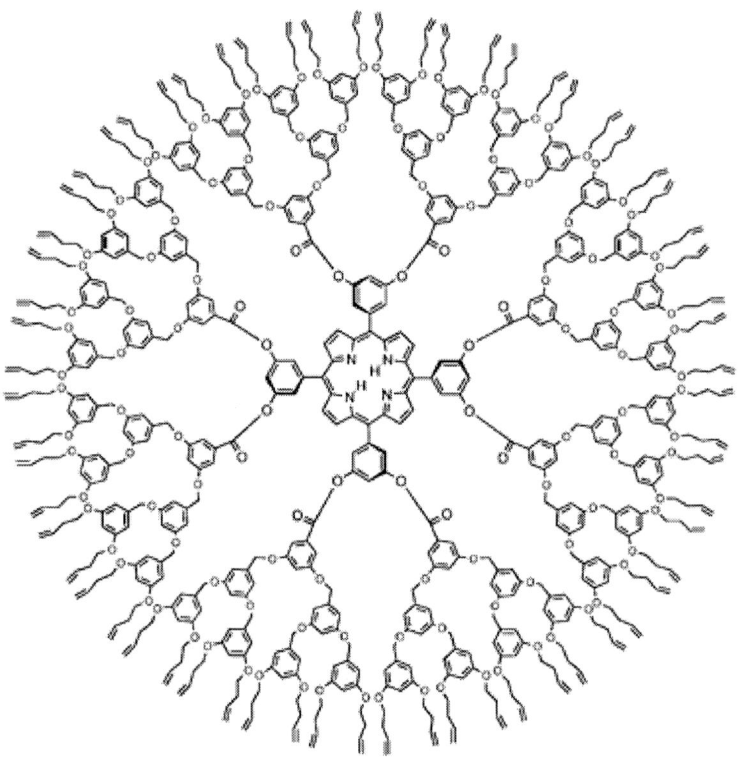

Figure 17. Porphyrin-core dendrimer.

In contrast to more traditional polymer imprinting, the approach ensures nearly homogeneous binding sites and quantitative template removal. Moreover, the hosts are soluble in common organic solvents and amenable to the incorporation of other functional groups, which should facilitate further development of this system for novel applications. When developing 'monomolecular imprinting' methodology, Neal A. Rakow et al selected dendrimers as macromolecular hosts because their molecular homogeneity and excellent solubility facilitate characterization. Ho-Cheol Kim et al [45] reported oriented nanoporous lamellar organosilicates templated from topologically unsymmetrical dendritic-linear block copolymers.

Jurriaan Huskens et al [46] reported heavyweight dendritic inks for positive microcontact printing. Poly(propylene imine) dendrimers with dialkyl sulfide end groups were prepared and developed as inks for positive microcontact printing ((+)μCP) on gold. Long (C10H21-S-C10H20-), medium (C3H7-S-C4H8-), and short (CH3- S-CH2-) dialkyl sulfide end groups were attached to second- and third-generation PPI dendrimers to create a family of dendritic sulfides.

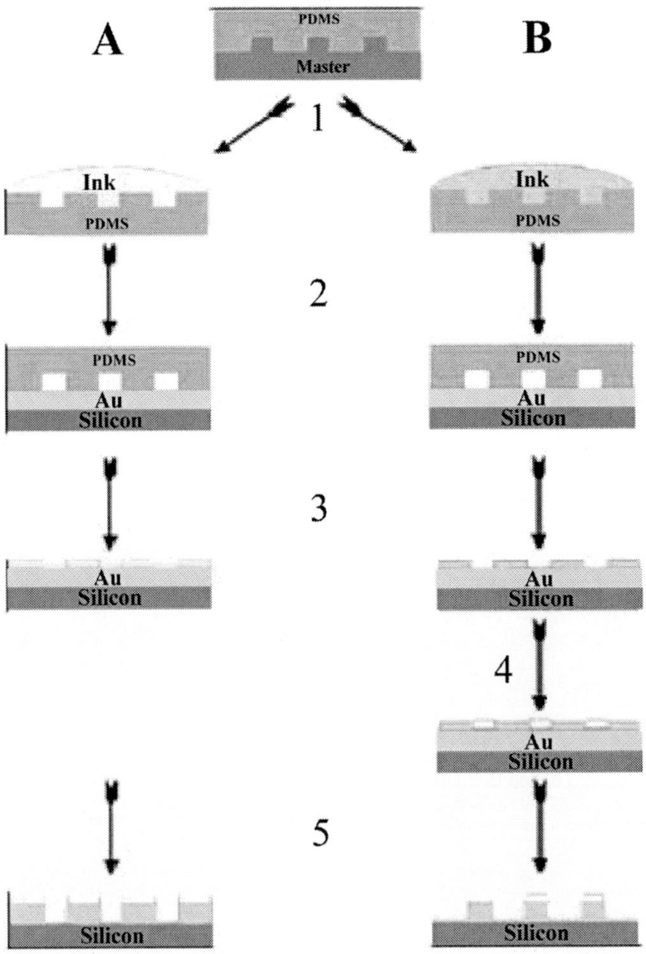

Figure 18. Heavyweight dendritic inks for positive microcontact printing.

The dendritic inks flatten upon adsorption and form monolayers on gold. (+)μCP was performed on gold using commercially available poly(dimethylsiloxane) as stamp material and *n*-octadecanethiol as etch resist. The gold beneath the dendrimers was selectively etched away with an acidic Fe(NO3)3/thiourea solution to give the positive copy of the original master pattern. The multivalent sulfide attachment and the relatively high molecular mass of these dendrimers ensured minimal lateral ink spreading and thus optimal feature reproducibility (Figure 18).

4. Research Advances in Biology and Medicine

4.1. Imaging Contrast Reagent

Walter H. Backes et al [47] reported the evaluation of Gd(III) DTPA-terminated poly(propylene imine) dendrimers as contrast agents for MR imaging. Serial and dynamic CE-MRI were performed to monitor the distribution of MRI contrast agent in the heart, arteries, renal system, liver, spleen, bladder and tumor periphery. Dynamic CE-MRI showed timely renal clearance for all contrast agents, but a stronger and a prolonged blood signal enhancement for the higher generations of the dendritic contrast agent. Moreover, G0 and G1 showed a rapid tumor wash-in and wash-out, whereas G3 and G5 displayed a more gradual and prolonged tumor wash-in. Both G0 and dendritic contrast agents G1, G3 and G5 are well suited for non-tissue-specific MRI of submillimeter- sized blood vessels and evaluating tumor microcirculatory characteristics in mice. Higher generations of dendritic contrast agents display lower concentration detection limits, which suggests their future use for molecular imaging (Figure 19).

Hisataka Kobayashi et al [48] reported MR lymphangiography using dendrimer-based contrast agents to a comparison at 1.5T and 3.0T. Most macromolecular contrast agents (CAs) show lower r1 and higher r2 relaxivities at 3.0T than at 1.5T (Figure 20). MR lymphangiography in mice using a macromolecular G6 dendrimer-based CA was serially performed and compared at both 1.5T and 3.0T. The r1 and r2 relaxivities of the G6 CA were 25 and 78/s/mM at 1.5T and 17 and 82/s/mM at 3.0T, respectively.

Robert C. Brasch et al [49] designed, synthesized, and initially characterized a representative set of novel constructs for large-molecular radiographic/computed tomography (CT) contrast agents, intended for a primarily intravascular distribution. A new assembly of well-known and biocompatible components consists of paired, symmetrical dendritic polylysines initiated from both ends of a poly(ethylene glycol) (PEG) core, yielding an array of multiple free amino groups to which were conjugated highly soluble and stable triiodophthalamide ("triiodo") moieties. An array of six dendritic contrast agents was synthesized originally, using three different PEG cores (3, 6, 12 kDa) with t-Boc lysine-generated dendrimer "amplifiers" (from three to five generations) containing 16 to 64 amino groups for conjugation with reactive triiodo moieties. A clinically used, nonionic, small molecular CT contrast agent, iobitridol, was derivatized via a hydroxyl protection/deprotection strategy, introducing a new carboxyl group available for conjugation to the lysine amino groups of dendrimers.

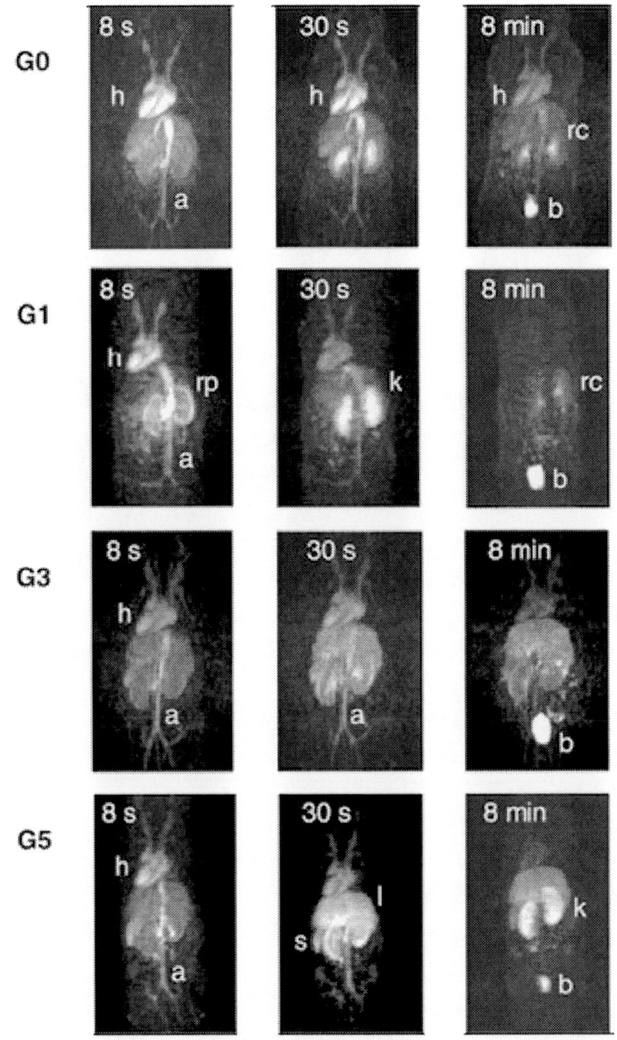

Figure 19. Digitally subtracted contrast-enhanced whole-body MR images.

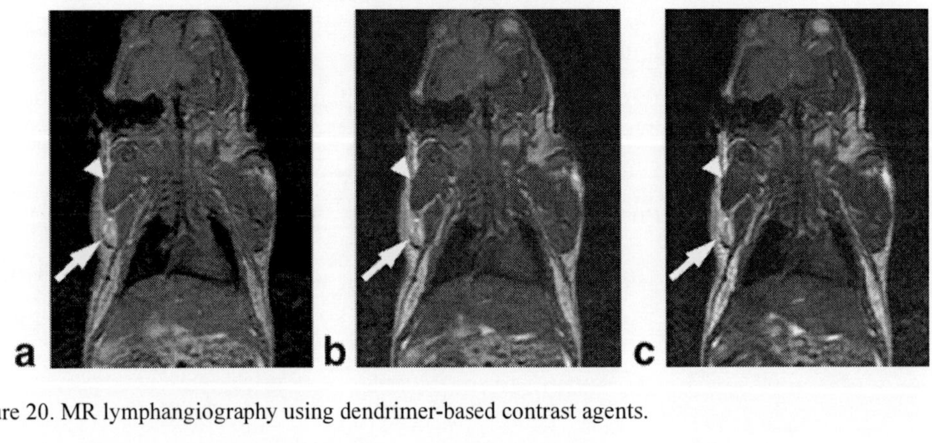

Figure 20. MR lymphangiography using dendrimer-based contrast agents.

The results showed the synthetic feasibility, desired basic characteristics, and potential utility for CT contrast enhancement achieved with a new type of iodinated, large-molecular PEG-core dendritic construct.

Peptide dendrimers have a variety of applications in biology such as the vehicles for drug and gene delivery, molecular inhibitors, protein mimics, and synthetic vaccines. The multiple antigenic peptide (MAP) system is a well-known example of a discrete, dendrimeric scaffold. Ching-Hsuan Tung et al [50] reported enzyme-targeted fluorescent imaging probes on a multiple antigenic peptide core. They explored a novel application of the MAP based scaffold by designing molecular probes that fluoresce only after enzymatic treatment. The probes, which were synthesized on solid support, incorporate a cathepsin S dipeptide substrate (Leu-Arg), and a poly(ethylene glycol) (PEG) chain in their dendritic arms. The fluorescence emission of the near-infrared fluorochromes attached to the N-termini of the dendritic arms was quenched. Mechanistic studies revealed formation of H-type dye aggregates within the tetravalent MAP system.

Vladimir S. Talanov et al [51] reported a dendrimer-based nanoprobe for dual modality magnetic resonance and fluorescence imaging. A novel PAMAM dendrimer-based nanoprobe for dual magnetic resonance and fluorescence imaging modalities was synthesized. Fluorescence studies revealed that Gd(III) complexation to the probe has no effect on the quantum yield; however, increases in the dye content resulted in partial quenching. The potential of the new nanoprobe as a dual modality imaging agent was demonstrated in vivo by the efficient visualization of sentinel lymph nodes in mice by both MRI and fluorescence imaging modalities (Figure 21).

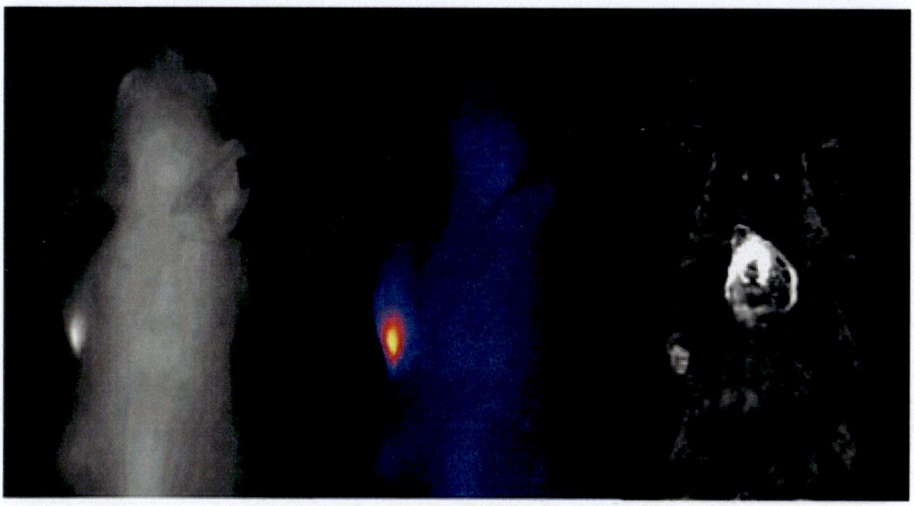

A B C

Figure 21. Dendrimer-based nanoprobe for dual modality magnetic resonance and fluorescence imaging.

4.2. Drug and Gene Delivery

The control of gene transfection in the body is a core issue in gene therapy. Photochemical internalization is a technology that allows light-induced delivery of DNA, drugs or other biological factors directly inside cells. Usually it requires that a photosensitizer be added to the drug-delivery system to photochemically destabilize the endosomal membrane. Nobuhiro Nishiyama et al [52] reported light-induced gene transfer from packaged DNA enveloped in a dendrimeric photosensitizer (Figure 22). A system for in vivo DNA delivery in which these two components are assembled into one structure was presented. This is a ternary complex composed of a core containing DNA packaged with cationic peptides and enveloped in the anionic dendrimer phthalocyanine, which provides the photosensitizing action. The ternary complex showed more than 100-fold photochemical enhancement of transgene expression in vitro with reduced photocytotoxicity. In an animal experiment, subconjuctival injection of the ternary complex followed by laser irradiation resulted in transgene expression only in the laser-irradiated site. Nobuhiro Nishiyama et al demonstrated a new biomedical application for dendrimers, and the first success in the photochemical-internalization-mediated gene delivery in vivo.

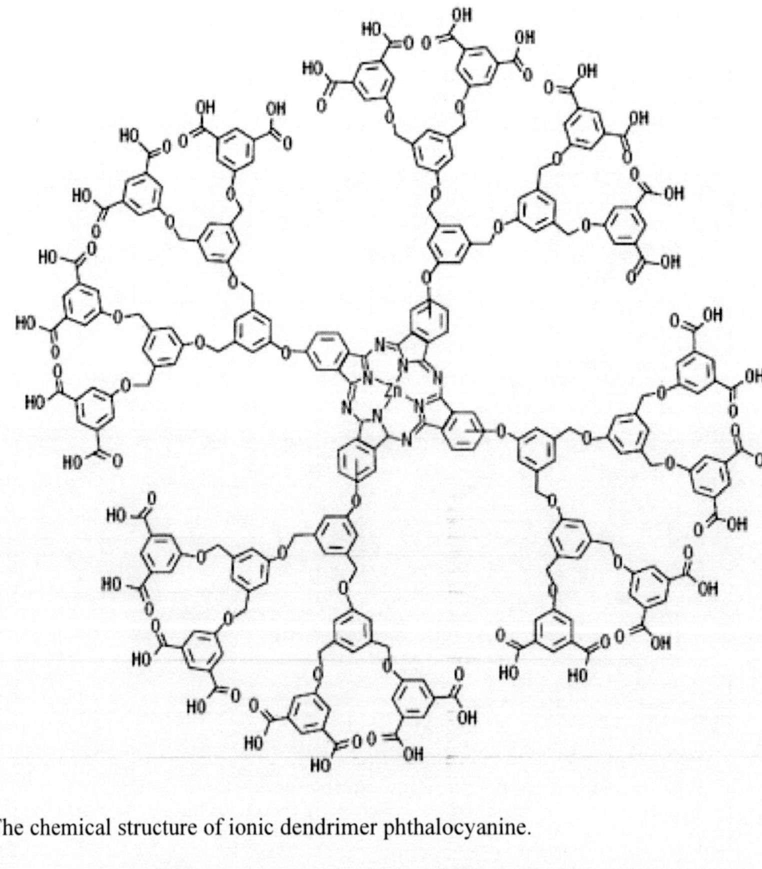

Figure 22. The chemical structure of ionic dendrimer phthalocyanine.

M. Adeli et al [53] reported linear-dendritic ABA triblock copolymers as nanocarriers. Linear-dendritic ABA triblock copolymers containing PEG were used for transport the small guest molecules such as 5,7-dibromo-8-hydoxy quinoline, ibuprofen, and Congo red. The linear-dendritic copolymers/ guest molecule complexes were stable at room temperature for about 10 months; during this time, guest molecules did not release from linear-dendritic copolymers/guest molecule complexes.

M. Adeli et al [54] reported multiarm star nanocarriers containing a poly(ethylene imine) core and polylactide arms. Star polymers (SPs) containing a hyperbranched poly(ethylene imine) (PEI; number-average molecular weight = 10,000) core and polylactide arms were synthesized via the ring-opening polymerization of lactide. PEI was used as a multifunctional macroinitiator for the ring-opening polymerization of lactide. Different lactide monomer/ amino-functional group (LA/NH_n; $n = 1$ or 2) ratios were used for preparing SPs with different molecular weights. SPs were able to encapsulate small guest molecules such as Rose Bengal; they also transported small, hydrophilic molecules from water to the organic phase. The transport capacity of all the nanocarriers depended on the LA/ NH_n ratio used for synthesizing the SPs. Nanocarriers with a higher LA/NH_n ratio had higher transport capacities.

Robert Langer et al [55] reported a family of hierarchically self-assembling linear-dendritic hybrid polymers for highly efficient targeted gene delivery.

James R. Baker Jr et al [56] have demonstrated the applicability of poly(amidoamine) (PAMAM) dendrimers as a platform for the targeted delivery of chemotherapeutic drugs both in vitro and in vivo. They showed the synthesis, characterization, and in vitro targeting of an engineered PAMAM nanodevice in which folic acid (FA) is conjugated as the targeting molecule and a caspase-specific FRET-based agent (PhiPhiLux G1D2) is conjugated as the apoptosis-detecting agent. This conjugate specifically targets FA-receptor-positive, KB cells. In these cells, the apoptosis-inducing agent staurosporine caused a 5-fold increase in the cellular fluorescence. These results show, for the first time, the potential applicability of a targeted apoptosis-measuring nanodevice, which could be used for simultaneously monitoring the apoptotic potential of a delivered drug.

James R. Baker Jr et al [57] reported the synthesis and human growth factor receptor-2 (HER2) specific tumor targeting properties of a dendrimer conjugated to anti-HER2 mAb (monoclonal antibody) conjugate. The polyamidoamine (PAMAM) dendrimer generation five (G5) was labeled with alexaFluor 488 and conjugated to anti-HER2 mAb. Uniquely, the conjugate demonstrated cellular uptake and internalization in HER2-expressing cells as compared to free antibody. The time course of internalization and blocking experiments with free antibody suggest that the rapid and efficient cellular internalization of the dendrimer-antibody conjugate was achieved without alterations in specificity of targeting. Animal studies demonstrated that the conjugate targets HER2-expressing tumors.

Rangaramanujam M. Kannan et al [58] demonstrated the activity of dendrimer-methotrexate conjugates on methotrexate-sensitive and -resistant cell lines. They have investigated the effect of dendrimer end-functionality on the activity of polyamido amine (PAMAM) dendrimer-methotrexate (MTX) conjugates in MTX-sensitive and MTX-resistant human acute lymphoblastoid leukemia (CCRF-CEM) and Chinese hamster ovary (CHO) cell lines (Figure 23). Two amide-bonded PAMAM dendrimer-MTX conjugates were prepared using a dicyclohexylcarbodiimide (DCC) coupling reaction: one between a carboxylic acid-terminated G2.5 dendrimer and the amine groups of the MTX (conjugate A) and another between an amine-terminated G3 dendrimer and the carboxylic acid group of the MTX (conjugate B). The results suggested that conjugate A showed an increased drug activity compared to an equimolar amount of free MTX toward both sensitive and resistant cell lines, whereas conjugate B did not show significant activity on any of the cell lines. These findings demonstrated the feasibility of using dendrimers as drug delivery vehicles for achieving higher therapeutic effects in chemotherapy, especially in drug-resistant cells.

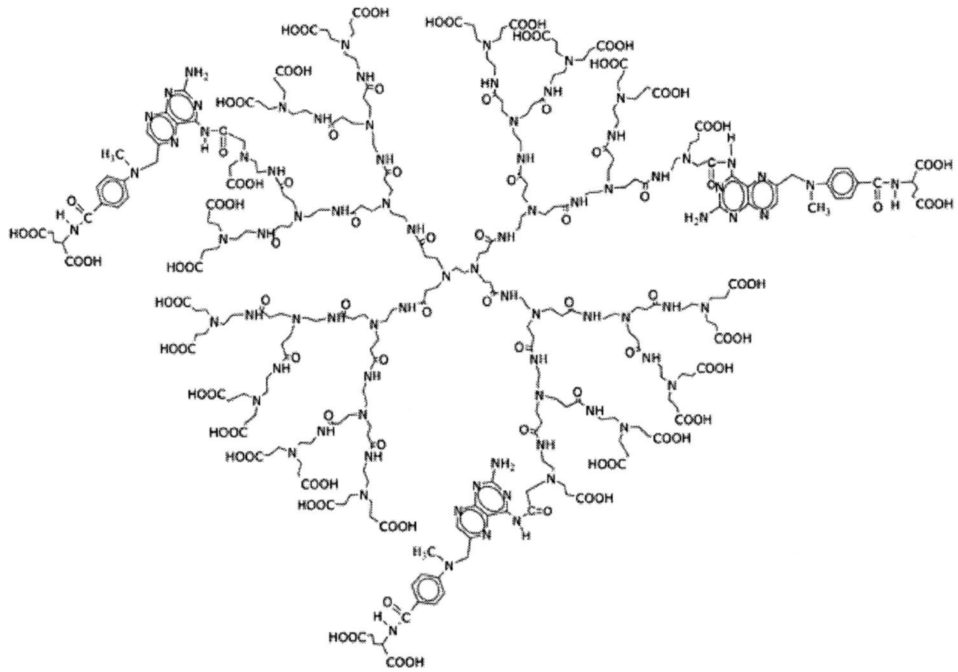

Figure 23. PAMAM-G2.5-COOH-MTX conjugate.

Patrice Hildgen et al [59] reported the synthesis and evaluation of novel dendrimers with a hydrophilic interior as nanocarriers for drug delivery. Novel polyester-co-polyether dendrimers consisting of a hydrophilic core were synthesized by a combination of convergent and divergent syntheses (Figure 24). The core was synthesized from biocompatible moieties, butanetetracarboxylic acid and aspartic acid, and the dendrons from PEO (poly(ethylene oxide)), dihydroxybenzoic acid or gallic acid, and PEG monomethacrylate. Dendrimers demonstrated good ability to encapsulate the guest molecule, with loading of 15.80 and 6.47% w/w for rhodamine and β-carotene, respectively. The release of the encapsulated compounds was found to be slow and sustained, suggesting that these dendrimers can serve as potential drug delivery vehicles.

István J. Majoros et al [60] have designed and synthesized poly(amidoamine) (PAMAM) dendrimer-based multifunctional cancer therapeutic conjugates. The primary amino groups on the surface of the generation 5 (G5) PAMAM dendrimer were neutralized through partial acetylation, providing enhanced solubility of the dendrimer (in conjugation of FITC (fluorescein isothiocyanate)) and preventing nonspecific targeting interactions (in vitro and in vivo) during delivery. The functional molecules fluorescein isothiocyanate (FITC, an imaging agent), folic acid (FA, targets overexpressed folate receptors on specific cancer cells), and paclitaxel (taxol, a chemotherapeutic drug) were conjugated to the remaining nonacetylated primary amino groups. These multifunctional dendrimer conjugates have been tested in vitro for targeted delivery of chemotherapeutic and imaging agents to specific cancer cells.

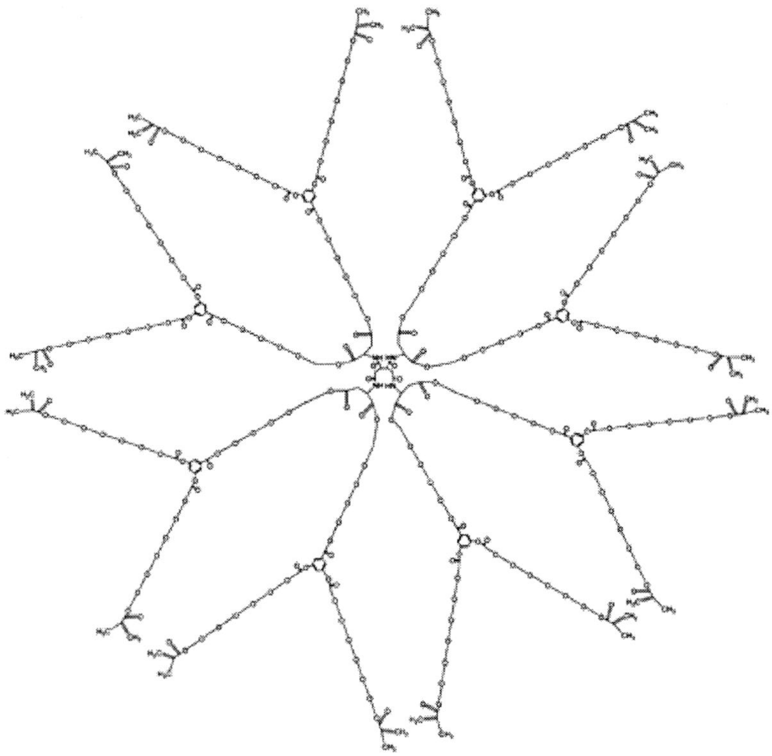

Figure 24. Novel dendrimers with a hydrophilic interior as nanocarriers for drug delivery.

Dendrimers, dendrons, and hyperbranched polymers are gaining popularity as novel drugs, imaging agents, and drug delivery systems. They present advantages of well-defined molecular weight, multivalent surfaces, and high drug carrying capacity. Moreover, it is emerging that such architectures can display unique endocytic properties. As poly(ethylene glycol) (PEG) is widely used for protein and drug conjugation, the aim of this study was for the first time to synthesize novel, branched PEG-based architectures, to define their cytotoxicity and, via preparation of Oregon green (OG) conjugates define the effect of structure on their cellular uptake. Ruth Duncan et al [61] reported novel monodisperse PEG-dendrons as new tools for targeted drug delivery. Five PEG-based dendrons were synthesized using monodisperse Fmoc-amino PEG propionic acid (Mw = 840) as a monomer, and cadaverine, tris(2-aminoethyl)amine or lysine as the branching moieties. These well-defined novel architectures have potential for further development as targetable drug delivery systems or tools for construction of structurally defined modified surfaces.

Jong-sang Park et al [62] synthesized arginine-conjugated polypropylenimine dendrimer G2 (DAB-8), PPI2-R for gene delivery systems. Results demonstrate that arginine-conjugation of PPI2 is successful in developing a low toxic and highly transfection efficient gene delivery carrier.

Prakash V. Diwan et al [63] developed folate-PAMAM dendrimer conjugates for targeted delivery of anti-arthritic drugs and studied their pharmacokinetics and biodistribution in arthritic rats. The aim of this study was to synthesize folate-dendrimer conjugates as suitable vehicle for site specific delivery of anti-arthritic drug (indomethacin) to inflammatory regions and to determine its targeting efficiency, biodistribution in adjuvant induced arthritic rats.

This study demonstrated the superiority of active targeting over dendrimer mediated passive targeting and also for the first time, folate-mediated targeting of an anti-arthritic drug to the inflammatory tissues.

Mark M. Banaszak Holl et al [64] reported the binding avidity of a nanoparticle-based multivalent targeted drug delivery platform. Dendrimer-based anticancer nanotherapeutics containing ~5 folate molecules have shown in vitro and in vivo efficacy in cancer cell targeting. This study provides quantitative and systematic evidence for multivalent interactions between these nanodevices and folate-binding protein (FBP). Qualitative evidence is also provided for a multivalent targeting effect to KB cells using flow cytometry. These data support the hypothesis that multivalent enhancement of KD, not an enhanced rate of endocytosis, is the key factor resulting in the improved biological targeting by these drug delivery platforms.

Mitsuru Hashida et al [65] synthesized a sixth generation lysine dendrimer (KG6) and two PEGylated derivatives thereof and evaluated their biodistribution characteristics in both normal and tumor-bearing mice. The intact KG6 showed a rapid clearance from the blood stream and nonspecific accumulation in the liver and kidney. In contrast, the PEGylated derivatives showed a better retention in blood and low accumulativeness in organs dependent of the rate of PEGylation. In addition, PEGylated KG6 with a high modification rate was accumulated effectively in tumor tissue via the enhanced permeability and retention (EPR) effect. Moreover, Mitsuru Hashida et al clarified that multiple administrations did not affect the biodistribution characteristics of a second dose of PEGylated KG6. PEGylated lysine dendrimer would be a useful material for a clinically applicable tumor-targeting carrier.

Luca Costantino et al [66] described a general method for incorporating target moieties in a well-defined arrangement into the surface of biocompatible polyester poly(D,L-lactic-co-glycolic acid) (PLGA) materials using dendrons. In this way it is possible to obtain nanoparticles (NPs) with a high degree of surface coverage. This new strategy was successfully applied to the preparation of peptide- and β-D-glucose-covered NPs. The first application is based on the discovery of NPs made of conjugates between PLGA and short peptidic sequences able to cross the blood–brain barrier (BBB) after systemic administration. Luca Costantino et al used a branched structure (dendron) in order to prepare a derivative of PLGA able to form, by simple nanoprecipitation, NPs with a high degree of surface coverage, characteristic that could influence the uptake by the liver and spleen. The same strategy, namely the use of a branched spacer (a dendron or a G1 dendrimer) inserted between one end of the PLGA chain and a derivatizing molecule, was also successfully applied to obtain β-D-glucose-covered NPs.

Dendrimers are emerging as a relatively new class of polymeric biomaterials withapplications in drug delivery, and imaging. Achieving a high drug payload in dendrimers, and understanding the therapeutic effect of the dendrimer–drug conjugates are receiving increasing attention. Rangaramanujam M. Kannan et al [67] reported the preparation, cellular transport, and activity of polyamidoamine-based dendritic nanodevices with a high drug payload. A high drug payload nanodevice was obtained by covalent conjugation of ibuprofen to a polyamidoamine (PAMAM-G4-OH) dendrimer. Significant amounts of the conjugate entered the cells rapidly within 15 min. Suppression of prostaglandin was noted within 30 min for the dendrimer–drug conjugates versus 1 h for the free ibuprofen. The results suggest that dendrimers with high drug payload improve the drug's efficacy by enhanced cellular delivery, and may produce a rapid pharmacological response. These dendrimer–drug conjugates can potentially be further modified by attaching antibodies and ligands for targeted drug delivery.

H. Namazi et al [68] reported the dendrimers of citric acid and poly (ethylene glycol) as the new drug-delivery agents. Citric acid–polyethylene glycol–citric acid (CPEGC) triblock dendrimers as biocompatible compounds containing G1, G2 and G3 were applied as the drug-delivery systems. Some of the small size molecules and drugs are trapped with the above-synthesized dendrimers. The guest molecules, which are hydrophobic when trapped into the suitable sites of dendrimers, are becoming soluble in aqueous solution. The quantity of trapped molecules and drugs such as 5-amino salicylic acid (5-ASA), pyridine, mefenamic acid, and diclofenac was measured. The drug/dendrimer complexes remained in room temperature for about 10 months and after this long time they were stable and the drugs were not released.

James R. Baker, Jr. et al [69] reported the synthesis and functional evaluation of DNA-assembled polyamidoamine dendrimer clusters for cancer cell-specific targeting. They sought to produce dendrimers conjugated to different biofunctional moieties (fluorescein [FITC] and folic acid [FA]), and then link them together using complementary DNA oligonucleotides to produce clustered molecules that target cancer cells that overexpress the high-affinity folate receptor (Figure 25). Amine-terminated, generation 5 polyamidoamine (G5 PAMAM) dendrimers are first partially acetylated and then conjugated with FITC or FA, followed by the covalent attachment of complementary, 5'-phosphate-modified 34-base-long oligonucleotides. Hybridization of these oligonucleotide conjugates led to the self-assembly of the FITC- and FA-conjugated dendrimers. In vitro studies of the DNA-linked dendrimer clusters indicated specific binding to KB cells expressing the folate receptor. Confocal microscopy also showed the internalization of the dendrimer cluster. These results demonstrate the ability to design and produce supramolecular arrays of dendrimers using oligonucleotide bridges. This will also allow for further development of DNA-linked dendrimer clusters as imaging agents and therapeutics.

Saghir Akhtar et al [70] reported a novel anionic dendrimer for improved cellular delivery of antisense oligonucleotides. The optimal design of hybridisation-competent antisense oligonucleotides (ODNs) coupled with an efficient delivery system appear to be important prerequisites for the successful use of antisense reagents for gene silencing.

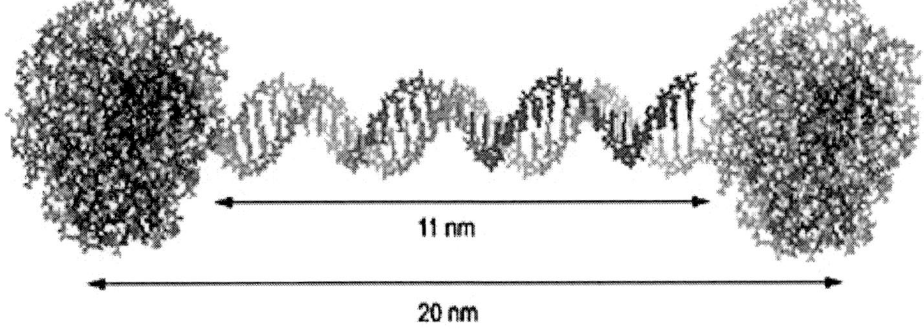

Figure 25. DNA-Assembled Polyamidoamine Dendrimer Clusters for Cancer Cell-Specific Targeting.

Saghir Akhtar et al selected an antisense ODN complementary to an accessible region of the epidermal growth factor receptor (EGFR) mRNA with the aid of an antisense oligonucleotide scanning array. This dendrimer conjugate was markedly more stable to serum nucleases compared to the free ODNs and the cellular uptake of ODN-dendrimer conjugates was up to 100-fold greater as compared to mannitol, a marker for fluid phase endocytosis, and up to 4-fold greater than naked ODN in cancer cells.

Eric E. Simanek et al [71] reported in vitro and in vivo evaluation of a melamine dendrimer as a vehicle for drug delivery. Cell-based and acute and subchronic in vivo toxicity profiles of a dendrimer based on melamine reveal that this class of molecules warrants additional study as vehicles for drug delivery (Figure 26).

Polymeric drug delivery systems are used not only to improve aqueous solubility of drug molecules but also to achieve desirable pharmacokinetics and an enhanced therapeutic index. New biodegradable polymers are needed to improve the biodistribution and targeting-ability of polymeric carriers. Chun Li et al [72] described the synthesis and characterization of branched poly(Lglutamic acid) (PG) containing multiple PG chains centered on a poly(amidoamine) (PAMAM) dendrimer or polyethyleneimine (PEI) cores. The branched PG polymers were obtained by ring-opening polymerization of benzyl ester of Lglutamic acid N-carboxyanhydride using PAMAM or PEI as the initiator. These polymers were degradable in the presence of the lysosomal enzyme cathepsin B, albeit more slowly than linear PG. Unlike conventional linear PG, each branched PG possessed multiple terminal amino groups.

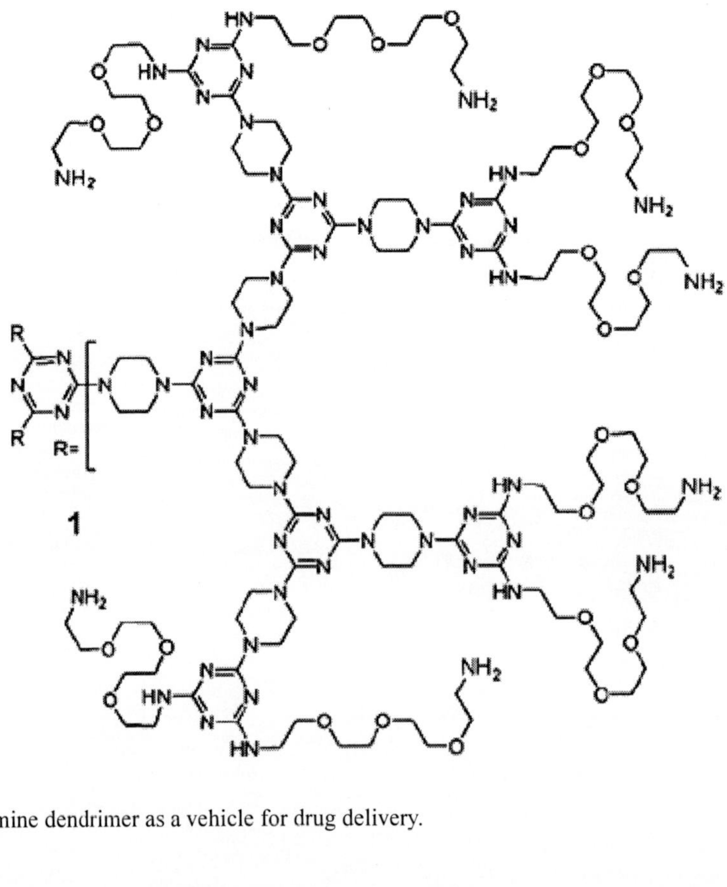

Figure 26. Melamine dendrimer as a vehicle for drug delivery.

This made it possible to attach multiple targeting moieties selectively to the termini of branched PG. Conjugation of monofunctional or heterodifunctional polyethylene glycol to the chain ends of branched PG was demonstrated in the presence of side chain carboxyl groups. Furthermore, folic acid, a model targeting moiety, and the nearinfrared dye indocyanine green, a model diagnostic agent, were successfully conjugated to the terminal amino groups and the side chain carboxyl groups of branched PG, respectively. The resulting conjugate had reduced nonspecific interaction and bound selectively to tumor cells expressing folate receptors (Figure 27).

To establish an effective nonviral gene transfer vector to hepatocytes, various oligo-carrier complexes were developed employing dendrimer (G4) and avidin–biotin systems (Av–bt), and their biodistribution were evaluated by Tsuneo Saga et al [73]. In-111-labeled-oligo, without any carriers, showed low uptake in normal organs other than the kidney. In contrast, 111In-oligo coupled with avidin through biotin (111In-oligo-bt-Av) showed very high accumulation in the liver. 111In-oligo complexed with G4 showed high uptake in the kidney and spleen, but its hepatic uptake was relatively low. When both G4 and Av–bt systems were employed, 111In-oligo/G4-bt-Av showed extremely high uptake in the lung. 111In-oligo-bt-Av, which exhibited the highest hepatic uptake in vivo, also showed high and rapid internalization into hepatocytes. The avidin–biotin system seems to have potential as a carrier of oligo-DNA to the liver.

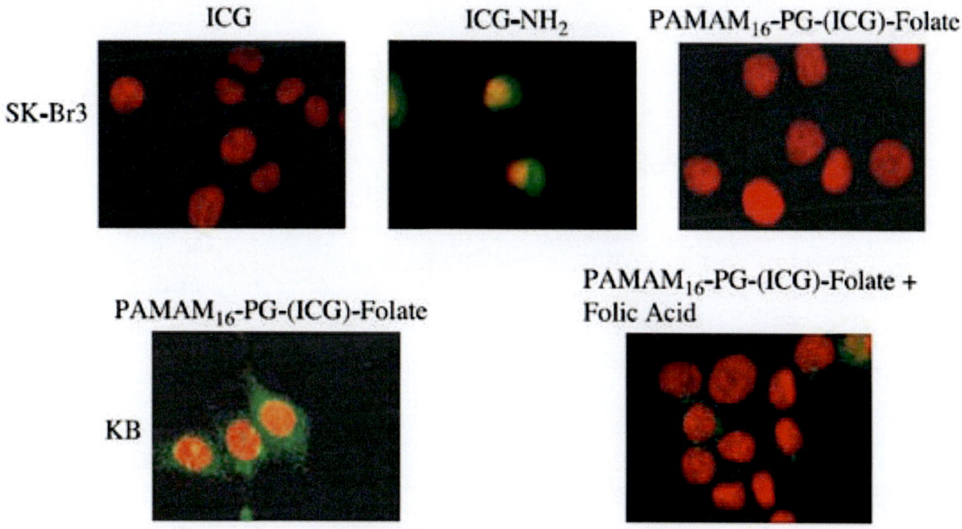

Figure 27. Branched poly(L-glutamic acid) as a biodegradable drug carrier.

4.3. Advances in Tissue Engineering

Sunil Shaunak et al [74] used anionic, polyamidoamine, generation 3.5 dendrimers to make novel water-soluble conjugates of D(+)-glucosamine and D(+)-glucosamine 6-sulfate with immuno-modulatory and antiangiogenic properties respectively (Figure 28).

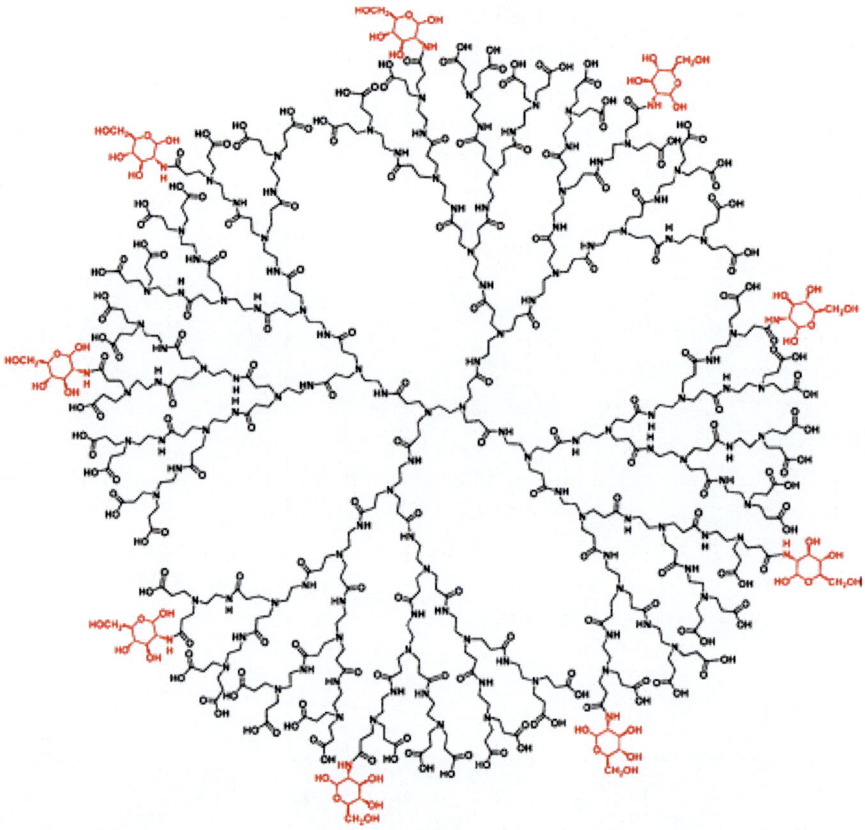

Figure 28. Polyvalent dendrimer glucosamine conjugates.

Dendrimer glucosamine inhibited Toll-like receptor 4–mediated lipopolysaccharide induced synthesis of pro-inflammatory chemokines (MIP-1α, MIP-1β, IL-8) and cytokines (TNF-α, IL-1β, IL-6) from human dendritic cells and macrophages but allowed upregulation of the costimulatory molecules CD25, CD80, CD83 and CD86. Sunil Shaunak et al concluded that synthetically engineered macromolecules such as the dendrimers described here can be tailored to have defined immuno-modulatory and antiangiogenic properties, and they can be used synergistically to prevent scar tissue formation.

Photo-crosslinkable dendritic macromolecules are attractive materials for the preparation of cartilage tissue engineering scaffolds that may be optimized for in situ formation of hydrated, mechanically stable, and wellintegrated hydrogel scaffolds supporting chondrocytes and chondrogenesis. Mark W. Grinstaff et al [75] reported biodendrimer-based hydrogel scaffolds for cartilage tissue repair. They designed and synthesized a novel hydrogel scaffold for cartilage repair, based on a multivalent and water-soluble tri-block copolymer consisting of a poly(ethylene glycol) core and methacrylated poly(glycerol succinic acid) dendrimer terminal blocks. The terminal methacrylates allow mild and biocompatible photo-crosslinking with a visible light, facilitating in vivo filling of irregularly shaped defects with the dendrimer-based scaffold. The multivalent dendrimer constituents allow high crosslink densities that inhibit swelling after crosslinking while simultaneously introducing biodegradation sites (Figure 29). The mechanical properties and water content of the hydrogel can easily be tuned by changing the biodendrimer concentration. In vitro chondrocyte encapsulation studies demonstrate significant synthesis of neocartilaginous material, containing proteoglycans and type II collagen.

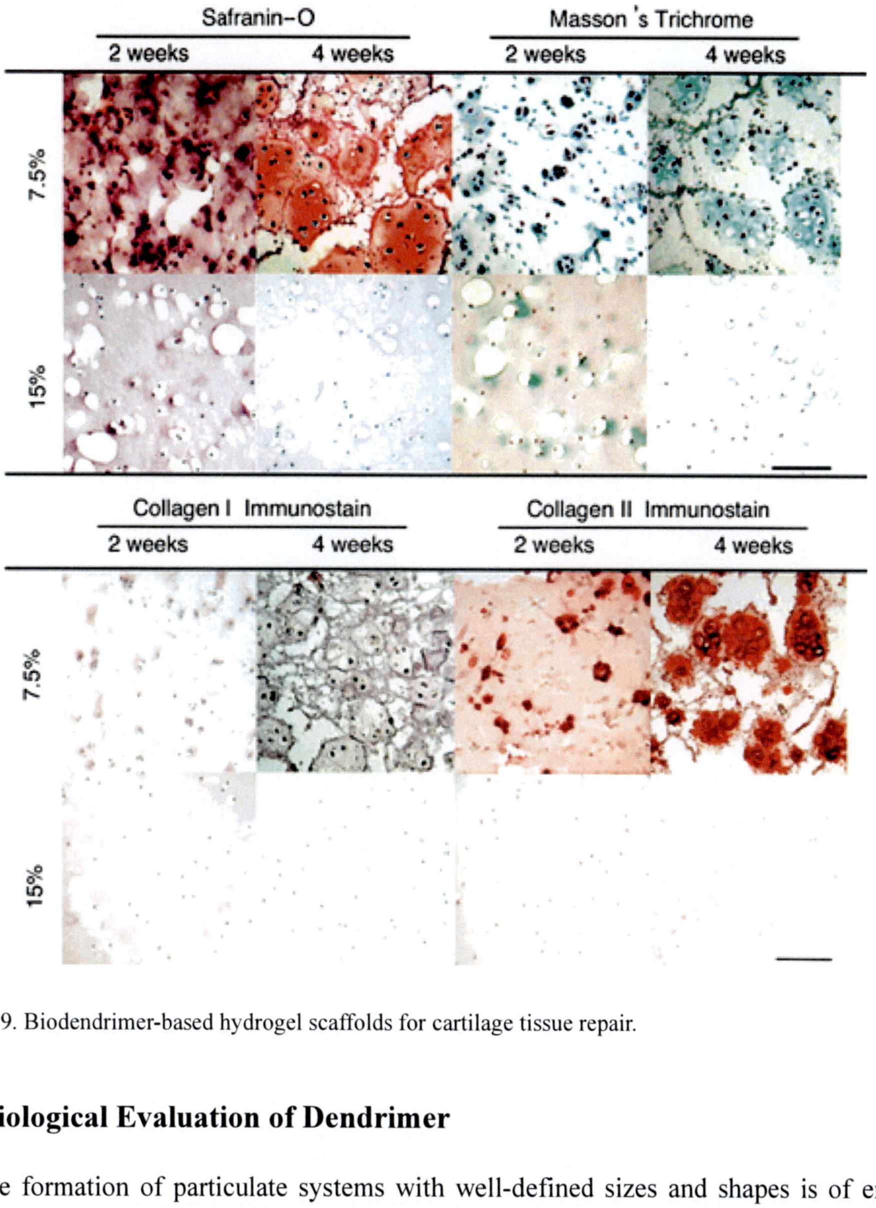

Figure 29. Biodendrimer-based hydrogel scaffolds for cartilage tissue repair.

4.4. Biological Evaluation of Dendrimer

The formation of particulate systems with well-defined sizes and shapes is of eminent interest in certain medical applications such as drug delivery, gene transfection, and imaging. The high level of control possible over the architectural design of dendrimers; their size, shape, branching length/density, and their surface functionality, clearly distinguishes these structures as unique and optimum carriers in those applications. The bioactive agents may be encapsulated into the interior of the dendrimers or chemically attached/physically adsorbed onto the dendrimer surface, with the option of tailoring the carrier to the specific needs of the active material and its therapeutic applications. In this regard, the high density of exo-presented surface groups allows attachment of targeting groups or functionality that may modify the solution behavior or toxicity of dendrimers. Quite remarkably, modified dendrimers have been shown to act as nano-drugs against tumors, bacteria, and viruses. Recent successes in simplifying and optimizing the synthesis of dendrimers such as the dlegoT and dclickT approaches, provide a large variety of structures while at the same time reducing the cost of their production. The reflections on biomedical applications of dendrimers given in this review clearly demonstrate the potential of this new fourth major class of polymer architecture and indeed substantiate the high hopes for the future of dendrimers [76].

The field of biomedical dendrimers is still in its infancy, but the explosion of interest in dendrimers and dendronised polymers as inherently active therapeutic agents, as vectors for targeted delivery of drugs, peptides and oligonucleotides, and as permeability enhancers able to promote oral and transdermal drug delivery makes it timely to review current knowledge regarding the toxicology of these dendrimer chemistries (currently under development for biomedical applications) [77]. Clinical experience with polymeric excipients, plasma expanders, and most recently the development of more dclassical polymerTderived therapeutics can be used to guide development of bsafeQ dendritic polymers. Moreover, in future it will only ever be possible to designate a dendrimer as bsafeQ when related to a specific application. The so far limited clinical experience using dendrimers make it impossible to designate any particular chemistry intrinsically bsafeQ or btoxicQ. Although there is widespread concern as to the safety of nano-sized particles, preclinical and clinical experience gained during the development of polymeric excipients, biomedical polymers and polymer therapeutics shows that judicious development of dendrimer chemistry for each specific application will ensure development of safe and important materials for biomedical and pharmaceutical use.

Dendrimers are branched, synthetic polymers with layered architectures that show promise in several biomedical applications. By regulating dendrimer synthesis, it is possible to precisely manipulate both their molecular weight and chemical composition, thereby allowing predictable tuning of their biocompatibility and pharmacokinetics. Advances in the understanding of the role of molecular weight and architecture on the in vivo behavior of dendrimers, together with recent progress in the design of biodegradable chemistries, has enabled the application of these branched polymers as anti-viral drugs, tissue repair scaffolds, targeted carriers of chemotherapeutics and optical oxygen sensors [78]. Before such products can reach the market, however, the field must not only address the cost of manufacture and quality control of pharmaceuticalgrade materials, but also assess the long-term human and environmental health consequences of dendrimer exposure in vivo.

Because of these well-defined characteristics, PAMAMs are finding utility in a variety of applications, many of which are biological in nature. Little is known, however, about the biological behavior of the PAMAMs, which is critical to their use in vivo. Jeanette C. Roberts et al [79] reported the preliminary biological evaluation of polyamidoamine (PAMAM) Starburst dendrimers. Generation 3, 5, and 7 PAMAMs were studied in V79 cells or in Swiss-Webster, mice for a number of biological properties, including (1) in vitro toxicity, (2) in vivo toxicity, (3) immunogenicity, and (4) biodistribution. Potential biological complications were observed only with G7 at the highest level tested. No evidence of immunogenicity was seen. The biodistribution properties of the Starburst dendrimers were rather unusual. G3 showed the highest accumulation in kidney tissue; G5 and G7 appeared to preferentially localize in the pancreas. In addition, G7 showed extremely high urinary excretion. In general, the dendrimers did not exhibit properties that would preclude their use in biological applications.

Muriel Blanzat et al [80] reported in vitro anti-HIV activity of phosphorus-containing dendrimers bearing Galβ1cer analogues. Two series of water-soluble dendritic catanionic assemblies, acting as multisite analogues of galactosylceramide (Galb1cer), have been prepared with the goal of blocking HIV infection prior to the entry of the virus into human cells. Trifunctional and hexafunctional cinnamic acid-terminated dendrimers have been synthesized from phosphorus-containing dendrimers bearing aldehyde end groups. A classical acid–base reaction performed in water between acid-terminated dendrimers and stoichiometric amounts of N-hexadecylamino-1-deoxylactitol provided the expected catanionic assemblies.

Koji Matsuoka et al [81] reported thiosialoside clusters using carbosilane dendrimer core scaffolds as a new class of influenza neuraminidase inhibitors. An efficient synthesis of a series of carbosilane dendrimers uniformly functionalized with α-thioglycoside of sialic acid was accomplished. The results of a preliminary study on biological responses against influenza virus sialidases using thiosialoside clusters showed that some of the glycodendrimers have inhibitory potencies against the sialidases.

Mitsuru Hashida et al [82] synthesized dendritic poly(L-lysine)s (DPKs), dendritic poly(L-ornithine)s (DPOs), which are constructed as novel amino acid dendrimers, and PEGylated KG6 (the sixth generation of DPKs), and evaluated the physicochemical properties and biodistribution characteristics of these dendrimers. After intravenous administration to mice, all tested dendrimers cleared rapidly from blood flow and mainly accumulated in the liver and kidney. The hepatic and renal accumulation changed in a generationdependent manner.

Costas Demetzos et al [83] reported the doxorubicin–PAMAM dendrimer complex attached to liposomes and the cytotoxic studies against human cancer cell lines. The low release rate of doxorubicin as well as the high incorporation efficiency of doxorubicin–PAMAM complex into liposomes are considered as beneficial factors concerning the activity of doxorubicin. Liposomal formulations composed of lipids and of a drug–dendrimer complex could be characterized as modulatory liposomal controlled release system, and could provide benefits to the delivery of drugs and modulate their release.

Natarajan Raju et al [84] reported the synthesis of a new scaffold derived from iminodipropionic acid for the preparation of peptide dimers and tetramers.

Toward the production of unimolecular nanocapsules with biocompatible surface, Kenji Kono et al [85] prepared a fourth generation polyamidoamine dendrimer having both a poly(ethylene glycol) (PEG) with the number average molecular weight of 2000 and a methacryloyl group at every chain end of the dendrimer through an L-lysine residue. Polymerization of the methacryloyl groups resulted in a significant reduction of the affinity of the dendrimer to a guest molecule, rose bengal, suggesting that the peripheral network formed by the linking of methacryloyl groups hides the dendrimer interior. In addition, when the methacryloyl groups of the dendrimer were polymerized in the presence of rose bengal, the guest molecule was found to be tightly associated with the dendrimer.

4.5. Pharmacokinetics of Dendrimers

Christopher J. H. Porter et al [86] reported the pharmacokinetics, biodistribution, and evidence for metabolism and bioresorption after intravenous administration to rats of cationic poly-L-lysine dendrimers. Cationic poly-L-lysine 3H-dendrimers with either 16 or 32 surface amine groups have been synthesized and their pharmacokinetics and biodistribution investigated after intravenous administration to rats. Plasma and whole blood pharmacokinetics were similar, precluding interaction with blood components as a causative factor in either the rapid removal or reappearance of radioactivity in plasma. For the L-lysine capped dendrimers, radiolabel was widely distributed throughout the major organs, with no apparent selectivity for organs of the reticuloendothelial system. In contrast, a greater proportion of the administered radiolabel was recovered in the organs of the reticuloendothelial system for the D-lysine capped system, as might be expected for a nondegrading circulating foreign colloid. This is the first data to demonstrate the biodegradation/bioresorption of poly-L-lysine dendrimers and has significant implications for the utility of these systems as drugs or drug delivery systems.

Single-triggered disassemble dendrimers were recently developed and introduced as a potential platform for a multi-prodrug. These unique structural dendrimers can release all of their tail units through a self-immolative chain fragmentation initiated by a single cleavage at the dendrimer's core. There are several examples for the bioactivation of first-generation self-immolative dendritic prodrugs. However, enzymatic activation failed for second-generation self-immolative dendrimers. The hydrophobic large molecular structure of the dendritic prodrugs results in aggregation under aqueous conditions and prevented the enzyme from reaching the triggering substrate. Doron Shabat et al [87] showed a simple solution for the enzymatic activation of second-generation self-immolative dendrimers. Poly(ethylene glycol) (PEG) was conjugated to the dendritic platform via click chemistry (Figure 30). The poly(ethylene glycol) tails significantly decreased the hydrophobic properties of the dendrimers and thereby prevented aggregate formation. Doron Shabat et al designed and synthesized a dendritic prodrug with four molecules of the anticancer agent camptothecin and a trigger that can be activated by penicillin-G-amidase. The PEG5000-conjugated, self-immolative dendritic prodrug was effectively activated by penicillin-G-amidase under physiological conditions and free camptothecin was released to the reaction media. Cell-growth inhibition assays demonstrated increased toxicity of the dendritic prodrug upon incubation with the enzyme.

Ijeoma F. Uchegbu et al [88] reported preferential liver gene expression with polypropylenimine dendrimers. Quaternisation of dendrimer proved to be critical in improving DNA binding, as evidenced by data from the ethidium bromide exclusion assay and dendrimer–DNA colloidal stability data. This improved colloidal stability had a major effect on vector tolerability, as dendrimer–DNA formulations were well tolerated on intravenous injection while a similar dendrimer–DNA dose was lethally toxic by the same route. Ijeoma F. Uchegbu et al concluded that the polypropylenimine dendrimers are promising gene delivery systems which may be used to target the liver and avoid the lung and also that molecular modifications conferring colloidal stability on gene delivery formulations have a profound effect on their tolerability on intravenous administration.

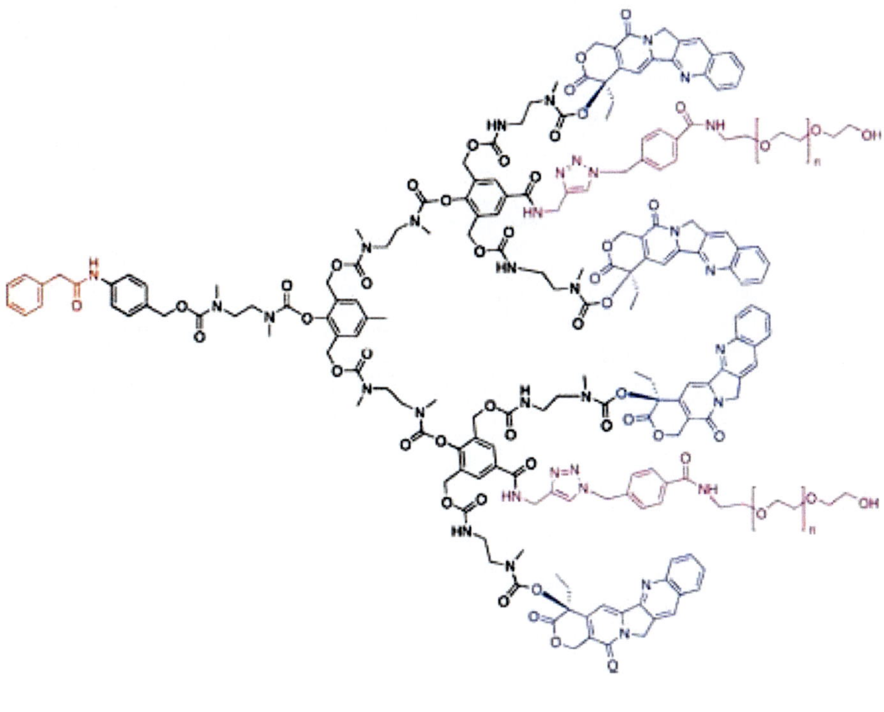

Figure 30. Conjugation of self-immolative dendrimers with poly(ethylene glycol) via click chemistry.

Antony D'Emanuele et al [89] reported the use of a dendrimer-propranolol prodrug to bypass efflux transporters and enhance oral bioavailability. The aim of the study was to determine the effects on the transport of propranolol across monolayers of the human colon adenocarcinoma cell line, Caco-2, of forming a prodrug by conjugating to generation 3 (G3) and lauroyl-G3 PAMAM dendrimers. The findings show that the A→B Papp of propranolol conjugates was reduced in the presence of the endocytosis inhibitor colchicines and was lower at 4 °C than at 37°C, suggesting that the enhancement mechanism involves endocytosis-mediated transepithelial transport. The A→B Papp of conjugated propranolol was not altered in the presence of the P-gp inhibitor cyclosporin A suggesting that conjugation of drug to dendrimer allows the bypassing of the efflux transporter. The results suggest that dendrimer-drug prodrugs may be used to increase drug solubility and bypass drug efflux transporters, therefore increasing drug bioavailability.

4.6. Peptides Dendrimers

The main stress is given to peptide dendrimers, namely, multiple antigenic peptides (MAPs). Their synthesis, physicochemical properties, biological activities, etc. have been described with many examples. MAPs can be used as diagnostics, mimetics, for complexation of different cations, as vaccines against parasites, bacteria, viruses, etc [90].

PEGylated polyamidoamine (PAMAM) dendrimers as drug carriers have been a topic of interest because of their biomedically favorable features, including minimal toxicity, reduced immunogenicity, and excellent solubility in aqueous and most organic solutions. A PEG shell on dendrimer surface may provide steric hindrance, known as stealth properties of PEG, to stabilize drug molecules to be delivered. Hu Yang et al [91] reported in vitro enzymatic stability of dendritic peptides. The effects of PEG and coupling sequence of drug, PEG, and dendrimer in modulating the stability of delivered drug molecules were evaluated. Nsuccinyl-Ala-Ala-Pro-Phe-p-nitroanilide was chosen as a model peptide. Dendritic peptides, that is, peptide-dendrimer, peptide-PAMAM-PEG, and peptide-PEG-dendrimer, were constructed based on Starburst G3.0 PAMAM dendrimer. The enzymatic stability of dendritic peptides was peptide-PAMAM-PEG > peptide- PAMAM > free peptide > peptide-PEG-PAMAM. The ratio of PEG/peptide could be reduced for increasing peptide loading while maintaining the delivered peptides' relatively high enzymatic stability.

Jean-Louis Reymond et al [92] reported peptide dendrimers as artificial enzymes, receptors, and drug-delivery agents. The dendritic architecture applied to peptides provides a practical entry into globular macromolecules resembling proteins (Figure 31).

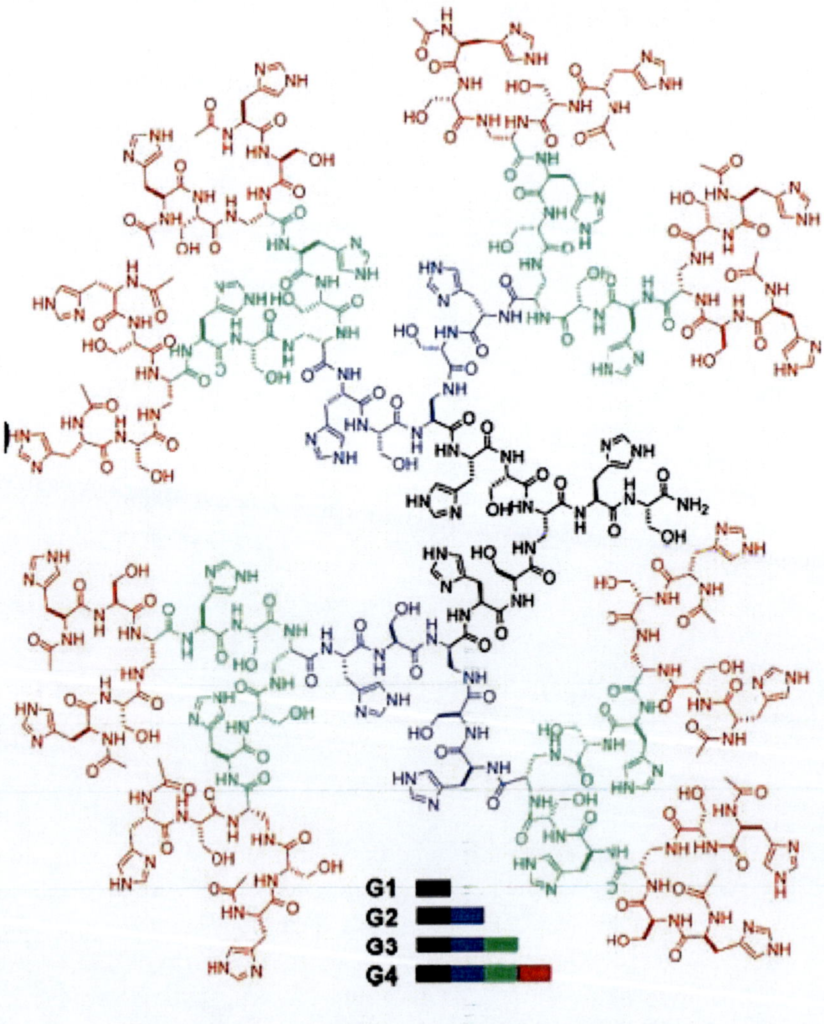

Figure 31. Peptide dendrimers G1-G4 with iterations of a $(His\text{-}Ser)_2Dap$ dendron.

A modular design was chosen using a divergent synthesis on solid support alternating proteinogenic α-amino acids with branching diamino acids, producing peptide dendrimers with a molecular weight of 3-5 kDa. Initial studies focused on models for hydrolases and produced esterase peptide dendrimers featuring histidine as the key catalytic residue. Variations of amino acid composition and the branching diamino acid led to enantioselective catalysts. Rate accelerations of kcat/kuncat = 90 000 were obtained when the design was changed to monomeric peptide dendrimers alternating two amino acids with the branching unit. A combinatorial approach was developed allowing for the preparation of large libraries (>60 000 members), which were screened for B12 binding and catalytic activity. Glycopeptide dendrimers conjugated to colchicines selectively inhibited the proliferation of targeted cells, whereas colchicine alone displayed high toxicity.

M. Bryszewska et al [93] reported the biological properties of low molecular mass peptide dendrimers. A series of new, low molecular mass, lysine-based peptide dendrimers with varying distribution of cationic and aromatic groups in the structure were synthesized. They expressed antimicrobial activity against Gram-positive (*Staphylococcus aureus*) and Gram-negative (*Escherichia coli*) bacteria as well as against fungal pathogens (*Candida albicans*). Their cytotoxic, haematotoxic, and genotoxic effects were studied. It appears that degree of branching and steric distribution and types of hydrophobic (aromatic) groups and cationic centres are important components of dendrimeric structure and influence both antimicrobial potency and toxicity. Such 3D structure of our dendrimers mimics that of the natural antimicrobial peptides and can be achieved by application of dendrimer chemistry.

5. Advances of Metal Dendrimers

5.1. Nano Dendritic Catalysis

Interest in catalysis by metal nanoparticles (NPs)is increasing dramatically, as reflected by the large number of publications in the last five years. This field, "semi-heterogeneous catalysis", is at the frontier between homogeneous and heterogeneous catalysis, and progress has been made in the efficiency and selectivity of reactions and recovery and recyclability of the catalytic materials [94]. Usually NP catalysts are prepared from a metal salt, a reducing agent, and a stabilizer and are supported on an oxide, charcoal, or a zeolite. Besides the polymers and oxides that used to be employed as standard, innovative stabilizers, media, and supports have appeared, such as dendrimers, specific ligands, ionic liquids, surfactants, membranes, carbon nanotubes, and a variety of oxides. Ligand-free procedures have provided remarkable results with extremely low metal loading. The Review presents the recent developments and the use of NP catalysis in organic synthesis, for example, in hydrogenation and C-C coupling reactions, and the heterogeneous oxidation of CO on gold NPs.

Dirk E. De Vos et al [95] reported the synthesis and application of porphyrin-functionalized dendrimers for the application in recyclable photocatalysts in a nanofiltration membrane reactor. The convergent synthesis of a series of porphyrin-functionalized pyrimidine dendrimers has been accomplished by a procedure involving the nucleophilic aromatic substitution (NAS) as a key reaction step. The resulting dendritic porphyrin catalysts show high activity in the light-induced generation of singlet oxygen (1O2) from ground-state oxygen. These materials are synthetically useful photosensitizers for the oxidation of various olefinic compounds to the corresponding allylic hydroperoxides.

Oriol Rossell et al [96] reported carbosilane dendrimers peripherally functionalized with P-stereogenic monophosphines and the catalytic behavior of their allylpalladium complexes in the asymmetric hydrovinylation of styrene. The first dendrimers containing P-stereogenic monophosphines as peripheral groups were obtained by reacting the protected BH3-LiCH2PPhR (R = 2-biphenylyl, 9-phenanthryl) with zeroth- and first-generation chlorocarbosilane dendrimers, followed by treatment with morpholine and subsequent chromatographic purification (Figure 32). The reaction of the chiral dendrimers with the dinuclear allyl complex [Pd(μ-Cl)(η3-2-MeC3H4)]2 allowed us to graft PdCl- (η3-2-MeC3H4) units on their surface. The catalytic properties of the new palladodendrimers were tested in the asymmetric hydrovinylation of styrene. The results in terms of activity, selectivity, and enantiomeric excess depend strongly on the nature of the phosphine and the halide abstractor.

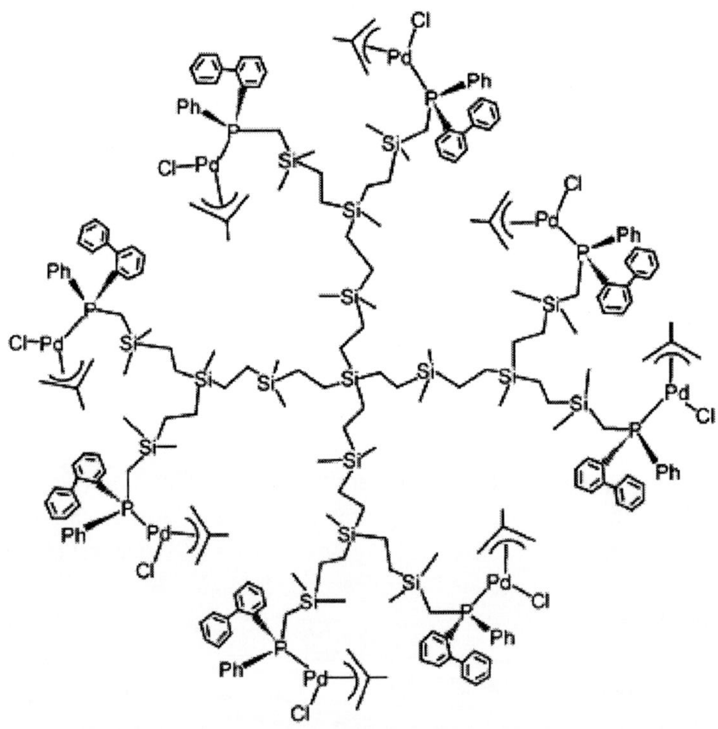

Figure 32. Carbosilane dendrimers peripherally functionalized with P-stereogenic monophosphines.

5.2. Metal Nanoparticles Preparation Using Dendrimer as Molecular Template

Dong-Mei Xu et al [97] reported the preparation of Ag nanoparticles in the presence of low generational poly(ester-amine) dendrimers. Stable Ag nanoparticles of 10–20 nm were prepared by reduction of AgNO3 with NaBH4 in water solution in the presence of low generational hydroxylterminated poly(ester-amine) dendrimer G1.0 (OH)16 and amino-terminated poly(ester-amine) dendrimer G1.5 (NH2)8 by optimizing preparation conditions. The results showed that the size of the Ag particles increased with Ag+/dendrimer molar ratio, and the size of Ag nanoparticles in Ag/G1.0 (OH)16 system was larger than that of Ag nanoparticles in Ag/G1.5 (NH2)8 system, while the polydispersities of two systems were similar.

Toyoko Imae et al [98] reported size-controlled in situ synthesis of metal nanoparticles on dendrimer-modified carbon nanotubes. Multiwalled carbon nanotubes (MWCNTs) were covalently modified with fourth-generation NH2-terminated poly(amido amine) (G4-PAMAM) dendrimers, named CNT/DEN. By simply tuning the metal ion-to-dendrimer terminal amine ratio, Au nanoparticles (NPs) with different sizes from 3 to 11 nm were in situ synthesized and deposited on the surface of CNT/DEN, as confirmed by a transmission electron microscopy (TEM) and an ultra violet-visible (UV-vis) absorption spectrometry. Furthermore, many metal NPs other than Au NPs, such as Ag, Cu, and Pt NPs, as well as binary NPs of Ag@Au, were synthesized in situ on CNT/DEN, yielding the water-dispersible and stable CNT/DEN/NPs (Figure 33).

M. P. Srinivasan et al [99] reported covalent molecular assembly in a supercritical medium and the formation of nanoparticles encapsulated in immobilized dendrimers. The work demonstrates the feasibility of forming metallic nanoparticles within the confines of dendrimerladen ultrathin films that are immobilized on a solid surface. The functional property of the poly amido amine dendrimer (PAMAM) as a sequestering agent for Cu nanoparticles was exploited by the introduction of copper acetylacetonate (Cu(acac)2) as a precursor. The nanoparticle precursors were introduced into the matrix using supercritical carbon dioxide and liquid tetrahydrofuran (THF) as processing media. The precursors were subsequently reduced to form the nanoparticles. Higher yield, a denser distribution of nanoparticles, and greater stability toward polar solvent attack were observed when the structures were assembled in supercritical CO2 than when THF was employed. TEM images revealed that the embedded nanoparticles are fairly tightly distributed in terms of size with an average diameter of 7 nm.

Richard M. Crooks et al [100] reported the synthesis, characterization, and magnetic properties of dendrimer-encapsulated nickel nanoparticles containing <150 atoms. Ni nanoparticles containing <150 atoms are prepared in an organic solvent using hydrophobic dendrimers as templates, which results in a high degree of size monodispersity. Magnetic (SQUID) studies indicate that the particles are ferromagnetic at 5 K with magnetic saturations that vary depending upon the particle size.

Hee-Tae Jung et al [101] prepared gold nanostructures from a taper-shaped supramolecular material (1-COOH) containing a carboxylic acid (-COOH) core, rigid biphenylene group, and three perfluorinated (-CF3) tails. The carboxylic core of the supramolecule serves as the site for reaction with the metal ions. Hee-Tae Jung et al showed that the self-assembling behavior of the material is strongly affected by substituting the carboxylic core of the supramolecule with a Na+ ion, followed by ion transfer of AuIII and subsequent reduction of AuIII with UV irradiation. This order and spacing are preserved during reduction of the metal salt with UV irradiation, although the intensities of the peaks increase. The size and shape of gold nanocrystals are controllable depending on UV irradiation time. These types of perfluorinated supramolecular templates have enhanced mesophase stability on account of the existence of strong intermolecular interactions, although many of these molecules are not traditional amphiphiles.

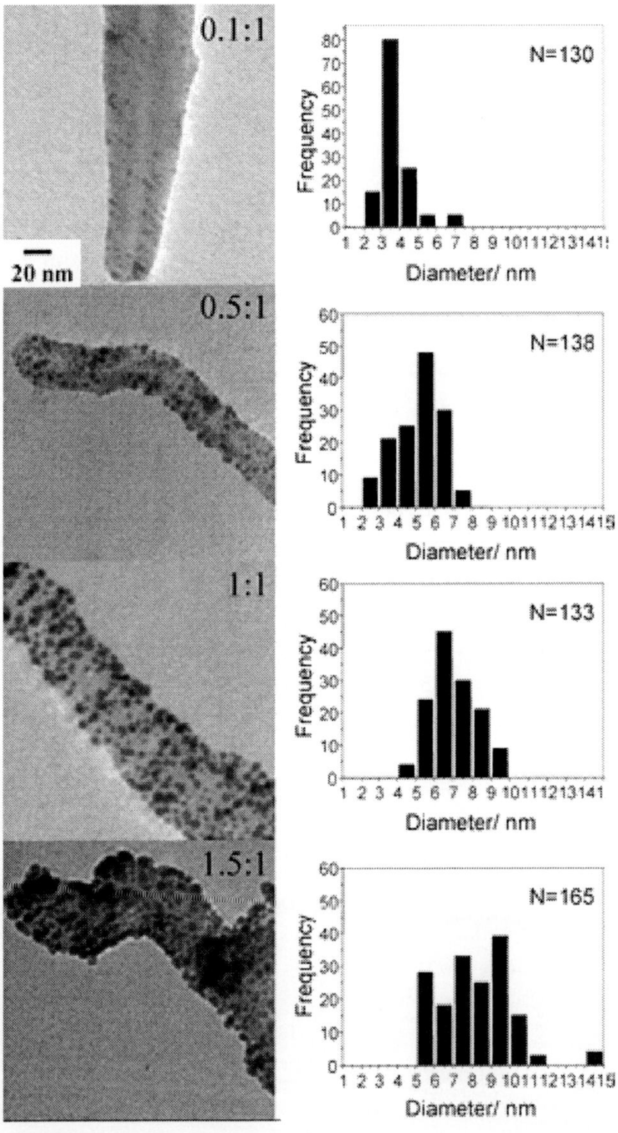

Figure 33. Size-controlled in situ synthesis of metal nanoparticles on dendrimer-modified carbon nanotubes.

Richard M. Crooks et al [102] reported the synthesis, characterization, and extraction of Au dendrimer-encapsulated nanoparticles (DENs) prepared in organic solvents. DENs composed of 31 and 55 Au atoms were prepared using organic solvents and poly(amidoamine) (PAMAM) dendrimer templates modified on their periphery with dodecyl groups. The spectral and microscopic properties of the resulting materials were identical to those prepared using water-soluble PAMAM dendrimers. It was possible to extract the organic-soluble DENs into water using the water-soluble thiols tiopronin and glutathione. The properties of the resulting monolayer-protected clusters were nearly identical to those of the precursor DENs. The synthetic methodology reported here provides a convenient method for preparing DENs of non noble metals such as Ni and Fe.

5.3. Complexation between Dendrimer and Metal Ions/Nanoclusters

Kimihisa Yamamoto et al [103] showed that tin ions, Sn2+, complex to the imine groups of a spherical polyphenylazomethine dendrimer in a stepwise fashion. This behaviour reflects a gradient in the electron density associated with the imine groups, with complexation in a more peripheral generation proceeding only after complexation in generations closer to the core has been completed. By attaching an electron-withdrawing group to the dendrimer core, Kimihisa Yamamoto et al are able to change the complexation pattern, so that the core imines are complexed last. By further extending this strategy, it should be possible to control the number and location of metal ions incorporated into dendrimer structures, which might find uses as tailored catalysts or building blocks for advanced materials (Figure 34).

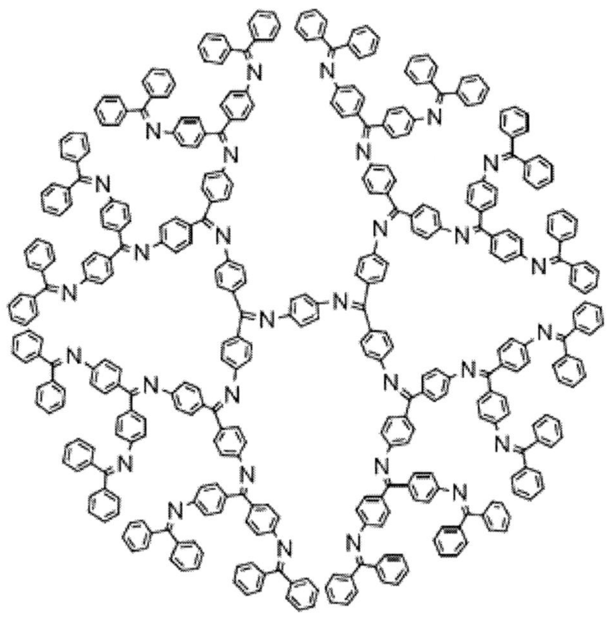

Figure 34. Chemical structures of the dendrimers and oligomers investigated by complexation with $SnCl_2$.

Jian Weng et al [104] synthesized a series of cyclens substituted with mixed stilbene and poly(ethylene glycol) dendritic arms. All dendrimers terminated with different peripheral groups had good solubility in common organic solvents, and dendrimers terminated with -CO2H groups (CO2H-dendrimers) were also soluble in alkaline solutions. The nickel coordination properties of these dendrimers were investigated in organic solvents. Dendrimers terminated with -CN groups (CN-dendrimers) and the second-generation CO2H-dendrimer [(CO2H)8L2] could produce pentacoordinated nickel complexes; the third-generation CO2H-dendrimer [(CO2H)16L3] could form tetra- and pentacoordinated nickel complexes, and the nickel complex of the fourth-generation CO2H-dendrimer [(CO2H)32L4] could not be obtained. The CN-dendrimers could not form hydrogen bonds on the surface, and the first- and secondgeneration CO2H-dendrimers could not form intramolecular hydrogen-bond networks.

Therefore, they had no selectivity for positive nickel ions and negative chloride ions. (CO2H)32L4 could not produce a nickel complex because it had a crammed backbone structure that could not penetrate nickel and chloride ions. Therefore, it was possible to control the ion access of cations and anions with the hydrogen-bond network of (CO2H)16L3.

Perla B. Balbuena et al [105] reported dendrimer-tetrachloroplatinate precursor interactions. Density functional theory is used to investigate the structure and energetics of the tetrachloroplatinate anion and its hydrolysis products at several degrees of hydration, as well as those of outer dendrimer pockets hosting such species.

Chunhui Huang et al [106] reported the synthesis and photopysical properties of polyaryl ether dendrimer with a 4-phenylacetyl-5-pyrazolone-based Terbium(III) complex as core. A series of novel dendritic β-diketone ligands, 1-phenyl-3-[G-*n*]-4-phenylacetyl-5-pyrazolone (*n* = 0-3, G stands for polyaryl ether), were synthesized by introducing Fréchet-type dendritic branches (Figure 35). These dendritic complexes, prepared from aqueous solution, exhibit high stability. Interestingly, the study of photophysical properties shows that the luminescence quantum yields of the dendritic Tb-complexes increase from 0.1 to 2.26% with an increase of the dendritic generation from 0 to 3. Importantly, an "energy-reservoir effect" was observed in the dendritic system using the method based on the resonance energy transfer from these complexes to rhodamine 6G. With the increase of the dendritic generation, the metal-centered luminescence quantum yield was almost the same, and the energy transfer (-transfer) from the ligand to Tb3+ increased.

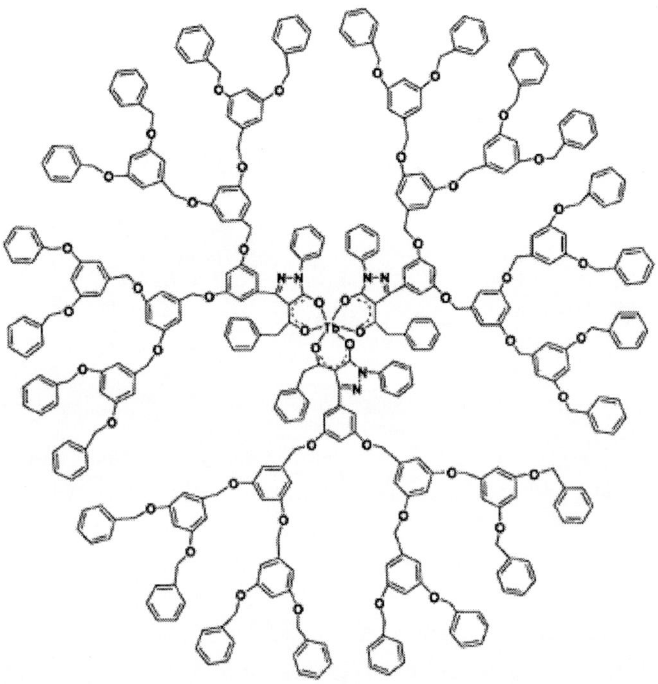

Figure 35. Polyaryl ether dendrimer with a 4-phenylacetyl-5-pyrazolone-based Terbium(III) complex as core.

Further measurements of the triplet state and oxygen quenching of these dendritic complexes verify that this enhancement of the energy transfer (-transfer) is attributed to both an “antenna effect” and a “shell effect”.

Robert C. Hider et al [107] reported the iron binding dendrimers for the treatment of haemochromatosis. A range of iron binding dendrimers terminated with hexadentate ligands formed from hydroxypyridinone, hydroxypyranone, and catechol moieties have been synthesized in order to investigate their potential as clinically useful iron(III)-selective chelators capable of removing dietary iron from the gastrointestinal tract and preventing the development of iron overload typical of haemochromatosis and thalassaemia intermedia. Hydroxypyridinone-terminated dendrimers were found to possess a high affinity and selectivity for iron(III). A hydroxypyridinone-based dendrimer was demonstrated to be highly efficient at reducing the absorption of iron(III) in rat intestine. This family of dendrimers may find an application in the treatment of iron overload.

Fritz Vögtle et al [108] reported cyclam cored luminescent dendrimers as ligands for Co(II), Ni(II) and Cu(II) ions. Fritz Vögtle et al have investigated the photophysical properties of two dendrimers containing a cyclam core decorated with 4 naphthyl units (G0), 12 dimethoxybenzene and 16 naphthyl units (G2). These dendrimers show fluorescence bands that can be assigned to naphthyl localized excited states (λ_{max} = 337 nm), naphthyl-amine exciplexes (λ_{max} = 470 nm) and, for G2, naphthyl excimers (λ_{max} ca. 400 nm). Cyclam is a very good ligand for transition metal ions and we have investigated complex formation between these dendritic ligands and Ni(II), Co(II) and Cu(II), added as nitrate salts (Figure 36).

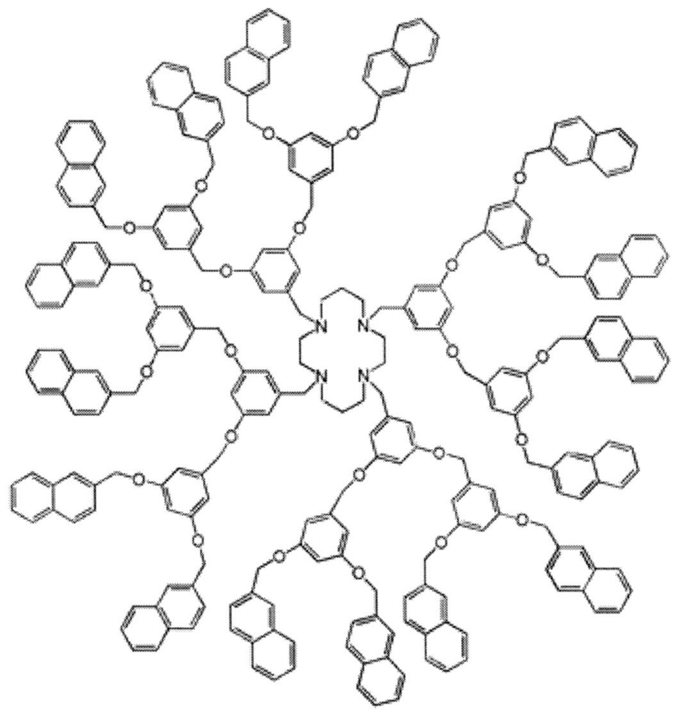

Figure 36. Cyclam cored luminescent dendrimers as ligands for Co(II), Ni(II) and Cu(II) ions.

This process can be monitored by the strong changes, both in shape and intensity, observed in the emission spectra of these dendrimers. Complexation with Cu(II) causes not only changes in the relative intensities of the fluorescence bands, but also the appearance of a new absorption band in the near UV spectral region. An analysis of the titration curves has allowed us to obtain clear evidence for the formation of not only 1:1 species, but also 1:2 metal to ligand species. G2 shows a clear preference, compared to G0, in forming complexes with a 1:2 metal-to-ligand stoichiometry, although it possesses very bulky dendrons appended to the cyclam central unit.

Wai-Yeung Wong et al [109] reported triphenylamine-dendronized pure red Iridium Phosphors with superior OLED efficiency/color purity trade-offs (Figure 37). Herbert Schumann et al [110] reported convergent synthesis and characterization of organotin dendrimers Sn{(CH2)nSn[(CH2)4SnPh3]3}4 (n=3,4). The reaction of the ω-haloalkyltin trihalides Br(CH2)3SnBr3 and Br(CH2)4SnBr3 with 3 equiv of but-3-enylmagnesium bromide yielded Br(CH2)3Sn(CH2CH2CH=CH2)3 and Br(CH2)4Sn(CH2CH2- CH=CH2)3. Both dendritic branches can be converted into their corresponding Grignard reagents, whose consequent treatment with 0.25 M amounts of SnCl4 resulted in the formation of the dendrimers Sn[(CH2)3Sn(CH2CH2CH=CH2)3]4 and Sn[(CH2)4Sn(CH2CH2CH=CH2)3]4, respectively.

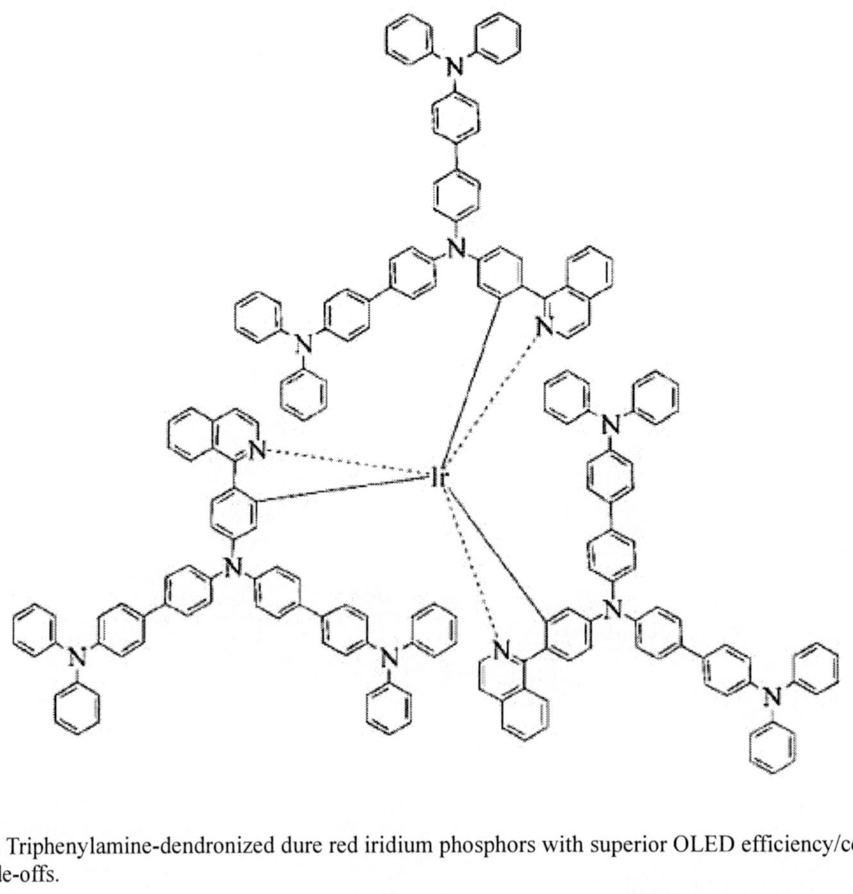

Figure 37. Triphenylamine-dendronized dure red iridium phosphors with superior OLED efficiency/color purity trade-offs.

Bradley D. Fahlman et al [111] reported facile synthesis of tin oxide nanoparticles stabilized by dendritic polymers. Kimihisa Yamamoto et al [112] reported the investigation of a molecular morphology effect on polyphenylazomethine dendrimers and their physical properties and metal-assembling processes. A series of novel dendritic polyphenylazomethines (DPA) with asymmetric morphologies was synthesized. Bis- and tris-substituted DPAs of the fourth generation also act as frameworks for stepwise assembly of a metal component (SnCl2), like the fully substituted symmetric DPA. The experimental results suggest a bent conformation of the dendrimer in which the core moiety is folded into the dendron structure. The structures of bis- and tris-substituted dendrimers which accurately follow the stepwise process are fixed. These observations provide a new insight into the finely controlled metal-assembly chemistry of dendritic macromolecules, and a rigid and fixed conformation is one of important factors for their unique properties.

Oriol Rossell et al [113] reported dendrimers containing two metallic layers, and the chloride migration from peripheral gold, palladium, or rhodium metals to internal ruthenium atoms (Figure 38). An efficient strategy for the synthesis of carbosilane dendrimers containing double metallic layers has been developed. The method consists of the attachment of the Ru(*p*-cymene)Cl2 unit to the branches of a phosphino-terminated dendrimer, followed by abstraction with AgOTf of the chloride ligands from the ruthenium and selective incorporation, through the nitrogen pyridine atom, of the bifunctional 4-pyPPh2 molecule.

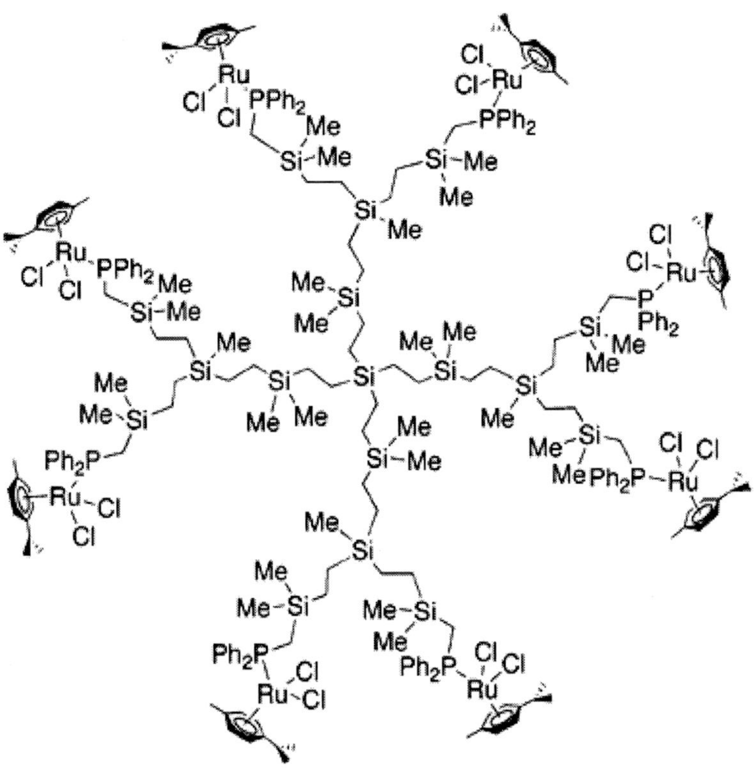

Figure 38. Dendrimers Containing Two Metallic Layers.

Finally, metalation of the resulting species with the complexes [(tht)AuCl], [PdCl(η3-2-MeC3H4)]2, and [RhCl(cod)]2 renders the targeted Ru/Au, Ru/Pd, and Ru/Rh dendrimers. An unexpected migration of the chloride ligand bonded to the peripheral metals to the internal ruthenium atoms was observed in terms of the results obtained from density functional calculations. In the case of the Ru/Au dendrimers, the different abilities for the ligand transfer process shown by the dendrimers in comparison with a mononuclear Ru/Au model compound are explained in terms of dendritic effects.

Novel Au-Pt bimetallic flower nanostructures fabricated on a polyamidoamine dendrimers-modified surface by electrodeposition are reported by Xiurong Yang et al [114]. These polyamidoamine dendrimers were stable, and they assisted the formation of Au-Pt bimetallic nanoflowers during the electrodeposition process. FE-SEM images showed that the bimetallic nanoflower included two parts: the "light" and the "pale" part. The two parts consisted of many small bimetallic nanoparticles, which was attributed to the progressive nucleation process. Moreover, the "light" part contained more bimetallic nanoparticles (Figure 39). The morphologies of bimetallic nanoflowers depended on the electrodeposition time and potential and the layer number of assembled dendrimers. The average size of nanoflowers increased with the increase in electrodeposition time.

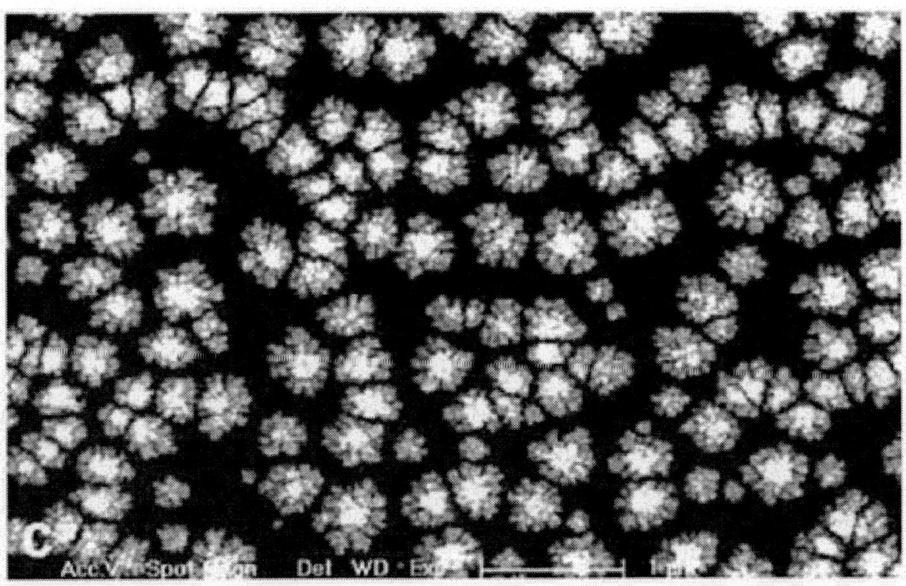

Figure 39. Polyamidoamine dendrimers-assisted electrodeposition of gold-platinum bimetallic nanoflowers.

Mark G. Humphrey et al [115] reported the electrochemical, spectroelectrochemical, and molecular quadratic and cubic nonlinear optical properties of alkynylruthenium dendrimers (Figure 40). Mo6X8i Nanocluster cores (X = Br, I): From inorganic solid state compounds to hybrids. The [(M6L8i)L6a] units (a = apical, i = inner) are the basic building blocks in the molybdenum, tungsten and rhenium octahedral cluster chemistry. Nano-sized metallic clusters are easily obtained by solid-state synthesis with transition elements associated with halogens or chalcogens (L). The intrinsic properties of M6 cluster units – one electron reversible redox process, magnetism and luminescence – depend on the nature of the metal and ligands. The solubilization of M6 solid state compounds provides [(M6L8i)L6a]n–

building blocks with individual properties that can be further used for the design of hybrid organic/inorganic materials [116].

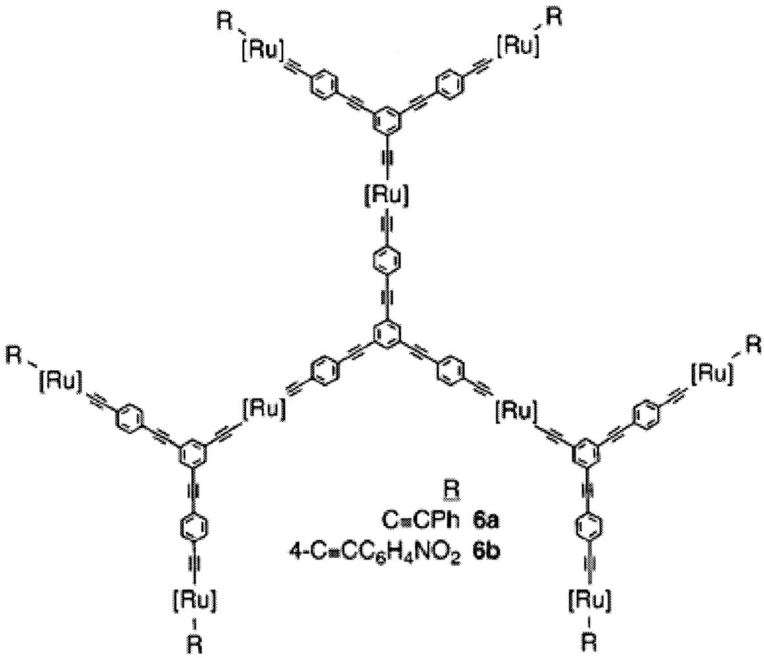

Figure 40. Alkynylruthenium dendrimers.

From Simple Monopyridine Clusters [Mo6Br13(Py-R)][*n*-Bu4N] and Hexapyridine Clusters [Mo6X8(Py-R)6][OSO2CF3]4 (X = Br or I) to Cluster-Cored Organometallic Stars, Dendrons, and Dendrimers were reported by Didier Astruc et al [117]. Hexasubstitution of apical triflate ligands in the octahedral clusters [M]2[Mo6X8(CF3SO3)6] (M = *n*-Bu4N or Cs, X = Br or I) and monosubstitution in [*n*-Bu4N]2[Mo6Br13(CF3SO3)] was carried out in tetrahydrofuran at 60 °C with simple pyridines and then extended to organometallic pyridines, yielding cluster-cored stars, and to dendronic polyallyland polyferrocenylpyridines, yielding cluster-cored polyallyl and polyferrocenyl dendrimers and dendrons (Figure 41).

Christiane Perrin et al [118] reported Mo6Br8-Cluster-cored organometallic stars and dendrimers. The octahedral molybdenum cluster [*n*-Bu4N]2[Mo6Br8(CF3SO3)6] undergoes substitution of all six terminal triflate ligands with the organometallic pyridine ligands [RuCp(PPh3)2(η1-C2-4-pyridinyl)] and 1-ferrocenyl-2-(4-pyridinyl) acetylene to give the new light and air sensitive hexa-functionalized Mo6 clusters, and with the dendronic phenolate ligand *p*-NaO–C6H4C{CH2CH2CH2Si(Me)2Fc}3 to give the air-sensitive Mo6-cluster-cored octadecylferrocenyl dendrimer that discloses a single CV wave in CH2Cl2 and recognizes the biologically important adenosyl triphosphate di-anion (ATP2–). The organometallic pyridines were also coordinated to Ag+ to give the new trinuclear AgRu2 and AgFc2 cationic complexes for comparison of the structures and electronic delocalization with those of the clusters.

Multiporphyrin dendrimers are among the most promising architectures to mimic the oxygenic lightharvesting complex because of their structural similarities and synthetic convenience. The overall geometries of dendrimers are determined by the core structure, the type of dendron, and the number of generations of interior repeating units. The rigid core and bulky volume of exterior porphyrin units in multiporphyrin dendrimers give rise to well-ordered three-dimensional structures. As the number of generations of interior repeating units increases, however, the overall structures of dendrimers become disordered and randomized due to the flexibility of the repeating units. Takuzo Aida et al [119] investigated the relationship between incoherent excitation energy migration processes and molecular structures in Zinc(II) Porphyrin dendrimers. To reveal the relationship between molecular structure and processes of excitation-energy migration in multiporphyrin dendrimers, Takuzo Aida et al calculated the molecular structure and measured the time-resolved transient absorption and fluorescence anisotropy decays for various hexaarylbenzeneanchored polyester zinc(II) porphyrin dendrimers along with three types of porphyrin dendrons as references.

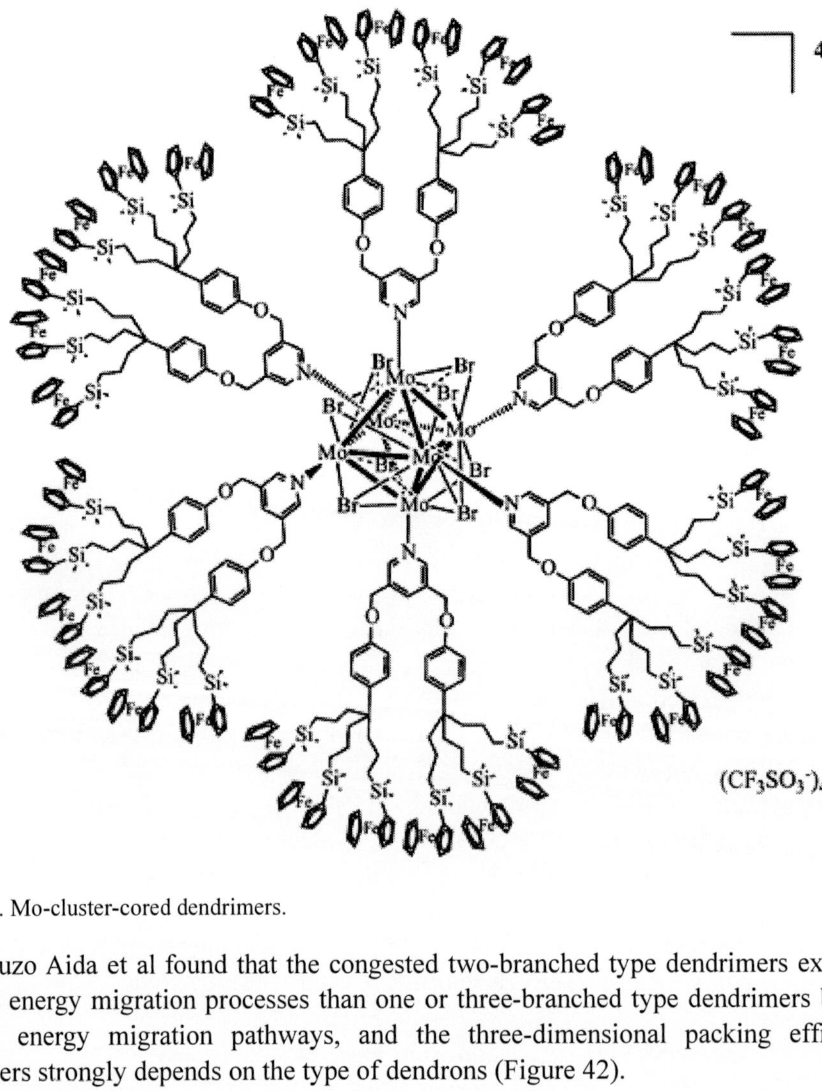

Figure 41. Mo-cluster-cored dendrimers.

Takuzo Aida et al found that the congested two-branched type dendrimers exhibit more efficient energy migration processes than one or three-branched type dendrimers because of multiple energy migration pathways, and the three-dimensional packing efficiency of dendrimers strongly depends on the type of dendrons (Figure 42).

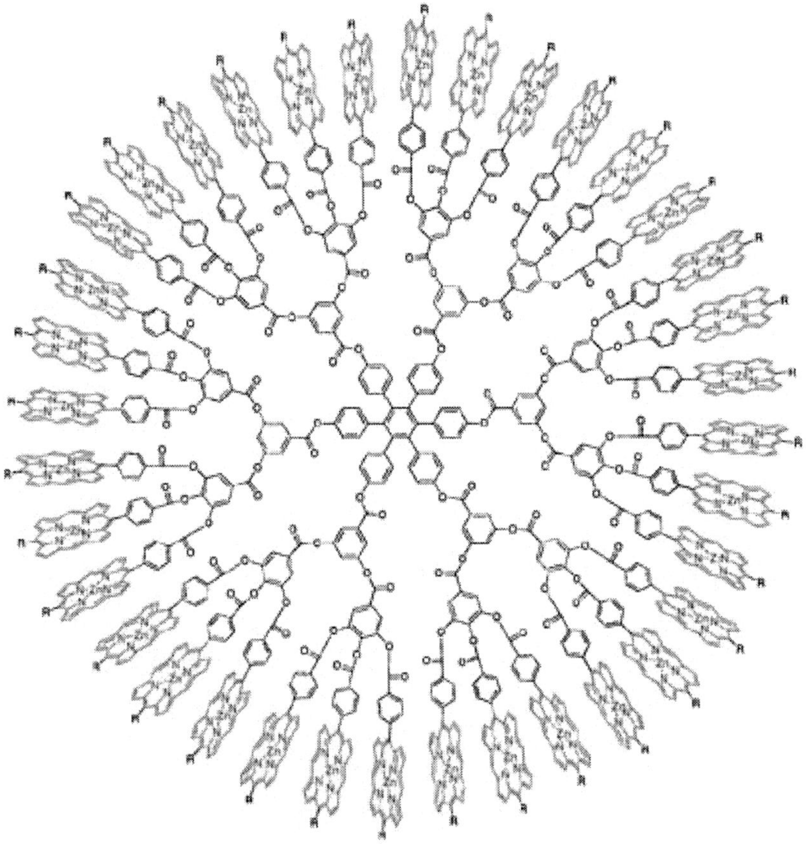

Figure 42. Zinc(II) porphyrin dendrimers.

6. Self-Assembly of Dendrimers

6.1. Biomolecular Self-Assembly and Recognition

Thomas Schrader et al [120] reported dendrimeric bisphosphonates for multivalent protein surface binding. A single weak-binding event is multiplied into an efficient receptor site for protein surfaces (<10–1 to >106M–1 in buffered aqueous solution) in a biomimetic fashion. This has hitherto been done with natural host/guest pairs, but not with artificial receptors. The organic reaction presented is one of very few that enable chemists to fuse multiple ionic building blocks covalently in highly polar solution; this one-pot reaction proceeds with virtually quantitative yield. According to this concept, other building blocks with aldehyde groups can likewise be multiplied into monodisperse functional dendrimers.

Mauri A. Kostiainen et al [121] reported multivalent dendrons for high-affinity adhesion of proteins to DNA. One of the keys for using deoxyribonucleic acid (DNA) as a nanomaterial relies on how the individual DNA chain can be aligned and how a multitude of DNA chains can be packed into ordered nanostructures. Hsin-Lung Chen et al [122] reported two-dimensional densely packed DNA nanostructure derived from DNA complexation with a low-generation poly(amidoamine). They presented a simple method for constructing a 2-D densely packed DNA nanostructure using the electrostatic complex of DNA with a poly(amidoamine) (PAMAM) dendrimer of generation two. Ordered DNA arrays are formed by drop-casting an aqueous solution containing positively overcharged complexes onto mica followed by a prolonged incubation. During the incubation, the complexes tend to adsorb onto the negatively charged mica surface through electrostatic attraction. The rodlike complexes organize to form ordered arrays to increase the surface density of the adsorbed complexes and hence the attractive free energy of adsorption. The densely packed nanostructure obtained here is distinguished from the previously reported spheroid or toroid structure derived from DNA complexations with the higher-generation dendrimers.

G. Floudas et al [123] reported the self-assembly and molecular dynamics of peptide-functionalized polyphenylene dendrimers. The self-assembly mechanism and the associated molecular dynamics are studied for a series of poly-L-lysine-functionalized polyphenylene dendrimer melts as a function of the core size (generation), functionality, and polypeptide length using X-rays, solid-state NMR, calorimetry, and dielectric spectroscopy. A striking dependence of the polyphenylene self-assembly on the poly-L-lysine length is shown. In addition, the type (α-helix/ β-sheet) of peptide secondary structure is controlled by the packing restrictions imposed by the polyphenylene core (Figure 43). G. Floudas et al showed that constrained poly-L-lysines can adopt different secondary structures from their linear analogues.

Cationic lipids and polymers are routinely used for cell transfection, and a variety of structure-activity relation data have been collected. Few studies, however, focus on the structural aspects of self-assembly as a crucial control parameter for gene delivery. Francois Diederich et al [124] reported the self-assembly, DNA complexation, and pH response of amphiphilic dendrimers for gene transfection. Francois Diederich et al presented here the observations collected for a set of cationic dendritic amphiphiles based on a stiff tolane core that are built from identical subunits but differ in the number and balance of their hydrophobic and cationic hydrophilic moieties. Francois Diederich et al found that the primary self-assembly of the amphiphile is of paramount importance and influences DNA binding, serum sensitivity, and pH response of the vector system.

Prabal K. Maiti et al [125] studed sequence-dependent complexation between oligonucleotides (single-strand DNA) and various generation ethylene diamine (EDA) cored poly amido amide (PAMAM) dendrimers through atomistic molecular dynamics simulations accompanied by free energy calculations and inherent structure determination. Simulations reveal formation of a stable complex and provide a detailed molecular level understanding of the structure and dynamics of such a complexation. The reaction free energy surface in the initial stage is found to be funnel-like, with a significant barrier arising in the late stage due to the occurrence of misfolded states of DNA. Complexation shows surprisingly strong sensitivity to the ssDNA sequence, which is found to arise from a competition between enthalpic versus entropic rigidity of ssDNA.

The formation of amyloid plaques is a key pathological event in neurodegenerative disorders, such as prion and Alzheimer's diseases. B. Klajnert et al [126] reported the pH Dependence of Molecular Interactions of Dendrimers with Amyloid Peptides. They have studied the effect of polypropyleneimine dendrimers on the formation of amyloid fibrils as a function of pH in order to gain further insight in the aggregation mechanism and its inhibition.

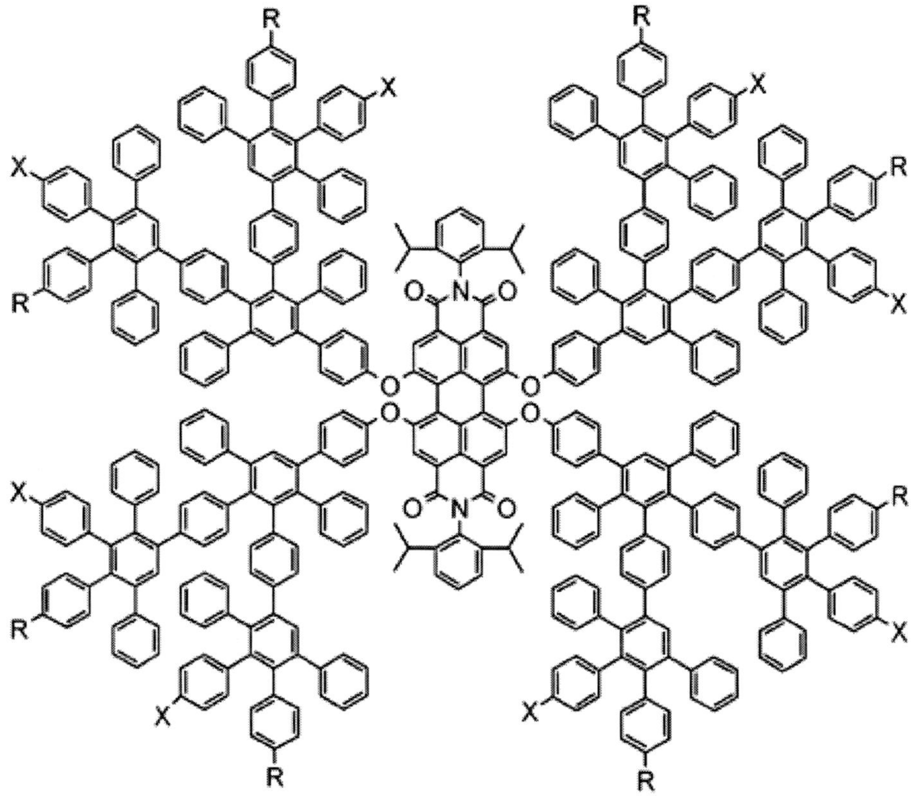

Figure 43. Polyphenylene scaffold of dendrimers.

The results showed that the level of protonation of His, Glu, and Asp residues is important for the final effect, especially at low dendrimer concentration when their inhibiting capacity depends on the pH. At the highest concentrations, dendrimers were very effective against fibril formations for both prion and Alzheimer's peptides.

The interactions between plasmid DNA and cationic polymers are of interest for their potential biological applications. Xiangliang Yang et al [127] reported studies on the interactions between DNA and PAMAM with fluorescent probe [Ru(phen)2dppz]2+. The interactions of DNA with polyamidoamine (PAMAM) dendrimer were studied by fluorescence spectroscopy and stopped-flow technique. A rapid and reproducible fluorescent assay method had been developed for assessing PAMAM and DNA interactions using [Ru(phen)2dppz]2+ as a probe. Xiangliang Yang et al further studied the kinetics of PAMAM binding to DNA and the reverse process of DNA dissociation from the complexes. The results indicated that DNA condensation was the rate-determining step during the complexation process, while DNA unfolding and expansion was the rate-determining step during the DNA dissociation process. And at N/P ratios above 2.0, DNA was fully condensed by PAMAM and dissociation was increasingly difficult. These results provided some useful instructions for self-assembling and disassembling of DNA as well as efficient gene delivery applications.

Hiroshi Tsukube et al [128] reported the proteo-dendrimers designed for complementary recognition of cytochrome c for preparing dendrimer architecture toward nanoscale protein complexation. "Proteo-dendrimers" in which polyanionic hepta(glutamic acids), fluorescent zinc porphyrinate cores, hydrophilic polyether surfaces, and nonpeptide hydrophobic dendrons are combined, were developed as a new series of synthetic receptors for protein recognition. They have polyanionic "patch" structures on their surfaces and undergo complementary electrostatic interactions with a positively charged cytochrome c patch, as observed in biological protein–protein recognition systems. These proteo-dendrimer receptors exhibited higher affinities with cytochrome c proteins in aqueous solutions than with biological cytochrome b5. Furthermore, they effectively blocked complexation of biological cytochrome b5 with cytochrome c, indicating that the proteo-dendrimers and cytochrome b5 similarly occupy the polycationic patch of cytochrome c.

Daiyo Terunuma et al [129] reported the syntheses and lectin-binding activities of lactotriaose-containing carbosilane dendrimers. Carbosilane dendrimers periphery-functionalized with lactotriaose (GlcNAcb1–3Galb1–4Glc) with valencies of three, four, six, and twelve were prepared for use in a lectin-binding assay. A lactotriaose derivative was prepared from D-glucosamine and D-lactose derivatives. The N-Troc-protected glucosamine glycosyl donor and 3'-O-unprotected lactose glycosyl acceptor were condensed in the presence of silver trifluoromethanesulfonate and methylsulfenyl bromide to provide corresponding trisaccharide with new β-1-3 linkages in 92% yield. The protection group of the trisaccharide was transformed into an acetyl group. The 4-pentenyl glycoside was prepared from the acetate via glycosyl bromide. The alkene was converted into acetyl sulfide by addition of thioacetic acid under radical conditions. The lactotriaose unit was linked with carbosilane dendrimers to afford acetyl-protected glycodendrimers. De-O-acetylation of the dendrimers was carried out in the presence of sodium methoxide and then aq NaOH to give the desired lactotriaose clusters using a carbosilane dendrimer backbone.

6.2. Supramoleclular Assembly

Dendrimers and dendrons are a unique class of molecular-scale building blocks for the construction of nanostructured materials. This makes these structures ideal for the construction of multiply crosslinked covalent assemblies, and also for the self-assembly of multiply bound non-covalent (supramolecular) assemblies [130].

Rigoberto Advincula et al [131] reported the conjugated polymer nanoparticles via intramolecular crosslinking of dendrimeric precursors.

Shiyong Liu et al [132] described the syntheses and solution behavior of model amphiphilic dendritic–linear diblock copolymers that self-assemble in aqueous solutions into micelles with thermoresponsive shells. The investigated materials are constructed of poly(benzyl ether) monodendrons of the second generation ([G-2]) or third generation ([G-3]) and linear poly(N-isopropylacrylamide) (PNIPAM). [G-2]-PNIPAM and [G-3]-PNIPAM dendritic–linear diblock copolymers have been prepared by reversible addition–fragmentation transfer (RAFT) polymerizations of N-isopropylacrylamide with a [G-2]- or [G-3]-based RAFT agent, respectively. The thermoresponsive collapse of the PNIPAM shell is a two-stage process; the first one occurs gradually in the temperature range of 20–29°C, which is much lower than the lower critical solution temperature of linear PNIPAM homopolymer, followed by the second process, in which the main collapse of PNIPAM chains takes place in the narrow temperature range of 29–31°C.

A new methodology for the preparation of well-defined core-shell nanoparticles was developed by Karen L. Wooley et al [133], based upon the employment of a multifunctional crosslinker to coincidently stabilize supramolecular polymer assemblies and imbed into the shell unique chemical functionalities. Amphiphilic diblock copolymers of poly(acrylic acid)80-*b*-poly(styrene)90 that had been assembled into micelles and partially functionalized throughout the corona with alkynyl groups were utilized as Click-readied nanoscaffolds for the formation of shell Click-crosslinked nanoparticles (SCCs). Divergently grown dendrimers of the zero, first, second, and third generations having increasing numbers of azide terminating groups ((N3)2-[G-0], (N3)4-[G-1], (N3)8- [G-2], and (N3)16-[G-3], respectively) were investigated as crosslinkers via Click reactions with the alkynyl groups to form covalent linkages throughout the block copolymer micelle corona, thus forming a crosslinked shell. The dendritic crosslinker is important as it also allows for the incorporation of excess functionality that can undergo complementary reactions.

Xin-Ru Jia et al [134] reported the self-assembly of a new class of amphiphilic poly(amidoamine) dendrimers and their electrochemical properties. Amphiphilic poly(amidoamine) (PAMAM) dendrimers consisting of a hydrophilic dendrimer core and hydrophobic aromatic dansyl or 1-(naphthalenyl)-2- phenyldiazene (NPD) shells have been synthesized. These amphiphilic dendrimers from the zero generation to the third generation self-assemble into vesicular aggregates in water. The self-assembly behavior of these dendrimers strongly depends on their generations. For the PAMAM–NPD aggregates, the photoisomerization process leads to a change in the aggregate size.

The solution behavior of spherical dendrimers as well as hybrid-linear dendritic diblock copolymers has been extensively studied, and the size, shape, and ability of these polymers to encapsulate small molecules have led to their comparison with traditional micelles. Paula T. Hammond et al [135] studied the solution behavior of linear-dendritic rod diblock copolymers in methanol. The diblock copolymers consist of a linear poly(ethylene oxide)-poly(ethylene imine) diblock copolymer backbone around which poly(amido amine) branches have been divergently synthesized from the poly(ethylene imine) block. The intrinsic viscosity of the amine and ester terminated polymers was found to follow two very different trends at low generation; however, at higher generations, they followed similar, yet slightly different, curves with the values for the amine terminated polymers only a little larger than those of the ester terminated polymers.

Daniel R. Blasini et al [136] studied the stepwise self-assembly of ordered supramolecular assemblies based on coordination chemistry. Dendrimer layers were deposited onto SiO2 wafers by alternately immersing the substrate into a solution of terpyridyl (tpy)-pendant poly(amido amine) (PAMAM) dendrimers (dend-n-tpy; n = 8, 16) dissolved in CH2Cl2, followed by the interfacial coordination reaction with cobalt (Co2+) from aqueous solution.

Virgil Percec et al [137] programmed the internal structure and stability of helical pores self-assembled from dendritic dipeptides via the protective groups of the peptide. The synthesis of dendritic dipeptides (4-3,4-3,5)12G2-CH2-X-L-Tyr-L-Ala-OMe with X = Boc, Moc, and Ac; their self-assembly in bulk and in solution; and the structural and retrostructural analysis of their supramolecular helical porous assemblies are reported. The dimensions, structure, internal order, thermal stability of the supramolecular helical pores, and conformations of the dendron and supramolecular dendrimer are programmed by the nature of the protective groups of the dipeptide. The ability of the protective groups to program the structure of the helical pore reveals the simplest design strategy that complements the more complex strategies based on the architecture of the dendron, the stereochemistry, and the structure of the dipeptide.

6.3. Liquid-Crystal Dendrimer

Carsten Tschierske et al [138] reported the first example of a third-generation liquid-crystalline carbosilane dendrimer with peripheral bent-core mesogenic units for understanding of "dark conglomerate phases".

Hee-Tae Jung et al [139] reported the large-area, highly aligned cylindrical perfluorinated supramolecular dendrimers using magnetic fields.

Milagros Pinol et al [140] reported the synthesis, characterization, and induction of stable anisotropy in liquid crystalline photo-addressable PPI dendrimers. A series of liquid crystalline azobenzene codendrimers with light-driven properties has been prepared by chemical modification of the third generation of an amine-terminated poly(propylene imine) (PPI) dendrimer. These photoactive dendrimers were obtained by grafting the photochromic 4-cyanoazobenzene and/or liquid crystalline 4-cyanobiphenyl units onto the periphery of PPI. The synthesized dendrimers include different percentages of each functional unit randomly distributed in the dendritic shell. Illumination with either blue (488 nm) or UV (350 nm) linearly polarized light at room temperature gave rise to a stable dichroism and moderate birefringence values when 80% (molar ratio) of the units are azobenzene. In addition, pre-irradiation of the films with unpolarized UV light gave rise to higher birefringence values.

6.4. Interfacial Behaviour of Dendrimer

Dendrimers are novel polymeric nanoarchitectures for solubility enhancement. Poor solubility and hydrophobicity of drugs/bioactives limit their possible applications in drug delivery and formulation development. Apart from conventional methods of solubility enhancement, there are some novel methods which can be used in solubilization. Dendrimers represent a novel type of polymeric material that has generated much interest in many diverse areas due to their unique structure and properties. Dendrimer-mediated solubility enhancement mainly depends on factors such as generation size, dendrimer concentration, pH, core, temperature, and terminal functionality [141]. Added advantage in solubilization can be achieved considering these factors. Available literature suggests that ionic interaction, hydrogen bonding, and hydrophobic interactions are the possible mechanisms by which a dendrimer exerts its solubilizing property.

A. I. Cooper et al [142] reported the extraction of a hydrophilic compound fromwater into liquid CO2 using dendritic surfactants.

Amphiphiles with a dendritic structure are attractive materials as they combine the features of dendrimers with the self-assembling properties and interfacial behavior of amphiphiles. Jean-Louis Gallani et al [143] have designed and synthesized three series of segmented amphiphilic block co-dendrimers (Janus-type) and studied their interfacial properties on the Langmuir trough. Various behaviors are observed with, as a rule, the lowest generation dendrimers behaving more or less like traditional amphiphiles while the larger molecules tend to exhibit more complicated isotherms, with a non-straighforward temperature dependence, one particular molecule seemingly forming supramolecular assemblies spontaneously (Figure 44).

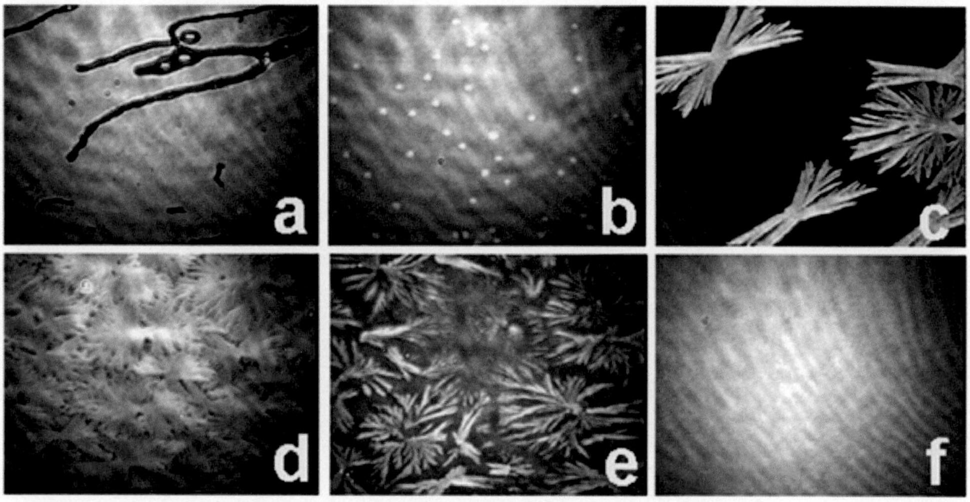

Figure 44. Interfacial behavior of a series of amphiphilic block co-dendrimers.

Dendritic folate rosettes as ion channels in lipid bilayers. Naomi Sakai et al [144] reported that π-stacked supramolecular rosettes can form synthetic ion channels. These esthetically pleasing supramolecules can self-assemble from heteroaromatic "sectors" with directional hydrogen-bond donor-acceptor arrays such as folate guanines, and so on. H-bond directed circular self-assembly of these sectors gives cyclic oligomers, which, in turn, may self-assemble on top of each other to give π-stacks. Templation of this process by cation binding to carbonyl lone pairs in the rosette centers accounts for the confirmed ionophoric properties of many π-stacked supramolecular rosettes in bulk liquid membranes.

6.5. Dendrimer Supramolecular Helix

Virgil Percec et al [145] reported the self-assembly of amphiphilic dendritic dipeptides into helical pores (Figure 45).

V. Percec et al [146] reported the self-organization of supramolecular helical dendrimers into complex electronic materials (Figure 46).

Virgil Percec et al [147] found that the internal structure of helical pores self-assembled from dendritic dipeptides is stereochemically programmed and allosterically regulated.

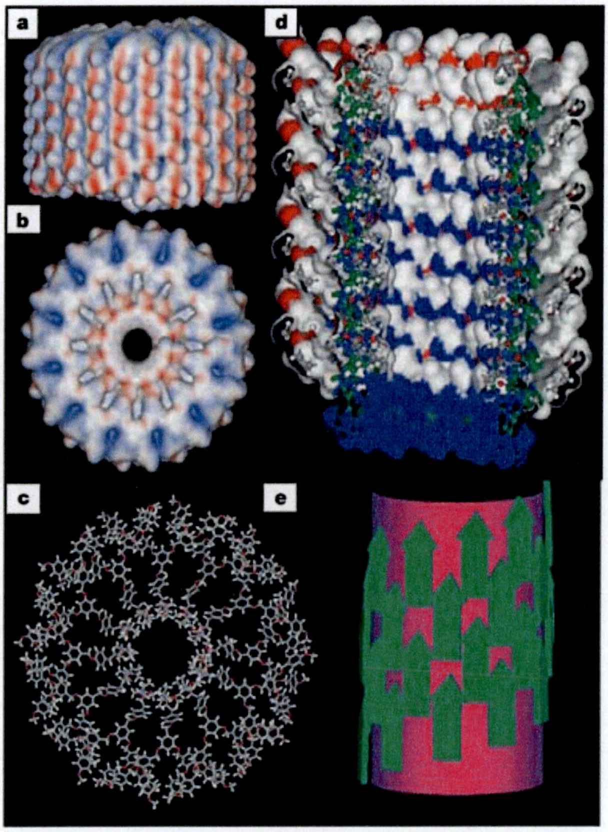

Figure 45. Self-assembly of amphiphilic dendritic dipeptides into helical pores.

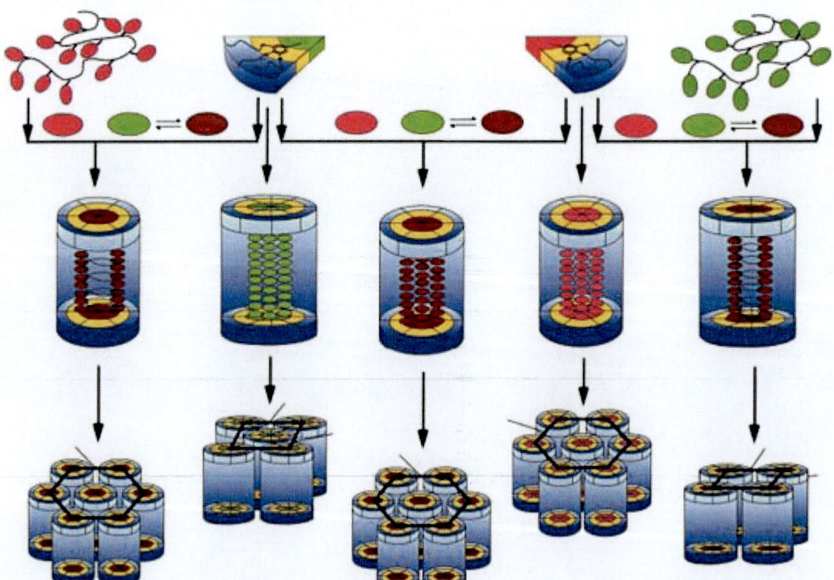

Figure 46. Self-organization of supramolecular helical dendrimers into complex electronic materials.

7. Charge and Energy Transfer

7.1. Fluorescence Properties

Kung-Hwa Wei et al [148] have been prepared highly luminescent dendron-substituted copolyfluorenes that incorporate surface-modified cadmium sulfide nanoparticles. A small percentage of these nanoparticles can be incorporated into the dendritic structures upon tailoring the interfaces between the ligands on the nanoparticles and the dendritic structures in the copolyfluorene. Both the photoluminescence and electroluminescence efficiencies of the polymer nanocomposites are dramatically enhanced—sometimes more than doubled—relative to the values of the pure polymer.

Donald A. Tomalia et al [149] have been synthesized Hydroxyl terminated, poly(alkyl ether) dendrons possessing single site phosphine functional groups at their focal points. These single site reactive dendrons are then used in ligand exchange reactions to produce stabilized surface capped CdSe/core CdS-shell quantum dots that exhibit high quantum yield, non-quenching photoluminescence (PL) properties. These enhanced features are in sharp contrast to quantum dots capped with analogous single site thiol functionalized dendrons that exhibit substantially quenched photoluminescence. These new dendron constructions begin with protected pentaerythritol moieties which are used as the initiator core as well as the branch cells. By protecting one of the four hydroxyl groups in the pentaerythritol moiety, the remaining three hydroxyl groups are utilized as divergent growth sites for the poly(ether) dendron construction. Desired generation levels (i.e., generations 1, 2) are synthesized, followed by deprotection of the focal point hydroxyl group to expose appropriate functional groups suitable for conversion to sulfhydryl, phosphine or other desired functional groups.

Toyoko Imae et al [150] studied the strong fluorescence emission from poly(amido amine) (PAMAM) dendrimers with different terminal groups or a poly(propylene imine) (PPI) dendrimer under different conditions by varying experimental parameters such as pH value, aging time, temperature, and concentration. The increase of fluorescence intensity was fast at low pH or high temperature but linear with respect to dendrimer concentration. It was reasonable that the formation of a fluorescence-emitting moiety had a close relation to protonated tertiary amine groups in PAMAM or PPI dendrimers. Furthermore, oxidation of the tertiary amines was confirmed to play an important role, which was evidently caused by oxygen in air. The results of fluorescence decay indicated that the deactivation of luminescence was raised with increasing temperature. Dendrimers emitted blue photoluminescence along fiber chain templates on a fluorescent microscope.

7.2. Energy Transfer

S. Thayumanavan et al [151] have designed and synthesized difunctionalized dendrimers containing two donors in the periphery and an acceptor at the core to serve as scaffolds for comparison with linear analogues to investigate the advantage of dendritic scaffolds for energy and charge transfer. Comparison of these dendrimers with the fully decorated dendrimers provides information on the advantage of chromophore density in energy/charge transfer from periphery to the core.

Jane Larsen et al [152] explored energy transfer processes within a series of Zn-porphyrin-appended dendrimers by means of excitation intensity dependent transient absorption measurements. Singlet-singlet annihilation on two distinct time scales of 18 ± 5 ps and 130 ± 10 ps in the dimer and the dendrimers is reported. The two distinct processes reflect the presence of two structural conformer distributions. Analysis of the singlet-singlet annihilation transient kinetics shows that sequential annihilation occurs within subunits up to four Zn-porphyrins in the dendrimers. The onset of the singlet-singlet annihilation process depending on the size of the molecule reveals a difference in the number of communicating Zn-porphyrins.

René M. Williams et al [153] described the synthesis and charge transfer properties of triarylamine-oligothiophene-perylenemonoimide dendrimers, TPA(T2-PMI)3 and TPA(T4-PMI)3. The fluorescence quantum yields indicate strong emission quenching by electron transfer [ΦTHF = 0.004 for TPA(T2-PMI)3, ΦTHF = 0.003 for TPA(T4-PMI)3, and ΦTHF = 0.8 for PMI]. Moreover, with the increase of the solvent polarity, the quantum yields decrease indicating that the A+•D–• (acceptor/donor) couple is more stabilized. The femtosecond transient absorption spectra show a very fast charge separation process (≈2 ps; kcs ≈ 5 × 1011 s–1) and a charge recombination of more than 1 order of magnitude slower (≈50 ps; kcr ≈ 2 × 1010 s–1), as observed from the rise time and decay of the radical anion and radical cation absorption bands. The analysis of the transient absorption spectroscopy and of the energetics of the process using Marcus theory indicates that in the electron transfer process the thiophene unit is the first electron donor. The presence of the triarylamine and/or the proximity of the oligothiophenes does improve the donor capabilities of the oligothiophene unit slightly and enhances its conjugation as seen in the absorption spectra and the transients of the radial cations.

7.3. Photon Absorption

Lucia Flamigni et al [154] reported the photoinduced processes in a dendritic Zn porphyrin structure with a free-base porphyrin core. The photophysical characterization of a nonameric porphyrin assembly made up of a central free-base and eight peripheral zinc porphyrins connected by flexible nucleosidic linkers, and the identification of the photoinduced processes taking place in the array was reported (Figure 47). The kinetic parameters of singlet–singlet energy transfer are interpreted by assuming the existence of several nonequilibrated conformations in the array due to the flexibility of the linkages. A fast quenching of the donor beyond the time detection limit of the instrument is also hypothesized and assigned to strong interactions between the zinc chromophores in close contact which can give self-quenching phenomena. For the zinc porphyrin triplet excited state, a triplet–triplet annihilation is detected at moderate photon flux as a consequence of multi excitation of the same array. Drastic reduction of the photon flux allows us to determine the triplet energy transfer rate from zinc porphyrin to free-base porphyrin, which is the final recipient of the energy collected by the array. The decay of the triplet state of the core free-base porphyrin is prolonged with respect to the model, indicating a shielding effect of the peripheral groups.

Theodore Goodson III et al [155] reported the entangled photon absorption in an organic porphyrin dendrimer. Two-photon absorption spectroscopy is an intensity dependent nonlinear effect related to the excitation of virtual intermediate states. The classical two-photon absorption has an extremely low efficiency which is quantified by its cross-section (δ ~ 10–48 cm4 s at 800 nm).

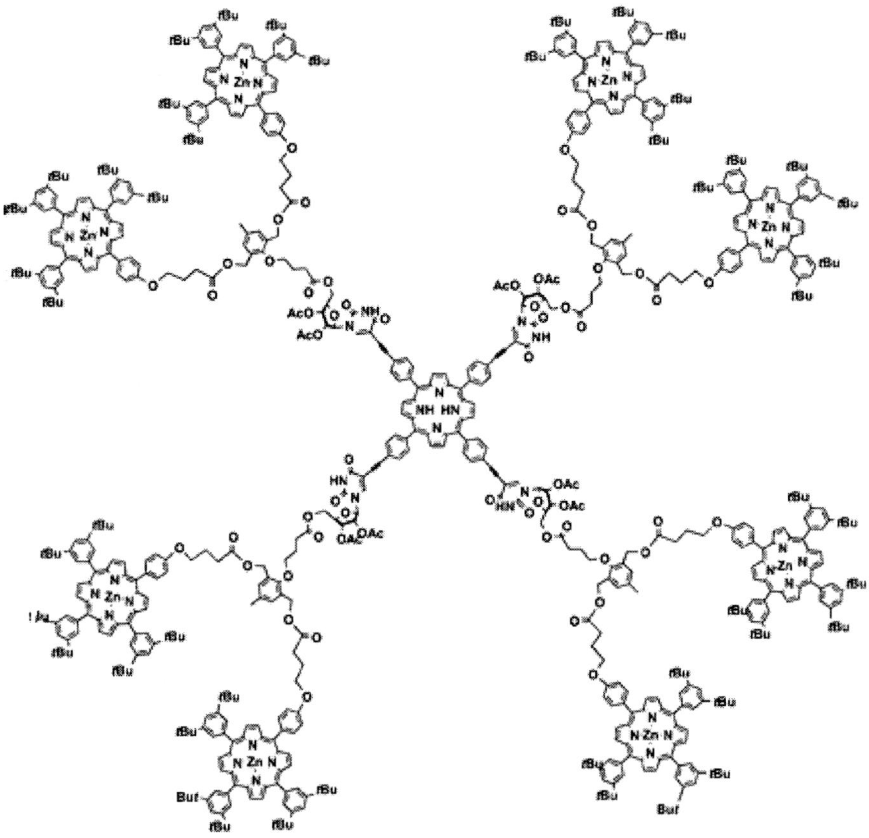

Figure 47. Dendritic Zn porphyrin structure with a free-base porphyrin core.

To overcome this limitation, Theodore Goodson III et al demonstrated a novel effect of the two-photon absorption method utilizing the high degree of quantum optical correlation between photon pairs created by the process of spontaneous parametric downconversion. A large entangled two-photon absorption cross-section (δ ~ 10–17 cm2 at 800 nm) was measured in an organic porphyrin dendrimer. This novel effect may open new avenues for ultrasensitive detection in chemical and biological systems. The use of entangled photons from a correlated source for the purpose of the two-photon effect is promising in this regard as one may obtain two-photon effects with very small numbers of photons.

Theodore Goodson, III et al [156] reported a strongly interacting new dendrimer system with an extended spectroscopic unit (coherent domain) beyond the trimer configuration. The report focused on the mechanism of enhancement of the two-photon absorption (TPA) response of this material as a function of the dendrimer generation number (Figure 48). This enhancement is strongly correlated to the size and geometry of the spectroscopic unit in a strongly interacting macromolecular system.

Uday Maitra et al [157] described a facile synthesis of bile acid-based dendritic structures with multiple naproxen groups at the periphery and a single anthracenyl moiety on the side chain. Absorption studies indicate that there are no groundstate interactions between the chromophores and the molar extinction coefficients increase linearly with increasing number of naproxen units. Upon excitation of the peripheral chromophores, there is efficient intramolecular energy transfer to the anthracenyl chromophore. These novel, non-conjugated dendrimers thus act as efficient molecular light harvesters.

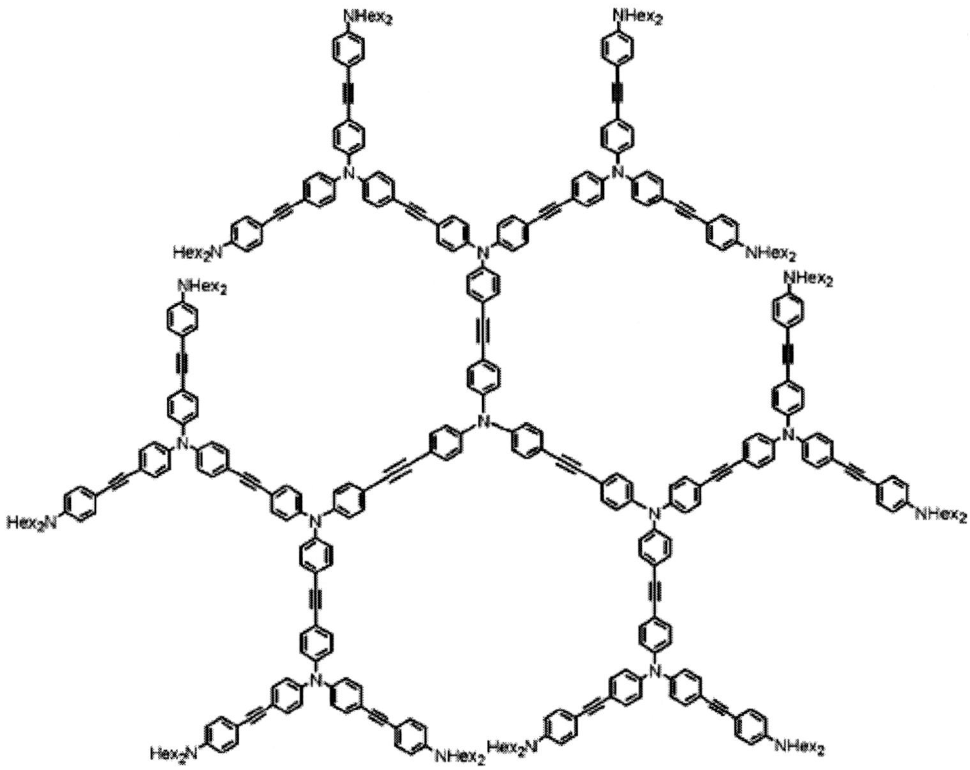

Figure 48. Organic conjugated dendrimers with enhanced two-photon absorption.

Utilization of two-photon (TP) excited fluorescence resonance energy transfer (FRET) within a lightharvesting dendrimer has proven to be a reliable method for the enhancement of the effective TP absorption efficiency of many FRET acceptor molecules. This light-harvesting approach has enabled TP photosensitization of singlet oxygen from a porphyrin in both aqueous and organic media using wavelengths more transmissive to human body tissue (750-1000 nm). This ability to utilize near-infrared irradiation to induce photochemical reactions is especially attractive for applications including in vivo photochemistry, oxygen sensing, and photodynamic cancer therapy. Jean M. J. Fréchet et al [158] reported light-harvesting chromophores with metalated porphyrin cores for tuned photosensitization of singlet oxygen via two-photon excited FRET. Jean M. J. Fréchet et al synthesized an array of novel photosensitizers by metalation of the porphyrin core with aluminum, silver, and zinc. Singlet oxygen photosensitization efficiency was found to be most efficient using aluminum followed by zinc and least efficient using silver.

7.4. Charge Transfer

Nicola Armaroli et al [159] reported the structure-dependent photoinduced electron transfer in fullerodendrimers with light-harvesting oligophenylenevinylene terminals. Oligophenylenevinylene (OPV)-terminated phenylenevinylene dendrons G1–G4 with one, two, four, and eight "side-arms", respectively, were prepared and attached to C60 by a 1,3-dipolar cycloaddition of azomethine ylides generated in situ from dendritic aldehydes and N-methylglycine. The relative electronic absorption of the OPV moiety increases progressively along the fullerodendrimer family C60G1–C60G4, reaching a 99:1 ratio for C60G4 (antenna effect). Regardless of solvent, all the fullerodendrimers exhibit ultrafast OPV→C60 singlet energy transfer. In CH2Cl2, the OPV→C60 electron transfer from the lowest fullerene singlet level (1C60*) is slightly exergonic (ΔGCS≈0.07 eV), but is observed, to an increasing extent, only in the largest systems C60G2–C60G4 with lower activation barriers for electron transfer (Figure 49). Monitoring of the lowest fullerene triplet state by sensitized singlet oxygen luminescence and transient absorption spectroscopy shows that this level is populated through intersystem crossing and is not involved in photoinduced electron transfer.

Clemens Krempner et al [160] reported hydroxy-substituted oligosilane dendrimers for controlling the electronic properties through hydrogen bonding. S. Thayumanavan et al [161] examined the photophysics of a series of molecules consisting of a benzthiadiazole core surrounded by a network of benzyl ether arms terminated by aminopyrene chromophores, which function as both energy and electron donors.

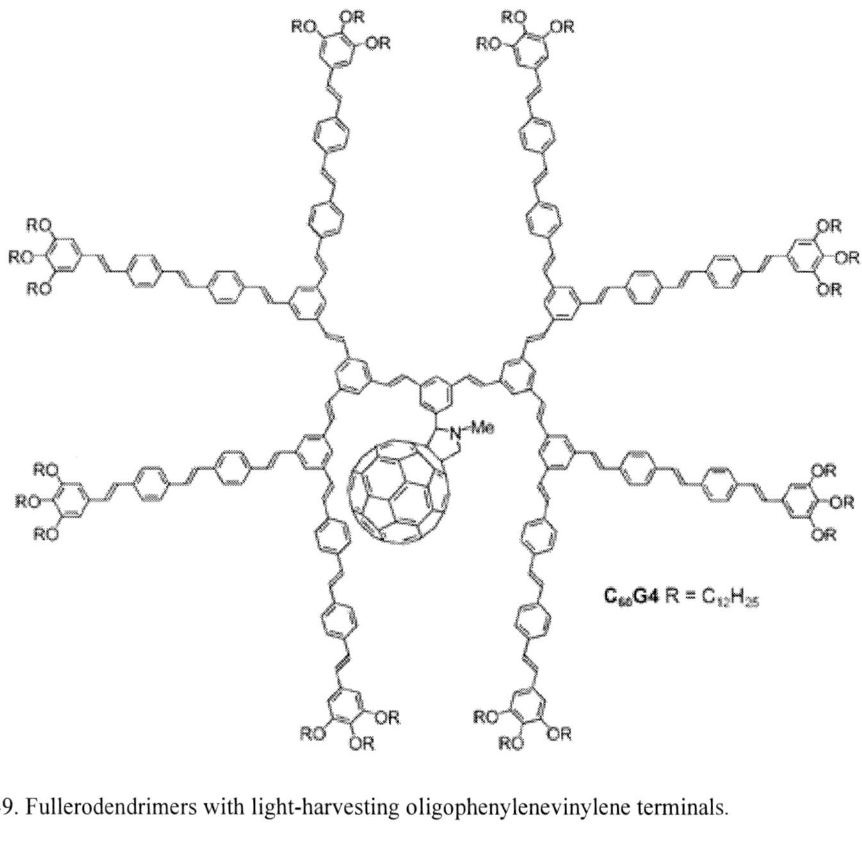

Figure 49. Fullerodendrimers with light-harvesting oligophenylenevinylene terminals.

Three classes of molecules are studied: dendrimers whose peripheries are fully decorated with aminopyrene donors (F), disubstituted dendrimers whose peripheries contain only two donors (D), and linear analogues in which a pair of benzyl ether arms link two donors to the central core (L). In all three types of molecules, the electronic energy transfer (EET) time scales as the square root of the generation number G, consistent with the flexible nature of the benzyl ether framework. Transient anisotropy measurements confirm that donor-donor energy hopping does not play a major role in determining the EET times. The charge transfer (CT) dynamics occur on the nanosecond time scale and lead to stretched exponential decays, probably due to conformational disorder. This work illustrated two important points about light-harvesting and charge-separation dendrimers. First, the use of a flexible dendrimer framework can lead to a more favorable scaling of the EET time (and thus the light-harvesting efficiency) with dendrimer size, relative to what would be expected for a fully extended dendrimer. Second, fully decorated dendrimers can compensate for the distancedependent slowdown in CT rate as G increases by providing additional pathways for the CT reaction to occur.

James K. Whitesell et al [162] reported intramolecular charge transfer in 1:1 Cu(II)/pyrenylcyclam dendrimer complexes. Two newly synthesized pyrenylcyclam dendrons exhibit a new emission band centered at 450 nm when coordinated with copper triflate. Observed fluorescence shifts induced by coordinative metalation indicate an unusual intramolecular charge transfer from a pyrenyl excited state to the coordinated metal ion that competes with pyrene excimer formation. This interaction likely proceeds by photoexcitation of a π-complex of the appended arene, followed by intramolecular charge transfer within the dendritic 1:1 cyclam/metal complex, effecting reduction of the bound Cu(II) metal ion. The appended dendritic groups not only decrease the equilibrium binding constant with Cu(II) but also participate in a new excited-state pathway as an alternative to energy-dissipative excimer formation.

Clifford P. Kubiak et al [163] reported charge gating and electronic delocalization over a denderimeric assembly of trinuclear Ruthenium clusters. A zeroth-order dendrimer was formed using a tridentate bridging ligand, 2,4,6-tri-4-pyridyl-S-triazine, and the redox active trinuclear ruthenium cluster Ru3O(OAc)6(CO)(py)(H2O). The electronic properties of this dendrimer were probed using cyclic voltammetry. IR spectroelectrochemistry was performed at both low (—30°C) and room temperature. The IR spectroelectrochemical response at –30°C was straightforward, but at room temperature, the dendrimer exhibits an unusual and complex series of electronic behaviors, including intramolecular cluster-to-bridging-ligand charge transfer, gated electron transfer, and dynamic exchange on the IR time scale.

Wei-Shi Li et al [164] reported the construction of segregated arrays of multiple donor and acceptor units using a dendritic scaffold for remarkable dendrimer effects on photoinduced charge separation. Dendritic molecules appended with multiple zinc porphyrin units (DPm,m [number of zinc porphyrin units] = 6, 12, and 24) trap bipyridine compounds carrying multiple fullerene units (Py2Fn, n [number of C60 units] = 1-3), affording coordination complexes DPm⊃Py2Fn having a photoactive layer consisting of spatially segregated donor and acceptor arrays on their surface. Complexes DPm⊃Py2Fn are stable enough to be isolated by gel permeation chromatography. UHV-STM microscopy enables clear visualization of a petal-like structure of DP12⊃Py2F3. Photoexcitation of the zinc porphyrin units in DPm⊃Py2Fn results in a zinc porphyrin-to-fullerene electron transfer to generate a charge separation (Figure 50).

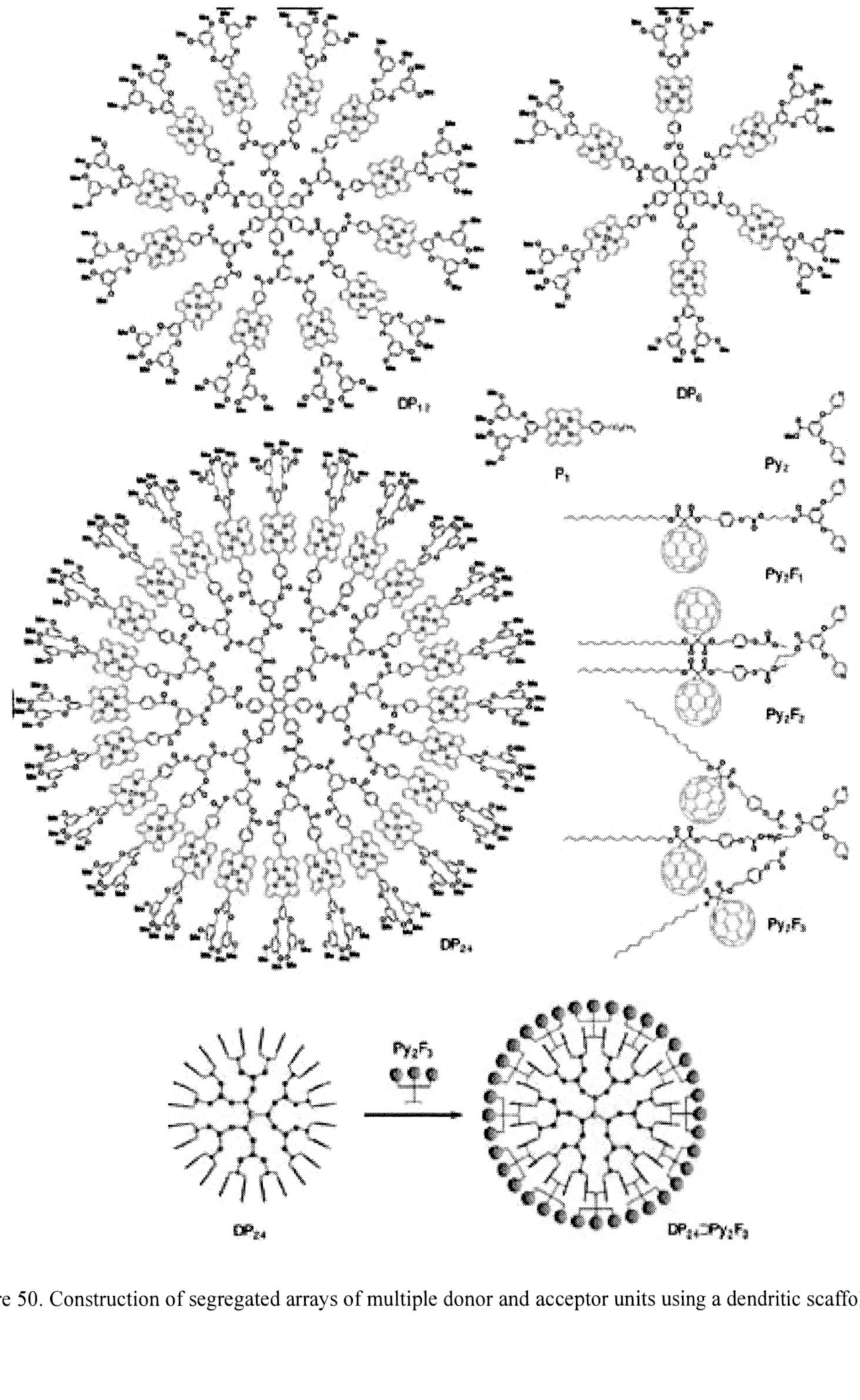

Figure 50. Construction of segregated arrays of multiple donor and acceptor units using a dendritic scaffold.

Chungkyun Kim et al [165] described the divergent synthesis of end-capped silole dendrimers Gn-2n+1Silole (n = 1-4) using 2,4,6,8-tetramethyl-2,4,6,8-tetravinylcyclotetrasiloxane [Me(CH2=CH)SiO]4 as a core molecule and allyl alcohol/ dichloromethylsilane as a building block. The reaction of the dichloride functionalities of the carbosiloxane dendrimers Gn-2n+2Cl (n = 1-4) with the 1,4-diphenylbutadiene- 1,4-dianions provided an effective route for the attachment of silole functionalities to a dendritic periphery. In addition, dendritic siloles GBn-2nSilole (n = 1-3) based on phenylethynylmethylsilane (PhC=C)2Me2Si as cores were synthesized. The building blocks of the dendrimers consisted of double bonds (-PhC=CHMeSi-) in the inner shell and silole groups on the outermost periphery. The generational limit of the dendrimer for the two-branching type core molecule was found to have eight silole groups for the third generation. Using this methodology, a series of silole-containing dendrimer systems showing green to greenish-blue fluorescence were synthesized.

7.5. Proton Harvest

π-conjugated dendrimers are molecular examples of tree-like structures known in physics as Bethe lattices. Electronic excitations in these systems can be spatially delocalized or localized depending on the branching topology. Without a priori knowledge of the localization pattern, understanding photoexcitation dynamics reflected in experimental optical spectra is difficult. 'Supramolecular'-like quantum-chemical calculations quickly become intractable as the molecular size increases. Dong-Lin Jiang et al [166] reported photoisomerization in dendrimers by harvesting of low-energy photons.

Vladmir Y. Chernyak et al [167] reported exciton scattering and localization in branched dendrimeric structures. They developed a reduced excitonscattering (ES) model, which attributes excited states to standing waves in quasi-one-dimensional structures, assuming a quasiparticle picture of optical excitations. Direct quantum-chemical calculations of branched phenylacetylene chromophores are used to verify the model and to derive relevant parameters. Complex and non-trivial delocalization patterns of photoexcitations throughout the entire molecular tree can then be universally characterized and understood using the proposed ES method, completely bypassing 'supramolecular' calculations. This allows accurate modelling of excited-state dynamics in arbitrary branched structures.

8. Advances in Dendrimer Sensors

8.1. Electrochemistry Biosensors

Rigoberto Advincula et al [168] have achieved nanomolar detection and specific recognition of pinacolyl methylphosphonate (PMP), a hydrolysis product and an analog of a relatively persistent class of toxic nerve agents, In addition, picomolar sensitivity is observed with methylphosphonic acid (MPA), an end-hydrolysis product for all organophosphate-based nerve agents. This is achieved using a combined surface plasmon resonance (SPR) and potentiometry setup. A modified polyamidoamine (PAMAM) carbazole/Cu2+ dendrimer, which is electrochemically crosslinked on a self-assembled monolayer (SAM) modified Au substrate, is used as an active sensing element for trapping the nerve-agent analogs. The ultrathin films have been used to study the anchoring of nerve agents via non-covalent interactions. The carbazole to amine ratio is optimized to ensure free primary amines are available to interact with the analyte and the Cu2+ ions present in the system, which further enhances the selectivity. The carbazole group on the periphery serves a dual purpose: crosslinking the dendrimers to form a conjugated network film, and generating the potentiometric response. The adsorption kinetics are monitored by using an in situ SPR/potentiometric setup. This technique not only offers a real-time dual detection of highly toxic nerve-agent analogs, but also shows viability for future sensor-device applications.

Chongmok Lee et al [169] reported electrochemical deposition of end-capped triarylamine and carbazole dendrimers as the alternate technique for the manufacture of multilayer films (Figure 51).

Haesik Yang et al [170] reported a nanocatalyst-based assay for proteins based on DNA-free ultrasensitive electrochemical detection using catalytic reduction of p-nitrophenol by gold-nanoparticle labels (Figure 52). Lawrence D. Margerum et al [171] reported the effect of PAMAM dendrimer size and pH on the electrostatic binding of metal complexes using cyclic voltammetry. Cyclic voltammetry was employed to investigate the extent of electrostatic binding for metal complexes such as ferrocyanide, [Fe(CN)6]4−, to full generations of polyamidoamine (PAMAM) dendrimers (G1.0-G5.0) as a function of size and pH. Addition of dendrimer solutions to metal complexes caused a large net positive shift in *E*1/2 and reduced peak currents that varied with the size of the dendrimer. Binding ratios, *K*4−/*K*3−, calculated from the potential shifts changed from a ratio of five with small dendrimers (G1.0-G3.0) to 10 for G4.0 and G5.0, consistent with known structural changes in PAMAM dendrimers. Conditional binding constants for [Fe(CN)6]4−, calculated from the diffusion-limited oxidation currents, also increased from G3.0 (8.8 ×104 M−1) to G4.0 (31 × 104 M−1). At pH 5, the binding ratios increased dramatically at all generations, suggesting that the interior amine sites in the dendrimer are easily accessible to metal complexes and that dendrimers show some charge selectivity.

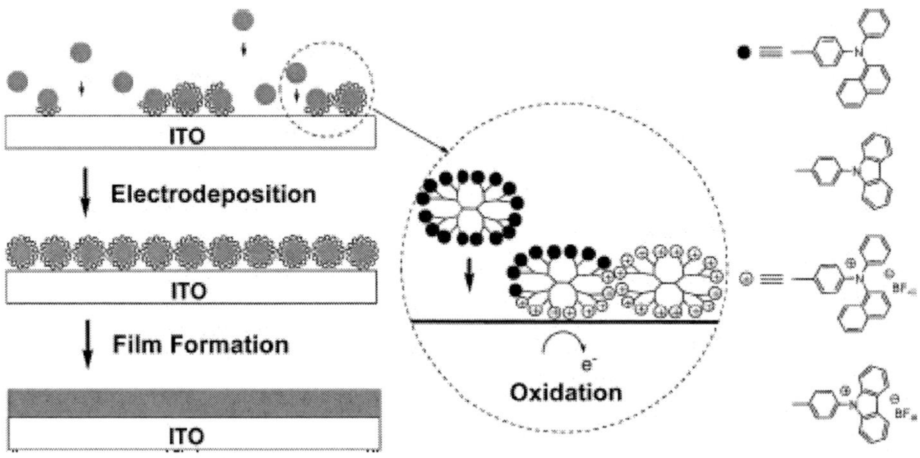

Figure 51. Electrochemical deposition of end-capped triarylamine and carbazole dendrimers.

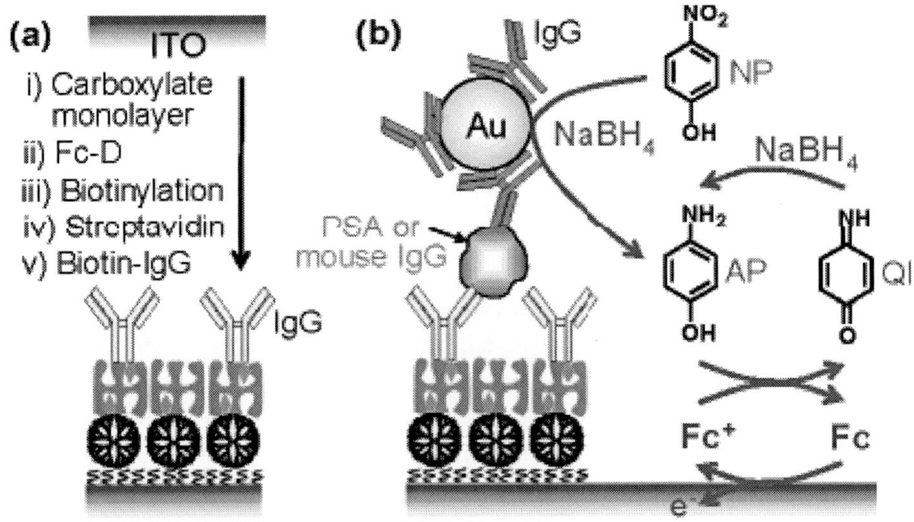

Figure 52. Preparation of an immunosensing layer for electrochemical detection of mouse IgG or PSA.

Francisco C. Nart et al [172] exploited the molecular engineering capability of the layer-by-layer (LbL) method to immobilize layers of gold nanoparticles on indium tin oxide (ITO) substrates, which exhibit enhanced charge transfer and may incorporate mediating redox substances. Polyamidoamine (PAMAM generation 4) dendrimers were used as template/stabilizers for Au nanoparticle growth, with PAMAM-Au nanoparticles serving as cationic polyelectrolytes to produce LbL films with poly(vinylsulfonic acid) (PVS) (Figure 53).

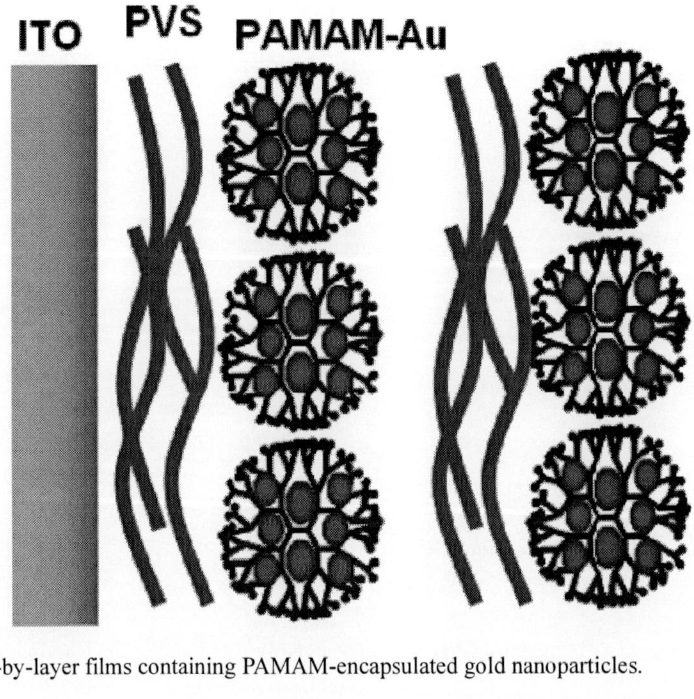

Figure 53. Layer-by-layer films containing PAMAM-encapsulated gold nanoparticles.

The cyclic voltammetry (CV) of ITO-PVS/PAMAM-Au electrodes in sulfuric acid presented a redox pair attributed to Au surface oxide formation. Electron hopping can be considered as the charge transport mechanism between the PVS/ PAMAM-Au layers within the LbL films. This charge transport was faster than that for nonmodified electrodes, shown by employing hexacyanoferrate(III) as the surface reaction marker.

Shaojun Dong et al [173] reported the assembly process of CuHCF/MPA multilayers on gold nanoparticles modified electrode and characterization by electrochemical SPR. Gold nanoparticles were deposited onto 2-mercaptoethylamine (MEA)-assembled planar gold thin film to construct gold nanoparticles modified electrode by virtue of a solution-based self-assembly strategy. Subsequently, 3-mercaptopropionic acid (MPA)-bridged copper hexacyanoferrate (CuHCF) multilayers were constructed on the as-prepared gold nanoparticles modified electrode. Compared with the multilayer system on a planar gold electrode, the different electrochemical and optical properties might result from higher curvature effect and extraordinary surface-to-volume ratio characteristic of gold nanoparticles and the nanoparticle- selective growth of CuHCF. A dendrimer-like assembly process was proposed to explain the experiment results. This new motif of multilayer on the gold nanoparticles modified electrode was different from that of on a planar gold electrode, indicating a potential application of EC-SPR technique in the study of nanocomposite materials.

Functionalization of multi-walled carbon nanotubes with poly(amidoamine) dendrimer for mediator-free glucose biosensor is of great interest. Yun-Long Zeng et al [174] have successfully prepared novel carbon nanotube (CNT) nano-composites by grafting with poly(amidoamine) (PAMAM) dendrimer. The novel functionalized matrix with plenty amino groups circumvents the troublesome solubility problem of CNTs in solvents, especially in water, greatly expanding the scope of the application of carbon nanotubes. The GOx and HRP immobilized CNT-PAMAM based on the functional CNTs was synthesized. The bi-enzymatic CNT-PAMAM nano-composites are highly dispersible in water and show very promising applications in the fabrication of mediator-free bi-enzymatic biosensors for sensitive detection of glucose. The cooperation of nano-composite between CNT and high dense GOx and HRP results in very high sensitivity to glucose with a current response of 2200 nA mM−1 and fast response (~1 s). The modified electrode exhibits a wide linear response range for glucose from 4.0 μM to 1.2 mM (R = 0.9971, N = 15), with a detection limit of 2.5 μM. The negative electrode potential of −0.34 V is favorable for glucose detection in real samples without interference caused by other biomolecules.

E. Lojou et al [175] used dendritic macromolecules of two adjacent (G3.5 and G4) generations to modify gold or carbon electrodes. Dendrimers have been shown to adsorb spontaneously on electrode materials as compressed macromolecular films. They are able to inhibit (G3.5) or promote (G4) electroactive anionic species such as Fe(CN)63−/4− used as a probe system. Mixed protein/dendrimer assemblies have been constructed with proteins differing in charge, nature of the prosthetic groups and sizes such as lysozyme, cytochrome c, polyhemic cytochrome c3 or glucose oxidase. Generally, the stability of adsorbed films seems to be limited to one dendrimer/protein bilayer. Owing to the satisfactory stability of composite cytochrome c3/G3.5 or glucose oxidase/G4 films, biosensing applications are described for metal bioremediation and glucose detection, respectively.

Christopher M.A. Brett et al [176] described a new approach to produce nanostructured electrocatalytic membranes using a combination of three methods. Using the layer-by-layer technique, PAMAM dendrimers with cobalt hexacyanoferrates-modified gold nanoparticles were alternated with poly(vinylsulfonic acid) layers on ITO (indium tin oxide) electrodes (Figure 54). This film was used as substrate for immobilization of glucose oxidase in the presence of bovine serum albumin and glutaraldehyde as cross-linker. The modified electrode was successfully applied as a biosensor for the amperometric measurement of glucose, using glucose oxidase enzyme, at 0.0 V vs. SCE.

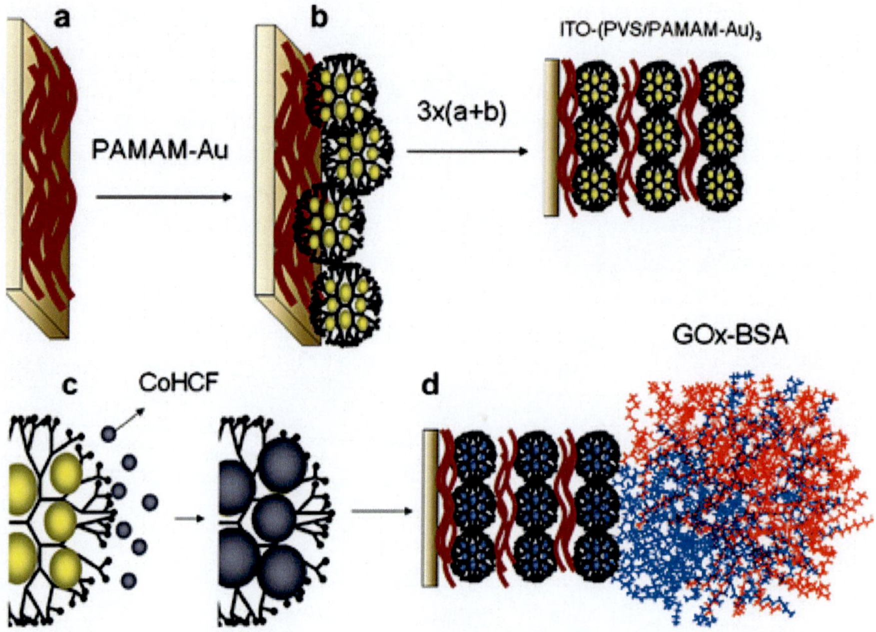

Figure 54. Fabrication of layer-by-layer films comprising PVS and PAMAM–Au.

M. Pilar García Armada et al [177] described the electrochemical characterization of polymethylferrocenyl dendrimers deposited onto a platinum electrode and their applications as hydrogen peroxide and glucose sensor. The redox dendrimers consist of flexible poly(propileneimine) dendrimer cores functionalised with octamethylferrocenyl units. Amperometric biosensors for glucose were prepared by immobilization of glucose oxidase onto these modified electrodes.

The detection of aromatic compounds from pesticides and industrial wastewater has become of great interest, since these compounds withstand chemical oxidation and biological degradation, accumulating in the environment. Valtencir Zucolotto et al [178] reported catechol biosensing using a nanostructured layer-by-layer film containing Cl-catechol 1,2-dioxygenase. A highly sensitive biosensor for detecting catechol was obtained with the immobilization of Cl-catechol 1,2-dioxygenase (CCD) in nanostructured films. CCD layers were alternated with poly(amidoamine) generation 4 (PAMAM G4) dendrimer using the electrostatic layer-by-layer (LbL) technique (Figure 55). Circular dichroism (CD) measurements indicated that the immobilized CCD preserved the same conformation as in solution. PAMAM/CCD 10-bilayer films were employed in detecting diluted catechol solutions using either an optical or electrical approach. Due to the mild immobilization conditions employed, especially regarding the pH and ionic strength of the dipping solutions, CCD remained active in the films for periods longer than 3 weeks.

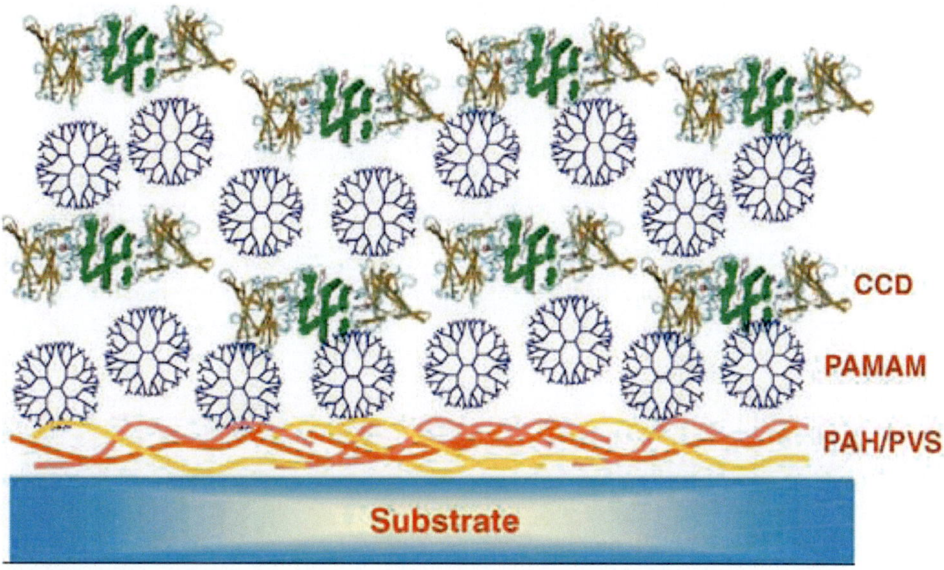

Figure 55. Catechol biosensing using a nanostructured layer-by-layer film containing Cl-catechol 1,2-dioxygenase.

Ru-Qin Yu et al [179] reported a hydrogen peroxide biosensor based on nano-Au/PAMAM dendrimer/cystamine modified gold electrode. Horseradish peroxidase (HRP) to a nano-Au monolayer, which was supported by PAMAM dendrimer/cystamine modified gold electrode. The high affinity of PAMAM dendrimer for nano-Au with its amine groups was used to realize the role of nano-Au as an intermediator to retain the high bioactivity of HRP. A hydrogen peroxide biosensor was fabricated based on the effective immobilization of the high bioactivity of HRP. In the presence of the hydroquinone mediator in the solution, the immobilized HRP exhibited excellent electrocatalytical response to the reduction of H2O2. The biosensor exhibited high sensitivity and good stability and reproducibility. Performance of the biosensor was evaluated with respect to possible interferences and the application to real sample analysis.

T. Hianik et al [180] reported an acetylcholinesterase sensors based on gold electrodes modified with dendrimer and polyaniline. Potentiometric and amperometric enzyme sensors based on modified gold electrodes have been developed and compared in pesticide determination. PAMAM dendrimer (generation G4) stabilized with 1-hexadecanethiol was used for the immobilization of acetylcholinesterase from electric eel and choline oxidase from Alcaligenes species in the assembly of amperometric sensor. Polyaniline-doped with camphorsulfonic acid was used to obtain potentiometric response. Trichlorfon, carbofuran and eserine suppress the biosensor response due to their inhibitory effect. The difference in the biosensor behavior and the high sensitivity of the dendrimer modified sensor to the inhibitors is due to the specific organization of protein layer at charged surface of the modifier macromolecules.

T. Hianik et al [181] studied the properties of mixed alkanethiol–dendrimer layers on a gold support and their application in biosensing. They showed that properties of glucose sensor can be modified using a different ratio of 1-hexadecanethiol (HDT) and poly(amidoamine) dendrimer of first generation (G1). The cyclic voltammetry in the presence of the redox couple, Fe(CN)63−/Fe(CN)64−, was used for estimating how effectively the layer blocks the redox probe's access to the electrode surface. A scanning electrochemical microscope (SECM) was used to image the resulting distribution of the organic compounds. T. Hianik et al found that with increasing content of dendrimers, the integrity of the layers was improved.

Naifei Hu et al [182] reported a heme protein films with polyamidoamine dendrimer for direct electrochemistry and electrocatalysis. Biocompatible nanosized polyamidoamine (PAMAM) dendrimer films provided a suitable microenvironment for heme proteins to transfer electron directly with underlying pyrolytic graphite (PG) electrodes. Hemoglobin (Hb), myoglobin (Mb), horseradish peroxidase (HRP), and catalase (Cat) incorporated in PAMAM films exhibited a pair of well-defined, quasi-reversible cyclic voltammetric peaks, respectively, characteristic of the protein heme Fe(III)/Fe(II) redox couples. Oxygen, hydrogen peroxide, and nitrite were catalytically reduced at the protein-PAMAM film electrodes, showing the potential applicability of the films as the new type of biosensors or bioreactors based on direct electrochemistry of the proteins.

Tibor Hianik et al [183] reported a QCM Immunosensor Based on Polyamidoamine Dendrimers. The poly(amidoamine)(PAMAM) dendrimers of fourth generation (G4) and 1-hexadecanethiol (HDT) has been used for formation of self assembled monolayers on a gold surface for preparation of quartz crystal microbalance (QCM) immunosensor for detection of human IgG. The anti-IgG antibodies were immobilized on a surface of G4-HDT layers or those covered by Protein A (Figure 56).

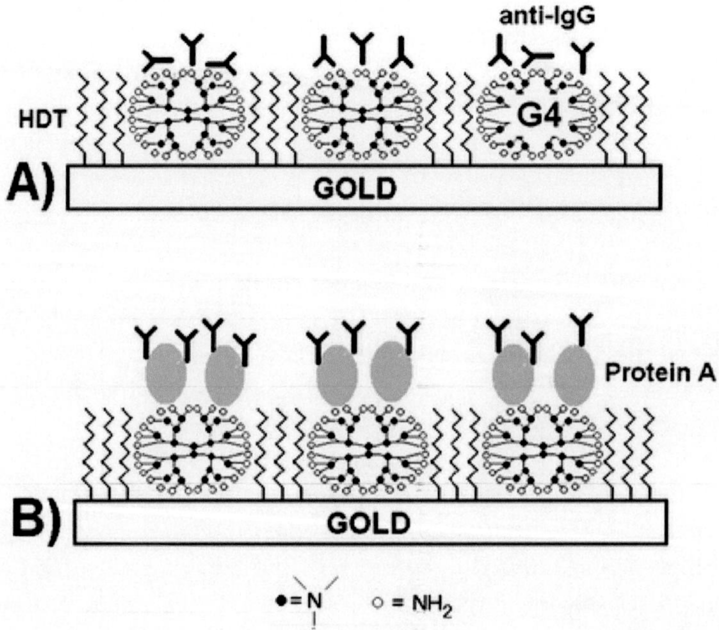

Figure 56. Quartz crystal microbalance immunosensor based on polyamidoamine dendrimers.

In later case the oriented immobilization of anti-IgG takes place due to preferred interaction of F_c portion of antibodies with the Protein A. The detection limit in a flow mode was 7 nM both for biosensor with and without Protein A. Increased concentration of IgG and formation of Ab-Ag complexes resulted in increase of motional resistance, that suggest increased contribution of viscoelasticity, as revealed from impedance analysis.

Mark H. Schoenfisch et al [184] reported the preparation and characterization of nitric oxide (NO)-releasing dendrimer conjugates. Generation 3 and 5 polypropylenimine dendrimers (DAB-Am-16 and DAB-Am-64) were modified at the exterior to impart different amine functionalities. The ability to store NO on a dendritic scaffold using N-diazeniumdiolate NO donors was examined via the reaction of primary amine, secondary amine, and amide functionalities with high pressures of NO (5 atm). The secondary amine dendrimer conjugates exhibited a high storage capacity for NO (up to 5.6 ímol NO/mg), greatly increasing the "payload" of released NO over existing macromolecular NO donors. The multivalent exterior of dendrimers allows for the future combination of NO donors and other functionalities on a single molecular scaffold, enabling diverse utility as NO storage/delivery systems.

8.2. Optical Biosensor

Andrei G. Kutateladze et al [185] reported photoinduced signal amplification through controlled externally sensitized fragmentation in masked sensitizers.

Atsushi Ikeda et al [186] used PAMAM dendrimers (G1-G4) as stationary phases to deposit porphyrinintercalated DNA onto ITO electrodes. The surface density of the external amino moieties of the dendrimers, i.e., the cationic charge density on the surface, increased with each generation, and correspondingly increased the surface coverage of DNA and porphyrin units through electrostatic interactions and intercalation. The resulting dendrimer-DNA-porphyrin membranes were sufficiently stable that they did not peel off the ITO electrodes. The resulting electrodes generated photocurrent waves in response to irradiation with visible light. The photocurrent increased exponentially upon increasing the generation of the dendrimer, i.e., a positive dendritic effect in these systems can be observed; actually, the photocurrent for the G4-DNA-porphyrin membrane (77.5 nA cm-2 at 440 nm) was ca. nine times larger than that for the G1-DNA-1 membrane (8.5 nA cm-2 at 440 nm), and the quantum yield for the former (1.0%) was five times larger than that for the latter (0.2%). Furthermore, the performance of the two-dimensional polymer G4 as a stationary phase was far superior to that of the one-dimensional polymer poly(allylamine hydrochloride) (PAH).

REFERENCES

[1] Christophe Galliot, Christophe Larre, Anne-Marie Caminade, Jean-Pierre Majoral. Regioselective Stepwise Growth of Dendrimer Units in the Internal Voids of a Main Dendrimer. *Science*, 1997, 277, 1981-1984

[2] Daniel Taton, Yves Gnanou, Rachid Matmour, Stephanie Angot, Sijian Hou, Raju Francis, Benedicte Lepoittevin, Delphine Moinard, and Jerome Babin. Controlled

polymerizations as tools for the design of star-like and dendrimer-like polymers. *Polym. Int.* 55:1138–1145 (2006)

[3] George R. Newkome, Kishore K. Kotta, and Charles N. Moorefield. Design, Synthesis, and Characterization of Conifer-Shaped Dendritic Architectures. *Chem. Eur. J.* 2006, 12, 3726 – 3734.

[4] Sandra Reemers, Ahmed Mourran, Helmut Keul, and Martin Moller. Novel Route to Dendritic Structures and Their Application for Surface Modification. *Journal of Polymer Science: Part A: Polymer Chemistry*, Vol. 44, 1372–1386 (2006).

[5] Chun-Yan Hong, Ye-Zi You, Jun Liu, Cai-Yuan Pan. Dendrimer-Star Polymer and Block Copolymer Prepared by Reversible Addition-Fragmentation Chain Transfer (RAFT) Polymerization with Dendritic Chain Transfer Agent. *Journal of Polymer Science: Part A: Polymer Chemistry,* Vol. 43, 6379–6393 (2005).

[6] Jianhua Zou, Yongbing Zhao, and Wenfang Shi. Encapsulation Mechanism of Molecular Nanocarriers Based on Unimolecular Micelle Forming Dendritic Core-Shell Structural Polymers. *J. Phys. Chem.* B 2006, 110, 2638-2642.

[7] Paola Carbone, Aurora Calabretta, Marco Di Stefano, Fabrizia Negri, and Klaus Mu1llen. Shape Persistence and Bistability of Planar Three-Fold Core Polyphenylene Dendrimers: A Molecular Dynamics Study. *J. Phys. Chem.* A 2006, 110, 2214-2224.

[8] Chae Kyu Kim, Eun Seok Song, Hyung Joo Kim, Chiyoung Park, Jai Kyeong Kim, Young Chul Kim, Jae Woong Yu, Chulhee Kim. Conjugated Dendrimers with Triazine Peripheries and a Distyrylanthracene Core. *Journal of Polymer Science: Part A: Polymer Chemistry,* Vol. 44, 5855–5862 (2006).

[9] Jin-Seong Lee, June Huh, Cheol-Hee Ahn, Minhyung Lee, Tae Gwan Park. Synthesis of Novel Biodegradable Cationic Dendrimers. Macromol. *Rapid Commun.* 2006, 27, 1608–1614.

[10] Stefan Bernhardt, Martin Baumgarten, and Klaus Müllen. Dendritic Encapsulation – "Postsynthetic" Functionalizations of a Single Benzophenone Shielded by Shape-Persistent Polyphenylene Dendrons. *Eur. J. Org. Chem.* 2006, 2523–2529.

[11] Mackay B. Steffensen, Emily Hollink, Frank Kuschel, Monika Bauer, Eric E. Simanek. Dendrimers Based on [1,3,5]-Triazines. *Journal of Polymer Science: Part A: Polymer Chemistry*, Vol. 44, 3411–3433 (2006).

[12] Jianfeng Pan, Weihong Zhu, Shangfeng Li, Jun Xu, and He Tian. Synthesis of Carrier-Transporting Dendrimers with Perylenebis(dicarboximide)s as a Luminescent Core. *Eur. J. Org. Chem.* 2006, 986–1001.

[13] Michael L. Szalai, Dominic V. McGrath, David R. Wheeler, Thomas Zifer, and James R. McElhanon. *Dendrimers Based on Thermally Reversible Furan-Maleimide Diels-Alder Adducts*. 2007; 40(4); 818-823.

[14] Kohei Hosomizu, Hiroshi Imahori, Uwe Hahn, Jean-Francüois Nierengarten, Andrea Listorti, Nicola Armaroli, Takashi Nemoto, and Seiji Isoda. Dendritic Effects on Structure and Photophysical and Photoelectrochemical Properties of Fullerene Dendrimers and Their Nanoclusters. *J. Phys. Chem.* C. 2007; 111(6); 2777-2786.

[15] Su-Hoon Yim, June Huh, Cheol-Hee Ahn, and Tae Gwan Park. Development of a Novel Synthetic Method for Aliphatic Ester Dendrimers. *Macromolecules* 2007, 40, 205-210.

[16] Zi-Fa Shi, Le-Jia Wang, Hong Wang, Xiao-Ping Cao, and Hao-Li Zhang. Synthesis of Oligo(phenylene ethynylene)s with Dendrimer "Shells" for Molecular Electronics. *Organic Letters*. 2007; 9(4); 595-598.

[17] Jianping Lu, Ping Fang Xia, Pik Kwan Lo, Ye Tao, and Man Shing Wong. Synthesis and Properties of Multi-Triarylamine-Substituted Carbazole-Based Dendrimers with an Oligothiophene Core for Potential Applications in Organic Solar Cells and Light-Emitting Diodes. *Chem. Mater*. 2006, 18, 6194-6203.

[18] Herbert Schumann, and Yilmaz Aksu. Synthesis and Characterization of Silicon-Based Group 14 Dendrimers Si{(CH2)2Sn[(CH2)4MPh3]3}4 (M = Ge, Sn). *Organometallics* 2007, 26, 397-402.

[19] Stephane Campidelli, Chloe Sooambar, Enrique Lozano Diz, Christian Ehli, Dirk M. Guldi, and Maurizio Prato. Dendrimer-Functionalized Single-Wall Carbon Nanotubes: Synthesis, Characterization, and Photoinduced Electron Transfer. *J. Am. Chem. Soc.* 2006, 128, 12544-12552.

[20] Rajesh Kumar Kainthan, Edward B. Muliawan, Savvas G. Hatzikiriakos, and Donald E. Brooks. Synthesis, Characterization, and Viscoelastic Properties of High Molecular Weight Hyperbranched Polyglycerols. *Macromolecules* 2006, 39, 7708-7717.

[21] Yang Jiang, Jie-Yu Wang, Yuguo Ma, Yu-Xin Cui, Qi-Feng Zhou, and Jian Pei. Large Rigid Blue-Emitting π-Conjugated Stilbenoid-Based Dendrimers: Synthesis and Properties. *Org. Lett*. 2004, 6, 3297-3300.

[22] Koji Matsuoka, Mikiko Terabatake, Atsushi Umino, Yasuaki Esumi, Ken Hatano, Daiyo Terunuma, and Hiroyoshi Kuzuhara. Carbosilane Dendrimers Bearing Globotriaoses: Syntheses of Globotrioasyl Derivative and Introduction into Carbosilane Dendrimers. *Biomacromolecules* 2006, 7, 2274-2283.

[23] Oleg Lukin, Volker Gramlich, Ramchandra Kandre, Igor Zhun, Thorsten Felder, Christoph A. Schalley, and Grigoriy Dolgonos. Designer Dendrimers: Branched Oligosulfonimides with Controllable Molecular Architectures. *J. Am. Chem. Soc*. 2006, 128, 8964-8974.

[24] Blanca Gonzalez, Beatriz Alonso, Jose Losada, M. Pilar Garcia-Armada, and Carmen M. Casado. Aza-Crown Ethers Attached to Dendrimers through Amidoferrocenyl Units. *Organometallics* 2006, 25, 3558-3561.

[25] Mark L. Wolfenden and Mary J. Cloninger. Carbohydrate-Functionalized Dendrimers To Investigate the Predictable Tunability of Multivalent Interactions. Bioconjugate Chem. 2006, 17, 958-966.

[26] Alessia Amore, Rieko van Heerbeek, Niek Zeep, Jan van Esch, Joost N. H. Reek, Henk Hiemstra, and Jan H. van Maarseveen. Carbosilane Dendrimeric Carbodiimides: Site Isolation as a Lactamization Tool. *J. Org. Chem*. 2006, 71, 1851-1860.

[27] Virgil Percec, Mihai Peterca, Monika J. Sienkowska, Marc A. Ilies, Emad Aqad, Jan Smidrkal, and Paul A. Heiney. Synthesis and Retrostructural Analysis of Libraries of AB3 and Constitutional Isomeric AB2 Phenylpropyl Ether-Based Supramolecular Dendrimers. *J. Am. Chem. Soc*. 2006, 128, 3324-3334.

[28] Vincenzo Balzani, Heinz Bandmann, Paola Ceroni, Carlo Giansante, Uwe Hahn, Frank-Gerrit Klarner, Ute Muller, Walter M. Muller, Carla Verhaelen, Veronica Vicinelli, and Fritz Vogtle. Host-Guest Complexes between an Aromatic Molecular Tweezer and Symmetric and Unsymmetric Dendrimers with a 4,4'-Bipyridinium Core. *J. Am. Chem. Soc.* 2006, 128, 637-648.

[29] Osamu Enoki, Hideaki Katoh, and Kimihisa Yamamoto. Synthesis and Properties of a Novel Phenylazomethine Dendrimer with a Tetraphenylmethane Core. *Organic Letters*. 2006, 8(4): 569-571.

[30] Julien Leclaire, Rodolphe Dagiral, Suzanne Fery-Forgues, Yannick Coppel, Bruno Donnadieu, Anne-Marie Caminade, and Jean-Pierre Majoral. Octasubstituted Metal-Free Phthalocyanine as Core of Phosphorus Dendrimers: A Probe for the Properties of the Internal Structure. *J. Am. Chem. Soc*. 2005, 127, 15762-15770.

[31] Zinaida B. Shifrina, Marina S. Rajadurai, Nina V. Firsova, Lyudmila M. Bronstein, Xinlei Huang, Alexander L. Rusanov, and Klaus Muellen. Poly(Phenylene-pyridyl) Dendrimers: Synthesis and Templating of Metal Nanoparticles. *Macromolecules* 2005, 38, 9920-9932.

[32] Natarajan Raju, Rama S. Ranganathan, Mike F. Tweedle and Rolf E. Swenson. A new dendrimer scaffold for preparing dimers or tetramers of biologically active molecules. *Tetrahedron Letters* 46 (2005) 1463–1465.

[33] Benoit Muls, Hiroshi Uji-i, Sergey Melnikov, Alain Moussa, Wendy Verheijen, Jean-Philippe Soumillion, Jacob Josemon, Klaus Mullen, and Johan Hofkens. Direct Measurement of the End-to-End Distance of Individual Polyfluorene Polymer Chains. *ChemPhysChem* 2005, 6, 2286 – 2294.

[34] Tong Zhou and Shing Bor Chen. A Simulation Study on Dynamics of Dendrimer-Polymer Conjugates. *Macromolecules* 2006, 39, 6686-6692.

[35] Neil Cumpstey, Raghu N. Bera, Paul L. Burn, and Ifor D. W. Samuel. Investigating the Effect of Steric Crowding in Phosphorescent Dendrimers. *Macromolecules* 2005, 38, 9564-9570.

[36] Ying Wang, Xinbing Xie, and Theodore Goodson, III. Enhanced Third-Order Nonlinear Optical Properties in Dendrimer-Metal Nanocomposites. *Nano Letters*. 2005, 5(12): 2379-2384.

[37] David Scanu, Natalia P. Yevlampieva, and Robert Deschenaux. Polar and Electrooptical Properties of [60]Fullerene-Containing Poly(benzyl ether) Dendrimers in Solution. *Macromolecules*. 2007; 40(4); 1133-1139.

[38] Winston Ong, and Robin L. McCarley. Chemically and Electrochemically Mediated Release of Dendrimer End Groups. *Macromolecules* 2006, 39, 7295-7301.

[39] Alena Krupkova, Jan Cermak, Zuzana Walterova, and Jiri Horsky. Quantitative Interpretation of MALDI-TOF Mass Spectra of Imperfect Carbosilane Dendrimers. *Analytical Chemistry*. 2007; 79(4); 1639-1645.

[40] Hak-Joon Seok, Mi-Young Hong, Young-Ja Kim, Min-Kyu Han, Dohoon Lee, Jung-Hwa Lee, Jong-Shin Yoo, Hak-Sung Kim. Mass spectrometric analysis of aYnity-captured proteins on a dendrimer-based immunosensing surface: investigation of on-chip proteolytic digestion. *Analytical Biochemistry* 337 (2005) 294–307.

[41] Sabine Rosenfeldt, Elena Karpuk, Matthias Lehmann, Herbert Meier, Peter Lindner, Ludger Harnau, and Matthias Ballauff. The Solution Structure of Stilbenoid Dendrimers: A Small-Angle Scattering Study. ChemPhysChem 2006, 7, 2097 – 2104.

[42] Kai Landskron and Geoffrey A. Ozin. Periodic Mesoporous Dendrisilicas. *Science*, 2004, 306: 1529-1532.

[43] Byeongdu Lee, Young-Hee Park, Yong-Taek Hwang, Weontae Oh, Jinhwan Yoon, and Moonhor Ree. Ultralow-k nanoporous organosilicate dielectric films imprinted with dendritic spheres. *Nature Mmaterials*, 2005, 4: 147-151.

[44] Steven C. Zimmerman, Michael S. Wendland, Neal A. Rakow, Ilya Zharov, Kenneth S. Suslick. Synthetic hosts by monomolecular imprinting inside dendrimers. *Nature*, 2002, 418: 399-403.

[45] Teddie Magbitang, Victor Y. Lee, Jennifer N. Cha, Hsiao-Lin Wang, W. Richard Chung, Robert D. Miller, Geraud Dubois, Willi Volksen, Ho-Cheol Kim, and James L. Hedrick. Oriented Nanoporous Lamellar Organosilicates Templated from Topologically Unsymmetrical Dendritic-Linear Block Copolymers. *Angew. Chem. Int. Ed.* 2005, 44, 7574 –7580.

[46] Andras Perl, Maria Peter, Bart Jan Ravoo, David N. Reinhoudt, and Jurriaan Huskens. Heavyweight Dendritic Inks for Positive Microcontact Printing. *Langmuir* 2006, 22, 7568-7573.

[47] Sander Langereis, Quido G. de Lussanet, Marcel H. P. van Genderen, E. W. Meijer, Regina G. H. Beets-Tan, Arjan W. Griffioen, Jos M. A. van Engelshoven, and Walter H. Backes. Evaluation of Gd(III)DTPA-terminated poly(propylene imine) dendrimers as contrast agents for MR imaging. *NMR Biomed.* 2006;19:133–141.

[48] Yukihiro Hama, Marcelino Bernardo, Celeste A.S. Regino, Yoshinori Koyama, Martin W. Brechbiel, Murali C. Krishna, Peter L. Choyke, and Hisataka Kobayashi. MR Lymphangiography Using Dendrimer-Based Contrast Agents: A Comparison at 1.5T and 3.0T. *Magnetic Resonance in Medicine* 57:431–436 (2007).

[49] Yanjun Fu, Danute E. Nitecki, David Maltby, Gerhard H. Simon, Kirill Berejnoi, Hans-Juergen Raatschen, Benjamin M. Yeh, David M. Shames, and Robert C. Brasch. Dendritic Iodinated Contrast Agents with PEG-Cores for CT Imaging: Synthesis and Preliminary Characterization. *Bioconjugate Chem.* 2006, 17, 1043-1056.

[50] Amit K. Galande, Scott A. Hilderbrand, Ralph Weissleder, and Ching-Hsuan Tung. Enzyme-Targeted Fluorescent Imaging Probes on a Multiple Antigenic Peptide Core. *J. Med. Chem.* 2006, 49, 4715-4720.

[51] Vladimir S. Talanov, Celeste A. S. Regino, Hisataka Kobayashi, Marcelino Bernardo, Peter L. Choyke, and Martin W. Brechbiel. Dendrimer-Based Nanoprobe for Dual Modality Magnetic Resonance and Fluorescence Imaging. *Nano Letters*. 2006, 6 (7): 1459-1463.

[52] Nobuhiro Nishiyama, Aya Iriyama, Woo-Dong Jang, Kanjiro Miyata, Keiji Itaka, Yuji Inoue, Hidenori Takahashi, Yasuo Yanagi, Yasuhiro Tamaki, Hiroyuki Koyama, and Kazunori Kataoka. Light-induced gene transfer from packaged DNA enveloped in a dendrimeric photosensitizer. *Nature Materials*. 2005, 4: 934-941.

[53] M. Adeli, Z. Zarnegar, A. Dadkhah, R. Hossieni, F. Salimi, A. Kanani. Linear-Dendritic ABA Triblock Copolymers as Nanocarriers. *Journal of Applied Polymer Science*, Vol. 104, 267–272 (2007).

[54] M. Adeli, R. Haag. Multiarm Star Nanocarriers Containing a Poly(ethylene imine) Core and Polylactide Arms. *Journal of Polymer Science: Part A: Polymer Chemistry*, Vol. 44, 5740–5749 (2006).

[55] Kris C. Wood, Steven R. Little, Robert Langer, and Paula T. Hammond. Family of Hierarchically Self-Assembling Linear-Dendritic Hybrid Polymers for Highly Efficient Targeted Gene Delivery. *Angew. Chem. Int. Ed.* 2005, 44, 6704 –6708.

[56] Andrzej Myc, Istvan J. Majoros, Thommey P. Thomas, and James R. Baker, Jr. Dendrimer-Based Targeted Delivery of an Apoptotic Sensor in Cancer Cells. *Biomacromolecules* 2007, 8, 13-18.

[57] Rameshwer Shukla, Thommey P. Thomas, Jennifer L. Peters, Ankur M. Desai, Jolanta Kukowska-Latallo, Anil K. Patri, Alina Kotlyar, and James R. Baker, Jr. HER2 Specific Tumor Targeting with Dendrimer Conjugated Anti-HER2 mAb. *Bioconjugate Chem.*, 2006, 17 (5): 1109-1115.

[58] Sezen Gurdag, Jayant Khandare, Sarah Stapels, Larry H. Matherly, and Rangaramanujam M. Kannan. Activity of Dendrimer-Methotrexate Conjugates on Methotrexate-Sensitive and -Resistant Cell Lines. *Bioconjugate Chem.* 2006, 17, 275-283.

[59] Renu Singh Dhanikula and Patrice Hildgen. Synthesis and Evaluation of Novel Dendrimers with a Hydrophilic Interior as Nanocarriers for Drug Delivery. *Bioconjugate Chem.* 2006, 17, 29-41.

[60] Istvan J. Majoros, Andrzej Myc, Thommey Thomas, Chandan B. Mehta, and James R. Baker, Jr. PAMAM Dendrimer-Based Multifunctional Conjugate for Cancer Therapy: Synthesis, Characterization, and Functionality. *Biomacromolecules* 2006, 7, 572-579.

[61] Manuela Berna, Daniele Dalzoppo, Gianfranco Pasut, Maria Manunta, Lorella Izzo, Arwyn T. Jones, Ruth Duncan, and Francesco M. Veronese. Novel Monodisperse PEG-Dendrons as New Tools for Targeted Drug Delivery: Synthesis, Characterization and Cellular Uptake. *Biomacromolecules* 2006, 7, 146-153.

[62] Tae-il Kim, Jung-un Baek, Cheng Zhe Bai, Jong-sang Park. Arginine-conjugated polypropylenimine dendrimer as a non-toxic and efficient gene delivery carrier. *Biomaterials* 28 (2007) 2061–2067.

[63] Durairaj Chandrasekar, Ramakrishna Sistla, Farhan J. Ahmad, Roop K. Khar, Prakash V. Diwan. The development of folate-PAMAM dendrimer conjugates for targeted delivery of anti-arthritic drugs and their pharmacokinetics and biodistribution in arthritic rats. *Biomaterials* 28 (2007) 504–512.

[64] Seungpyo Hong, Pascale R. Leroueil, Istvan J. Majoros, Bradford G. Orr, James R. Baker, Jr., and Mark M. Banaszak Holl. The Binding Avidity of a Nanoparticle-Based Multivalent Targeted Drug Delivery Platform. *Chemistry and Biology* 14, 107–115.

[65] Tatsuya Okuda, Shigeru Kawakami, Naomi Akimoto, Takuro Niidome, Fumiyoshi Yamashita, Mitsuru Hashida. PEGylated lysine dendrimers for tumor-selective targeting after intravenous injection in tumor-bearing mice. *Journal of Controlled Release* 116 (2006) 330–336.

[66] Luca Costantino, Francesca Gandolfi, Leila Bossy-Nobs, Giovanni Tosi, Robert Gurny, Francesco Rivasi, Maria Angela Vandelli, Flavio Forni. Nanoparticulate drug carriers based on hybrid poly(D,L-lactide-co-glycolide)-dendron structures. *Biomaterials* 27 (2006) 4635–4645.

[67] Parag Kolhe, Jayant Khandare, Omathanu Pillai, Sujatha Kannan, Mary Lieh-Lai, Rangaramanujam M. Kannan. Preparation, cellular transport, and activity of polyamidoamine-based dendritic nanodevices with a high drug payload. *Biomaterials* 27 (2006) 660–669.

[68] H. Namazi, M. Adeli. Dendrimers of citric acid and poly (ethylene glycol) as the new drug-delivery agents. *Biomaterials* 26 (2005) 1175–1183.

[69] Youngseon Choi, Thommey Thomas, Alina Kotlyar, Mohammad T. Islam, and James R. Baker, Jr. Synthesis and Functional Evaluation of DNA-Assembled Polyamidoamine Dendrimer Clusters for Cancer Cell-Specific Targeting. *Chemistry and Biology*, 12: 35–43.

[70] Majad Hussain, Mikhail S. Shchepinov, Muhammad Sohail, Ibrahim F. Benter, Andrew J. Hollins, Edwin M. Southern, Saghir Akhtar. A novel anionic dendrimer for improved cellular delivery of antisense oligonucleotides. *Journal of Controlled Release* 99 (2004) 139–155.
[71] Michael F. Neerman, Wen Zhang, Alan R. Parrish, Eric E. Simanek. In vitro and in vivo evaluation of a melamine dendrimer as a vehicle for drug delivery. *International Journal of Pharmaceutics* 281 (2004) 129–132.
[72] W. Tansey, S. Ke, X.-Y. Cao, Marites J. Pasuelo, Sidney Wallace, Chun Li. Synthesis and characterization of branched poly(L-glutamic acid) as a biodegradable drug carrier. *Journal of Controlled Release* 94 (2004) 39– 51.
[73] Marcelo Mamede, Tsuneo Saga, Takayoshi Ishimori, Tatsuya Higashi, Noriko Sato, Hisataka Kobayashi, Martin W. Brechbiel, Junji Konishi. Hepatocyte targeting of 111In-labeled oligo-DNA with avidin or avidin–dendrimer complex. *Journal of Controlled Release* 95 (2004) 133– 141.
[74] Sunil Shaunak, Sharyn Thomas, Elisabetta Gianasi, Antony Godwin, Emma Jones, Ian Teo, Kamiar Mireskandari, Philip Luthert, Ruth Duncan, Steve Patterson, Peng Khaw, and Steve Brocchini. Polyvalent dendrimer glucosamine conjugates prevent scar tissue formation. *Nature Biotechnology*, 2004, 22: 977-984.
[75] Serge H. M. Sontjens, Dana L. Nettles, Michael A. Carnahan, Lori A. Setton, and Mark W. Grinstaff. Biodendrimer-Based Hydrogel Scaffolds for Cartilage Tissue Repair. *Biomacromolecules* 2006, 7, 310-316.
[76] Sonke Svenson, Donald A. Tomalia. Dendrimers in biomedical applications—reflections on the field. *Advanced Drug Delivery Reviews* 57 (2005) 2106– 2129.
[77] Ruth Duncan, Lorella Izzo. Dendrimer biocompatibility and toxicity. *Advanced Drug Delivery Reviews* 57 (2005) 2215– 2237.
[78] Cameron C Lee, John A MacKay, Jean M J Fréchet, and Francis C Szoka. Designing dendrimers for biological applications. *Nature Biotechnology*. 2005, 12: 1517-1526.
[79] Jeanette C. Roberts, Mahesh K. Bhalgat, and Richard T. Zera. Preliminary biological evaluation of polyamidoamine (PAMAM) StarburstTM dendrimers. *Journal of Biomedical Materials Research*, 30, 53-65 (1996).
[80] Muriel Blanzat, Cedric-Olivier Turrin, Anne-Marie Aubertin, Christiane Couturier-Vidal, Anne-Marie Caminade, Jean-Pierre Majoral, Isabelle Rico-Lattes, and Armand Lattes. Dendritic Catanionic Assemblies: In vitro Anti-HIV Activity ofPhosphorus-Containing Dendrimers Bearing Galb1cer Analogues. *ChemBioChem* 2005, 6, 2207 – 2213.
[81] Jun-Ichi Sakamoto, Tetsuo Koyama, Daisei Miyamoto, Sangchai Yingsakmongkon, Kazuya I. P. J. Hidari, Wipawee Jampangern, Takashi Suzuki, Yasuo Suzuki, Yasuaki Esumi, Ken Hatano, Daiyo Terunuma and Koji Matsuoka. Thiosialoside clusters using carbosilane dendrimer core scaffolds as a new class of influenza neuraminidase inhibitors. *Bioorganic and Medicinal Chemistry Letters* 17 (2007) 717–721.
[82] Tatsuya Okuda, Shigeru Kawakami, Tadahiro Maeie, Takuro Niidome, Fumiyoshi Yamashita, Mitsuru Hashida. Biodistribution characteristics of amino acid dendrimers and their PEGylated derivatives after intravenous administration. *Journal of Controlled Release* 114 (2006) 69–77.

[83] Aristarchos Papagiannaros, Kostas Dimas, George Th Papaioannou, Costas Demetzos. Doxorubicin–PAMAM dendrimer complex attached to liposomes: Cytotoxic studies against human cancer cell line. *International Journal of Pharmaceutics* 302 (2005) 29–38.

[84] Natarajan Raju, Rama S. Ranganathan, Mike F. Tweedle and Rolf E. Swenson. A new dendrimer scaffold for preparing dimers or tetramers of biologically active molecules. *Tetrahedron Letters* 46 (2005) 1463–1465.

[85] Yasuhiro Haba, Atsushi Harada, Toru Takagishi, Kenji Kono. Synthesis of biocompatible dendrimers with a peripheral network formed by linking of polymerizable groups. *Polymer* 46 (2005) 1813–1820.

[86] Ben J. Boyd, Lisa M. Kaminskas, Peter Karellas, Guy Krippner, Romina Lessene, and Christopher J. H. Porter. Cationic Poly-L-lysine Dendrimers: Pharmacokinetics, Biodistribution, and Evidence for Metabolism and Bioresorption after Intravenous Administration to Rats. *Molecular Pharmaceutics* Vol. 3, NO. 5, 614-627.

[87] Anna Gopin, Sharon Ebner, Bernard Attali, and Doron Shabat. Enzymatic Activation of Second-Generation Dendritic Prodrugs: Conjugation of Self-Immolative Dendrimers with Poly(ethylene glycol) via Click Chemistry. *Bioconjugate Chem.* 2006, 17, 1432-1440.

[88] Andreas G. Schatzlein, Bernd H. Zinselmeyer, Adurrahim Elouzi, Christine Dufes, Ya Tsz A. Chim, Clive J. Roberts, Martyn C. Davies, Avril Munro, Alexander I. Grayb, Ijeom F. Uchegbu. Preferential liver gene expression with polypropylenimine dendrimers. *Journal of Controlled Release* 101 (2005) 247–258.

[89] Antony D'Emanuele, Rachaneekorn Jevprasesphant, Jeffrey Penny, David Attwood. The use of a dendrimer-propranolol prodrug to bypass efflux transporters and enhance oral bioavailability. *Journal of Controlled Release* 95 (2004) 447– 453.

[90] Petr Niederhafner, Jaroslav Sebestik and Jan Jezek. Peptide dendrimers. *J. Peptide Sci.* 11: 757–788 (2005).

[91] Hu Yang, Stephanie T. Lopina. In vitro enzymatic stability of dendritic peptides. *J. Biomed. Mater. Res.* 76A: 398–407, 2006.

[92] TAMIS DARBRE AND JEAN-LOUIS REYMOND. Peptide Dendrimers as Artificial Enzymes, Receptors, and Drug-Delivery Agents. *Acc. Chem. Res.* 2006, 39, 925-934.

[93] B. Klajnert, J. Janiszewska, Z. Urbanczyk-Lipkowska, M. Bryszewska, D. Shcharbin, M. Labieniec. Biological properties of low molecular mass peptide dendrimers. *International Journal of Pharmaceutics* 309 (2006) 208–217.

[94] Didier Astruc, Feng Lu, and Jaime Ruiz Aranzaes. Nanoparticles as Recyclable Catalysts: The Frontier between Homogeneous and Heterogeneous Catalysis. *Angew. Chem. Int.* Ed. 2005, 44, 7852 – 7872.

[95] Suhas A. Chavan, Wouter Maes, Lieven E. M. Gevers, Joos Wahlen, Ivo F. J. Vankelecom, Pierre A. Jacobs, Wim Dehaen, and Dirk E. De Vos. Porphyrin-Functionalized Dendrimers: Synthesis and Application as Recyclable Photocatalysts in a Nanofiltration Membrane Reactor. *Chem. Eur. J.* 2005, 11, 6754 – 6762.

[96] Lara-Isabel Rodriguez, Oriol Rossell, Miquel Seco, Arnald Grabulosa, Guillermo Muller, and Merce Rocamora. Carbosilane Dendrimers Peripherally Functionalized with P-Stereogenic Monophosphines. Catalytic Behavior of Their Allylpalladium

Complexes in the Asymmetric Hydrovinylation of Styrene. *Organometallics* 2006, 25, 1368-1376.

[97] Dong-Mei Xu, Ke-Da Zhang, Xiu-Lin Zhu. Preparation of Ag Nanoparticles in the Presence of Low Generational Poly(ester-amine) Dendrimers. *Journal of Applied Polymer Science,* Vol. 104, 422–426 (2007).

[98] Stephane Campidelli, Chloe Sooambar, Enrique Lozano Diz, Christian Ehli, Dirk M. Guldi, and Maurizio Prato. Dendrimer-Functionalized Single-Wall Carbon Nanotubes: Synthesis, Characterization, and Photoinduced Electron Transfer. *J. Am. Chem. Soc.* 2006, 128, 12544-12552.

[99] Sreenivasa Reddy Puniredd and M. P. Srinivasan. Covalent Molecular Assembly in a Supercritical Medium: Formation of Nanoparticles Encapsulated in Immobilized Dendrimers. *Ind. Eng. Chem. Res.* 2007, 46, 464-471.

[100] Marc R. Knecht, Joaquin C. Garcia-Martinez, and Richard M. Crooks. Synthesis, Characterization, and Magnetic Properties of Dendrimer-Encapsulated Nickel Nanoparticles Containing <150 Atoms. *Chem. Mater.* 2006, 18, 5039-5044.

[101] Seong A Kim, Dae-Hwan Jung, Yun-Ho Kim, Min Ah Kang, and Hee-Tae Jung. Core-Controlled Formation of Micellar Cubic Assemblies of Supramolecular Dendrimers: From Their Mechanism to Formation of Gold Nanostructures. *Macromolecules* 2006, 39, 6186-6190.

[102] Marc R. Knecht, Joaquin C. Garcia-Martinez, and Richard M. Crooks. Hydrophobic Dendrimers as Templates for Au Nanoparticles. *Langmuir* 2005, 21, 11981-11986.

[103] Kimihisa Yamamoto, Masayoshi Higuchi, Satoshi Shiki, Masanori Tsuruta and Hiroshi Chiba. Stepwiseradialcomplexationofimine groups in phenylazomethine dendrimers. *Nature* 2002, 415: 509-511.

[104] JIAN WENG, QIQING ZHANG. Syntheses and Nickel Coordination Properties of Cyclen Substituted with Mixed Stilbene/Poly(ethylene glycol) Dendritic Arms. *Journal of Polymer Science: Part A: Polymer Chemistry*, Vol. 43, 5414–5428 (2005).

[105] Francisco Tarazona-Vasquez and Perla B. Balbuena. Dendrimer-Tetrachloroplatinate Precursor Interactions. 1. Hydration of Pt(II) Species and PAMAM Outer Pockets. *J. Phys. Chem.* A 2007, 111, 932-944.

[106] Li Shen, Mei Shi, Fuyou Li, Dengqing Zhang, Xianghong Li, Enxian Shi, Tao Yi, Yukou Du, and Chunhui Huang. Polyaryl Ether Dendrimer with a 4-Phenylacetyl-5-pyrazolone-based Terbium(III) Complex as Core: Synthesis and Photopysical Properties. *Inorg. Chem.* 2006, 45, 6188-6197.

[107] Tao Zhou, Hendrik Neubert, Ding Yong Liu, Zu Dong Liu, Yong Min Ma, Xiao Le Kong, Wei Luo, Sykes Mark, and Robert C. Hider. Iron Binding Dendrimers: A Novel Approach for the Treatment of Haemochromatosis. *J. Med. Chem.* 2006, 49, 4171-4182.

[108] Giacomo Bergamini, Paola Ceroni, Mauro Maestri, Sang-Kyu Lee, Jeroen van Heyst, Fritz Vogtle. Cyclam cored luminescent dendrimers as ligands for Co(II), Ni(II) and Cu(II) ions. *Inorganica Chimica Acta* 360 (2007) 1043–1051.

[109] Guijiang Zhou, Wai-Yeung Wong, Bing Yao, Zhiyuan Xie, and Lixiang Wang. Triphenylamine-Dendronized Pure Red Iridium Phosphors with Superior OLED Efficiency/Color Purity Trade-Offs. *Angew. Chem. Int. Ed.* 2007, 46, 1149 –1151.

[110] Herbert Schumann, Yilmaz Aksu, and Birgit C. Wassermann. Convergent Synthesis and Characterization of Organotin Dendrimers Sn{(CH2)nSn[(CH2)4SnPh3]3}4 (n= 3, 4). *Organometallics* 2006, 25, 3428-3434.

[111] Venkateswarlu Juttukonda, Robert L. Paddock, Jeffery E. Raymond, Dan Denomme, Andrew E. Richardson, Laura E. Slusher, and Bradley D. Fahlman. Facile Synthesis of Tin Oxide Nanoparticles Stabilized by Dendritic Polymers. *J. Am. Chem. Soc.* 2006, 128, 420-421.

[112] Takane Imaoka, Reiko Tanaka, and Kimihisa Yamamoto. Investigation of a Molecular Morphology Effect on Polyphenylazomethine Dendrimers; Physical Properties and Metal-Assembling Processes. *Chem. Eur. J.* 2006, 12, 7328 – 7336.

[113] Inma Angurell, Oriol Rossell, Miquel Seco, and Eliseo Ruiz. Dendrimers Containing Two Metallic Layers. Chloride Migration from Peripheral Gold, Palladium, or Rhodium Metals to Internal Ruthenium Atoms. *Organometallics* 2005, 24, 6365-6373.

[114] Lei Qian and Xiurong Yang. Polyamidoamine Dendrimers-Assisted Electrodeposition of Gold-Platinum Bimetallic Nanoflowers. *J. Phys. Chem.* B 2006, 110, 16672-16678.

[115] Marie P. Cifuentes, Clem E. Powell, Joseph P. Morrall, Andrew M. McDonagh, Nigel T. Lucas, Mark G. Humphrey, Marek Samoc, Stephan Houbrechts, Inge Asselberghs, Koen Clays, Andre Persoons, and Takashi Isoshima. Electrochemical, Spectroelectrochemical, and Molecular Quadratic and Cubic Nonlinear Optical Properties of Alkynylruthenium Dendrimers. *J. Am. Chem. Soc.* 2006, 128, 10819-10832.

[116] Stephane Cordier, Kaplan Kirakci, Denise Mery, Christiane Perrin, Didier Astruc. Mo6Xi8 Nanocluster cores (X = Br, I): From inorganic solid state compounds to hybrids. *Inorganica Chimica Acta* 359 (2006) 1705–1709.

[117] Denise Mery, Lauriane Plault, Catia Ornelas, Jaime Ruiz, Sylvain Nlate, Didier Astruc, Jean-Claude Blais, Joao Rodrigues, Stephane Cordier, Kaplan Kirakci, and Christiane Perrin. From Simple Monopyridine Clusters [Mo6Br13(Py-R)][n-Bu4N] and Hexapyridine Clusters [Mo6X8(Py-R)6][OSO2CF3]4 (X = Br or I) to Cluster-Cored Organometallic Stars, Dendrons, and Dendrimers. *Inorg. Chem.* 2006, 45, 1156-1167.

[118] Denise Méry, Cátia Ornelas, Marie-Christine Daniel, Jaime Ruiz, João Rodrigues, Didier Astruc, Stéphane Cordier, Kaplan Kirakci, Christiane Perrin. Mo6Br8-Cluster-cored organometallic stars and dendrimers. *C. R. Chimie* 8 (2005) 1789–1797.

[119] Sung Cho, Wei-Shi Li, Min-Chul Yoon, Tae Kyu Ahn, Dong-Lin Jiang, Jiwon Kim, Takuzo Aida, and Dongho Kim. Relationship between Incoherent Excitation Energy Migration Processes and Molecular Structures in Zinc(II) Porphyrin Dendrimers. *Chem. Eur. J.* 2006, 12, 7576 – 7584.

[120] Markus Arendt, Wei Sun, Jens Thomann, Xiulan Xie, and Thomas Schrader. Dendrimeric Bisphosphonates for Multivalent Protein Surface Binding. *Chem. Asian J.* 2006, 1, 544 – 554.

[121] Mauri A. Kostiainen, Geza R. Szilvay, David K. Smith, Markus B. Linder, and Olli Ikkala. Multivalent Dendrons for High-Affinity Adhesion of Proteins to DNA. *Angew. Chem. Int. Ed.* 2006, 45, 3538 –3542.

[122] Chun-Jen Su, Yi-Chun Liu, Hsin-Lung Chen, Yen-Cheng Li, Hsien-Kuang Lin, Wen-Liang Liu, and Chain-Shu Hsu. Two-Dimensional Densely Packed DNA Nanostructure Derived from DNA Complexation with a Low-Generation Poly(amidoamine) Dendrimer. *Langmuir* 2007, 23, 975-978.

[123] M. Mondeshki, G. Mihov, R. Graf, H. W. Spiess, and K. Mullen. Self-Assembly and Molecular Dynamics of Peptide-Functionalized Polyphenylene Dendrimers. *Macromolecules* 2006, 39, 9605-9613.

[124] Marine Guillot-Nieckowski, Derk Joester, Meike Stohr, Myriam Losson, Marc Adrian, Bjoern Wagner, Manfred Kansy, Harry Heinzelmann, Raphael Pugin, Francois Diederich, and Jean-Louis Gallani. Self-assembly, DNA Complexation, and pH Response of Amphiphilic Dendrimers for Gene Transfection. *Langmuir* 2007, 23, 737-746.

[125] Prabal K. Maiti and Biman Bagchi. Structure and Dynamics of DNA-Dendrimer Complexation: Role of Counterions, Water, and Base Pair Sequence. *Nano Letters* 2006, 6 (11): 2478-2485.

[126] B. Klajnert, J. Cladera, and M. Bryszewska. Molecular Interactions of Dendrimers with Amyloid Peptides: pH Dependence. *Biomacromolecules* 2006, 7, 2186-2191.

[127] Linzhu Zhou, Lu Gan, Haibing Li, Xiangliang Yang. Studies on the interactions between DNA and PAMAM with fluorescent probe [Ru(phen)2dppz]2+. *Journal of Pharmaceutical and Biomedical Analysis* 43 (2007) 330–334.

[128] Dharam Paul, Hiroyuki Miyake, Satoshi Shinoda, and Hiroshi Tsukube. Proteo-Dendrimers Designed for Complementary Recognition of Cytochrome c: Dendrimer Architecture toward Nanoscale Protein Complexation. *Chem. Eur. J.* 2006, 12, 1328 – 1338.

[129] Akihiro Yamada, Ken Hatano, Tetsuo Koyama, Koji Matsuoka, Naonori Takahashi, Kazuya I. P. J. Hidari, Takashi Suzuki, Yasuo Suzukic, and Daiyo Terunuma. Lactotriaose-containing carbosilane dendrimers: Syntheses and lectin-binding activities. *Bioorganic and Medicinal Chemistry* 15 (2007) 1606–1614.

[130] David K. Smith. Dendritic Gels—Many Arms Make Light Work. *Adv. Mater*. 2006, 18, 2773–2778.

[131] Prasad Taranekar, Jin-Young Park, Derek Patton, Timothy Fulghum, Glenn J. Ramon, and Rigoberto Advincula. Conjugated Polymer Nanoparticles via Intramolecular Crosslinking of Dendrimeric Precursors. *Adv. Mater*. 2006, 18, 2461–2465.

[132] ZHISHEN GE, SHIZHONG LUO, SHIYONG LIU. Syntheses and Self-Assembly of Poly(benzyl ether)-b- Poly(N-isopropylacrylamide) Dendritic–Linear Diblock Copolymers. *J. Polym. Sci. Part A: Polym. Chem.* 44: 1357–1371, 2006.

[133] Maisie J. Joralemon, Rachel K. O'Reilly, Craig J. Hawker, and Karen L. Wooley. Shell Click-Crosslinked (SCC) Nanoparticles: A New Methodology for Synthesis and Orthogonal Functionalization. *J. Am. Chem. Soc*. 2005, 127, 16892-16899.

[134] BING-BING WANG, XIN ZHANG, XIN-RU JIA, ZI-CHEN LI, YAN JI, YEN WEI. Self-Assembly of a New Class of Amphiphilic Poly(amidoamine) Dendrimers and Their Electrochemical Properties. *J. Polym. Sci. Part A: Polym. Chem*. 43: 5512–5519, 2005.

[135] Catherine M. B. Santini, T. Alan Hatton, and Paula T. Hammond. Solution Behavior of Linear-Dendritic Rod Diblock Copolymers in Methanol. *Langmuir* 2006, 22, 7487-7498.

[136] Daniel R. Blasini, Samuel Flores-Torres, Detlef-M. Smilgies, and Hector D. Abruna. Stepwise Self-Assembly of Ordered Supramolecular Assemblies Based on Coordination Chemistry. *Langmuir* 2006, 22, 2082-2089.

[137] Virgil Percec, Andres E. Dulcey, Mihai Peterca, Monica Ilies, Monika J. Sienkowska, and Paul A. Heiney. Programming the Internal Structure and Stability of Helical Pores Self-Assembled from Dendritic Dipeptides via the Protective Groups of the Peptide. *J. Am. Chem. Soc.* 2005, 127, 17902-17909.

[138] Harald Hahn, Christina Keith, Heinrich Lang, Ramaiahgari Amaranatha Reddy, and Carsten Tschierske. First Example of a Third-Generation Liquid-Crystalline Carbosilane Dendrimer with Peripheral Bent-Core Mesogenic Units: Understanding of "Dark Conglomerate Phases". *Adv. Mater.* 2006, 18, 2629–2633.

[139] Dong Ki Yoon, Su Rim Lee, Yun Ho Kim, Sung-Min Choi, and Hee-Tae Jung. Large-Area, Highly Aligned Cylindrical Perfluorinated Supramolecular Dendrimers Using Magnetic Fields. *Adv. Mater.* 2006, 18, 509–513.

[140] Rafael Alcala, Raquel Gimenez, Luis Oriol, Milagros Pinol, Jose Luis Serrano, Belen Villacampa, and Ana I. Vinuales. Synthesis, Characterization, and Induction of Stable Anisotropy in Liquid Crystalline Photo-addressable PPI Dendrimers. *Chem. Mater.* 2007, 19, 235-246.

[141] Umesh Gupta, Hrushikesh Bharat Agashe, Abhay Asthana, and N. K. Jain. Dendrimers: Novel Polymeric Nanoarchitectures for Solubility Enhancement. *Biomacromolecules,* 2006, 7 (3): 649-658.

[142] I. Cooper, J. D. Londono, G. Wignall, J. B. McClain, E. T. Samulski, J. S. Lin, A. Dobrynin, M. Rubinstein, A. L. C. Burke, J. M. J. Frechet, and J. M. DeSimone. Extraction of a hydrophilic compound fromwater into liquidCO2 using dendritic surfactants. *Nature*, 1997, 389: 368-371.

[143] Izabela Bury, Bertrand Donnio, Jean-Louis Gallani, and Daniel Guillon. Interfacial Behavior of a Series of Amphiphilic Block Co-dendrimers. *Langmuir* 2007, 23, 619-625.

[144] Naomi Sakai, Yuko Kamikawa, Masayuki Nishii, Toru Matsuoka, Takashi Kato, and Stefan Matile. Dendritic Folate Rosettes as Ion Channels in Lipid Bilayers. *J. Am. Chem. Soc.* 2006, 128, 2218-2219.

[145] Virgil Percec, Andres E. Dulcey, Venkatachalapathy S. K. Balagurusamy, Yoshiko Miura, Jan Smidrkal, Mihai Peterca, Sami Nummelin, Ulrica Edlund, Steven D. Hudson, Paul A. Heiney, Hu Duan, Sergei N. Magonov, and Sergei A. Vinogradov. Self-assembly of amphiphilic dendritic dipeptides into helical pores. *Nature,* 2004, 430: 764-768.

[146] V. Percec, M. Glodde, T. K. Bera, Y. Miura, I. Shiyanovskaya, K. D. Singer, V. S. K. Balagurusamy, P. A. Heiney, I. Schnell, A. Rapp, H.-W. Spiess, S. D. Hudsonk, and H. Duank. Self-organization of supramolecular helical dendrimers into complex electronic materials. *Nature,* 2002, 419: 384-387.

[147] Virgil Percec, Andres E. Dulcey, Mihai Peterca, Monica Ilies, Janine Ladislaw, Brad M. Rosen, Ulrica Edlund, and Paul A. Heiney. The Internal Structure of Helical Pores Self-Assembled from Dendritic Dipeptides is Stereochemically Programmed and Allosterically Regulated. *Angew. Chem.* 2005, 117, 6674 –6679.

[148] Chia-Hung Chou, Hsu-Shen Wang, Kung-Hwa Wei, and Jung Y. Huang. Thiophenol-Modified CdS Nanoparticles Enhance the Luminescence of Benzoxyl Dendron-Substituted Polyfluorene Copolymers. *Adv. Funct. Mater.* 2006, 16, 909–916.

[149] Baohua Huang, Donald A. Tomalia. Poly(ether) dendrons possessing phosphine focal points for stabilization and reduced quenching of luminescent quantum dots. *Inorganica Chimica Acta* 359 (2006) 1961–1966.

[150] Dongjun Wang, Toyoko Imae, Masao Miki. Fluorescence emission from PAMAM and PPI dendrimers. Journal of Colloid and Interface, *Science* 306 (2007) 222–227.

[151] Arpornrat Nantalaksakul, Raghunath Reddy Dasari, Tai-Sang Ahn, Rabih Al-Kaysi, Christopher J. Bardeen, and S. Thayumanavan. Dendrimer Analogues of Linear Molecules to Evaluate Energy and Charge-Transfer Properties. *Organic Letters* 2006, 8(14): 2981-2984.

[152] Jane Larsen, Ben Brulggemann, Tomas Polivka, Villy Sundstrolm, Eva Akesson, Joseph Sly, and Maxwell J. Crossley. Energy Transfer within Zn-Porphyrin Dendrimers: Study of the Singlet-Singlet Annihilation Kinetics. *J. Phys. Chem.* A 2005, 109, 10654-10662.

[153] Andrea Petrella, Jens Cremer, Luisa De Cola, Peter Baluerle, and Rene M. Williams. Charge Transfer Processes in Conjugated Triarylamine-Oligothiophene-Perylenemonoimide Dendrimers. *J. Phys. Chem.* A 2005, 109, 11687-11695.

[154] Lucia Flamigni, Anna Maria Talarico, Barbara Ventura, Chloé Sooambar, and Nathalie Solladié. Photoinduced Processes in a Dendritic Zn Porphyrin Structure with a Free-Base Porphyrin Core. *Eur. J. Inorg. Chem.* 2006, 2155–2165.

[155] Dong-Ik Lee and Theodore Goodson III. Entangled Photon Absorption in an Organic Porphyrin Dendrimer. *J. Phys. Chem.* B, 2006, 110 (51): 25582-25585.

[156] Oleg Varnavski, Xingzhong Yan, Olivier Mongin, Mireille Blanchard-Desce, and Theodore Goodson, III. Strongly Interacting Organic Conjugated Dendrimers with Enhanced Two-Photon Absorption. *J. Phys. Chem.* C 2007, 111, 149-162.

[157] N. Vijayalakshmi and Uday Maitra. Multiple Chromophore Labeled Novel Bile Acid Dendrimers for Light Harvesting. *Macromolecules* 2006, 39, 7931-7940.

[158] Michael A. Oar, William R. Dichtel, Jason M. Serin, Jean M. J. Frechet, Joy E. Rogers, Jonathan E. Slagle, Paul A. Fleitz, Loon-Seng Tan, Tymish Y. Ohulchanskyy and Paras N. Prasad. Light-Harvesting Chromophores with Metalated Porphyrin Cores for Tuned Photosensitization of Singlet Oxygen via Two-Photon Excited FRET. *Chem. Mater.* 2006, 18, 3682-3692.

[159] Nicola Armaroli, Gianluca Accorsi, John N. Clifford, Jean-FranÅois Eckert, and Jean-Francois Nierengarten. Structure-Dependent Photoinduced Electron Transfer in Fullerodendrimers with Light-Harvesting Oligophenylenevinylene Terminals. *Chem. Asian J.* 2006, 1, 564 – 574.

[160] Ulrike Jager-Fiedler, Martin Kockerling, Ralf Ludwig, Alexander Wulf, and Clemens Krempner. Hydroxy-Substituted Oligosilane Dendrimers: Controlling the Electronic Properties through Hydrogen Bonding. *Angew. Chem. Int. Ed.* 2006, 45, 6755 –6759.

[161] Tai-Sang Ahn, Arpornrat Nantalaksakul, Raghunath R. Dasari, Rabih O. Al-Kaysi, Astrid M. Mulller, S. Thayumanavan, and Christopher J. Bardeen. Energy and Charge Transfer Dynamics in Fully Decorated Benzyl Ether Dendrimers and Their Disubstituted Analogues. *J. Phys. Chem.* B 2006, 110, 24331-24339.

[162] Tao Gu, James K. Whitesell, and Marye Anne Fox. Intramolecular Charge Transfer in 1:1 Cu(II)/Pyrenylcyclam Dendrimer Complexes. *J. Phys. Chem.* B 2006, 110, 25149-25152.

[163] Benjamin J. Lear and Clifford P. Kubiak. Charge Gating and Electronic Delocalization over a Denderimeric Assembly of Trinuclear Ruthenium Clusters. *Inorg. Chem.* 2006, 45, 7041-7043.

[164] Wei-Shi Li, Kil Suk Kim, Dong-Lin Jiang, Hiroyuki Tanaka, Tomoji Kawai, Jung Ho Kwon, Dongho Kim, and Takuzo Aida. Construction of Segregated Arrays of Multiple Donor and Acceptor Units Using a Dendritic Scaffold: Remarkable Dendrimer Effects on Photoinduced Charge Separation. *J. Am. Chem. Soc.* 2006, 128, 10527-10532.

[165] Ho-Jin Son, Won-Sik Han, Hyojeong Kim, Chungkyun Kim, Jaejung Ko, Chongmok Lee, and Sang Ook Kang. End-Capped Silole Dendrimers on a Carbosilane Periphery: Potential Electroluminescent Materials. *Organometallics* 2006, 25, 766-774.

[166] Dong-Lin Jiang, and Takuzo Aida. Photoisomerization in dendrimers by harvesting of low-energy photons. *Nature,* 1997, 388: 454-456.

[167] Chao Wu, Sergey V. Malinin, Sergei Tretiak, and Vladimir Y. Chernyak. Exciton scattering and localization in branched dendrimeric structures. *Nature Physics*. 2006, 2: 631-635.

[168] Prasad Taranekar, Akira Baba, Jin Young Park, Timothy M. Fulghum, and Rigoberto Advincula. Dendrimer Precursors for Nanomolar and Picomolar Real-Time Surface Plasmon Resonance/Potentiometric Chemical Nerve Agent Sensing Using Electrochemically Crosslinked Ultrathin Films. *Adv. Funct. Mater.* 2006, 16, 2000–2007.

[169] Ho-Jin Son, Won-Sik Han, Kuk Hwa Lee, Ha Jin Jung, Chongmok Lee, Jaejung Ko, and Sang Ook Kang. Electrochemical Deposition of End-Capped Triarylamine and Carbazole Dendrimers: Alternate Technique for the Manufacture of Multilayer Films. *Chem. Mater.*, Vol. 18, No. 25, 2006.

[170] Jagotamoy Das, Md. Abdul Aziz, and Haesik Yang. A Nanocatalyst-Based Assay for Proteins: DNA-Free Ultrasensitive Electrochemical Detection Using Catalytic Reduction of p-Nitrophenol by Gold-Nanoparticle Labels. *J. Am. Chem. Soc.* 2006, 128, 16022-16023.

[171] Agnieszka Kulczynska, Tony Frost, and Lawrence D. Margerum. Effect of PAMAM Dendrimer Size and pH on the Electrostatic Binding of Metal Complexes Using Cyclic Voltammetry. *Macromolecules* 2006, 39, 7372-7377.

[172] Frank N. Crespilho, Valtencir Zucolotto, Christopher M. A. Brett, Osvaldo N. Oliveira, Jr., and Francisco C. Nart. Enhanced Charge Transport and Incorporation of Redox Mediators in Layer-by-Layer Films Containing PAMAM-Encapsulated Gold Nanoparticles. *J. Phys. Chem.* B 2006, 110, 17478-17483.

[173] Fuan Wang, Jianlong Wang, Hongjun Chen, Shaojun Dong. Assembly process of CuHCF/MPA multilayers on gold nanoparticles modified electrode and characterization by electrochemical SPR. *Journal of Electroanalytical Chemistry* 600 (2007) 265–274.

[174] Yun-Long Zeng, Yu-Fang Huang, Jian-Hui Jiang, Xiao-Bing Zhang, Chun-Ran Tang, Guo-Li Shen, Ru-Qin Yu. Functionalization of multi-walled carbon nanotubes with poly(amidoamine) dendrimer for mediator-free glucose biosensor. *Electrochemistry Communications* 9 (2007) 185–190.

[175] E. Lojou, P. Bianco. Assemblies of dendrimers and proteins on carbon and gold electrodes. *Bioelectrochemistry* 69 (2006) 237–247.

[176] Frank N. Crespilho, M. Emilia Ghica, Monica Florescu, Francisco C. Nart, Osvaldo N. Oliveira Jr., Christopher M.A. Brett. A strategy for enzyme immobilization on layer-by-

layer dendrimer–gold nanoparticle electrocatalytic membrane incorporating redox mediator. *Electrochemistry Communications* 8 (2006) 1665–1670.

[177] M. Pilar García Armada, José Losada, Magdalena Zamora, Beatriz Alonso, Isabel Cuadrado, Carmen M. Casado. Electrocatalytical properties of polymethylferrocenyl dendrimers and their applications in biosensing. *Bioelectrochemistry* 69 (2006) 65–73.

[178] Valtencir Zucolotto, Andressa P.A. Pinto, Tathyana Tumolo, Marli L. Moraes, Maur´ıcio S. Baptista, Antonio Riul Jr., Ana Paula U. Araujo, Osvaldo N. Oliveira Jr. Catechol biosensing using a nanostructured layer-by-layer film containing Cl-catechol 1,2-dioxygenase. *Biosensors and Bioelectronics* 21 (2006) 1320–1326.

[179] Zhi-Min Liu, Yu Yang, Hua Wang, Yan-Li Liu, Guo-Li Shen, Ru-Qin Yu. A hydrogen peroxide biosensor based on nano-Au/PAMAM dendrimer/cystamine modified gold electrode. *Sensors and Actuators* B 106 (2005) 394–400.

[180] M. Snejdarkova, L. Svobodova, G. Evtugyn, H. Budnikov, A. Karyakin, D.P. Nikolelis, T. Hianik. Acetylcholinesterase sensors based on gold electrodes modified with dendrimer and polyaniline A comparative research. *Analytica Chimica Acta* 514 (2004) 79–88.

[181] L. Svobodova, M. Snejdarkova, K. Toth, R.E. Gyurcsanyi, T. Hianik. Properties of mixed alkanethiol–dendrimer layers and their applications in biosensing. *Bioelectrochemistry* 63 (2004) 285– 289.

[182] Li Shen, Naifei Hu. Heme protein films with polyamidoamine dendrimer: direct electrochemistry and electrocatalysis. *Biochimica et Biophysica Acta* 1608 (2004) 23–33.

[183] Lenka Svobodova, Maja Snejdarkova, Vladimira Polohova, Igor Grman, Peter Rybar, Tibor Hianik. QCM Immunosensor Based on Polyamidoamine Dendrimers. *Electroanalysis* 2006, 18 (19-20) 1943 – 1949.

[184] Nathan A. Stasko and Mark H. Schoenfisch. Dendrimers as a Scaffold for Nitric Oxide Release. *J. Am. Chem. Soc.* 2006, 128, 8265-8271.

[185] Rudresha Kottani, Janaki R. R. Majjigapu, Alexei Kurchan, Kavitha Majjigapu, Tiffany P. Gustafson, and Andrei G. Kutateladze. Photoinduced Signal Amplification through Controlled Externally Sensitized Fragmentation in Masked Sensitizers. *J. Am. Chem. Soc.* 2006, 128, 14794-14795.

[186] Shin Ogasawara, Atsushi Ikeda, and Jun-ichi Kikuchi. Positive Dendritic Effect in DNA/Porphyrin Composite Photocurrent Generators Containing Dendrimers as the Stationary Phase. *Chem. Mater.* 2006, 18, 5982-5987.

In: Nanotechnology Research Developments
Editor: R. Jiménez-Contreras, pp. 95-136
ISBN: 978-1-60021-899-6

Chapter 2

FABRICATION OF CORE-SHELL STRUCTURED PEROVSKITE NANOPARTICLES WITH HETEROGENEOUS OXIDE SHELLS AND THEIR CERAMICS

***Jian Quan Qi*[*, 1], *Helen Lai Wah Chan*[2], *Hu Yong Tian*[2], *Wan Ping Chen*[2] and *Long Tu Li*[1]**

[1] Department of Materials Sciences and Engineering and
State Key Laboratory of Fine Ceramics and New Processing,
Tsinghua University, Beijing 100084, China
[2] Department of Applied Physics and Materials Research Center,
The Hong Kong Polytechnic University, Hong Kong, China

ABSTRACT

The perovskite nanoparticles such as BaTiO3, SrTiO3, BaZrO3 and their solid solutions can be prepared using a new method, DSS, near room temperature. Hydrogen interstitials are introduced into the lattice of perovskite nanoparticles synthesized by DSS method so that the nanoparticles take positive charges and adsorb the hydroxyl group with negative charges on the surface. Hydroxyl groups adsorbed on the surface of nanoparticles can be ionized freely, and the nanoparticles exhibit an alkalescent property. When this kind of perovskite nanopaticles disperses in a certain solution of inorganic salt such as Mg(NO3)2•6H2O, Zn(NO3)2•6H2O, Al(NO3)3•9H2O, Y(NO3)3•6H2O etc., the hydroxide related to these salts spontaneously deposit on the surface of the nanoparticles, as a result, a perovskite structure nanoscaled core materials are successfully self-wrapped by a non-ferroelectric hydroxide derived from a nitrate salt solution. After annealing at high temperature, core-shell structured perovskite nanoparticles with heterogeneous oxide shells can be synthesized. When the core-shell structured perovskite nanoparticles with heterogeneous oxide shells are composited with doped shell oxide materials, the pinning effect during sintering can be avoided, as a result, the sintering temperature are decreased and dense ceramics are obtained. The core-shell structures are analyzed by a

[*] E-mail: jianquanqi@mail.tsinghua.edu.cn

conjunction of XRD, HRTEM and FE-SEM. The lattice match of shell and core in our ceramics of MZO wrapped BST nanoparticles probably follows the second match mode. The tunability and FOM values of the ceramics of MZO wrapped BST nanoparticles have been improved significantly compared with that of pure BST ceramics. The ceramics of MZO wrapped BST nanoparticles are promising materials for tunable microwave device applications and high energy density ceramic capacitors for their distinct dielectric loss improvement.

Core-shell structured nanoparticles received much attention for their unique physical and chemical properties, as well as their technological applications.[1-7] With oxide shells, they are also interesting from a number of perspectives.[8-11] The shells offer protection to the cores as well as introduce new properties to the hybrid structures.[12] This kind of hybrid structures can have advantages of both the core and the shell. Perovskite oxides such as BaTiO3, SrTiO3, PbTiO3, BaZrO3, PbZrO3 etc. is a series of important functional materials with a formula of ABO3, and most of them can form solid solution with each other while keeping a perovskite structure. The perovskite ceramics are important for their ferroelectricity and anti-ferroelectricity, which find a wide variety of applications in nonlinear optics, thin-film capacitors, pyroelectric detectors, optical memories, and electrooptic modulators [13]. However, with the reduction in grain size, the ratio of grain boundary to grain of the ceramics will increase and degenerate their electrical properties. In order to improve the properties of the ceramics, introduction of heterogeneous oxide shells which have outstanding properties, such as very low dielectric loss, is a wonderful choice. Self-wrapping MgO shell to BST nanoparticles is a successful example which can be applied for microwave materials.[14-15] Fabrication of this kind of core-shell structured perovskite nanoparticles with heterogeneous oxide shells and their ceramics will be discussed here and self-wrapping MgO shell to $Ba_{1-x}Sr_xTiO_3$ (BST) nanoparticles will be used as an example.

1. Synthesis of Perovskite Nanoparticles from Solution

In order to fabricate core-shell structured perovskite nanoparticles with heterogeneous oxide shells, it is a key to synthesize perovskite nanoparticles with suitable physical and chemical properties. However, most perovskite oxides are still prepared by conventional solid-state reactions between the corresponding oxides or oxides and carbonates at temperatures above 1000oC. Aiming at a better understanding of the structure-property relationship and the development of novel electronic devices, the study of the nano-scaled ferroelectric materials has attracted immense attention in recent years. Novel synthesis techniques have been developed. Among them, wet-chemical methods provide promising alternatives because they can be better controlled from the molecular precursor to the final material to give highly pure and homogeneous materials, and allow low reaction temperatures to be used, the size and morphology of the particles to be controlled, and metastable phases be prepared. There are two kinds of wet-chemical routes for the synthesis of perovskite nanocrystals. One is precipitation method which produces precursors such as carbonate, hydroxide, oxalate, dry gel etc. by solution method and then calcines them to obtain the perovskite nanocrystals. Although the methods are very simple and the reaction in solution can be even operated at room temperature, agglomeration and grain-growth are inevitable during the drying and calcination processes. The other one is direct synthesis of perovskite nanostructures in solution and much effort has been made on it in recent years. Hydrothermal method is one of the most popular approaches to obtain the perovskite nanostructures directly from solution, but the synthesis processes are often conducted at elevated temperatures (typically 100-280oC) and/or under relatively high pressure to improve the crystallinity of the products.[16-21] To further simplify the process and lower the processing temperature and pressure is very important and rather challenging from the technical point of view. Fortunately, some most important perovskite oxides have been directly synthesized from solution successfully near room temperature in recent years. This kind of method, which was called direct synthesis from solution (DSS),[22] was developed to synthesize BaTiO3,[23] SrTiO3,[24] BaZrO3[25] and their solid solutions near room temperature. The nanoparticles synthesized by DSS method are very suitable for the fabrication of core–shell structured perovskite nanoparticles with heterogeneous oxide shells.

1.1. Synthesis of $BaTiO_3$ Nanoparticles from Solution

The synthesis of fine ceramic powders has become an important part of modern ceramic research, so that there is a strong trend toward the application of chemical method for powder synthesis in electronic ceramics. Using fine ceramic powders can lower the sintering temperature, achieve uniform fine ceramic grain size and save energy. BaTiO3 is one of the most typical and important perovskite materials which has been widely used in multilayer ceramic capacitors (MLCC) and positive temperature coefficient of resistivity (PTCR) ceramics.[26-32] Recent advances in microelectronics and communications have led to the miniaturization of MLCCs. Higher capacitance in smaller case sizes requires a reduction of the thickness of the ceramic layers below 2-3 μm and the increase of the number of layers (>200).[33-34] To achieve this goal, powders with improved quality with small and uniform size are required. The synthesis of BaTiO3 nanostructure is also a basic method that many other perovskite nanostructures can be obtained similarly. There are several methods to synthesize BaTiO3 nanostructures from solution. For example, the precursors, barium titanium isopropoxide (BaTi[OCH(CH3)2]6), were used to synthesize BaTiO3 nanostructures at temperatures over 100oC by solution-phase decomposition in the presence of coordinating ligands followed by heat-treating at about 280oC.[16-17] Another synthesis protocol is based on the dissolution of alkali or alkaline earth metals in benzyl alcohol and subsequent reaction with transition metal alkoxides at relatively low temperatures of 200-220oC to synthesize BaTiO3 nanoparticles.[21] However, DSS is one of the most convenient methods to obtain BaTiO3 nanoparticles in a large-scale because the synthesis temperature can be as low as near room temperature, such as 45oC.[22-23] It may be a good replacement of the well known oxalate method.[35] Furthermore, improved DSS method can be operated even at room temperature or lower to synthesize BaTiO3 nanoparticles directly.

1. BaTiO3 Nanoparticle Synthesis Process

The chart of DSS method is shown in Figure 1. A typical method to synthesize BaTiO3 nanoparticles using DSS method is as follows: The reagents Ba(OH)2•8H2O and tetrabutyl titanate are adopted as starting raw materials to prepare nanoscaled BaTiO3 powders.

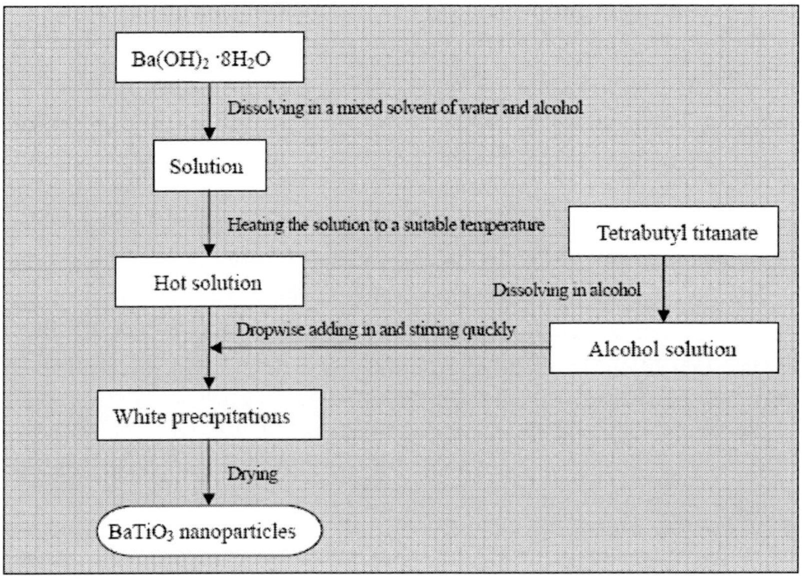

Figure 1. The chart of DSS method.

The base solution is prepared by dissolving Ba(OH)2•8H2O in deionized water (DIW) with the concentration of 0.4M ethanol. Tetrabutyl titanate is dissolved in ethanol absolute to obtain a solution with concentration of 0.4M ethanol. Different reaction temperatures of 45, 60, 70 and 85oC are adopted in order to study the temperature effect on the properties of prepared powders. When the base solution reaches the desired reaction temperature, the same amount of tetrabutyl titanate solution is added drop by drop to the base solution under normal pressure and is stirred quickly. An instantaneous formation of white precipitation can be observed. After the reaction has been completed, the white precipitation is filtered and dried at 60oC for 24h to obtain the powder.

2. Characterization of Nanoparticles

Modern analytic instruments such as X-ray diffraction (XRD), scanning electron microscopy (SEM), and transmission electron microscopy (TEM) etc can be used to characterize the features of the as-prepared nanoparticles. The powders prepared by conventional chemical method often are hardened easily due to agglomeration during the drying period. Fortunately, if the precipitations have been crystallized, agglomeration can be suppressed. The powders prepared by DSS method are all soft after drying at 60oC when the synthesis temperature is lower than or equal to 70oC. The powders synthesized at 85oC are hardened slightly, but it also can be crushed into soft powders easily. All of the samples are crystalline BaTiO3 powders as shown in the XRD spectra (Figure 2).

The line broadening effect of XRD profile is caused by the size reduction of the grains. The crystallite size of BaTiO3 can be estimated by the Scherrer's equation: [23]

$$d_x = K\lambda/(\beta\cos\theta) \quad (1)$$

where d_x is crystallite size, K a constant and usually taken as 0.89, λ the wavelength of diffraction X-ray (0.15418 nm for CuKα), β the line width at half-maxiumum height after subtraction of broadening caused by equipment and θ is half the diffraction angle of the center of the peak (in degrees). The XRD profiles of (111) plane of the samples synthesized at different temperatures are shown in Figure 3. The XRD peak shift with the synthesis temperature can be observed from this figure. It is revealed that the powders have different crystallite lattice parameters when synthesized at different temperatures. The lattice parameter and crystallite size of the powders synthesized at different temperatures are shown in Figure 4.

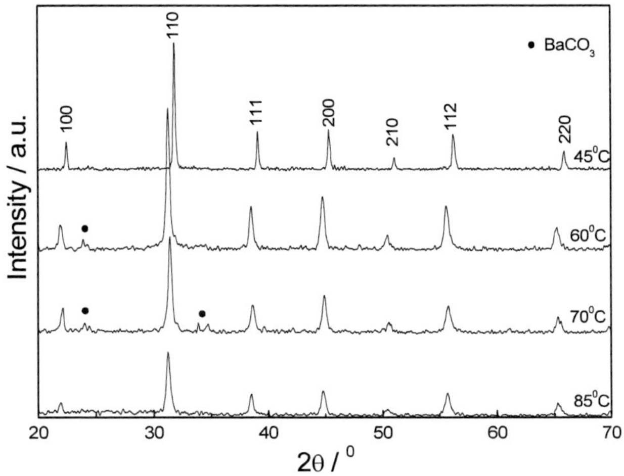

Figure 2. XRD profiles of $BaTiO_3$ powders synthesized at different temperatures [23].

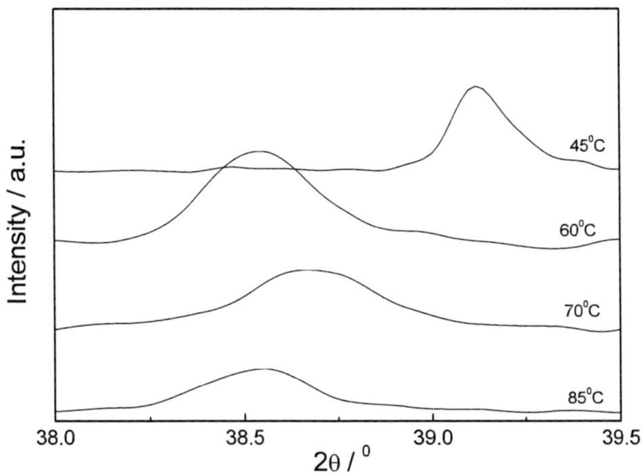

Figure 3. XRD profiles of (111) plane of $BaTiO_3$ powders synthesized at different temperatures [23].

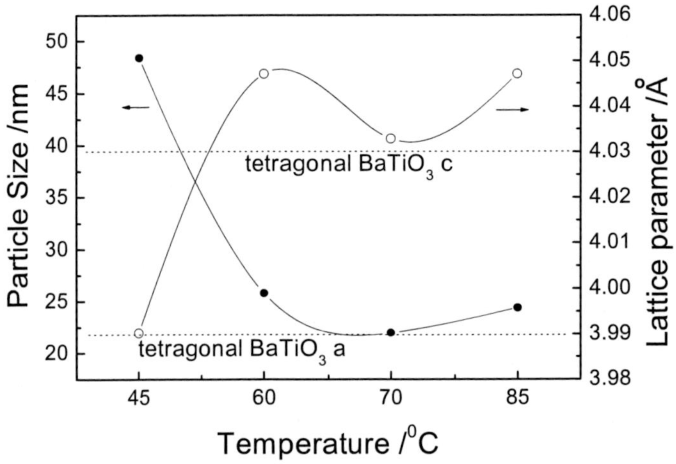

Figure 4. The lattice parameter and crystallite size of the $BaTiO_3$ powders synthesized at different temperatures [23].

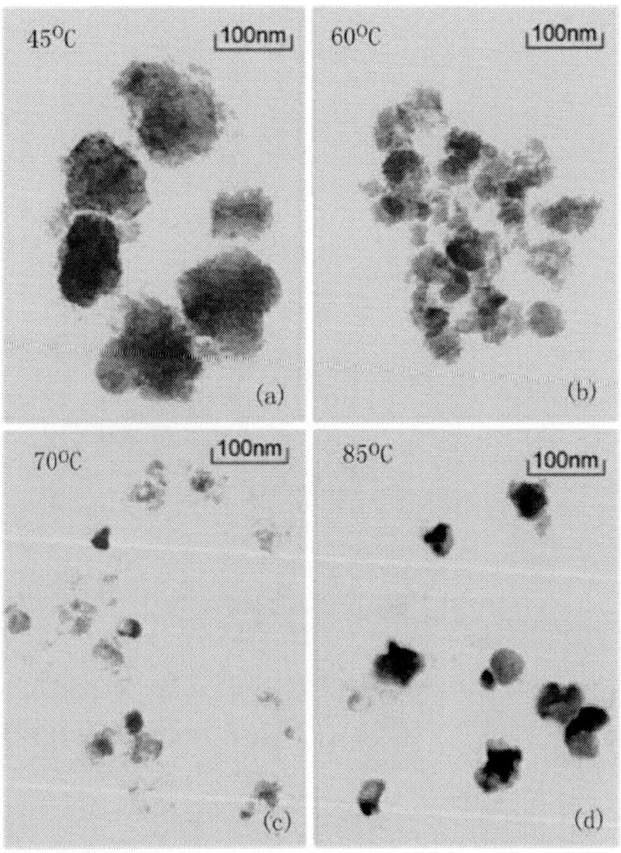

Figure 5. The TEM images of the samples [23].

The crystallite size of as-prepared powders decreases rapidly from 50nm to 20nm with the increase in the synthesized temperature from 45oC up to 70oC. After that, the crystallite size increases slightly. The TEM images of the samples are shown in Figure 5. The change tendency of the particle size with the synthesis temperature is similar to that of the crystallite size measured by XRD. Figure 6 shows the image of a single crystal diffraction spots obtained through focusing the transmission electron beam on one particle shown in Figure 5. Moreover, the shape of the particles in Figure 5 is almost spherical. Therefore, one particle is a single-crystal particle and the crystallite size is the average particle size of the powders.

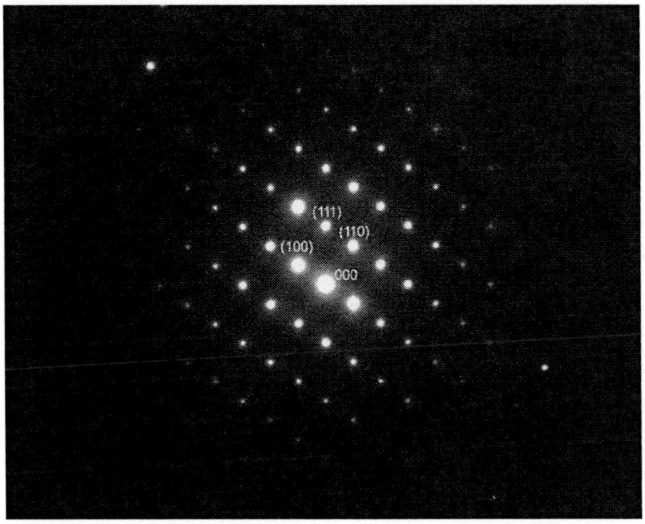

Figure 6. Transmission electron diffraction spot images of a single particle [22].

1.2. Synthesis of $SrTiO_3$ and BST Nanoparticles

Both SrTiO3 and Ba1-xSrxTiO3 (BST) are important functional materials. SrTiO3 ceramics can be used in grain boundary layer capacitor and voltage varistor.[36-38] The Curie temperature (TC) of the bulk BST varies almost linearly with the composition, x, in the solid solution Ba1−xSrxTiO3. For a particular application temperature, the composition is adjusted to move the ferroelectric phase transitions to the range of the device application temperature so that BST ceramics can be also used as low temperature PTCR and high dielectric constant capacitors.[39-40] BST films are being extensively investigated as the charge-storage dielectric for Gbit dynamic random access memory and other integrated capacitor structures, and more recently as electric-field tunable elements for high frequency devices, and infrared detectors.[41-51] The nano-scaled BST is also a good candidate for the fundamental research of nano-ferroelectricity because of its excellent compositional and structural modifiability.

The method of synthesis of SrTiO3 nanoparticles is very similar to that of BaTiO3 nanoparticles because of the very similar chemical properties between Ba and Sr, such as both Ba(OH)2 and Sr(OH)2 have strong alkalescence in water solution. However, their solubilities are quite different and Sr(OH)2 has a lower solubility than Ba(OH)2 especially at low temperature as shown in Figure 7.

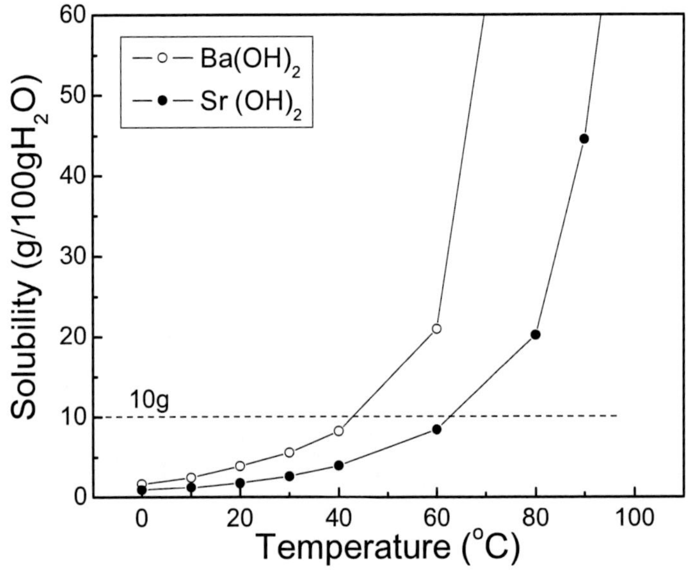

Figure 7. Solubility of $Ba(OH)_2$ and $Sr(OH)_2$, the number of grams of substance which can be dissolved in 100g of water at a certain temperature [22].

For example, the solubility of Ba(OH)2 reaches 10g at 43oC, while the solubility of Sr(OH)2 can only reach that number at 63oC. Thus, synthesis of the same amount of SrTiO3 nanoparticles needs a higher temperature than that of BaTiO3 nanoparticles in one batch so that the alkaline earth metal hydroxide can be dissolved completely at the beginning of the reaction. Therefore, the lower solubility of Sr(OH)2 than Ba(OH)2 should be noted during synthesis of Sr-included perovskite nanoparticles.

The process of DSS method for synthesizing BST nanoparticles is the same as that of synthesizing BaTiO3 and SrTiO3 nanoparticles as shown in Figure 1 except that the starting materials, Ba(OH)2·8H2O, is replaced with Ba(OH)2·8H2O and Sr(OH)2·8H2O according to a designated ratio in BST. The synthesis process may be divided into two steps.[24] The first step is to separately prepare a base solution and a titanium solution. The base solution is prepared by dissolving Ba(OH)2·8H2O (28.392 g) and Sr(OH)2·8H2O (2.658 g) into DIW (500 ml). The concentration of the base solution is ~ 0.2M. The solution was then heated to 60°C in a water bath. The titanium solution was prepared by dissolving Ti(OBu)4 (34.0 g) into absolute ethanol (250 ml). These solution concentrations are for the composition of (Ba0.9Sr0.1)TiO3; for other compositions, the solution concentration can be changed accordingly. The second step is to allow the reaction to take place by dripping the titanium solution into the base solution (60°C) under vigorous stirring. An instantaneous formation of white precipitation can be observed. After the reaction has been completed and followed by an ageing for 3 h, the white precipitation is obtained by filtration and baked at 60°C for 24 h in an oven.

Figure 8 shows the representative XRD pattern of the as-prepared powder that has been synthesized and baked at 60°C. Except for one small peak appearing at about 24°, all other peaks are identified as from perovskite BST. The peak sharpness and intensity indicate that the BST is well crystallized. The particles are found to have a cubic symmetry with lattice parameter ~0.402 nm, which is different from the symmetry of bulk BST.

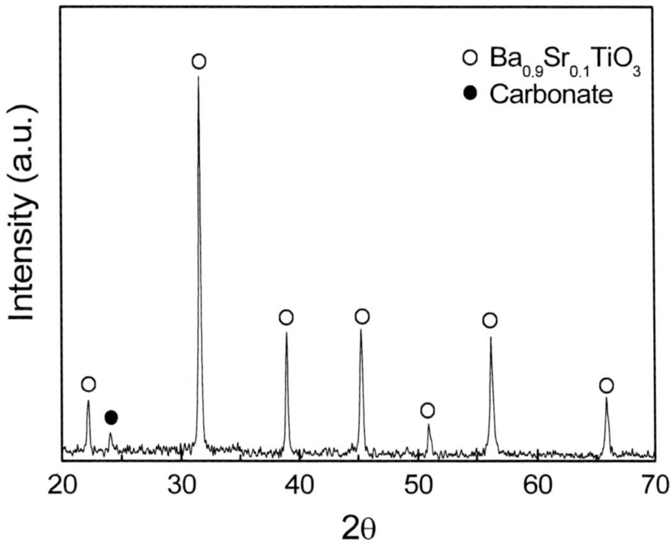

Figure 8. X-ray diffraction pattern of BST(90/10) nanoparticles [24].

In literature, BST ceramic of the same composition has a tetragonal phase at room temperature with lattice parameters a=0.399 nm and c=0.401 nm.[52] This symmetry change is a typical size effect that has been observed in other nano-scaled ferroelectric materials as well. The peak at 24° in the XRD pattern is identified as from barium carbonate. The formation of barium carbonate is believed to be a result of the reactions between carbon dioxide dissolved into the solutions from air and reacted with the alkaline earth hydroxides during the process. The impure phase can either be avoided by undertaking the whole process in a protective atmosphere (such as argon or nitrogen) or be removed by washing the powders using dilute acids.

Another microstructural factor that possibly affects the lattice symmetry is interstitial hydrogen. Synthesized from an aqueous solution, the particles are expected to contain hydrogen that may exist in the interstitial sites of BST lattices. Despite direct observation of such interstitial hydrogen is not yet available, some indirect evidences can support the assumption, such as the as-synthesized BST particles have a larger lattice than bulk BST because the interstitial hydrogen is responsible for such slight lattice expansion in the as-synthesized BST. For the particles subjected to post-synthesis thermal treatment, the interstitial hydrogen may be partially/completely removed.

The DTA-TG analysis has helped to study the release of hydrogen from the lattices. Figure 9 shows the weight and DSC of the as-synthesized powders as a function of temperature. From room temperature to about 220°C, there is a weight loss of ~1.5%. Such weight loss is mainly due to the removal of residual alcohol and adsorbed water in the particles. From 220°C to ~830°C, another weight loss, up to ~3%, is observed. This weight loss is believed to be corresponding to the release of interstitial hydrogen and the adsorbed hydroxyl groups. At higher temperatures, the sample weight is almost a constant value, meaning that the release of the interstitial hydrogen and hydroxyl groups is completed at ~830°C (i.e., the particles subjected to >830°C thermal treatment are stoichiometric BST). In the DSC curve, no sharp endothermic or exothermal peak is observed. This is because the as-synthesized particles are already well crystallized. A rather broadened exothermic curve is observed, which is likely corresponding to the crystallite size increase at higher temperatures.

The microstructure of the as-synthesized BST particles has been further characterized by means of scanning electron microscopy (SEM) and transmission electron microscopy (TEM). A representative SEM image of the as-prepared BST nanoparticles is shown in Figure 10. It is observed that the particles are uniform, with ~30 nm in diameter. The result is in good agreement with the XRD estimation.

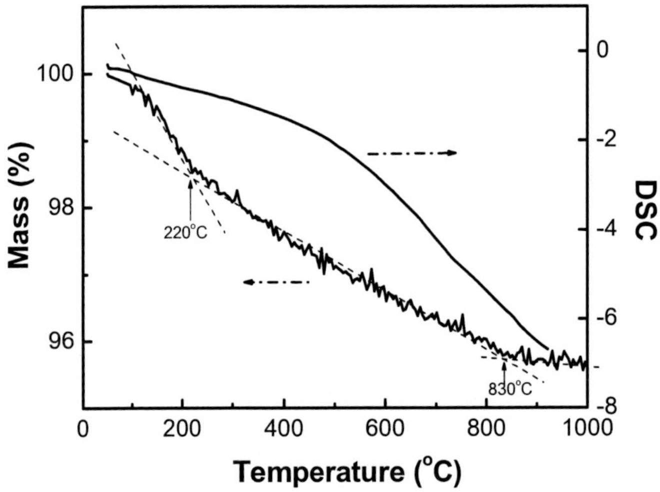

Figure 9. DTA-TG analysis of as-prepared BST(90/10) powder [24].

Figure 10. BST(90/10) nanoparticles observed under a field-emission scanning electron microscope (FE-SEM) [24].

The TEM observation results are shown in Figure 11. Figure 11(a) shows the morphology of the powder. Consistent with the SEM observation, the particle size is found under TEM to be ~50 nm in diameter. A selected-area electron diffraction (SAED) pattern with indexed atomic planes is shown in the inset of Figure 11(a).

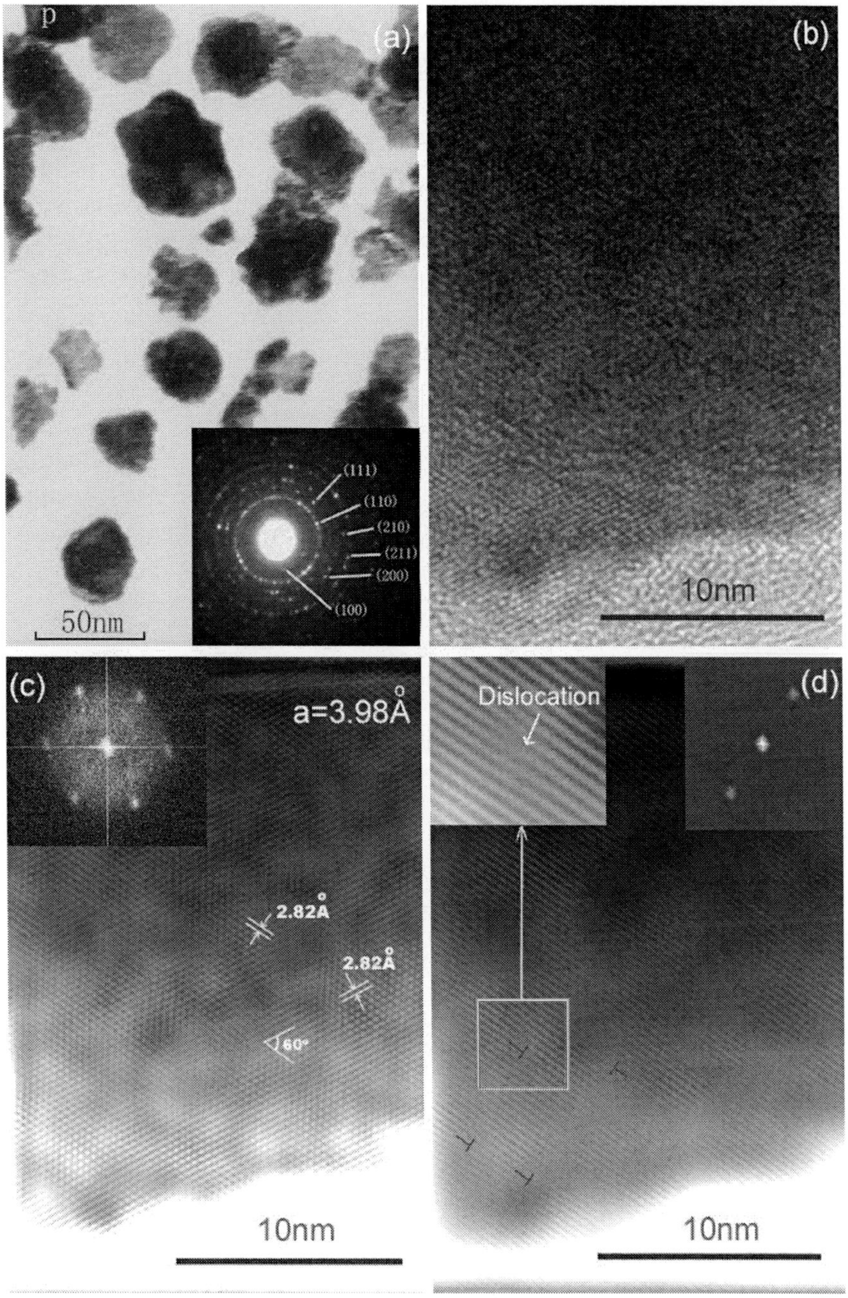

Figure 11. TEM observation of BST(90/10) nanoparticles: (a) Morphology, (b) High resolution image, (c) FFT image indicating relationship of atomic planes and (d) FFT image revealing the dislocations in the nano-structure [24].

Figure 11(b) is a high-resolution TEM image of "particle p" in Figure 11(a). Regularly arranged patterns can be observed in the darker region of the photo, indicating that the particles under observation are crystallized. A Fast Fourier Transform Algorithm (FFT) has yielded structural information of the nano-particles, as shown in Figures 11(c) and 11(d). The lattice parameter of the material is found to be ~0.398 nm. Dislocations have also been identified in the FFT image shown in Figure 11(d). Obviously, the low synthesis temperature is the main reason for the formation of the dislocations.

Further experiments have indicated that the particle size can be controlled through modifying the solution concentration, the synthesis temperature, the stirring rate and/or the drying temperature. For example, finer particles can be obtained in thinner solution. Although, as a typical example, the synthesis in this paper has been conducted at 60°C, it does not mean that crystallized nanoparticles could not be obtained via this technique at other temperatures. As a matter of fact, BST nanoparticles with acceptable crystallinity have been obtained via the same process conducted at several different temperature points (~25°C, 40°C, 80°C etc). Another point to mention is that this synthesis technique is applicable to BST with any Ba:Sr ratio, despite only the composition of Ba : Sr = 0.9 : 0.1 is discussed here.

It is difficult to enlarge the production scale of nanoparticles for many synthesis methods, especially for hydrothermal method in an autoclave. With DSS method, products can be synthesized in a large scale easily. In the laboratory, using a 2000 ml beaker as reaction container, about 100 g of nanoparticles can be obtained in one batch. Therefore, the method is ideal for industrial production.

The BST nanoparticles prepared by the above process are very useful for making high-quality ceramics and composites. The microstructural features are very small in size (hence large specific area) and already crystallized without any high temperature treatment mean a better sinterability of the BST nano-particles than powders prepared via regular solid-state reactions. Based on this, we have managed to prepare BST ceramics by a simplified ceramic processing in which the calcination step was omitted and the sintering temperature was lowered.

1.3. Synthesis of $BaZrO_3$ and BZT Nanoparticles

Barium zirconate (BZ) is another typical of perovskite oxide that can form solid solution with barium titanate completely, which is expressed as BZT generally. The substitution of small amounts of titanium by zirconium in barium titanate modifies the ferroelectric properties e.g., the thermal stability of the dielectric properties which is desirable for many technological applications [53, 54]. Also, the BZT ceramic [Ba(Zr, Ti)O3] shows excellent bulk properties of a high dielectric material (dielectric constant >20,000)[55, 56]. BZT ceramics have been established as one of the most important compositions for dielectrics in multilayer ceramic capacitors (MLCCs) of the EIA Y5V specification.[56-58] With the change of Zr/Ti ratio, the microstructure and properties of the material change, as already observed in BZT ceramics.[52] Due to the excellent and easily modifiable dielectric properties, BZT is useful for multilayer capacitors and frequency-agile microwave devices.[59] In the literature, the synthesis of perovskite BZT nano-powders is usually through the sol-gel method which requires post-annealing, typically at 400°C or above, to crystallize the powders.[51,61] In DSS method, however, by employing appropriate chemical precursors (barium hydroxide, zirconium isopropanol and tetrabutyl titanate, etc), perovskite BZT nanoparticles can be prepared under much milder conditions.

The synthesis process of BZT and BZ is very simple and similar to that of BST and BT [25] as shown in Figure 1. However, some starting materials are different, such as isopropanol zirconium is adopted to obtain the element zirconium in BZ and BZT. Because isopropanol zirconium is sensitive to moisture and even can form precipitates in absolute ethanol, the solvent of the alcohol solution has to be replaced by isopropanol. More details of our experiments are as follows: Firstly, a base solution is prepared by dissolving barium hydroxide into warm water (60°C). The concentration of the solution is usually 0.4 M. After dissolving, a small amount of ammonia is added into the solution. In the meantime, a solution is prepared by dissolving tetrabutyl titanate and zirconium isopropanol into isopropanol. In this solution, the concentration of titanium and zirconium has been adjusted in terms of the desired composition of the final products. To prepare Ba(Zr0.50Ti0.50)O3 powder, for example, the Ti and Zr concentration are both 0.2 M. Then the organic solution is slowly added into the base solution (~60°C) under vigorous stirring. An instantaneous formation of white precipitation can be observed. After the reaction has been completed, the white precipitation is filtered out of the solution and baked at ~60°C for 24h in an oven. Dried white powders obtained via the above process are the final products.

X-ray diffraction spectra of the BZT powders of different compositions (x = 0, 0.1, 0.2, 0.5, 0.7, and 1.0) are shown in Figure. 12.

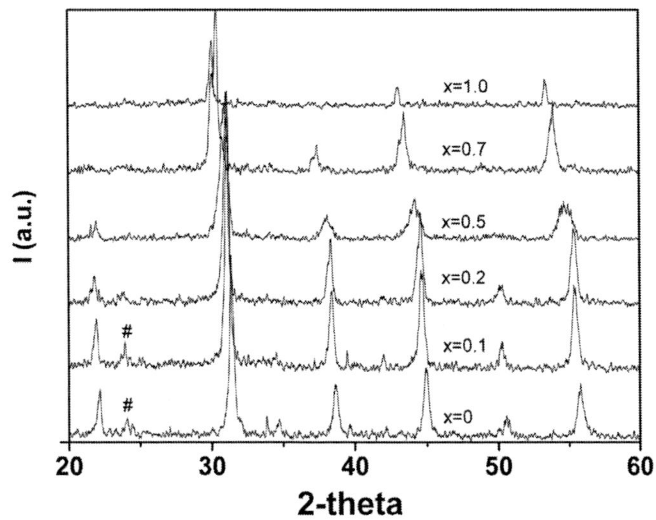

Figure 12. X-ray diffraction spectra of $Ba(Zr_xTi_{1-x})O_3$ powders. The peak marked with "#" is identified as $BaCO_3$ peak. All other peaks are perovskite peaks [25].

All the powders have been prepared at 60°C. Except for minor barium carbonate peaks (located at about 25°) occasionally appear in the plot, all major peaks shown in the patterns can be identified as from perovskite phase. With the increase in Zr/Ti ratio, the peak positions exhibit a shift towards lower angles, indicating an increase in the lattice constant. This trend is in good consistency with that in bulk BZT solid solutions,[52] suggesting a good compositional control in the as-prepared powders.

Figure 13 shows a TEM micrograph of the Ba(Zr0.50Ti0.50)O3 powders prepared in 60°C solutions. It is found that the powders are quite uniform in size, with an average diameter of about 40 nm. TEM observations and X-ray diffractions on the powder samples prepared under other conditions reveal that the particle size may vary as the processing parameters change. Typically, lowering the solution concentration and raising the solution temperature are helpful for making smaller-sized powders. Nevertheless, under the conditions we have investigated (i.e., solution temperature from room temperature to 80°C and concentration from 0.1 M to 1 M), the average diameter value of the powders is seldom beyond the range of 20 nm to 120 nm.

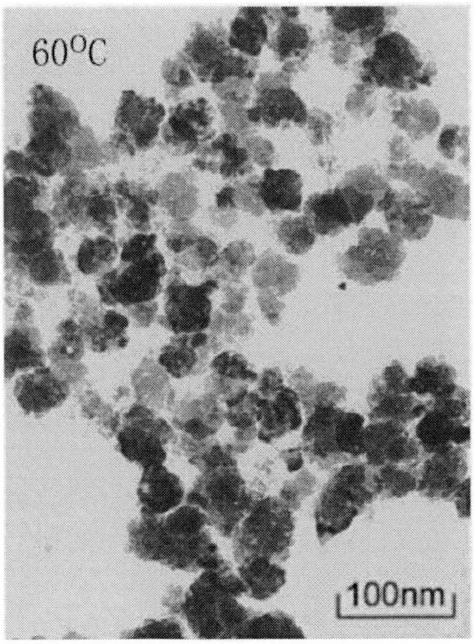

Figure 13. TEM observation of $Ba(Zr_{0.50}Ti_{0.50})O_3$ powders [25].

2. Synthesis of Core-Shell Structured Perovskite Nanoparticles with Heterogeneous Oxide Shells

Core-shell structured perovskite nanoparticles with heterogeneous oxide shell may have advantages of both the core and the shell. This kind of hybrid structure has potentials for some important applications, such as microwave devices and high energy density ceramic capacitors. Paraelectric phase Ba1-xSrxTiO3 (BST) at room temperature can be obtained by adjusting the ratio of Ba and Sr and/or doping foreign ions to substitute A or B sites in the ABO3 perovskite system, which is a promising material for electric tunable devices such as phase shifters, delay lines, tunable capacitors, and oscillators [62,63]. Driven by the applications of electric materials in telecommunications, the optimum material parameters have been focused on a low dielectric constant (ε), large tunability, a low loss dissipation factor (tanδ), and low temperature dependence of ε. Composite materials, including a ferroelectric BST phase and non-ferroelectric oxides with low loss (MgO, Ta2O3, Al2O3, etc.), have attracted much attention [63-67]. To obtain the size of ferroelectric materials in the nanometer scale, more attention has also been paid to chemical routes, like sol-gel procedures

[68], to the thermal decomposition of precursors [69], the hydrothermal process [70], selected coprecipitation techniques [71], as well as to the solvothermal method [72]. As noted previously, it is a key to synthesize perovskite nanoparticles with suitable physical and chemical properties. The special feature of the nanoparticles by DSS method is ideal in this respect, making them a favorable candidate for self-wrapping heterogeneous oxide shell during synthesis of core-shell structured perovskite nanoparticles. This strategy for creating the core-shell BST@MgO particles is described in Section 2.2. The nanoscaled BST powders synthesized by DSS method are used as core materials and the core-shell structure is formed using the chemical solution method, by adding magnesium nitrate solution into the BST base solution with an ultrasonic dispersion process.

2.1. The Special Feature of Perovskite Nanoparticles Prepared by DSS Method

The special feature of perovskite nanoparticles prepared by DSS method is that they can carry positive charge and attract free hydroxyl group on their surfaces. The hydroxyl group can often exist in perovskite lattice of compound such as BaTiO3, to form $Ba1-_x[VBa'']_xTiO3-2_x[OHO\bullet]2_x$ when the BaTiO3 powders are synthesize from water solution[23]. Hydroxyl group in lattice can also be considered as interstitial hydrogen ion as shown in Equation (2)

$$V_O^{\bullet\bullet} + H_2O \rightarrow 2H_i^{\bullet} + O_O^{\times} \tag{2}$$

or

$$H_2O \rightarrow 2H_i^{\bullet} + O_O^{\times} + V_{Ba}^{''} \tag{3}$$

For a low synthesis temperature, it seems difficult to form vacancy defects and Equations (3) and (2) are not quite reasonable. A more reasonable equation may be revised as:

$$H_2O \rightarrow H_i^{\bullet} + OH^{-}(free) \tag{4}$$

The hydroxyl group with negative charge should locate on the surface of the nanoparticles and can be ionized freely. Because the perovskite nanoparticles by DSS method are synthesized from solution near room temperature, hydrogen interstitials could be introduced into the perovskite lattice and carry positive charges. When hydrogen interstitials are introduced into perovskite lattice, the lattice cell of the powders will expand and change the crystal symmetry. In our experiments, all the lattice cell parameters of the nanoparticles synthesized by DSS method are larger than that of the bulk materials and have cubic symmetry. This indicates that the lattice cell has expanded and hydrogen interstitials have been introduced into the perovskite lattice.

It is reasonable to expect that the hydrogen interstitial will introduce positive charges into the lattice and therefore, the particles have to adsorb negatively charged hydroxyl groups to compensate and neutralize such positive charges as shown in Equation (4). If such particles are dispersed in water, then the solution must show an alkalescent property. On the other hand, if stoichiometric perovskite particles are suspended in water, the solution should keep neutral. In Figure 14, DSS-prepared BaTiO3 nanoparticles have been dispersed in water and the pH value is checked. Alkalescent property of the suspension is proved and this means the hydrogen interstitials in perovskite lattice are actually compensated by surface-charged hydroxyl groups. It should be noted that around the nanoparticles, alkalinity is much stronger than far away from them. This is due to the adsorption effect of hydrogen interstitials with positive charges in the lattice which attract hydroxyl group with negative charges on the surface.

Another test has been adopted to prove this assumption. BST nanoparticles prepared by DSS method have been divided into three parts and subjected to different annealing processes. They are i) the as-synthesized, ii) the 500°C-annealed and iii) the 1000°C-annealed powders, respectively. With dispersion of the particles in water, the three suspensions have been found to have three different pH values: i) 10, ii) 8 and iii) 7. While samples i) and ii) are obviously the cases discussed above, sample ii) actually corresponds to a case where the interstitial hydrogen has been partially removed by the thermal treatment.

Figure 14. Checking the pH value of DSS-prepared $BaTiO_3$ nanoparticles dispersed in water.

The hydrogen interstitials have been removed in sample iii). All of these experiments are guided by the DSC-TG experiments of BST shown in Figure 9. Here we should note that hydrogen interstitials can be introduced into the lattice of perovskite nanoparticles synthesized by DSS method and behave as alkalescence. This can also happen in oxide particles prepared by other chemical routes such as hydrothermal method. However the alkalescence of perovskite nanoparticles by DSS method is much stronger than those by other methods because DSS method operates at much lower temperature compared to other methods.

2.2. Synthesis of Core-shell Structured BST Nanoparticles with MgO Shells

The perovskite nanoparticles prepared by DSS method behave as alkalescence, and can react with the solution of many salts. Because hydrogen interstitials will introduce positive charges into the lattice of perovskite nanoparticles obtained directly from solution and therefore the particles have to absorb negatively charged hydroxyl groups to compensate and neutralize such positive charges as shown in Equation (4). When this kind of perovskite nanopaticles are dispersed in a certain solution of inorganic salt such as Mg(NO3)2•6H2O, Zn(NO3)2•6H2O, Al(NO3)3•9H2O, Y(NO3)3•6H2O etc., the hydroxide related to these salts can spontaneously deposit on the surface of the nanoparticles. As a result, a perovskite structured nanoscaled core material will be successfully self-wrapped by a non-ferroelectric hydroxide derived from a nitrate salt solution. After annealing at high temperature, core-shell structured perovskite nanoparticles with heterogeneous oxide shells can be synthesized. Here nitrate salts are proposed in order to avoid some impurity ions, such as Cl-, SO42-, which are difficult to remove from the final products.

The flow chart of processing of self-wrapped core-shell structured perovskite nanoparticles with heterogeneous oxide shell is shown in Figure 15. Here we use BST(50/50) nanoparticles and magnesium nitrate hexahydrate [Mg(NO3)2·6H2O] as an example of perovskite nanoparticles and nitrate salt to illustrate the process. The BST nanoparticles prepared by DSS method is dispersed in ethanol by ball milling for 2 h. Mg(NO3)2·6H2O is dissolved into deionized water to form Mg-solution. This solution is added into the suspension and then subjected to ultrasonic agitation for 1 h. Ammonia is subsequently added to adjust the acidity of the mixture to near PH~11. The resulting MgO-wrapped BST mixture is filtered out, carefully washed several times by deionized water, and then dried and calcined at different temperatures.

A simultaneous thermal analyzer (NETZSCH STA 449C) was used for thermogravimetric analysis and differential scanning calorimetry (TGA and DSC). The weight of the sample is about 15 mg. The range in which the temperature increases is 30-1100oC. The measurements are made under oxygen (20 ml/min) with protective argon gas (10 ml/min) at a heating rate of 10oC/min. A phase analysis of the powder has been performed by X-ray diffraction (XRD) using a Rigaku D/Max-B diffractometer with CuKα radiation.

The powder sample is pressed on a glass plate and mounted vertically on the sample table, and diffraction patterns are recorded using CuKα (0.15406 nm) radiation. Data are collected in the 2θ range of 20o to 100o in steps of 0.02o using a scintillation detector. The morphology is observed by field emission scanning electron microscopy (SEM, JSM-6335F) and transmission electron microscopy (TEM, Philips CM20) working at 200 kV. The Fourier transform infrared spectroscopy (FTIR, Magna 760) has been used to determine the phase transitions that occur qualitatively during the syntheses. These spectra are recorded in the range of 400~4000 cm-1 at room temperature.

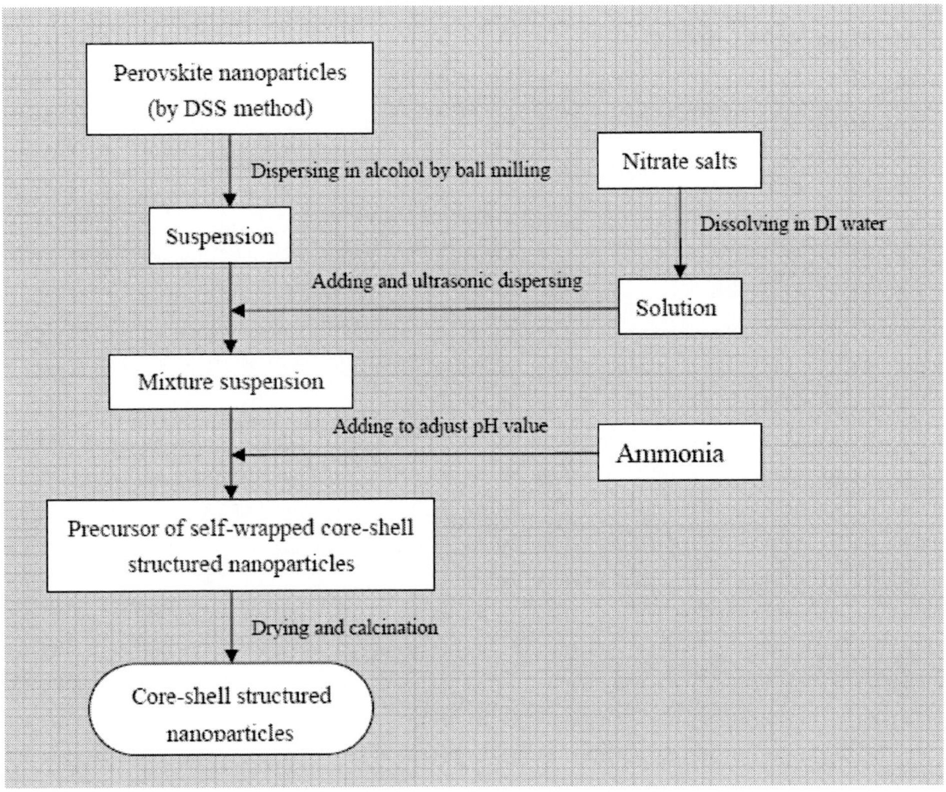

Figure 15. Processing of self-wrapped core-shell structured perovskite nanoparticles with heterogeneous oxide shell.

TGA and DSC are often employed as analysis of the temperature point of chemical reaction or phase transformation of a certain system. Figure 16 shows the TGA and DSC curves of as-prepared BST@MgO powder in flowing oxygen. The TGA trace indicates a three-step loss in the sample mass, corresponding to the evaporation of water and organic solvents, the reaction between metal ions and the surface of the BST core, and decomposition and crystallization of precursors. During the first step, the weight loss is close to 26.13%, which can be attributed to the evaporation of residual water and a small amount of organic solvent. Water can be found in the powder, even in the powder that has been dried at a higher temperature, probably because atmospheric H2O is trapped at the surface of the particles. It is reasonable to get the FTIR spectra belonging to the vibration bands of hydrogen bonding in the 500oC dried BST@MgO powder, although the spectra should disappear after the mixture is heated to over 100oC. It is speculated that the water does not only come from the process of synthesis. A sharp endothermic peak near 230.8oC is found in the DSC curve, which corresponds to the relatively large loss of mass in this step.

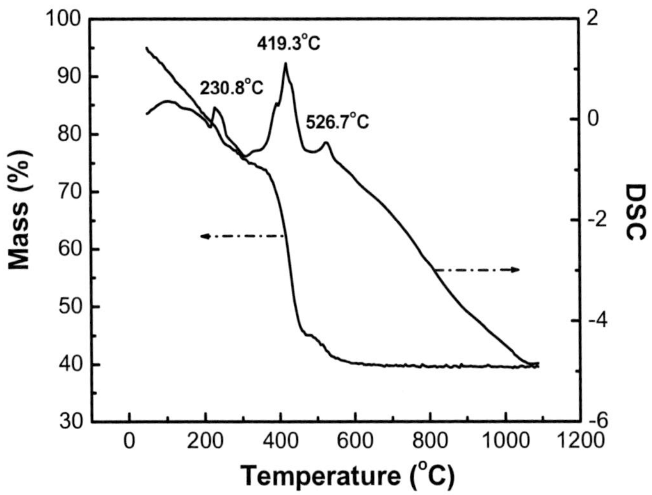

Figure 16. TGA and DSC curves of as-prepared BST(50/50)@MgO powder in flowing oxygen [15].

There is a sharp loss of mass from the onset temperature of 409oC, with the loss of mass being close to 29.09% in the narrow temperature range of 30oC. There is a sharp peak around 419.3oC, accompanied by two shoulder peaks in the DSC curve. We believe that the main chemical reactions occur in this temperature region. During synthesis process of core-shell structure in the suspension, the core with alkalescence can react with the acidic salt solution, and the hydroxide precursor can be self-wrapped onto the surface of the core. The shell will form after heating. As the temperature increases, one exothermic peak around 488.6oC and a weak endothermic peak around 526.7oC are shown in the DSC curve in the third step corresponding to a mass loss of 4.93%. This is attributed to decomposition and crystallization in the forming of MgO as a final shell.

Figure 17 shows the FTIR reflectivity spectra of: (a) as-prepared BST, (b) as-prepared BST@MgO, and (c) 500oC annealed BST@MgO between 4000~400 cm-1 at room temperature. The main bands are clearly shown in the spectra as follows: Metal-oxygen (M-O) bonding between 400 and 800 cm-1; Carbon-oxygen bonding between 1300 and 1700 cm-1 and around 2400 cm-1; C-H stretching vibrations between 2800 and 3000 cm-1 and weak nitrate peaks at about 1774 cm-1; and hydrogen bonding at around 3500 cm-1. The broad peak at 3522 cm-1 and one intensive peak at 1502 cm-1, shown in Figure 17(a), are due to the excitation of stretching and to the deformation vibration modes of hydroxide ions. These peaks are found at around 3536 cm-1, 3330 cm-1, 1502 cm-1, and 3499 cm-1, 1466 cm-1, 1643 cm-1 in the as-prepared BST@MgO powder and BST@MgO powder annealed at 500oC, as shown in Figure 17 (b) and (c). These peaks are assigned to the hydrogen bond between H2O and ethanol, which will become much weaker as the temperature increases because of the evaporations of the solvents. In Figure 17 (b) and (c), the absorption peak at 1678, 1774 cm-1 in curve (b), and only one peak at 1774 cm-1 in curve (c) are assigned to the nitrates group ($O-O-NO_2$ or $O-N=O$) because a small amount of residue of nitrate radicals has not been thoroughly removed after the mixtures have been washed several times.

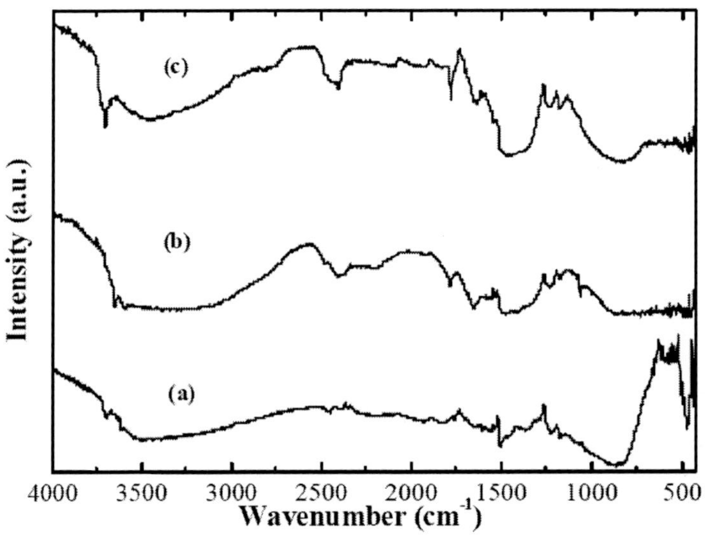

Figure 17. FTIR reflectivity spectra of the samples. (a) as-prepared BST(50/50), (b) as-prepared BST(50/50)@MgO, and (c) 500°C annealed BST(50/50)@MgO [15].

The bands at 2800-3000 cm-1 are due to C-H stretching vibrations, whereas the intensities are weaker in the as-prepared BST powder and as-prepared BST@MgO powder compared to those in the 500oC annealed BST@MgO powder. This is an intriguing phenomenon because the intensity should decrease as the ethanol and other organic solvents in these mixtures evaporate. Carbon-oxygen bonding presents between 1300 and 1700 cm-1, and entrapped atmospheric CO2 at 2455 cm-1 in curve (a), 2357 cm-1 in curve (b) and 2405 cm-1 in curve (c). The peaks 847 cm-1 in curve (a), 802 cm-1 in curve (b), and 825 cm-1 in curve (c) are typical deformation and stretching vibration modes of carbonate ions [73,74], although Liu et al. assigned these peaks to the N-O bond vibrations of NO3-[75]. It is reasonable to rule out the N-O bond vibrations of NO3- in this range, and N-O bonding vibrations should not be found in the as-prepared BST powder and BST@MgO powder after being annealed at 500oC. Naturally, these peaks around 802~847cm-1 should be caused by other bonding groups.

XRD investigations have shown that the as-precipitated BST powder is crystallized in a single phase (Ba0.5Sr0.5TiO3), as shown in Figure 18(a). No other phases, such as BaO, SrO, and TiO2 have been detected except for a weak peak of (Ba,Sr)CO3. BaCO3 is also found in the nano-BaTiO3 powder prepared using the DSS method [23]. This result is consistent with the FTIR spectra, a peak at 2455 cm-1 for the as-prepared BST powder, which is assigned to C-O bonding by entrapping atmospheric CO2 in the mixtures. The composition of the powder has been checked by energy dispersive X-ray spectroscopy (EDX) during the SEM and TEM measurements. The EDX analysis of the powder shows no traces of other ions, as the stoichiometry of as-prepared BST is Ba:Sr:Ti=10:9.16:19.22, near (Ba+Sr)/Ti=1 within the allowed measuring errors. From XRD patterns, the lattice parameters of BST decrease as the treating temperature increases. This is partly because of the release of hydrogen ions from the lattice and/or formation of vacancy defects. The lattice parameter of the as-prepared BST is about 3.9535 Å, decreasing to 3.9443 Å for BST in the 500oC annealed BST@MgO. Other results for sintering discs are derived from the as-prepared BST@MgO powder at higher temperatures, from 1100oC to 1300oC. The lattice parameters decrease from 3.9457 Å (1100oC) to 3.9363 Å (1300oC) (they are not shown here). In Figure 18(b), some peaks of NH4NO3 are presented in the XRD pattern. It is obvious that the nitrates are present, although the precipitations are washed several times with water. Naturally, we find the vibration bands of the nitrate group in the FTIR spectra (as shown in Figure 17 (b,c)).

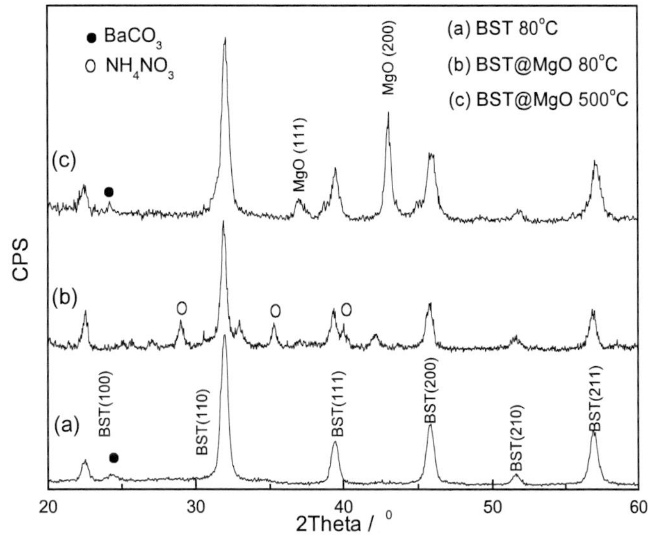

Figure 18. XRD patterns of : (a) as-prepared BST(50/50), (b) as-prepared BST(50/50)@MgO, and (c) 500°C annealed BST(50/50)@MgO [15].

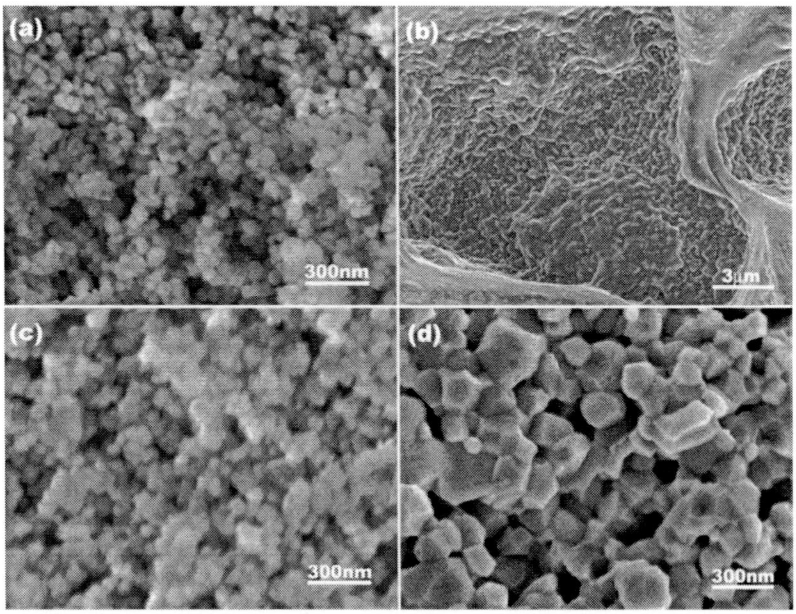

Figure 19. FE-SEM image, (a) BST(50/50) nanoparticles prepared at 70°C, (b) Interim composite of MgO precursor wrapped BST nanoparticles, (c) MgO wrapped BST(50/50) nanopaticles annealed at 500°C, and (d) MgO wrapped BST(50/50) nanopaticles annealed at 1100°C.

These peaks will become much weaker, and even disappear when the powders are heated at higher temperatures. MgO phase will become much stronger as the powders are heated at higher temperatures. Figure 19 shows the surface morphology of the nanoscaled BST powder and MgO-wrapped BST heat treated at different temperatures. The SEM observation shows that the as-prepared BST powder has an average particle size of 20 nm (Figure 19(a)). The particle size of BST powder calculated by the Scherrer's equation (Equation (1)) is 17.3 nm (as-prepared BST powder), while the particle size of BST in the composites, i.e., MgO-wrapped BST, shows a slight decrease from 17.9 nm (as-prepared BST@MgO) to 16 nm (BST@MgO annealed at 500oC). However, the size of the particle in the BST@MgO ceramic sintered at 1100oC increases to 22.6 nm. The ambiguous structure of the as-prepared BST@MgO powder is shown in Figure 19(b), possibly due to the amorphous structure of MgO caused by hydrolyzation and some organic residues or nitrates in it. Crystallized particles are shown to be embedded in an amorphous shell [65]. As temperature increases, all solvents and organic residues disappear, and well-crystallized MgO-wrapped BST will form. As shown in Figure 19(c) and (d), aggregated particles of about 100~150 nm (500oC) and several hundred nanometers (1100oC) consisting of more than one core-shell structure will form as crystalline growth is driven by the thermodynamic effect.

The fine details of the particles and their morphology have been investigated using a Philips TEM. The bright field electron micrograph of the BST powder is shown in Figure 20 (a). The average particle diameter is found to be approximately 20 nm. Figure 20 (b) shows the TEM images of BST@MgO annealed at 500oC. The particles from these images are not homogeneous in size, being between 40~200 nm. A big particle and its surroundings have been chosen to measure the EDX spectra to confirm the composition and its variations from core to shell. The spots are marked by "I," "II," and "III," as shown in Figure 20 (c). The results of the atomic percentage of the elements in BST@MgO are listed in Table I.

Table 1. Atomic percentage of the elements in BST@MgO[15]

Selected spot	Atomic %				Molar ratio of Mg/Ti
	Mg	Ti	Sr	Ba	
I	21.83	0.01	0.03	0.01	Without BST
II	13.9	10.4	4.2	7.4	1.34
III	8.12	19.22	10	9.16	0.42

According to the molar ratio of these spots, there is a radial chemical gradient from the outer spot (marked "I") to the boundary (marked "II") and then to the core of the grains (marked "III"). It is clear that a ferroelectric nanoscaled core, BST, in a thin surrounding shell, MgO, has been prepared successfully. As mentioned above, the outer shell consists of MgO. Figure 20(d) shows the SAED pattern recorded from an area containing a core-shell structure (BST@MgO) such as spot (III) in Figure 20(c). Two sets of diffraction patterns are found, which correspond to MgO and BST, respectively. There is a spot of diffraction plane of MgO beside that of the same diffraction plane of BST, such as labeled as M(220) for MgO and P(220) for BST, respectively in the figure, showing that the lattice cells are matched between the core and the shell. Therefore, BST nanoparticles synthesized by DSS method can be self-wrapped by the precursor of MgO such as amorphous Mg(OH)2 in magnesium nitrate solution and then form core-shell structures after annealed at 500oC.

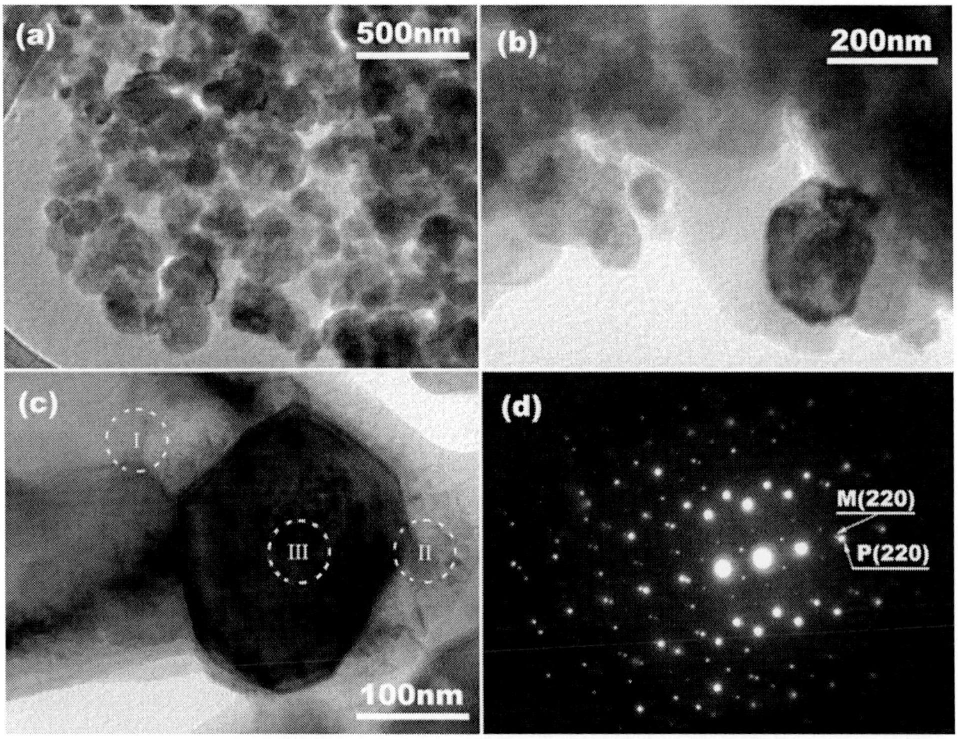

Figure 20. TEM image (a) BST(50/50) nanoparticles, (b) BST@MgO annealed at 500°C, (c) BST@MgO (a big particle and its surroundings), (d) SEAD of the BST@MgO (focus on (III) in (c)).

2.3. Other Patterns of Core-shell Structured Perovskite Nanoparticles with Heterogeneous Shells

Other patterns of core-shell structured perovskite nanoparticles with heterogeneous shell can be synthesized similar to that of BST@MgO. If the core is changed to BT or BZT, all the processing is the same as shown in Figure 15. If we use Mg(NO3)2 and Zn(NO3)2 as starting nitrates, the heterogeneous shell will change to Zn-doped MgO (MZO) and this kind of shell will improve the sintering characteristics of MgO. Thus, it can be expected that BT@MgO, BZ@MgO, ST@MgO, BZT@MgO and BST@MZO etc. can be synthesized by this method. If Mg(NO3)2 and Al(NO3)3 are used as the starting nitrates, spinel heterogeneous shell can be formed. Figure 21 shows the XRD pattern of BST@MgAl2O4 calcined at 1100oC. It is clearly revealed that spinel MgAl2O4 shell has been synthesized after calcined at 1100oC. There are two minor impurity phases indexed in this system, one is BaAl2O4 and the other is MgO. These two phases do not affect the electrical properties of the system and can be eliminated after a higher temperature annealing.

The shell with a spinel phase is not as sensitive to moisture as pure MgO. If the devices are fabricated using this kind of materials, much higher reliability can be expected than MgO wrapped materials.

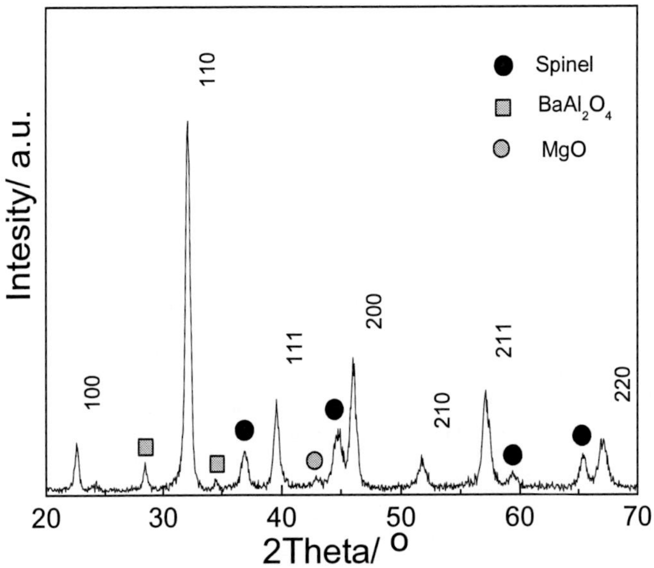

Figure 21. XRD pattern of BST(50/50)@$MgAl_2O_4$ calcined at 1100°C.

3. Ceramics of Core-Shell Structured Perovskite Nanoparticles with Heterogeneous Oxide Shells

It is very important to tailor either the grain size of the ferroelectric perovskite phase or the shell thickness of heterogeneous oxide shell. The ceramic materials with this kind of core-shell structures can be obtained by sintering of core-shell structured perovskite nanoparticles with heterogeneous oxide shells and can be applied to many electrical devices. However, it is difficult to obtain the dense ceramics if the powders with this kind of structures are directly sintered. Because inert oxide such as MgO etc. is often used as the heterogeneous shell in order to obtain very low dielectric loss. However, this kind of inert oxides with high melting point is very difficult to be sintered. In the conventional composite, it is well known that the sintering characteristics depend on the easily sintered phase, e.g., in the MgO-BST system, BST can be easily sintered at much lower temperatures than MgO. In MgO-wrapped BST, unfortunately, the sintering behavior can be dominated by MgO, making it difficult to be sintered at a low temperature like BST. Another reason of making the core-shell structured materials difficult to be sintered is that nanoparticles exert a pinning effect during sintering and thus the mass transfer can be suppressed from shell to shell. BST core in MgO can be considered as inclusions and increase the energy necessary for the movement of a grain boundary and inhibit grain growth during sintering. Here some efforts are focused on decreasing the sintering temperature and obtaining good ceramics.

Core-shell structured 40%wt MgO wrapped BST nanoparticles have been prepared following procedures shown in Figure 15. The powders of MgO wrapped BST nanoparticles have been pressed into disks with 1 mm thickness and 11.78 mm diameter under the pressure of 1000 kg/cm2. Then the samples are sintered at different temperature from 1250 to 1450oC to study their sintering characteristics. The firing shrinkages are listed in Table 2 and all of them are less than 10%. All the samples are porous and can be easily crushed. Their strength is bad.

Table 2. Firing shrinkage of samples (diameter of green ware is 11.78mm)

Sintering temperature	1250°C	1350°C	1450°C
Diameter of sintered samples	11.10mm	10.92mm	10.74mm
Firing shrinkage of samples	5.8%	7.3%	8.8%

In general, sintering aids will help in the densification of difficult-to-sinter ceramics by introducing SiO2 related secondary phase inclusion during sintering as reactive liquid. However, the results are not as good as expected when we use SiO2 as sintering aid in the system of core-shell structured MgO wrapped BST nanoparticles. The sintering has not been improved after adding the sintering aids which results in frothy samples. Firing shrinkage is only improved slightly. Another strategy is to improve the sintering characteristics of MgO shell. We try to substitute MgO with doped MgO. Of course the dopants should be restricted to a certain range.

3.1. ZnO Doped MgO

MgO and ZnO have rocksalt and wurtzite structures, respectively.[76] Since they have different coordination numbers and crystal symmetry, it is expected that Zn will replace Mg without changing the original structure up to a certain amount when alloying. According to the phase diagram, MgO allows a maximum of 56% ZnO (wt %, 40 at.%) at 1600 oC, and maintains its NaCl structure with the lattice constant staying close to that of pure MgO (4.208 Å). The bandgaps of ZnO doped MgO are listed in Table 3.[77] Being a kind of material with a wide bandgap, the slight doping will not change the electrical properties greatly. Thus, substitute of MgO by ZnO doped MgO can be accepted as a methodology for our applications.

Table 3. Bandgaps of ZnO doped MgO [77]

Mg atomic fraction	Bandgap energy (eV)
1	~7.7
0.88	>5.31
0.85	5.31
0.65	4.70
0.5	4.00
0.45	3.87
0.38	3.70
0.30	3.42
0.15	3.37
0	3.30

The conventional ceramic processing has been adopted to fabricate MgO solid solution. We use x=0.1 and 0.3, calcination temperature is 800oC. From Figure 22, we can see that x=0.1 at 800oC, only the pure rocksalt phase can be obtained, but the solution is not completed when x=0.3 at this temperature.

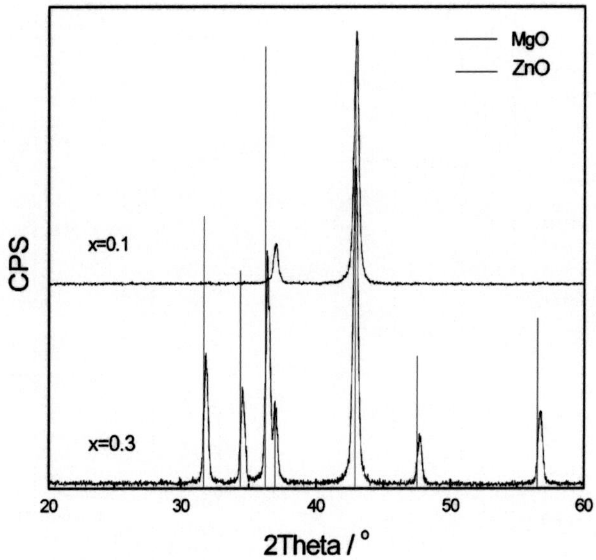

Figure 22. ZnO doped MgO calcined at 800°C.

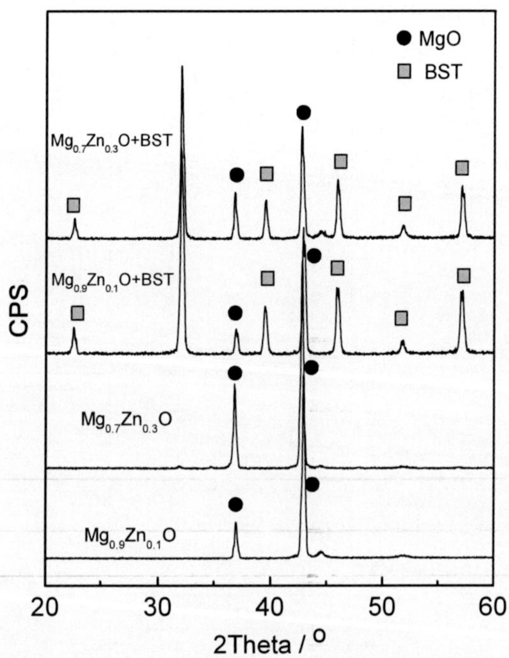

Figure 23. MZO and MZO-BST(50/50) composite ceramics sintered at 1250°C.

A long calcination time and a high temperature will promote the formation of solid solution. When the temperature increases to 1250oC, we obtain almost pure rocksalt MZO. There is no ZnO phase left. MZO is inert to BST, so that they do not react at high temperature. Composite of MZO and BST can be sintered into dense ceramics and keep their own phase respectively as shown in Figure 23. Firing shrinkage is listed in Table 4 and all the samples are dense.

Table 4. Firing shrinkage of MZO and composite of MZO-BST

Temperature	$Mg_{0.9}Zn_{0.1}O$	$Mg_{0.7}Zn_{0.3}O$	60%$Mg_{0.9}Zn_{0.1}O$+40%BST
1250°C	9.22mm/21.7%	8.97mm / 23.9%	9.64mm / 18.2%
1350°C	9.05mm / 23.2%		

3.2. Ceramics with Core-Shell Structured BST Nanoparticles with ZnO Doped MgO Shells

The advantages of core shell structures are that their properties can be conveniently tuned and tailored by designing the chemical compositions as well as sizes of the core and the shell. A strategy is synthesis of MgO wrapped BST core-shell structured nanoparticles using nanotechnology and obtaining their ceramics or films to improve the properties. In order to improve the sintering characteristics of this kind of materials, ZnO dopant has been added into the shell.

The flow chart of fabrication of MZO wrapped BST nanoparticles is similar to that of MgO wrapped one as shown in Figure 15. The BST nanoparticles are dispersed in distilled water by ball-grinding for 2 h. The solutions of magnesium nitrate hexahydrate (Mg(NO3)2·6H2O) and zinc nitrate hexahydrate (Zn(NO3)2·6H2O) with certain ratio to BST are added to the BST suspension under vigorous stirring, and then subject to ultrasonic agitation for 1 h. Ammonia is subsequently added to adjust the acidity of the mixture to near pH~11. The resulting MZO-wrapped BST mixture is filtered out, carefully wash several times by deionized water, and then dry at 80oC. In order to keep high resistivity and low electric loss of the shell but improve sintering distinctly, the fraction of ZnO in MgO is adopted as low as possible (here just 10%). The powders with the core of BST nanoparticles and the shell of MZO precursors are annealed at different temperatures from 500~1000oC in order to obtain the crystallized MZO shell. The MZO wrapped BST nanoparticle is pressed into disks with 1 mm thickness and 10 mm diameter and sintered at 1350oC to obtain the ceramics.

Table 5 lists the firing shrinkage of the samples of MZO wrapped BST nanoparticles with different ratio of MZO:BST. The sintering characteristics of MZO is improved distinctly after doped with ZnO, but that of core-shell structured powders is improved slightly (see Table 3). Here BST nanoparticles exert a pinning effect during sintering and thus the mass transfer can be suppressed from shell to shell. BST core in MZO can be considered as inclusions and increases the energy necessary for the movement of a grain boundary and inhibit grain growth during sintering.

In order to improve the sintering characteristics of this kind of materials, a strategy as shown schematically in Figure 24 has been adopted. When the core-shell structured MgO wrapped BST nanoparticles are composited with MZO powders as shown in Figure 24(b), the mass transfer can proceed from MZO to the shell MgO (or MZO) of BST nanoparticles without pinning effect as shown in Figure 24(c). Thus, dense ceramics can be obtained under this strategy as shown in Figure 24(d). Here is a typical example, core-shell structured MZO wrapped BST nanoparticles (5%wtMZO) are composited with 35%wt MZO conventional powders, firing shrinkage reach 25% when the sintering temperature is 1350oC (diameter shrinks to 8.87 from 11.78 mm) and thus obtained dense ceramics. At the same time, when all MZO are wrapped on the BST nanoparticles as shells, firing shrinkage is only 11% as shown in Table 5.

Table 5. Firing shrinkage of MZO(90/10) and its wrapped BST(50/50) nanoparticles(sintering at 1350oC, diameter of green disk is 11.78 mm)

Compositions	Pure MZO	20%wt MZO+80%BST	40%wtMZO+60%BST
Diameter/mm	9.05	10.35	10.50
Firing shrinkage	23%	12%	11%

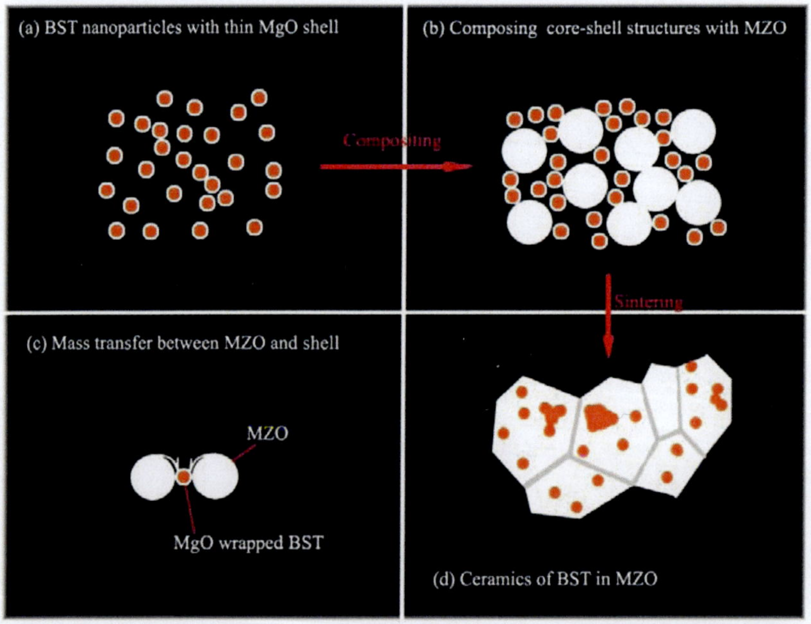

Figure 24. Schematic diagram showing the sintering process in MZO wrapped BST nanoparticles.

3.3. Analyzing the Core-Shell Structured BST Nanoparticles with ZnO Doped MgO Shells

The FE-SEM fractograph image of the ceramics of MZO wrapped core-shell structures prepared using the strategy shown in Figure 24, is shown in Figure 25. Both fractured styles of fracture along grain boundaries and fracture through grains, can be observed in the image. Some un-grown BST nanoparticles which keep their size similar to the as-prepared ones can also be observed in fractured grains, but on the boundary, it can hardly be observed. This confirmed our sintering strategy for MZO wrapped BST nanoparticles. On the other hand, it is also confirmed that the inert MZO shell can protect the core of BST nanoparticles and inhibit the grain growth.

Figure 25. Fractograph of the ceramics of MZO(90/10) wrapped BST(50/50) core-shell structures sintered at 1350°C [81].

The phases of the sample can be checked by XRD analysis as shown in Figure 26. In order to check the MgO phase clearly, The XRD profile of halite structured MZO ceramics is also given in Figure 26 (a). As shown in Figure 26 (b), because the MZO precursor is amorphous, there are no halite phase peaks in the as-prepared powders. Many other peaks are indexed as ammonia nitrate except for BST peaks After annealing or sintering at high temperature, the MZO precursor crystallized in halite structure as shown in Figure 26 (b), (c), (d), and (e), respectively.

The lattice cell parameters can be estimated from the peak positions and for a cubic crystal,

$$a = d\sqrt{h^2 + k^2 + l^2} \tag{5}$$

where a is lattice parameter and d is the distance between a certain diffraction crystalline plane and (h k l) crystalline plane index d can be calculated from Bragg's equation:

$$\lambda = 2d\sin\theta \tag{6}$$

where λis X-ray wavelength and θ is the diffraction angle.

The crystallite size of the powders can be estimated from the broadening of corresponding X-ray spectral peaks by the Scherrer's formula as shown in Equation (1). In the calculation the XRD peak of (200) plane of halite structured MZO (2θ~43) and (200) plane of perovskite structured BST (2θ~46) have been used.

The XRD profiles of (200) plane of two phases in different samples annealed at different temperatures, which have been regressed by Gaussian fitting, are shown in Figure 27. The annealing temperature dependence of the lattice cell parameters and crystallite size estimated by XRD data from Figure 27 are shown in Figures 28 and 29, respectively.

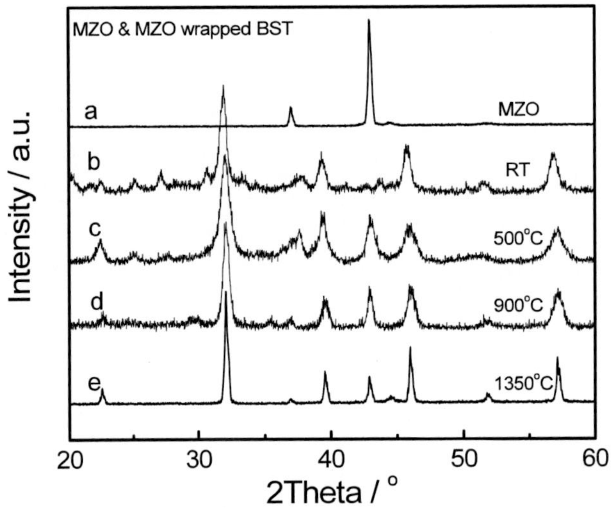

Figure 26. XRD profile of MZO(90/10) wrapped BST(50/50) nanoparticles annealing at different temperatures, (a)MZO ceramics, (b) as-prepared MZO precursor wrapped BST nanoparticles, (c) MZO wrapped BST nanoparticles annealed at 500°C, (d) MZO wrapped BST nanoparticles annealed at 900°C, (e) Ceramics of MZO wrapped BST nanoparticles sintered at 1350°C [81].

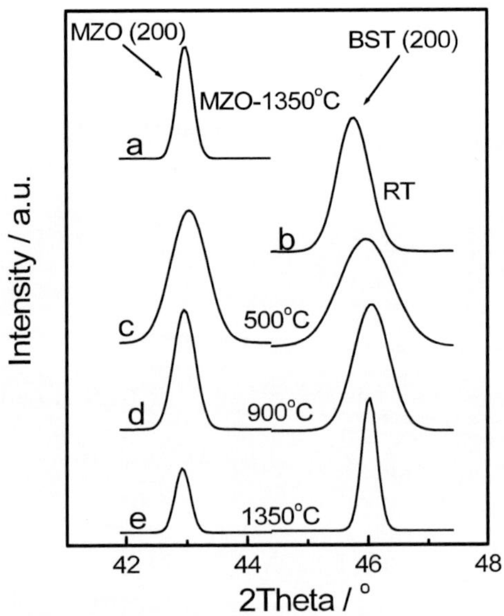

Figure 27. XRD profiles of (200) crystalline plane of (a) MZO(90/10) ceramics, (b) perovskite peak in MZO(90/10) precursor wrapped BST(50/50) nanoparticles, (c) MZO(90/10) wrapped BST(50/50) naoparticles annealed at 500°C (d) MZO(90/10) wrapped BST(50/50) nanoparticles annealed at 900°C, (e) Ceramics of MZO(90/10) wrapped BST(50/50) nanoparticles sintered at 1350°C [81].

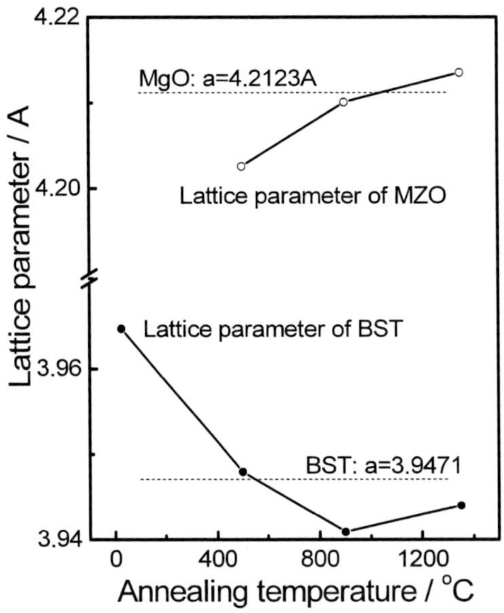

Figure 28. Lattice parameters of BST(50/50) core and MZO(90/10) shell as a function of annealing temperature.[81]

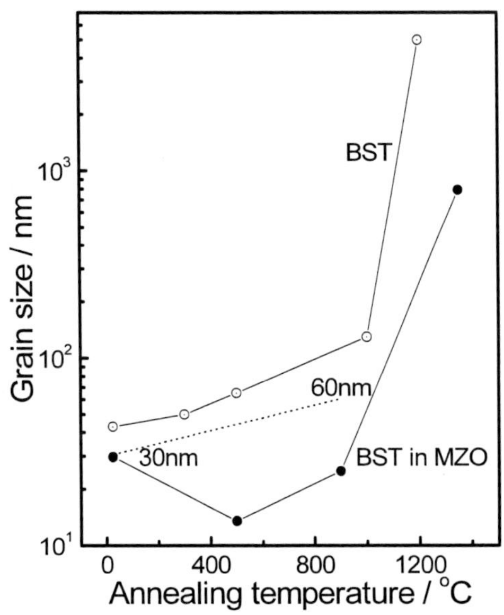

Figure 29. Annealing temperature dependance of nominal crystallite size of BST(50/50) as core in MZO(90/10) shell (solid circles) and that of pure BST(50/50) nanoparticles (hollow circles) [81].

The lattice cell parameter of MZO, aMZO, increases with annealing temperature. Because Zn ion is bigger (0.74A) than the Mg ion (0.72A) and cannot be dissolved into the MgO lattice completely at low temperature, the lattice cell parameter is small at 500oC. When the annealing temperature increases, more Zn ions are introduced into the lattice, so that the lattice cell of MZO expands. It is noted that the BST lattice cell is smaller than that in JCPDS cards (#39-1395) at high temperature because the lattice cell of BST core is subjected to inner stress caused by the MZO shell crystallization from amorphous to halite (rocksalt) structure.

It is easily understood that the crystallite size of pure BST increases with the annealing temperature for the grain growth even at moderately high temperature because nanoparticles have much greater chemical activity than conventional powders as shown in Figure 29. However, it should be noted that the nominal crystallite size of BST as core in MZO estimated from XRD data decreases before 500oC and then increases with annealing temperature and the size for the sample annealed at 900oC is smaller than the initial value. After the BST nanoparticles have been wrapped in MZO and the MZO crystallized from amorphous to halite structure, the lattice mismatch can cause inner stress and lead to a variation of lattice cell parameter which can cause XRD peaks broadening. Therefore, we should consider this factor when we estimate the crystallite size by XRD peak broadening effects. Different mathematical methods exist (e.g. Langford,[78] Balzar and Ledbetter,[79] Warren-Averbach[80]), which take into account strain in the analysis of XRD data and these methods use the integral breadth or the entire peak profile. However, here we focus much attention on the lattice parameter variation and the half width by Gaussian peak fits is used to estimate it in order to simplify the model. Given a variation of the lattice cell parameter δa, value of d changes by δd. According to Bragg's equation (Equation (6)),θ can change by δθ as Equation (7).

$$\delta d = -\frac{\lambda \cos\theta}{2\sin^2\theta}\delta\theta \tag{7}$$

Here, the minus sign implies that when the lattice cell is subjected to inner stress and shrank, the peak position shifts to higher angles. Because a cubic lattice cell parameter can be calculated using Equation (8):

$$\frac{1}{d^2} = \frac{h^2 + k^2 + l^2}{a^2} \qquad (8),$$

the variation of lattice cell parameter can be given in Equation (9):

$$\delta a = \delta d\sqrt{h^2 + k^2 + l^2} \qquad (9).$$

Using Equation (7), the broadening of XRD peak caused by lattice cell variation can be estimated using Equation (10):

$$\delta\theta = -\frac{2\sin\theta tg\theta}{\lambda\sqrt{h^2 + k^2 + l^2}}\delta a \tag{10}$$

or the variation of lattice cell parameter can be given by Equation (11):

$$\delta a = -\frac{\lambda\sqrt{h^2 + k^2 + l^2}}{2\sin\theta tg\theta}\delta\theta \tag{11}$$

where $\delta\theta$ is the line width at half-maxiumum height of the XRD peak after subtraction of equipment broadening and nano effect broadening β (refer to Scherrer's formula, Equation (1)). Because in our core-shell structure, the MZO shells form inert barriers around the BST nanoparticles, it prohibits the grain growth of BST nanoparticles before sintering started at low annealing temperatures, e.g., <1000°C. This conclusion can be confirmed by our observation of HRTEM. Therefore, the change of XRD peak broadening of BST caused by grain growth at low annealing temperature can be ignored. The variation of the lattice cell parameters is responsible for the increase in broadening of XRD peak compared with the initial as-prepared BST nanoparticles so that we can estimate it according to Equation (11) [81] as shown in Figure 30.

The maximum lattice cell variation is estimated to be 0.056Å, almost about 1.5% of 3.958Å. It means that the BST lattice parameter of some lattice cell in the nanosized BST core can decrease to 3.902Å from 3.958Å (calculated by the XRD peak position, refer to Figure 30). Therefore, the lattice mismatch in MZO wrapped BST nanoparticles is great and cause inner stress leading to shrinkage and variation in the lattice cell of BST core, which can be estimated by the XRD peak position and broadening effect.

There are two possible modes of lattice matching between MgO and BST lattice as shown in Figure 31. The bright bars in the Figure represent a certain match plane. The first mode, (I), is that the BST lattice matches MgO lattice one by one and insert dislocations when mismatch is great enough. The BST cell will expand in this mode. The second mode, (II), is that the BST lattice mismatches the MgO lattice and inserts dislocation to match MgO. The BST cell should shrink in this mode. From XRD analyzing results, the core-shell structure probably follows the second mode because the lattice cell parameter of BST nanoparticles decreases after wrapped by MZO.

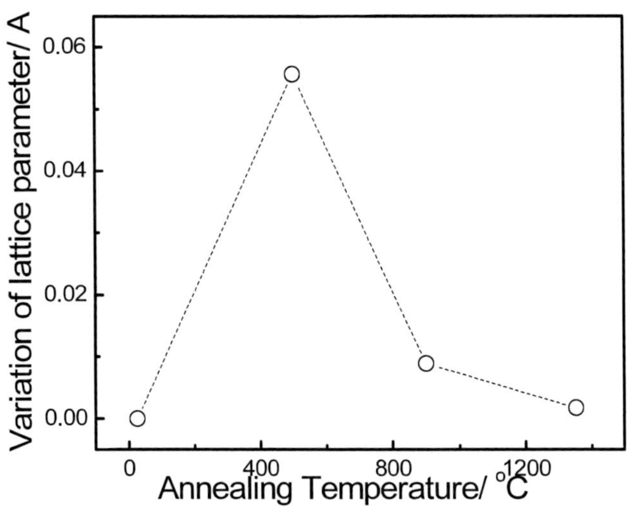

Figure 30. Variation of lattice cell parameters of BST core as a function of annealing temperature [81].

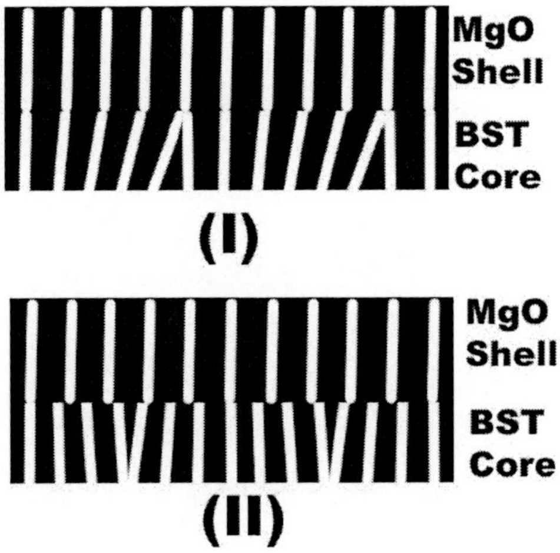

Figure 31. The lattice match modes between BST core and MZO shell [81].

The interface of BST core and MZO shell is particularly important because of its sharpness, lattice mismatch, and chemical gradient are critical for electron transfer and coupling. An epitaxial orientation relationship between the core and the shell is favorable, but an epitaxial growth is determined by their crystal structures.[1] The HRTEM image of MZO wrapped BST core-shell structure is shown in Figure 32. The interface of the core-shell structure is clearly shown in Figure 32(a). The lattice structure of MZO shell and BST core framed in Figure 32(a) are shown in Figure 32(b) and (c), respectively.

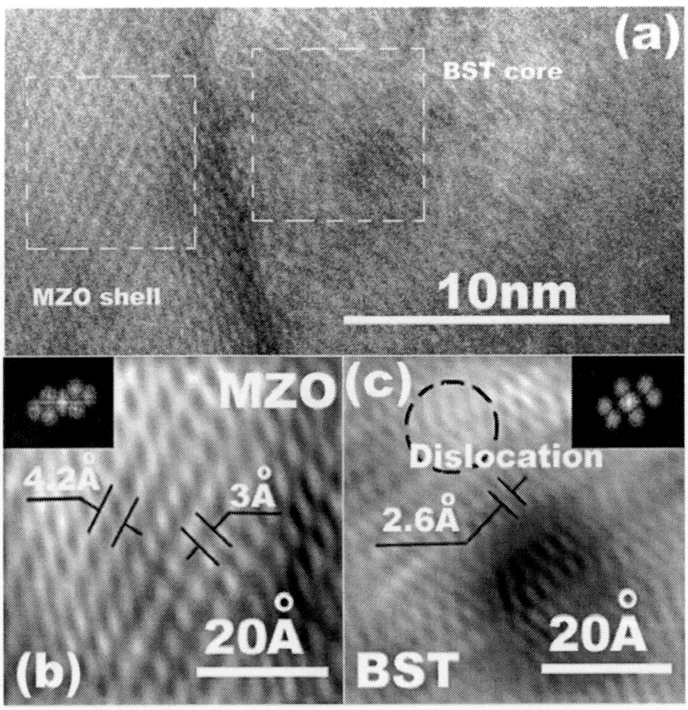

Figure 32. HRTEM image of MZO(90/10) wrapped BST(50/50) core-shell structure. (a) Interface of core-shell structure, (b) lattice of MZO shell, (c) lattice of BST core [81].

Figures 32(b) and (c) are obtained by fast Fourier transform (FFT) and smoothes the edge using a mask (as shown in insets of Figure 32(b) and (c), respectively) and then inverse FFT. It can be observed clearly that the MZO shell and BST core match along their (110) planes. The BST lattice cell is smaller than the normal one especially near the match boundary. There are many dislocations in BST lattice cell, and the integrity of MZO lattice is better than that of BST core. The distance between (110) planes is estimated as 2.6Å and the lattice cell parameter of BST is only 3.7Å. Therefore, the lattice cell has been depressed by the MZO shell as confirmed by XRD analysis.

3.4. Characterizing Electrical Properties of the Ceramics of the Core–shell Structured BST Nanoparticles with ZnO Doped MgO Shells

Relative permittivity and losses can be measured using a Stanford Research Systems model SR 720 LCR meter with the bias input by a Keithley 6517A electrometer/high resistance meter. In our experiments, the sample was fixed in an Oxford Cryostat with an intelligent temperature controller from 77 K to 400 K. High-frequency dielectric properties were measured using HP 4291B (1 MHz ~ 1.86 GHz) analyzer fitted with HP 16453 dielectric material testing fixture [14].

Core-shell structured perovskite nanoparticles with heterogeneous oxide shell may have advantages of both the core and the shell and have potentials for some important applications, such as microwave devices and high energy density ceramic capacitors. For high-frequency tunable device applications, a high dielectric constant is not desired for device impedance matching purposes, which leads to less efficient transfers of power to the device, thereby degrading the performance of the device. Moreover, a high dielectric loss also results in inferior device performance. Thus, the field dependence of the dielectric constant and loss should be evaluated. The tunability is expressed as:

$$T = \frac{\varepsilon(0) - \varepsilon(V)}{\varepsilon(0)} \times 100\% \tag{12}$$

where ε(V) and ε(0) denote the field-dependent and zero-field dielectric constants, respectively. The tunability as a function of temperature is shown in Figure 33. The figure of merit (FOM) is an important factor to demonstrate the quality of the ceramics. The FOM (K) is calculated based on the expression given below [82]:

$$K = \frac{T}{\tan\delta(0)} \tag{13}$$

where $\tan\delta(0)$ denotes the dielectric loss without a biased field on the samples.

In order to determine the temperature dependence of dielectric constant, we measured the dielectric constant and loss of the samples at 100 kHz from -200 °C to 50 °C under a biased field of 5 kV/cm. The tunability and FOM have also been calculated using Equations. (12) and (13). Figure 33 shows the tunability and FOM as functions of temperature. Relatively higher tunability has been obtained below the Curie temperature (T_c). The tunability is between 0.2% and 3.2% at room temperature (20~50°C), and the FOM value is between 0.8 and 12.8. The highest value in the tunability is close to 32% at -195°C, and the highest FOM value is close to 75 around the Curie temperature, as shown in Figure 33(a). The tunability of BST(50/50)-40%MZO(90/10) is very stable except for a sharp variation around the Curie temperature. The tunability is between 5.2% and 6.2% at room temperature, and the FOM value is between 12 and 13.9. The highest value of the tunability is close to 28%, below-100°C, and the highest FOM value is close to 65 around the Curie temperature, as shown in Figure 33(b). It is seen that the tunability and FOM of BST(50/50)-40%MZO(90/10) have improved compared with the pure BST(50/50) ceramic. Considering the applications in the microwave devices, a tunability larger than 10% is preferred. The tunability of BST(50/50)-40%MZO(90/10) is only 5.2~6.2% at room temperature (20~50°C), although it shows improved dielectric properties. Another composition, $Ba_{0.75}Sr_{0.25}TiO_3$ [BST(75/25)], has been chosen as the core material because its Curie temperature is closer to room temperature.

The temperature dependence of the dielectric constant and loss of the BST(75/25) ceramic is shown in Figure 34. The sample has been sintered at 1250oC for 2 h and measured at 100 kHz from 20oC to 120oC. The Curie temperature is near 55oC, and the loss is about 0.03 around the Curie temperature. The tunability is between 3.40% and 37.47% in the whole range of test temperatures with the highest value of 37.47% at 47.36oC, as shown in Figure 34(c). It increases from the lowest value of 3.43% at 20.25oC to 37.47% at 47.36 oC. The FOM value is between 1.75 and 36.41, and the largest value is 36.41 around 72.95oC. The FOM value is only ~7.5 at 30oC because of the high dielectric loss of ~0.024.

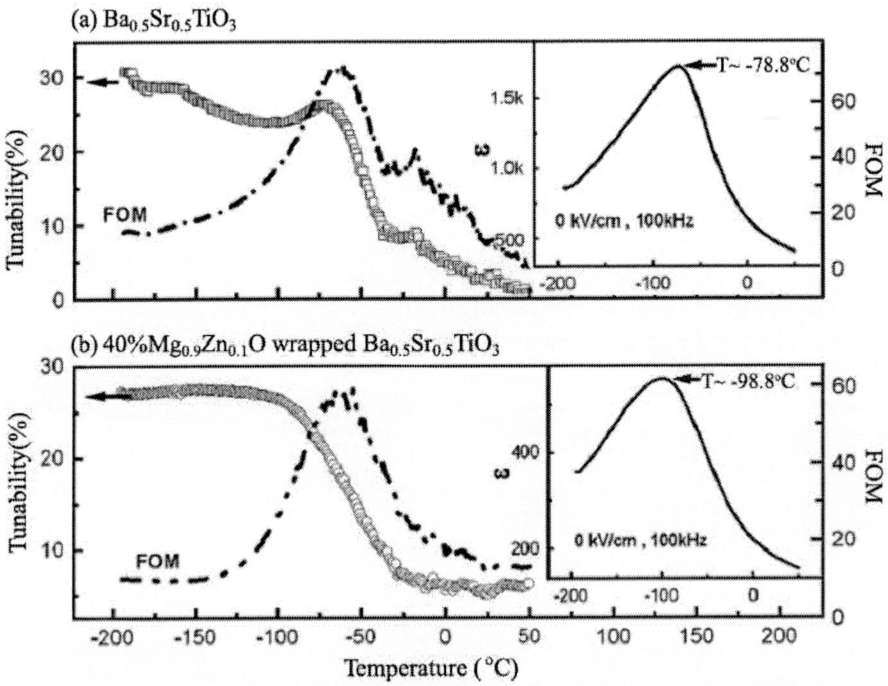

Figure 33. The tunability and the figure of merit as functions of measuring temperature. The insets show that the Curie temperatures are -78.8°C for BST(50/50) and -98.8°C for BST(50/50)-40%MZO(90/10), respectively [14].

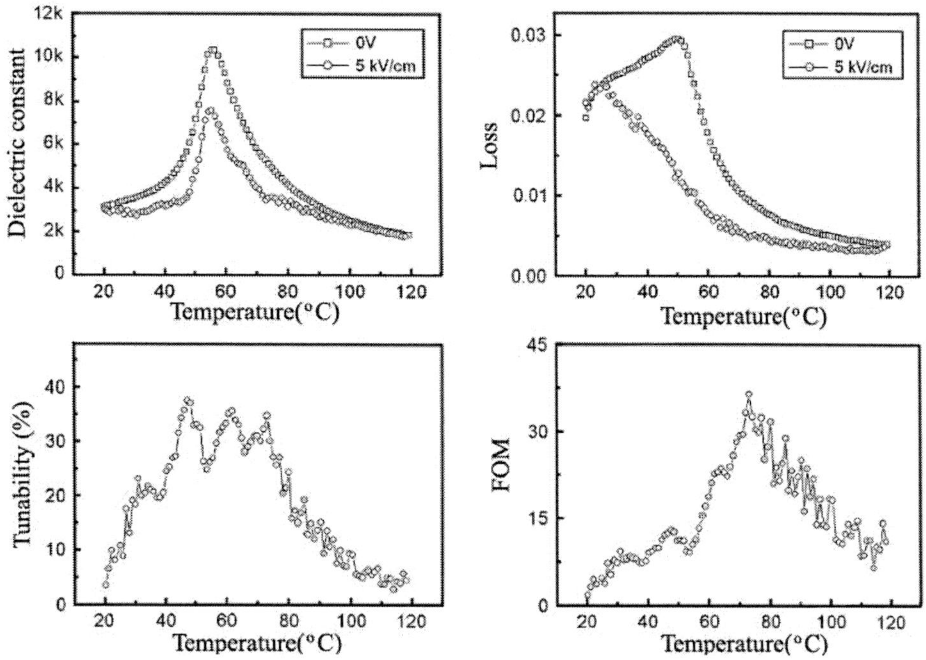

Figure 34. The dielectric properties of BST(75/25) ceramic sintered at 1250°C for 2 h.[14].

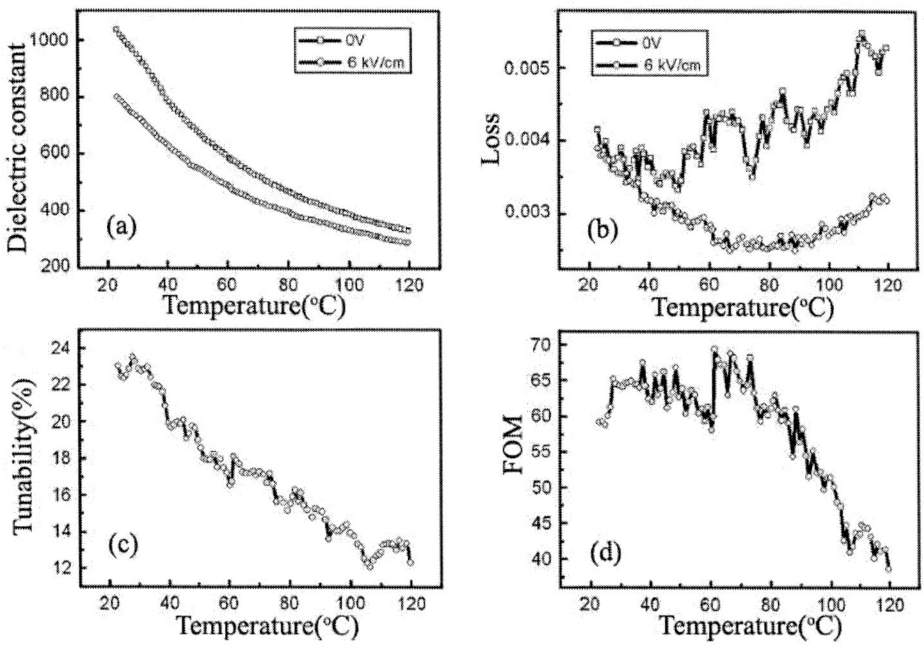

Figure 35. The dielectric properties of 40%$Mg_{0.9}Zn_{0.1}O$ wrapped $Ba_{0.75}Sr_{0.25}TiO_3$ ceramic sintered at 1350°C for 2h.[14].

The temperature dependence of the dielectric constant and loss of BST(75/25)-40%MZO(90/10) ceramic sintered at 1350oC for 2 h and measured at 100 kHz from 20oC to 120oC is shown in Figure 35. The dielectric constant decreases as the temperature increases, showing an obvious nonlinear dependence with regard to temperature. This meets an exponential decay law. For the sample without a bias:

$$\varepsilon=1370.78 \exp(-T/44.03)+243.97 \quad (14)$$

and for the sample with a bias of 5 kV/cm:

$$\varepsilon=961.62 \exp(-T/48.45)+210.20 \quad (15).$$

Here T is test temperature. As shown in Figure 35(b), the dielectric loss decreases significantly compared with that of pure BST(75/25), being between 0.002 and 0.005. The tunability has a monotonic decrease from 23.49% at 20 °C to 12.12% at 120 °C.

It is obvious that the Curie temperature of BST(75/25)-40%MZO shifts to a lower temperature than 25°C and cannot be observed in the measured temperature range. The FOM value is relatively stable between 60 and 70 from 20 °C to 80 °C, and decreases to 38.66 at 120 °C. High-frequency dielectric properties have also been measured by an HP 4291B (1 MHz ~ 1.86 GHz) analyzer at room temperature. The dielectric constant and loss of MZO(90/10) wrapped-BST and pure BST have been characterized and listed in Table 6.

Table 6. Room-temperature dielectric constant and loss at 100 MHz and 10 kHz

Samples	100MHz		10kHz	
	ε	Tanδ(%)	ε	Tanδ(%)
BST(50/50)	1994	2.0	1748	0.6
BST(50/50)-20%MZO	458	1.3	408	3.5
BST(50/50)-40%MZO	203	0.5	181	4.2
BST(75/25)	4655	8.0	4090	4.0
BST(75/25)-5%MZO	3220	1.2	2937	0.2
BST(75/25)-40%MZO	850	0.3	1001	0.4

It can be seen that the tunability and FOM value of the MZO wrapped-BST ceramics have been improved. These results show that BST-based composite ceramics derived from the core-shell structured nanoparticles can improve the dielectric properties, making it possible for use in tunable microwave device applications.

SUMMARY

In summary, perovskite nanoparticles such as BaTiO3, SrTiO3, BaZrO3 and their solid solutions can be fabricated using DSS near room temperature. When synthesized from solution directly, hydrogen interstitials will be introduced into the lattice of perovskite nanoparticles and carry positive charges, and they adsorb the hydroxyl group with negative charges on the surface of the nanoparticles. Because hydroxyl groups adsorbed on the surface of nanoparticles can be ionized freely, this kind of DSS nanoparticles exhibits an alkalescent property. When this kind of perovskite nanopaticles disperses in a certain solution of inorganic salt such as Mg(NO3)2•6H2O, Zn(NO3)2•6H2O, Al(NO3)3•9H2O, Y(NO3)3•6H2O etc., the hydroxide related to these salts spontaneously deposit on the surface of the nanoparticles, as a result, a perovskite structured nanoscaled core materials are successfully self-wrapped by a non-ferroelectric hydroxide derived from a nitrate salt solution. After annealing at high temperature, core-shell structured perovskite nanoparticles with heterogeneous oxide shells can be synthesized.

Because the sintering behavior are dominated by the inert oxide shells such as MgO and they often have high melting point, the core-shell structured perovskite nanoparticles with heterogeneous oxide shells are very difficult to be sintered directly at a low temperature like perovskite nanoparticles. Another reason of making the core-shell structured materials difficult to be sintered is that nanoparticles exert a pinning effect during sintering and thus the mass transfer can be suppressed from shell to shell. Thus it is difficult to obtain their dense ceramics using conventional method. When the core-shell structured perovskite nanoparticles with heterogeneous oxide shells are composited with doped shell oxide materials, the sintering temperature are decreased and pinning effect during sintering can be avoided. A good example is, when the core-shell structured MgO wrapped BST nanoparticles are composited with MZO powders, the mass transfer proceed from MZO to the shell MgO (or MZO) of BST nanoparticles. Thus, dense ceramics can be obtained.

From XRD analysis, the lattice cell parameters and crystallite size and the variation of lattice parameter can be estimated. When hydrogen interstitials are introduced into perovskite lattice, the lattice cell of the powders will expand and change the crystal symmetry. The lattice mismatch between core and shell causes the variation and shrinkage of lattice cell parameter of the perovskite core. The details of lattice match of shell and core can be observed and analyzed by HRTEM and SAED. In our ceramics of MZO wrapped BST nanoparticles, the core-shell structure probably follows the second match mode (Figure 31) because the lattice cell parameter of BST nanoparticles decreases after being wrapped by MZO. The morphology of ceramics can be observed by FE-SEM. Un-grown BST nanoparticles can also be observed in fractured grains of the ceramics of core-shell structured MZO wrapped BST nanoparticles.

The Curie temperatures of BST-MZO composite ceramics shift to a lower temperature compared with that of pure BST ceramics. The dielectric constant and loss decrease sharply depending on the percentage of the shell material (MZO) in the composites. The tunability and FOM values have been improved significantly compared with that of pure BST ceramics. It is thus seen that the ceramics of MZO wrapped BST nanoparicles are promising materials for tunable microwave device applications and high energy density ceramic capacitors for their distinct electric loss improvement.

REFERENCES

[1] J. Li, H. Zeng, S. H. Sun, J. P. Liu, Z. L. Wang, *J. Phys. Chem. B*, 108, 14005(2004).

[2] M. Danek, K. F. Jensen, C. B. Murray, M. G. Bawendi, *Chem. Mater.*, 8, 173(1996).

[3] X. Peng, M. C. Schlamp, A. V. Kadavanich, A. P. Alivisatos, *J. Am. Chem. Soc.*, 119, 7019(1997).

[4] M. A. Malik, P. O'Brien, N. Revaprasadu, *Chem. Mater.*, 14, 2004(2002).

[5] P. Reiss, J. Bleuse, A. Pron, *Nano Lett.*, 2, 781(2002).
[6] V. Skumryev, S. Stoyanov, Y. Zhang, G. Hadjipanayis, D. Givord, J. Nogues, *Nature*, 423, 850(2003).
[7] H. Zeng, J. Li, Z. L. Wang, J. P. Liu, S. H. Sun, *Nano Lett.*, 4, 187(2004).
[8] N. Kakuta, K. H. Park, M. F. Finlayson, A. Veno, A. J. Bard, A. Campion, M. A. Fox, S. E. Webber, J. M. White, *J. Phys. Chem.*, 89, 3828(1985).
[9] C. Nasr, S. Hotchandani, W. Y. Kim, R. H. Schmehl, P. V. Kamat, *J. Phys. Chem. B*, , 101, 7480(1997).
[10] S. S. Davis, *Trends. Biotechnol.*, 15, 217(1997).
[11] V. Suryanarayanan, N. A. Sreekumaran, R. T. Tom, T. Pradeep, *J. Mater. Chem.*, 14, 2661(2004).
[12] I. Pastoriza-Santos, D. S. Koktysh, A. A. Mamedov, M. Giersig, N. A. Kotov, L. M. Liz-Marzan, *Langmuir* 16, 2731(2000).
[13] M. E. Lines, A. M. Glass, *Principles and Applications of Ferroelectrics and Related Materials*, Oxford University Press, Oxford, 2001.
[14] H. Y. Tian, J. Q. Qi, Y. Wang, H. L. W. Chan, C. L. Choy, *Progress in Solid State Chemistry* 33, 207(2005)
[15] H. Y. Tian, J. Q. Qi, Y. Wang, J. Wang, H. L. W. Chan, C. L. Choy, *Nanotechnology*, 16, 47(2005).
[16] J. J. Urban, W. S. Yun, Q. Gu, H. K. Park, *J. Am. Chem. Soc.*, 124, 1186 (2002).
[17] S. O'Brien, L. Brus, C. B. Murray, *J. Am. Chem. Soc.*, 123, 12085 (2001).
[18] B. A. Hernandez, K. S. Chang, E. R. Fisher, P. K. Dorhout, *Chem. Mater.*, 14, 480 (2002).
[19] L. A. Perez-Maqueda, M. J. Dianez, F. J. Gotor, M. J. Sayagues, C. Real, J. M. Criado, *J. Mater. Chem.*, 13, 2234 (2003).
[20] J. J. Urban, J. E. Spanier, L. Ouyang, W. S. Yun, H. K. Park, *Adv. Mater.*, 15, 423 (2003).
[21] M. Niederberger, N. Pinna, J. Polleux, M. Antonietti, R. *Angew. Chem., Int. Ed.*, 43, 2270(2004).
[22] J. Q. Qi, "Direct Synthesis Perovskite Nanoparticles from Solution", Editor: George V. Karas, "*New Developments in Crystal Growth Research*", 2006, ISBN: 1-59454-539-1, Nova Science Publishers, NY
[23] J. Q. Qi, L. T. Li, Y. L. Wang, Z. L. Gui, *J. Crys. Growth*, 260, 551(2004).
[24] J. Q. Qi, Y. Wang, W. P. Chen, L. T. Li, H. L. W. Chan, *J. Sol. State Chem.*, 178, 279(2005).
[25] J. Q. Qi, Y. Wang, W. P. Chen, L. T. Li, H. L. W. Chan, *J. Nanoparticle Res.* 8, 959(2006).
[26] U. Weber, G. Greuel, U. Boettger, *J. Am. Ceram. Soc.*, 84, 759 (2001).
[27] J. Q. Qi, W. P. Chen, Y. Wang, H. L. W. Chan, L. T. Li, *J. Appl. Phys.*, 96, 6937(2004).
[28] J. Tennery, R. L. Cook, *J. Am. Ceram. Soc.*, 44, 187(1961).
[29] Q. Li, J. Q. Qi, Y. L. Wang, Z. L. Gui, L. T. Li, *J. Eu. Ceram. Soc*, 21, 2217(2001).
[30] W. Li, J. Q. Qi, Y. L. Wang, L. T. Li, Z. L. Gui, *Mater. Lett.*, 57,1(2002).
[31] J. Q. Qi, Z. L. Gui, W. Li, Y. L. Wang, Y. J. Wu, L. T. Li, *Mater. Lett.*, 56, 507(2002).
[32] P. W. Haayman, R. W. Dam, H. A. Klasens, Method of Producing a Semiconding Materials, German Pat. No. 929,350, June 23, 1955.
[33] S. Venigalla , *Am. Ceram. Soc. Bull.*, 6, 63(2001).

[34] M. Viviani, M. T. Buscaglia, A. Testino, V. Buscaglia, P. Bowen, P. Nanni, *J. Eu. Ceram. Soc.* 23, 1383(2003).
[35] R. T. Kristóf, *Építôanyag* 58, 34(2006).
[36] P. E. C. Franken, M. P. A. Viegers, A. P. Gehring, *J. Am. Ceram. Soc.* 64, 687(1981).
[37] P. A. Fang, H. Gu, H. Shen, *J. Eu. Ceram. Soc.* 24, 2509(2004).
[38] J. Y. Li, S. T. Li, F. Y. Liu, *J. Mater. Sci.-Mater. in Electron.* 14, 483(2003).
[39] V. Hornebecq, C. Huber, M. Maglione, *Adv. Funct. Mater.* 14, 899(2004).
[40] J. Q. Qi, W. P. Chen, Y. J. Wu, L. T. Li, *J. Am. Ceram. Soc.*, 81, 437(1998).
[41] S. R. Summerfelt, in Ferroelectric Thin Films, edited by R. Ramesh (Kluwer Academic, Netherlands, 1997), Chap. 1, p. 1.
[42] Kingon, S. K. Streiffer, C. Basceri, S. R. Summerfelt, *Mater. Res. Bull.* 21, 18(1995).
[43] J. M. Ponds, S. W. Kirchoefer, W. Chang, J. S. Horwitz, and D. B. Chrisey, *Integr. Ferroelectr.* 22, 317(1998).
[44] F. A. Miranda, F. W. Van Keuls, R. R. Romanofsky, and G. Subramanyam, *Integr. Ferroelectr.* 22, 269(1998)
[45] C. S. Hwang, *Mater. Sci. Eng. B,* 56, 178 (1998).
[46] W. J. Kim, W. Chang, S. B. Qadri, J. M. Pong, S. W. Kirchoefer, D. B. Chrisey, J. S. Honwitz, *Appl. Phys. Lett.*, 76, 1185 (2000).
[47] J. B. L. Rao, D. P. Patel, V. Krichevsky, *IEEE Trans Annte. Prog.*, 47, 458 (1999).
[48] P. Padmini, T. R. Taylor, M. J. Lefevre, A. S. Nagra, R. A. York, J. S. Speck, *Appl. Phys. Lett.*, 75, 3186 (1999).
[49] Y. Gim, T. Hudson, C. Kwon, A. T. Findikoglu, B. J. Gibbons, B. H. Park, Q. X. Jia, *Appl. Phys. Lett.*, 77, 1200 (2000).
[50] Kingon, J. P. Maria, S. K. Streiffer, *Nature*, 406, 1032 (2000).
[51] E. Ngo, P. C. Joshi, M. W. Cole, C. W. Hubbard, *Appl. Phys. Lett.*, 79, 248 (2001).
[52] Landolt-Bornstein Numerical Data and Functional Relationship in Science and Technology, Springer-Verlag Berlin 1981.
[53] T. R. Armstrong, L. E. Morgens, A. K. Maurice, and R. C. Buchanan, *J. Am. Ceram. Soc.* 72, 605 (1989).
[54] D. M. Tahan, A. Safari, and L. C. Klein, *J. Am. Ceram. Soc.* 79, 1593 (1996).
[55] N. Kamehara, M. Tsukada, J. S. Cross, and K. Kurihara, *J. Ceram. Soc. Jpn.* 105, 801 (1997).
[56] J. Q. Qi, Z. L. Gui, Y. L. Wang, Q. Li, L. T. Li, *J. Mater. Sci. Lett.*, 21, 405(2002).
[57] D. Hennings, A. Schnell, and G. Simon, *J. Am. Ceram. Soc.* 65, 539 (1982).
[58] H. Shizuno, S. Kusumi, H. Saito, and H. Kishi, *Jpn. J. Appl. Phys.* 32, 4380 (1993).
[59] Z. Yu, C. Ang, R. Guo, and A.S. Bhalla, *Appl. Phys. Lett.*, 81, 1285(2002).
[60] M. Veith, S. Mathur, N. Lecerf, V. Huch, and T. Decker, *J. Sol-gel Sci Tech.*, 17,145(2000).
[61] T. Tohma, H. Masumoto, and T. Goto, *Jpn. J. Appl. Phys.*, 41[11B] 6643(2002).
[62] A. Outzourhit, J. U. Trefny, T. B. Kito Yarar, *J. Mater. Res.* 10, 6(1995).
[63] X. F. Liang, W. B. Wu, Z. Y. Meng *Mater. Sci. and Eng. B* 99, 366(2003).
[64] W. Chang, L. Sengupta, *J. Appl. Phys.,* 92, 3941(2002).
[65] C. Huber, M. Treguer-Delapierre, C. Elissalde, F. Weill, M. Maglione, *Ferroelctrics* 294 13(2003).
[66] C. Huber, M. Treguer-Delapierre, C. Elissalde, F. Weill, M. Maglione, *J. Mater. Chem.* 13 650(2003).

[67] K. B. Chong, L. B. Kong, L. F. Chen, L. Yan, C. Y. Tan, C. K. Ong, T. Osipowicz, *J Appl. Phys.,* 95, 1416(2004).
[68] L. H. Hench, J. K. West, *Chem. Rev*. 90, 33(1990).
[69] H. P. Baldus, M. Jansen, *Angew. Chem.,* 109, 338(1997).
[70] S. Wada, H. Chikamori, T. Noma, T. Suzuki, *J Mater. Sci. Lett.* 19, 245(2000).
[71] W. E. Rhine, R. B. Hallock, W. M. Davis, W. Wong-Ng, *Chem. Mater*. 4, 1208(1992).
[72] J. F. Bocquet, K. Chhor, C. Pommier, *Mater. Chem. Phys*., 57, 273(1999).
[73] C. H. Chen, W. G. Zhu, T. Yu, X. F. Chen, X. Yao, R. G. Krishman, *Surf. and Coating Technology* 167, 245(2003).
[74] K. Nakamoto, 1986 *Infrared and Raman Spectra of Inorganic and Coordination Compounds* (Wiley, New York) p106.
[75] J. B. Liu, H. Wang, Y. D. Hou, M. K. Zhu, H. Yan, M. Yoshimura, *Nanotechnology,* 15, 777(2004).
[76] C. Kim, *Phys. Rev. B*, 66, 113404 (2002).
[77] G. J. Fang, *Phys. Stat. Sol.,* 2, 200(2003).
[78] M. Rand, J. I. Langford, A. Drake, J. S. Abell, *Cryogenics,* 33, 291(1993).
[79] D. Balzar, H. Ledbetter, *Advances in X-ray Analysis*, 38, 397(1995).
[80] B. E.Warren, B. L. Averbach, *J. Appl. Phys.,* 21, 595(1950).
[81] J. Q. Qi, H. Y. Tian, Y. Wang, G. K. H. Pang, L. T. Li, H. L. W. Chan, *J.Phys. Chem. B*, 109, 14006 (2005).
[82] X. X. Xi, H. C. Li, W. D. Si, A. A. Sirenko, I. A. Akimov, J. R. Fox, *J. Electroceram.,* 4, 393(2000).

In: Nanotechnology Research Developments
Editor: R. Jiménez-Contreras, pp. 137-171
ISBN: 978-1-60021-899-6

Chapter 3

THERMODYNAMIC AND KINETIC THEORIES OF NANOWIRE GROWTH

G. W. Yang
State Key Laboratory of Optoelectronic Materials and Technologies,
Institute of Optoelectronic and Functional Composite Materials,
School of Physics Science and Engineering, Zhongshan University,
Guangzhou 510275, China

ABSTRACT

The universal models of nucleation thermodynamics and growth kinetics have been established for nanowire growth upon metal-catalyst-assisted thermal chemical vapor transport (CCVT) on the basis of vapor-liquid-solid (VLS) mechanism, with respect to the size-effects induced by the curvature of critical nuclei and nanowire in the nucleation and growth. Based on the proposed thermodynamic and kinetic analyses on nanoscale, the thermodynamic and kinetic size limit of the nanowire growth was addressed. It was found that the critical size of the nanowire nucleation is suggested to be the thermodynamic size limit, i.e., the theoretical minimum size of the nanowire grown by CCVT both thermodynamically and kinetically, and the size of the catalyst droplet could limit the growing size, i.e., the theoretical possible or maximum size of the nanowire to a certain extent in kinetically. For the issue of catalyst nanoparticles on nanowires tip or substrate, a nucleation thermodynamic criterion and a diffusion kinetic criterion were deduced for the selection growth of catalyst nanoparticles on nanowires tip or substrate upon CCVT. Importantly, theoretical predictions on the basis of the proposed models are in well agreement with the experimental data. Additionally, a growth kinetic approach was developed to theoretically elucidate the growth of one-dimensional nanostructures inside carbon nanotubes upon vapor deposition. Theoretical analysis showed that the growth rate of one-dimensional nanostructures would go much higher once nuclei formed inside carbon nanotubes, due to the effect of surface tension induced by the nano-sized curvature of carbon nanotubes. Theoretical results are in well agreement with the experimental data. The size-dependence of the shape evolution of nanowires such as SnO2 was pursued in theoretical. It was found that the thermodynamic driving force, the Gibbs free energy difference between two phases, is always responsible for the shape choice in the different growth stages such as nucleation, growth, and structural transition of 1D nanostructures, implying that the thermodynamic driving force is one of the physical origins causing the interesting morphological transformation in the growth of 1D nanostructures.

1. INTRODUCTION

One-dimensional (1D) nanostructures such as wires, rods, belts, and tubes have become the focus of intensive research owing to their unique applications in mesoscopic physics and fabrication of nanoscaled devices. They not only provide a good system to study the electrical and thermal transport in one-dimensional confinement, but also are expected to play an important role in both interconnection and functional units in fabricating electronic, optoelectronic, and magnetic storage devices with nanoscale dimension [1]. Therefore, to attain various nanometer sized building blocks, a lot of self-assembly and synthesis processes have emerged in recent years [2]. Importantly, these assemblies and syntheses at the nanometer scale have revealed many unusual thermodynamic and kinetic behaviors of microphase growth [3-11], which plays an important role in the development of the classical thermodynamics [12]. A big challenge is how to flexibly control the self-assembly and synthesis processes of 1D nanostructure for these purposes of fundamental researches and potential applications. Basically, self-assembly processes require atomic or clustering interactions as the thermodynamic and kinetic driving forces to organize those atoms or clusters to form nanosized domain patterns [13]. Therefore, in order to control the growth of nanostructures, it is essential to clarify the thermodynamic and kinetic driving forces of the self-assembly process of nanostructures and develop new theoretical tools to address the nano-sized fabrications [14]. For this issue, in theoretical, we have developed a series of thermodynamic and kinetic tools to pursue the nucleation and growth of 1D nanostructures upon metal-catalyst-assisted thermal chemical vapor transport (CCVT) on the basis of vapor-liquid-solid (VLS) mechanism, with respect to the size-effects induced by the curvature of critical nuclei and nanowire in the nucleation and growth.

It is well known that CCVT has been one of leading methods to synthesize most semiconductor 1D nanostructures [2], and the process of VLS has been generally recognized to be the fundamental understanding of CCVT [15-16]. However, the VLS process is just an outline for the formation of 1D nanostructure upon CCVT, and many crucially physical and chemical issues involved in the nucleation and growth of nanowire have not been addressed in the present securable literatures, up to now. For instance, the nucleation position of nanowire on the surface of catalyst particles, the size limit of nanowire growth, the growth behavior of nanowire confined a nanotube, the size-dependence of nanowire growth rate, and the size-dependent shape evolution during nanowire growing. Importantly, these basic physical and chemical processes play key roles in controlling the fabrications of 1D nanostructure.

In this review chapter, we introduced many new theoretical tools to address the nucleation and growth of nanowire upon CCVT. The thermodynamic and kinetic theories of nanowire nucleation were described in section 2, in which thermodynamic and kinetic criteria are established to determine the selection nucleation of nanowire on the surface of catalyst. In section 3, we introduced the thermodynamic and kinetic theory of nanowire growth, which predicts the size limit of nanowire upon CCVT. Usually, the anomalous growth behavior would happen when a nanowire grows inside a carbon nantube. Therefore, a growth kinetic model was established to elucidate the formation of the nanowire confined nanotubes in section 4. The size-dependent growth behaviors of nanowire are theoretically pursued in section 5, in which a kinetic model of size-dependent growth rate of nanowire was introduced, and the thermodynamic and kinetic aspects of shape evolution of 1D nanostructure were addressed. Additionally, some remarked conclusions are summarized in section 6.

2. Thermodynamic and Kinetic Theory of Nanowire Nucleation

Based on the experimental data, it is not difficult to find that there mainly exist two general models of nanowire formations on substrates upon VLS process. One is that catalyst nanoparticles are on the bottom of nanowires, and another one that catalyst nanoparticles are on the tip of nanowires [1,17-21]. However, which physical and chemical origins result in two inverse growths of nanowires on catalyst nanoparticles? In other words, why would catalyst nanoparticles be on the tip of nanowires these times and on the bottom those times? To our knowledge, few studies have been reported to study the issue mentioned above [1,22]. We proposed the two universal theoretical criteria for the selection growth of nanowires upon CCVD in thermodynamical and kinetic, respectively, on the basis of the thermodynamical theory on nanoscale [10]. Importantly, taking Si nanowires as an example, the theoretical predictions are well consistent with experimental data. Noted that the thermodynamical theory on nanoscale (the capillary approximation) is valid when it is used to address the microphase growth, in which sizes of clusters are usually in the range of a couple of nanometers [8-10].

2.1. Thermodynamic Criteria of Selection Nucleation

Generally, the Gibbs free energy is an adaptable measure of the stability of a state in phase transition formation. Thermodynamically, the phase transformation is promoted by the difference of the free energy. Gibbs free energy of a phase can be expressed as a function of the pressure-temperature condition, and determined by a general coordinate or reactive coordinate [24]. Therefore, we will propose a thermodynamical criterion for the selection growth of nanowires by comparing the Gibbs free energy difference of the nanowire nucleation in two cases. The Gibbs free energy difference of a cluster can be expressed as

$$\Delta G = \Delta G_1 + \Delta G_2 \tag{2-1}$$

where the two terms on the right-hand side are the energy differences from surface and volume. According to the nucleation thermodynamics [24], the Gibbs energy difference aroused from the formation of spherical clusters in the gas phase can be expressed as follows [25]

$$\Delta G(r) = V\times\Delta g_v + A\sigma_{nv} \tag{2-2}$$

where σ_{nv}, V, and A are the interface free energy density of nucleus-vapor, volume, and area of spherical clusters, respectively, and Δg_v is the Gibbs free energy difference per unit volume, which can be expressed by $-RT/V_m \ln(P/P_e)$, where P, T, R, and V_m are the pressure, the temperature, the gas constant, and the mole volume of the nanowire, respectively. P_e is the vapor phase pressure in thermal equilibrium coexistence with the liquid of composition in a flat surface. Based on Tan's model [26] when the droplet is assumed to be an idea solution, Δg_v can be expressed as

$$\Delta g_v = -\frac{RT}{V_m}\ln\frac{p}{p_e} = -\frac{RT}{V_m}\ln\left(\frac{C}{C^{eq}}\right) \tag{2-3}$$

where C and Ceq are X concentrations X means nanowires materials_ on the solid and liquid lines of the eutectic catalyst-X phase diagram, respectively. Based on the Kelvin equation, taking the capillary effect of the nanosized curvature of the nanowire nuclei on the Gibbs free energy of the nucleation into account, Δg_v can be defined as

$$\Delta g_v = -\frac{1}{2}\left[\frac{RT}{V_m}\ln\left(\frac{C}{C^{eq}}\right) + \frac{2\sigma_{nv}}{r}\right] = \frac{1}{2}\Delta g_v^0 - \frac{\sigma_{nv}}{r} \tag{2-4}$$

where $\Delta g_v 0$ is defined as

$$\Delta g_v^0 = -\frac{RT}{V_m}\ln\left(\frac{C}{C^{eq}}\right) \tag{2-5}$$

The schematic illustration of the X nanowire growths by two models upon CCVT is shown in Figure 1. We will calculate the Gibbs free energies of the nanowire nucleation in two cases on the basis of the thermodynamic approach on nanoscale.

For the first case, i.e., the nanowire nucleation takes place on the surface of catalyst nanoparticles as shown in Figure 2 (a). When an atom cluster condensed with catalyst being on the substrate, the surface energy $\Delta G1_S$ can be calculated as follows

$$\Delta G_{1S} = 2\sigma_{nv}\pi r_1^2(1-\cos\theta_1) + \pi r_1^2\sin^2\theta_1(\sigma_{cn}-\sigma_{cv}) \tag{2-6}$$

where $r1$ is the radius of the cluster, σ_{cn} and σ_{cv} are the interface energies of the catalyst-nucleus, the catalyst-vapor, respectively, and $\theta 1$ is the contact angle between the catalyst and nucleus. When the interface between nanowire and catalyst liquid droplet is assumed to be incoherent interface, the values of σ_{cn} can be denoted as

$$\sigma_{cn} = \frac{\sigma_{cv}+\sigma_{nv}}{2} \tag{2-7}$$

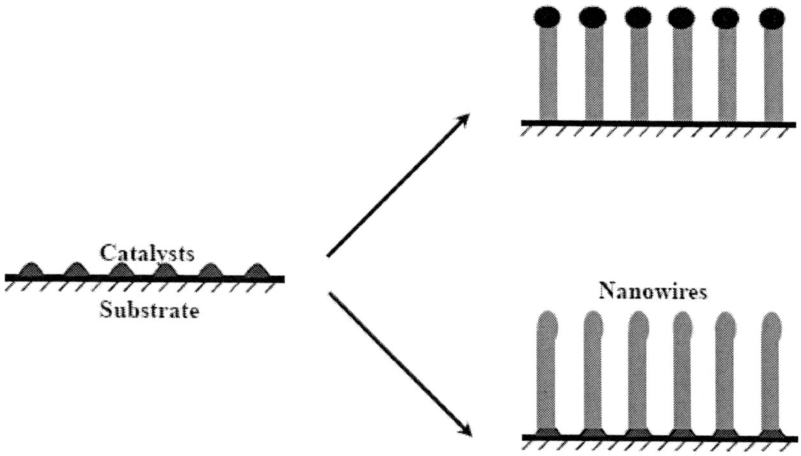

Figure 1. The selection growth of nanowires: catalyst nanoparticles on the tip and bottom of nanowires.

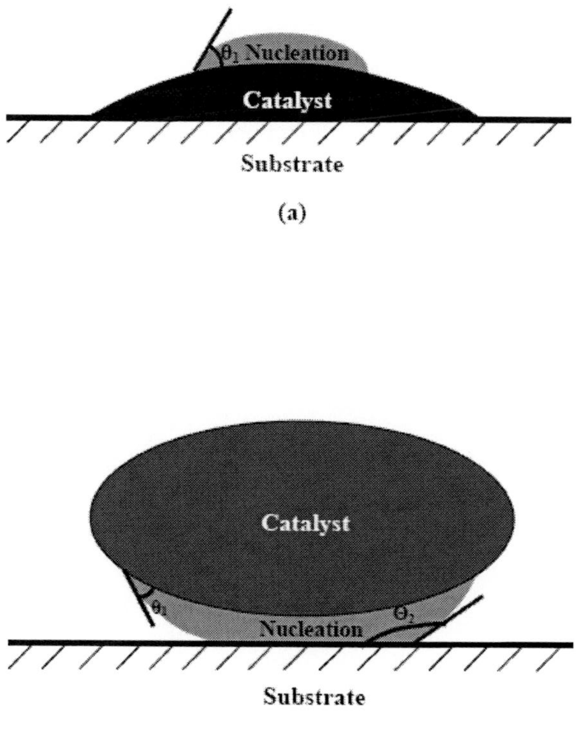

Figure 2. The schematic illustration of the nanowire nucleation by two models upon CCVD: nucleation on the surface of catalyst nanoparticles (a) and at the interface between catalyst nanoparticles and substrate (b).

The volume energy $\Delta G2_S$ can be given by

$$\Delta G_{2S} = \frac{4\pi r_1^3 \Delta g_v}{3} \frac{2 - 3\cos\theta_1 + \cos^3\theta_1}{4} \tag{2-8}$$

In this way, we obtain the Gibbs free energy can be expressed as

$$\Delta G_S = 2\sigma_{nv}\pi r_1^2(1-\cos\theta_1) + \pi r_1^2 \sin_2\theta_1(\sigma_{cn}-\sigma_{cv}) + \frac{4\pi r_1^3\Delta g_v}{3}\frac{2-3\cos\theta_1+\cos\theta_1^3}{4} \quad (2\text{-}9)$$

According to the balance condition of forces, we have $\sigma_{cv} = \sigma_{nv}\cos\theta 1 + \sigma_{cn}$. Thus, the Gibbs free energy can be expressed as

$$\Delta G_S = 2\sigma_{nv}\pi r_1^2(1-\cos\theta_1) - \pi r_1^2 \sin^2\theta_1\,\sigma_{nv}\cos\theta_1 + \frac{4\pi r_1^3\Delta g_v}{3}\frac{2-3\cos\theta_1+\cos\theta_1^3}{4} \quad (2\text{-}10)$$

Based on the condition of the critical nucleus formation, $\partial\Delta G_s/\partial r1=0$, when the atomic cluster transforms into the critical nucleus, the radius of the critical nucleus can be calculated to be

$$r_1^* = -\frac{8}{3}\frac{\sigma_{nv}}{\Delta g_v^0} \quad (2\text{-}11)$$

Accordingly, the formation energy $\Delta G_s^*(r1^*)$ of the critical nucleus can be given by

$$\Delta G_s^* = \left(\frac{256\pi\sigma_{nv}^3}{9(\Delta g_v^0)^2} + \frac{2048\pi\sigma_{nv}^3}{81(\Delta g_v^0)^3}\right)\left(\frac{2-3\cos\theta_1+\cos^3\theta_1}{4}\right) \quad (2\text{-}12)$$

For the second case, i.e., the nanowire nucleation takes place on the interface between the catalyst nanoparticle and the substrate as shown in Figure 2 (b). The surface energy $\Delta G1_T$ can be calculated as

$$\Delta G_{1T} = -2\sigma_{nv}\pi r_2^2(\cos\theta_1+\cos\theta_2) - \pi r_2^2\sin\theta_1^2\cos\theta_1\sigma_{nv} - \pi r_2^2\sin_2^2\cos\theta_2\sigma_{nv}$$
$$= -\pi r_2^2\sigma_{nv}f_1(\theta) \quad (2\text{-}13)$$

where r_2 is the radius of the cluster, σ_{nv} is the interface energy of the nucleus-vapor, and θ1 and θ2 are the contact angles of catalyst-nucleus and nucleus-substrate, respectively. In addition, $f1(\theta)$ is defined as 2(cos θ1+cosθ2)+sin2θ1cosθ1+sin2θ2cosθ2. The volume Gibbs free energy ΔG_{2T} can be given by

$$\Delta G_{2T} = \frac{4\pi r_2^3\Delta g_v}{3}\frac{\cos^3\theta_1+\cos^3\theta_2-3\cos\theta_1-3\cos\theta_2}{4} \quad (2\text{-}14)$$

$$= \frac{4\pi r_2^3 \Delta g_v}{3} f_2(\theta)$$

where $f2(\theta)$ is defined as

$$f_2(\theta) = (\frac{\cos^3\theta_1 + \cos^3\theta_2 - 3\cos\theta_1 - 3\cos\theta_2}{4} \tag{2-15}$$

Summing the surface energy and the volume one, the whole Gibbs free energy difference can be obtained. Similarly, based on the condition of the critical nucleus formation, $\partial\Delta G_T / \partial r2 = 0$, when the atomic cluster transforms into the critical nucleus, the radius of the critical nucleus can be calculated to be

$$r_2^* = \frac{\sigma_{nv}}{2\Delta g_v^0}\left(\frac{2f_1(\theta)}{f_2(\theta)} + \frac{8}{3}\right) \tag{2-16}$$

Accordingly, the formation energy $\Delta G_T*(r2*)$ of the critical nucleus can be expressed as

$$\Delta G_T^* = -\pi\sigma_{nv}\left[\frac{\sigma_{nv}}{2\Delta g_v^0}\left(\frac{2f_1(\theta)}{f_2(\theta)} + \frac{8}{3}\right)\right]^2 f_1(\theta) + \frac{4\pi}{3}\left[\frac{\sigma_{nv}}{2\Delta g_v^0}\left(\frac{2f_1(\theta)}{f_2(\theta)} + \frac{8}{3}\right)\right]^3$$
$$\times\left(\frac{1}{2}\Delta g_v^0 - \frac{2\Delta g_v^0}{\frac{2f_1(\theta)}{f_2(\theta)} + \frac{8}{3}}\right) f_2(\theta) \tag{2-17}$$

From the deduction above, one can see that the critical energy only depends on these interfacial parameters of σ_{cv}, σ_{nv}, and θ2, and nanowire materials-catalyst phase diagram, without any adjustable parameters. Thus, we can obtain that the nanowire nucleation on the surface of catalyst nanoparticles (Figure 2 (a)) would be preferable when $\Delta G_S* > \Delta G_T*$. On the other hand, the nucleation on the interface between catalyst nanoparticles and substrates (Figure 2 (b)) would be preferable when $\Delta G_S* < \Delta G_T*$ from the point of view of thermodynamics. Therefore, we naturally deduce a thermodynamical criterion for the selection growth of nanowires upon CCVD as follows. Catalyst nanoparticles would be on the tip of nanowires when $\Delta G_S* > \Delta G_T*$, and on the bottom when $\Delta G_S* < \Delta G_T*$, thermodynamically. Additionally, from the classic thermodynamical nucleation [24], the value of θ2 shows the interaction between nuclei and substrates, and different interactions between nuclei and substrates would lead different values of θ2. For example, the value of θ2 should be different when substrates are crystalline and amorphous, respectively. Therefore, we could compare ΔG_S* with ΔG_T* based on the thermodynamic criterion above if the value of θ2 could be determined in experimental or theoretical.

2.2. Kinetic Criteria of Selection Nucleation

On the basis of the Louchev's model [27], we suggest the diffusion kinetic mechanism of nucleation and growth of nanowires upon CCVT. However, there exists the distinct difference between the diffusion kinetic mechanism in our case and the Louchev's model, as our model takes the size effect of nanostructures into account, which did not appear in Louchev's work. The schematic illustration of the *X* nanowire nucleation and growth by two models upon CCVT is shown in Figure 3. Two kinds of diffusion processes, surface diffusion and body diffusion, of depositing atoms on the surface of nanoparticles are important for the selection nucleation stage, defining whether the nanoparticle remains at the substrate, or on the tip of the growing *X* nanowires. For the nucleation of nanowires upon CCVT, the surface diffusion of adatoms on the surface of nanoparticles is express by the surface saturation time of t_S, meaning that the increase of the adatom content to the saturation concentration. Thus, the surface diffusion time of order is suggested [27]

$$t_s = \frac{C^{*2} D_b(r,T)}{Q^2} \tag{2-18}$$

where C^* is a concentration threshold triggering the precipitation of the growing atoms directly on the upper surface of nanoparticle. The $D_b(r,T)$ is the size-dependent body diffusion coefficient of atoms, and it can be defined as [28]

$$D_b(r,T) = D_0 \exp\left[\frac{-E(\infty)}{RT} \exp\left(\frac{-2S_{vib}(\infty)}{3R} \frac{1}{\frac{r}{r_0}-1}\right)\right] \tag{2-19}$$

where $D0$, $E(\infty)$, $Svib(\infty)$, r, and $r0$ denote a pre-exponential constant, diffusion activation energy of bulk material, the activation entropy of bulk material, the radius of the nanoparticle, and three times the atomic diameter, respectively. In addition, $Q=[Pc/(2mkBT)1/2$ is the atom flux from gas upon CCVT on a unit area in a unit time, in which m, Pc, k_B, and T are the atom mass, partial pressure of Si, Boltzmann constant, and deposition temperature, respectively. On the other hand, the body diffusion is expressed by the body diffusion time of t_d, meaning that the adatoms penetrate to the nanoparticle base from surface. According to the diffusion theory [29], the diffusion distance of atom l is expressed as $\sqrt{6D_b(r,T)t}$, where t is the diffusion time. Assuming the diameter of the nanoparticle being d, the diffusion distance on the average is $l' = \sqrt[3]{\frac{4}{3}\pi(d/2)^3}$. Therefore, we get the average diffusion time t_d as follows

$$t_d = \frac{1}{6D_b(r,T)} \sqrt[3]{\left[\frac{4\pi}{3}\left(\frac{d}{2}\right)^3\right]^2} \tag{2-20}$$

Definitely, when $t_d >> t_s$ as shown in Figure 3 (a), the nanoparticle surface saturates with atoms much faster than one penetrates to its base, suggesting that the nucleation of adatoms would take place on the upper surface of catalyst nanoparticles, and then provide a nanoscale template for the nanowire growth. In contrast, when $t_d << t_s$ as shown in Figure 3 (b), atoms would penetrate to the base from the upper surface of nanoparticles much faster than adatoms reach the saturation threshold on the upper surface, implying that growing atoms will precipitate at the bottom of nanoparticles, and then result in the nucleation at the interface between the nanoparticle and the substrate. Thus, nanoparticles are lifted in the later nuclei growing and maintained on the nanowire tip.

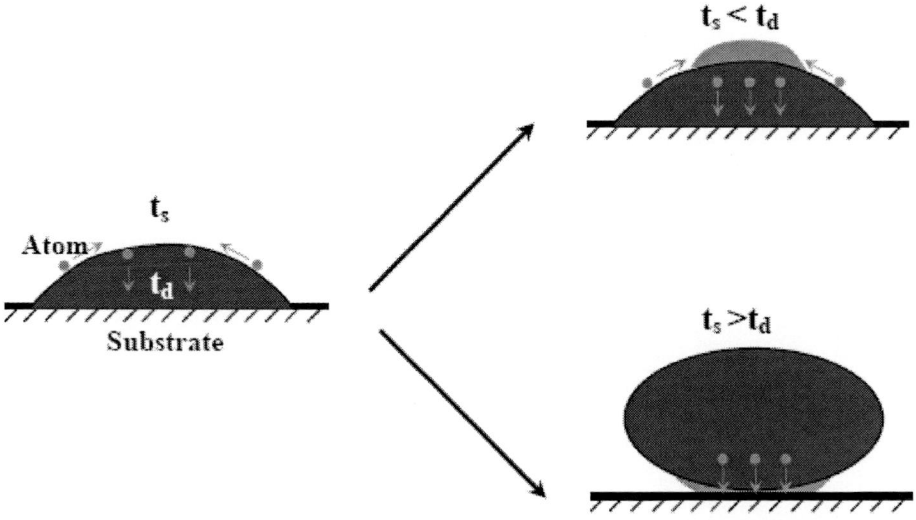

Figure 3. The schematic illustration of the nucleation kinetics of nanowires: the cluster forming on the surface of catalyst nanoparticles (a) when $t_d >> t_s$, and at the interface between catalyst nanoparticles and substrate (b) when $t_d << t_s$.

2.3. Comparison between Theory and Experiment

Note that CCVT on the basis of VLS mechanism is the steady-state process, implying that CCVT could be treated as a near-equilibrium process. In fact, the equilibrium thermodynamics has been validly employed to address the nucleation and growth upon CVD [9], especially, for the growth of 1D nanostructures upon CCVT [25,26]. Accordingly, it is convincing for our case to employ the equilibrium thermodynamic method to deal with the nanowire nucleation upon CCVT. Taking the Si nanowire growth upon CCVT as an example, we will compare the theoretical prediction with experimental data.

For the case of the Si nanowire nucleation on a Si substrate using Fe as catalyst, the typical growth conditions are T=1473 K and Pc=1.4 Torr [30]. The surface energy density of the silicon nanowire is σ_{nv}=1.610 J /m2 [31] and the liquid-catalyst surface energy density σ_{cv} is 0.885 J /m2 [32]. Note that the used Fe-Si phase diagram is from Ref. 33. Therefore, based on Eqs. (8) and (12), we can attain the relationship curves between θ2 and the critical radius of the two cases, as shown in Figure 4. Obviously, one can find that the critical radius $r2^*$ increases quickly with θ2 decreasing when θ2 is less than 80°, and it is close to $r1^*$ when θ2 is larger than 150°. In addition, we can obtain the dependence of θ2 and the critical energy on the basis of Eqs. (9) and (13), as shown in Figure 5. Similarly, we can see that the value of critical energy ΔG_T^* increases quickly with θ2 increasing when θ2 is more than 80°, and ΔG_S^* is much larger than ΔG_T^* when θ2 < 150°. When θ2 > 150°, ΔG_T^* is close to ΔG_S^*. Importantly, one can see that ΔG_T^* is always less than ΔG_S^* at given experiment parameters and θ2 values from 80° to 150° in Figure 5. However, comparing Figure 5 with Figure 6, we can see that the dependencies of the critical radius and critical energy of nuclei on the value of θ2 are inverse, i.e., the critical radius of nuclei decreases with increasing the value of θ2, and the critical energy of nuclei increases with increasing the value of θ2. Considering the physical process of nanowires nucleation shown as Figure 2, we propose that this anomalous behavior should be original from the interaction between nuclei and substrates. The large interface area between nuclei and substrates would reduce the formation energy of nuclei, while it would result in the critical radius of nuclei large.

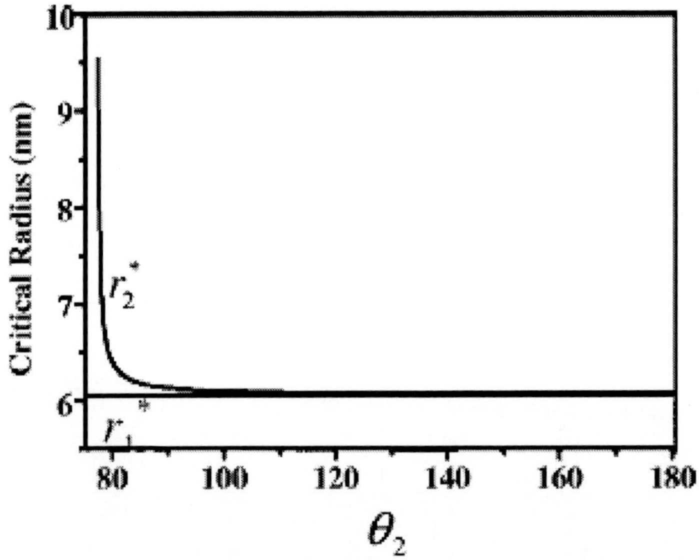

Figure 4. The dependence relationship between the contact angle θ2 and the critical nucleus $r1^*$, $r2^*$.

Therefore, these results indicate that the Gibbs free energy of the critical nuclei in the case of catalyst nanoparticles on the substrate is usually larger than that for catalyst nanoparticles on the tip of nanowires when θ2 is less than 180°. According to the thermodynamic criterion mentioned above, for the Si-Fe system, Fe catalyst nanoparticles would be on the tip of Si nanowires upon CCVT. More importantly, most of experimental cases of Si nanowires growing on Si substrates using Fe as catalysts by CCVT show that Fe nanoparticles are always on the tip of Si nanowires. [17,33–35].

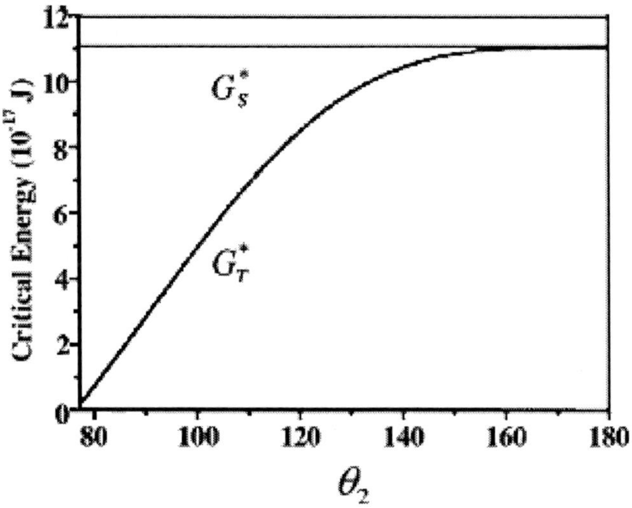

Figure 5. The relationship curves between the contact angle θ2 and the critical energy ΔG_S*, ΔG_T*.

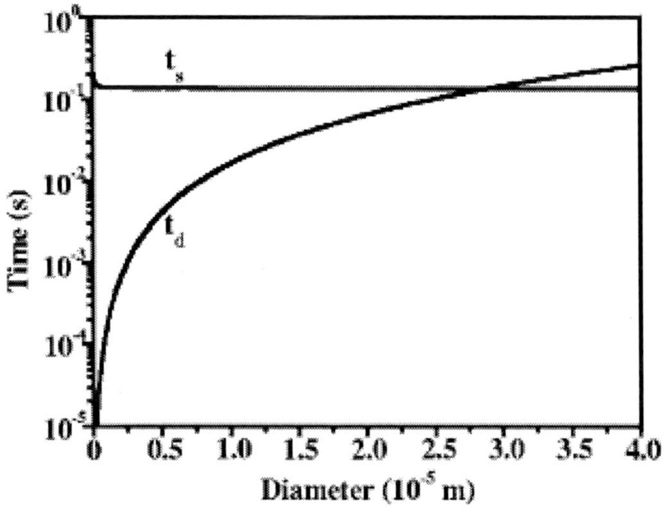

Figure 6. The comparison between the surface saturation time of t_s and the body diffusion time of t_d.

It is well known that the thermodynamical criterion just provides the probability for the selection growth of nanowires. On the other hand, the kinetic nucleation will play a key role in the achievement of the selection growth of nanowires upon CCVT, when the thermodynamic criterion operates. For instance, in the Si nanowire nucleation upon CCVT, when $t_d > t_s$, the Si atoms saturation on the catalyst surface is much faster than one penetrates to its base. Then, Si precipitates on the catalyst surface provides a nanoscaled template for the Si nanowire growth. In contrast, when $t_d < t_s$, the Si penetration to the base is much faster than the Si atoms saturation on the catalyst surface. Thus, Si precipitates at the bottom will lift the catalyst and maintain it on the nanowires tip later. In order to study the kinetics of Si nanowires growth, the diffusion time and saturation time are calculated at the given temperature and pressure in the Si-Fe system mentioned above on the basis of Eqs. (14–16).

The kinetics parameters are as follows: the value of $D0$, $E(\infty)$, Svib(∞), and $r0$ being 6.2×10−3 cm2/s [32], 8.4×104 J /mol [36], 34.32 J /mol K−1 [37] and 0.7449 nm [38], respectively, for the system of Si diffusion in Fe. Based on the Si-Fe phase diagram [39], Fe saturates at ≈50 at % of Si and the saturation concentration of Si is about 3.5×1022 cm−3 [40], we can calculate t_d and t_s under conditions of T=1473 K and P =1.4 Torr. Figure 6 show the dependence of t_d and t_s on the size of the Fe catalyst. From Figure 6, we can see that the diffusion time increases with the size increasing of Fe catalyst, while the saturation time hardly change with the diameter increasing of the nanoparticles. Very importantly, $t_d < t_s$_(more two orders of magnitude) when the size of the Fe catalyst is less than 25 μm. According to the kinetic criterion above, this result displays that Si penetrates to the base much faster than Si on the catalyst surface reaches saturation threshold when the size of the nanoparticles is less than 25 μm. In other words, Si will precipitate at the bottom, lift catalysts, and later maintain catalyst droplet on nanowires tips when the size of the Fe catalyst is less than 25 μm.

Gösele *et al.* pointed out that the size of the Si nanowires is limited by the size of the metallic catalyst droplet, and the diameter of the Si nanowires is usually much smaller than that of the catalyst droplet [26]. Accordingly, the size of Si nanowires should be smaller than that of the catalyst droplet when catalyst droplets are on nanowires tips. Interestingly, up to date, almost all experimental data at given temperatures and pressures in the Si-Fe system show that the diameter of Si nanowires is less than 200 nm, and catalyst droplets are definitely located on the tip of the Si nanowires [17,33–35], which corresponds with the theory predictions above. On the other hand, from Figure 6, one can see that there exists a critical diameter of 30 μm for catalyst droplets, and the catalyst surface-saturation and Si penetration-base time is nearly the same order of magnitude when the size of the nanoparticle is larger than 30 μm. Namely, when the radius of droplet is larger than 30 μm, no solid conclusions can be drawn from the above model predictions, and maybe, the selection growth of nanowires would be determined by chance. Assuredly, in experiment, Wagner and Ellis have prepared Si whiskers with 200 μm-sized diameters, which always had the catalytic particle on top. In the experimental case of Wagner and Ellis, they selected Au as catalyst, and the deposition temperature is about 1223 K. One can see that their experimental parameters are very different from the parameters of our calculating cases. Thus, the theoretical prediction is not comparable with their experimental results. Naturally, we would have tried to compare their experimental case with our calculations by using their experimental parameters. However, we did not in this case due to the lack of the detailed experimental parameters, e.g., the partial pressure of SiCl4, and so on. In fact, since the diameters of Si nanowires reported in experiment are small, the catalyst nanoparticles would be lifted on the nanowires tip in most cases from the proposed thermodynamical and kinetic criterion.

Therefore, these comparisons show that the universal thermodynamical and kinetic criterion of the selection growth of nanowires forming upon CCVT seems to operate to determine catalyst nanoparticles on the tip or bottom of nanowires. Note that, in some Si nanowire synthesis by laser ablation [41,42], no catalyst nanoparticles were observed anywhere on nanowire tip or substrate even when metals were mixed in the target. Actually, this nucleation and growth of nanowires is different from the VLS mechanism. Moreover, Wang *et al.* [42] pointed out that metal was not necessary for Si nanowire synthesis by laser ablation. Instead, oxides such as SiO2 are the special and effective catalyst that largely enhanced the Si nanowire growth in these cases [41,42].

In summary, the general thermodynamical and kinetic approaches on nanoscale are introduced to elucidate the nanowires nucleation taking place in CCVT, with respect to the capillary effect of the nanosized curvature of the nanowire nuclei in the initial growing. A thermodynamic criterion and a diffusion kinetic criterion on the basis of the thermodynamic and kinetic analyses on nanoscale are deduced for the selection growth of catalyst nanoparticles on nanowires tip or substrate upon CCVT. Taking Si nanowires as the example, we compared the experimental data with the theoretical predictions and found that they are nicely consistence. We thus expected that the thermodynamic and kinetic criteria on the nanometer scale to be a general approach applicable to understand nanowires formation in the occasion of CCVT.

3. Thermodynamic and Kinetic Theory of Nanowire Growth

Up to date, most semiconductor nanowires have been synthesized using metal catalyst assistant thermal chemical vapor transport on the basis of so-called vapor-liquid-solid mechanisms [43-46]. Actually, VLS mechanisms for applicable to the growth of silicon crystal whiskers was initially proposed and demonstrated by Wagner and Ellis in 1964 [15-16], and has subsequently been used to address the growth of semiconductor nanowires by CCVT [18, 47-49]. Based on these studies above, for the nanowire formation upon CCVT, the VLS mechanism reveals that (i) reactants are supplied in the vapor phase and nanowires are formed by extracting of nanowires materials from liquid droplets of metal alloy, (ii) the nanowire growth results from the difference of sticking coefficients of liquid droplets and nanowires, namely, sticking coefficients of nanowires are orders of magnitude smaller than that of liquid droplets. As a result, the droplet and the solid nanowire can capture and reject nearly all the constituents of the growing material from the vapor phase, respectively, as shown in Figure 1(a). Generally, the intrinsic properties of 1D nanostructures are mainly determined by their size, shape, composition, and crystalline structure. In principle, one can control any one of these characters to fine-tune the properties of this nanostructure. It is therefore essential to pursue the basic physics and chemistry involved in the formation of nanowires. However, the size limit of the nanowire growth upon CCVT still remains much less understood in theory, even though there have been a lot of relevant experimental studies to try to control the nanowire growth [49-51]. For instance, in both theoretical and experimental, we haven't known how thin or thick nanowires could be synthesized using CCVT yet. Accordingly, we proposed the universal nucleation thermodynamics and growth kinetics for the nanowire growth upon CCVT based on the VLS mechanism, and then theoretically deduced the size limit of the nanowire growth. Importantly, taking the Si nanowire (SiNW) growth as an example, we found that the theoretical predictions are in excellent agreement with experimental cases.

3.1. Thermodynamic Size Limit of Nanowire Growth

In general, the minimal size of the nanowires is both dependent on nucleation thermodynamics and growth kinetics (energy stability theory), meanwhile, it is also dependent on the Rayleigh instability of nanowire itself (linearized stability theory) [31]. However, when the linearized system is symmetric, the prediction of the stability based upon the linearized stability theory and the energy stability theory are consistent [52-53]. In other words, linearized equations do not yield much information on the energy stability range. Namely, we could have an instability (usually called a subcritical instability) below the critical values of Rayleigh numbers determined by the linearized theory, which can be in the energy stable region. Thermodynamically, the phase transformation is promoted by the difference of Gibbs free energies. The Gibbs free energy of a phase can be expressed as a function of the pressure and temperature, and determined by a general coordinate or reactive coordinate. The schematic illustration of the X nanowire (XNW) nucleation upon CCVT is shown in Figure 7(b). The Gibbs free energy difference of a cluster can be expressed as [24]

$$\Delta G = (\sigma_{lx} - \sigma_{lv})S_1 + \sigma_{xv}S_2 + \Delta g_v V \tag{3-1}$$

where σ_{lx}, σ_{lv}, and σ_{xv} are the liquid-XNW, the liquid-vapor, and the XNW-vapor interface energy density, S_1 and S_2 are the corresponding interface areas (see Figure1b) expressed as: $S_1 = 2\pi Rh$ and $S_2 = 2\pi rH$, respectively. When the interface between nanowire and catalyst liquid droplet is assumed to be incoherent interface, the values of σ_{lx} can be denoted as: $\sigma_{lx} = \frac{\sigma_{lv} + \sigma_{xv}}{2}$. V is the volume of X clusters defined as: $V = \frac{1}{6}\pi H\left(H^2 + 3L^2\right) - \frac{1}{6}\pi h\left(h^2 + 3L^2\right)$. Note that the theoretical model in this case is formulated based on the assumptions of (i) the catalyst droplet and X cluster are perfectly spherical, and (ii) X nuclei are mutually non-interaction. From Figure 7(b) and considering $r<R_0$, h, H, and $L2$ can be defined respectively as:

$$h = R_0 - \sqrt{R_0^2 - r^2 + \frac{r^2(R_0\cos\theta - r)^2}{R_0^2 + r^2 - 2R_0 r\cos\theta}} \tag{3-2}$$

$$H = r\left(1 - \frac{R_0\cos\theta - r}{\sqrt{R_0^2 + r^2 - 2R_0 r\cos\theta}}\right) \tag{3-3}$$

$$L^2 = r^2\left(1 - \frac{\left(R_0\cos\theta - r\right)^2}{R_0^2 + r^2 - 2R_0 r\cos\theta}\right) \tag{3-4}$$

where R_0, r, θ denote the curvature radius of the liquid droplet, the curvature of the cluster, and the contact angle between cluster and liquid droplet, respectively. Meanwhile, one can obtain $\cos\theta = \frac{\sigma_{lv} - \sigma_{xv}}{2\sigma_{xv}}$, and Δg_v is the Gibbs free energy difference per unit volume from vapor to solid, which is expressed by $-RT/V_m \ln(P/P_e)$, where P, T, R, and V_m are the pressure, the temperature, the gas constant, and the mole volume of the XNW, respectively. P_e is the X vapor phase pressure in thermal equilibrium coexistence with the liquid of composition in a flat surface. Based on Tan's model [26], when the droplet is assumed to be an idea solution, Δg_v can be expressed as

$$\Delta g_v = -\frac{RT}{V_m}\ln\frac{p}{p_e} = -\frac{RT}{V_m}\ln\left(\frac{C}{C^{eq}}\right) \tag{3-5}$$

where C and C^{eq} are X concentrations on the solid and liquid lines of the eutectic catalyst-X phase diagram, respectively.

On the basis of the above deductions, the Gibbs free energy of an X cluster is expressed reasonably as:

$$\begin{aligned}\Delta G = {} & \pi R(\sigma_{xv} - \sigma_{lv})\left(R - \sqrt{R^2 - r^2 + \frac{r^2(R\cos\theta - r)^2}{R^2 + r^2 - 2Rr\cos\theta}}\right) + \\ & 2\pi\sigma_{xv} r^2\left(1 - \frac{R\cos\theta - r}{\sqrt{R^2 + r^2 - 2Rr\cos\theta}}\right) \\ & + \frac{1}{6}\pi\left(\frac{RT}{V_m}\ln\left(\frac{C}{C^{eq}}\right)\right)\left(\left(R - \sqrt{R^2 - r^2 + \frac{r^2(R\cos\theta - r)^2}{R^2 + r^2 - 2Rr\cos\theta}}\right)^3\right. \\ & + 3r^2\left(R - \sqrt{R^2 - r^2 + \frac{r^2(R\cos\theta - r)^2}{R^2 + r^2 - 2Rr\cos\theta}}\right)\left(1 - \frac{(R\cos\theta - r)^2}{R^2 + r^2 - 2Rr\cos\theta}\right) - \\ & r^3\left(1 - \frac{R\cos\theta - r}{\sqrt{R^2 + r^2 - 2Rr\cos\theta}}\right)^3 - \\ & \left.3r^3\left(1 - \frac{R\cos\theta - r}{\sqrt{R^2 + r^2 - 2Rr\cos\theta}}\right)\left(1 - \frac{(R\cos\theta - r)^2}{R^2 + r^2 - 2Rr\cos\theta}\right)\right)\end{aligned} \tag{5-6}$$

From Eq. (6), one can see that the Gibbs free energy of a cluster of XNWs is related with two factors: one is the liquid droplet composition, and another one is the size of the liquid droplet. Therefore, when $\frac{\partial \Delta G(r)}{\partial r} = 0$, the critical size of the nuclei of XNWs can be attained

[10]. Distinctly, the critical size of the XNW nuclei is the thermodynamic size limit of the nanowire growth, the theoretical minimum size, from the view of point of thermodynamics.

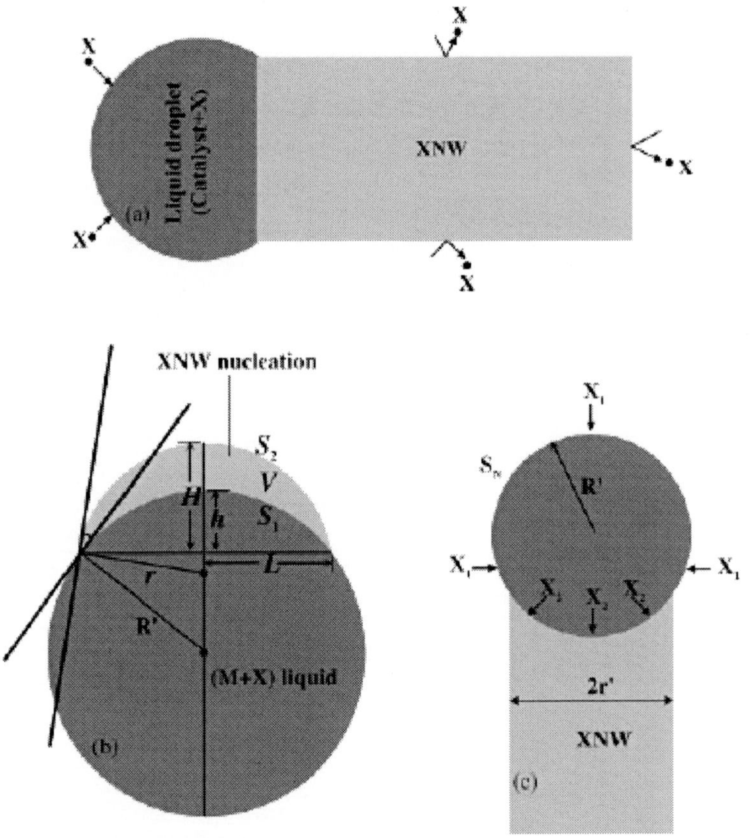

Figure 7. The schematic diagram showing the VLS mechanism, and XNW thermodynamic nucleation and kinetic growth processes. (a) The droplet and the solid nanowire capturing and rejecting nearly all the constituents of the growing material from the vapor phase, respectively, due to the sticking coefficient immense difference between them. (b) The thermodynamic nucleation case. (c) The kinetic growth case.

3.2. Kinetic Size Limit of Nanowire Growth

From the growth kinetics, when X atoms reach saturation in the liquid droplet at the given thermodynamic parameters, the number of X atoms of the droplet capturing (X_1) and extracting (X_2+ X_2^N) should be equal, as shown in Figure 7(c). Then, only X_2^N atoms contribute to the growth of XNWs in the extracting X atoms. The number of X_2^N atoms can be approximately equal to the net flux of impinging on the surface of the area whose value is equal to S_N. Reasonably, the number N of X_2^N atoms is expressed as

$$N = 2\pi R'\left(R' - \sqrt{R'^2 - r'^2}\right)\left(\theta - \theta_r\right) = \frac{10^6 \pi r'^2 S \rho N_m}{M} \tag{5-7}$$

where R', r', ρ, N_m, and M are the radius of droplet, the radius of XNW, the density of single crystal X, the Avogadoro constant, and the mole mass of single crystal X, respectively. θ is the number of X atoms from vapor impinging on a unit plane surface in an unit time. For the VLS progress of vapor growth or vapor deposition, an ideal gas should be a good approximation due to the low gas density. According to Maxwell's velocity distribution equation, at the given temperature T and pressure P, θ can be expressed as $\theta = \int_{-\infty}^{+\infty} du_x \int_{-\infty}^{+\infty} du_y \int_{0}^{+\infty} n u_z P(u) du_z = n\left(\frac{kT}{2\pi m}\right)^{1/2} = \frac{P}{\sqrt{2\pi mkT}}$ where $P(u)$ is the Maxwell's velocity distribution, n is the molecular number of the vapor phase per unit volume, and k is the Boltzmann constant, and m is the mass of X atom. θ_r is the number of desorbing flux from the surface of the droplet on an unit area in an unit time at a finite temperature T. Note that the desorbing flux (θ_r) is independent of the deposition flux from the vapor phase, but is a temperature function, and given by $\theta_r = \frac{P_r}{\sqrt{2\pi mkT}}$ where P_r is the saturation pressure in the liquid droplet surface. On the basis of the Gibbs-Thompson formula, we have

$$\ln\frac{P_r}{P_e} = \frac{2\sigma_{lv}\Omega}{r'kT} \tag{5-8}$$

where Ω is the volume of the X atom. According to Tan's model [26] and Eq. (8), one can obtain

$$P_r = \frac{PC^{eq}}{C}\exp\left(\frac{2\sigma_{lv}\Omega}{R'kT}\right) \tag{5-9}$$

In Eq. (7), S is the growth velocity of XNW, and can be expressed as [30]

$$S = h\nu\exp(-E_\alpha / RT)[1-\exp(-|\Delta g_m| / RT)] \tag{5-10}$$

where h, ν, E_α, R, and T are the lattice constant of the XNW in the growth direction, the thermal vibration frequency, the mole adsorption energy of adatoms attached of X atom at surface sites, gas constant, and the substrate temperature, respectively. Based on Eq. (5), the Gibbs free energy difference per mole, Δg_m, and it can be expressed as

$$\Delta g_m = -RT\ln\left(\frac{C}{C^{eq}}\right) \tag{5-11}$$

According to Eq. (7) - (11), we can obtain the relationship between the size ($2R'$) of catalyst and the diameter ($2r'$) of XNW as follows:

$$2r' = 4R'\left(\frac{M\left(P-\frac{PC^{eq}}{C}\exp\left(\frac{2\sigma_{lv}\Omega}{R'kT}\right)\right)}{10^{6}\sqrt{2\pi mkT}\,\rho N_{m} h\nu\exp\left(-\frac{E_{\alpha}}{RT}\right)\left(1-\exp\left(-\frac{\left|-RT\ln\left(\frac{C}{C^{eq}}\right)\right|}{RT}\right)\right)} - \left(\frac{M\left(P-\frac{PC^{eq}}{C}\exp\left(\frac{2\sigma_{lv}\Omega}{R'kT}\right)\right)}{10^{6}\sqrt{2\pi mkT}\,\rho N_{m} h\nu\exp\left(-\frac{E_{\alpha}}{RT}\right)\left(1-\exp\left(-\frac{\left|-RT\ln\left(\frac{C}{C^{eq}}\right)\right|}{RT}\right)\right)}\right)^{2}\right)^{1/2} \tag{5-12}$$

Clearly, from Eq. (12), one can see that, the diameter of nanowires is related with the size of droplet. On other words, the growing size of nanowires, the theoretical possible or maximum size, depends on the droplet size from the view of point of kinetics.

3.3. Comparison between Theory and Experiment

In order to validate our model, we take the Si nanowire (SiNW) growth upon CCVT as an example to check its operation, as sufficient experiments or theoretical parameters of Si-Au system are presently available [51]. From Leiber [51] and other references, the thermodynamic parameters of our calculation by Eqs. (6) and (12) are as follows: the liquid droplet-gas surface energy density σ_{lv} is 1.12 J/m2 [32], the solid silicon-gas surface energy density σ_{Siv} is 1.610 J/m2 [31], the mole adsorption energy of adatoms attached of Si atom at surface sites E_{α} can be evaluated to be about 118 kJ/mol [58], the thermal vibration frequency of Si atom ν is about 2×1013 Hz [59], the lattice constant of the SNW in the growth direction (111) h is 0.312 nm, the mole volume V_m of single crystal Si is 1.2×10-5 kg/m3, the mole mass M and density ρ of single crystal Si are 28.09 g/mol and 2.33 g/cm3, respectively. The experimental temperature T and pressure P are 713 K and 13.332 Pa, respectively.19 Note that the used Au-Si phase diagram is from Ref. [34]. On the basis of Eq. (6), we give the dependence of the radius of a cluster of Si (r) and the Gibbs free energy (ΔG) at the given temperature (713 K), pressure (13.332 Pa), and various Au-catalyst radius (R_0), as shown in Figure 8. We can see that the Gibbs free energy of Si clusters (ΔG) hardly changes with increasing of the Au-catalyst radius. In detail, the Gibbs free energy of Si clusters slightly little increases with the Au-catalyst radius decreasing. Namely, the nucleation barrier of nanowires does not actually depend on the size of metal catalyst. Based on the nucleation thermodynamics [10], the radius of Si clusters (r) corresponding to the peak value should be the critical radius (r') of the SiNW nucleation. Furthermore, from Eq. (6), the relationship curves between the Au-catalyst radius (R') and the critical radius of the SiNW nucleation (r') are given at the given temperature (713 K) and pressure (13.332 Pa) by numerical calculation, as shown in Figure 9. Definitely, the critical radius of the SiNW nucleation basically keeps unchanged with increasing the Au-catalyst radius when the catalyst radius more than 20 nm, and the critical radius decreases with increasing the catalyst radius when the catalyst radius less than 20 nm. However, the decreasing scope of the critical radius is very little when the catalyst radius is less than 20 nm, implying that the affect of the droplet size on the critical radius of the nanowire nucleation can be actually neglected. These results suggest that the critical radius of the nanowire nucleation seems a relatively independent value for a given system from thermodynamic viewpoint, e.g., Si-Au system. In other words, there is a certainly thermodynamic limit, a theoretical minimum value, for the nanowire growth size upon CCVT. For example, the thermodynamic limit is 1.4 nm in radius for the Si-Au system from Figure 9. Very interestingly, we haven't found any experiments that reported SiNWs less than 1.4 nm in radius was synthesized by CCVT yet, so far.

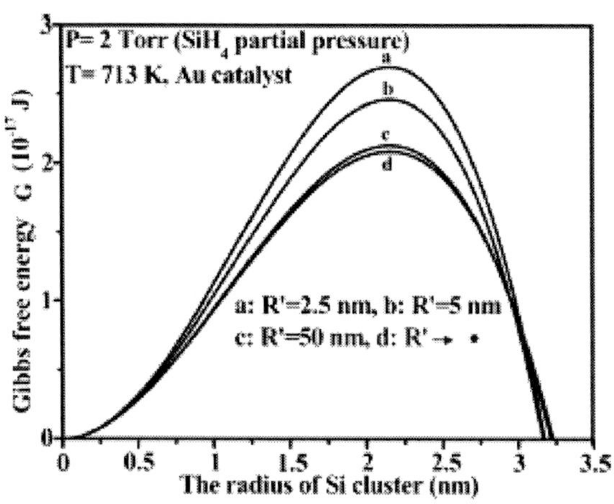

Figure 8. Dependence of the radius of the SiNW nucleation and the Gibbis free energy on the Si cluster size under conditions of various Au-catalyst sizes, temperature (713 K), and pressure (SiH_4 partial pressure of 2 Torr).

The dependence of the diameter of SiNWs on the Au-catalyst size at the given temperature (713 K) and pressure (13.332 Pa) can been attained from Eq. (12) as shown in Figure 10. One can see, the diameter of SiNWs increases with increasing the Au-catalyst diameter, and the SiNW diameter is always smaller than that of the Au-catalyst. Thus, these theoretical results are consistent with experimental cases [43,60]. Remarkably, the quantitative calculations are in excellent agreement with Lieber's experimental data.19 Similarly, Buhro *et. al.* [61] simulated the relationship curve between the diameter of catalyst and the size of GaAs nanowires on the basis of their empirically linear expression and experimental data. Interestingly, our theoretical predictions are consistent with Buhro's results. It is noticed that, in Figure 10, considering the oxidation effect after the SiNW growth, the data of Ref. [51] have been corrected on the basis of authors stated the thickness of oxidation layer. Accordingly, the size of the catalyst droplet can limit the possible growing size of the nanowire synthesized by CCVT to a certain extend from the kinetic viewpoint.

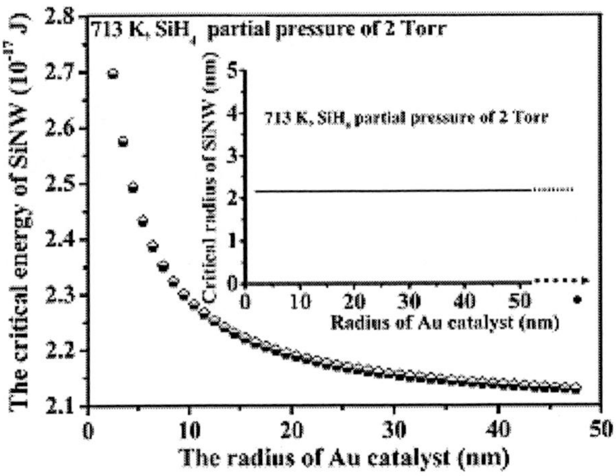

Figure 9. Relationship curve between the Au-catalyst radius and the critical energy of SiNW nucleation; the inset shows the dependence of the Au-catalyst radius and the critical nucleation of SiNW nucleation.

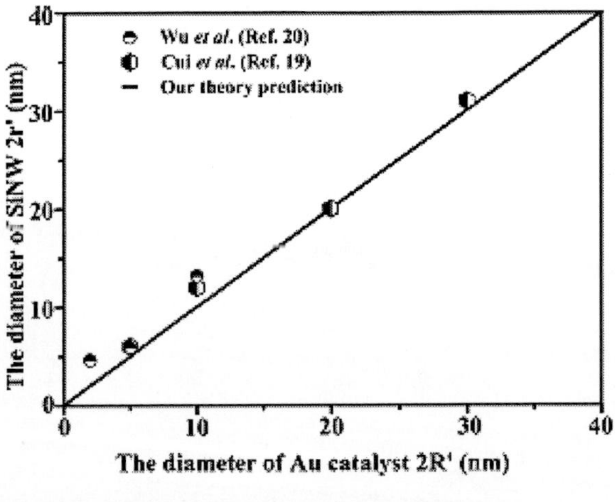

Figure 10. Experimental data of relationship between the size of the Au catalyst and the diameter of SiNW compared to the calculated curve by using eq 10 and Au-Si phase diagram, in which the Au-Si phase diagram from Ref [39] is used.

In summary, the critical size of the nanowire nucleation is suggested to be the thermodynamic size limit, the theoretical minimum size, of the nanowire grown by CCVT in both thermodynamic and kinetic, while the size of the catalyst droplet could limit the growing size, the theoretical possible or maximum size, of the nanowire to a certain extend in kinetics.

4. A Growth Kinetic Model of Nanowire Inside Nanotube

Carbon nanotubes (CNTs) regarded as a template have been demonstrated to be an effective route to fabricate one-dimensional nanostructures of metals and semiconductors [62-64]. Generally, fabrications of one-dimensional nanostructures by used CNTs regarded as template can be divided two cases. One is mechanic filled by capillarity of CNTs immersed liquid, and another one is vapor grown inside CNTs in gases by chemical and physical vapor depositions. In detail, the former is mainly employed to prepare metallic nanowires, and the latter is used to fabricate semiconductors nanowires [55-56,65-71]. We herein focus our attention on the last issue. Basically, it is important to control the growth of one-dimensional nanostructures inside CNTs both experimentally and theoretically for applications of one-dimensional nanostructures. However, compared with many studies of the experimental synthesis of nanowires by CNTs templates [71-72], few theoretical studies of nucleation thermodynamics and growth kinetics of one-dimensional nanostructures forming inside CNTs are found in the literature. To gain a better understanding of the formation of one-dimensional nanostructures inside CNTs in theory, we have proposed a nanothermodynamic nucleation theory to elucidate the nuclei formation of one-dimensional nanostructures inside CNTs [12]. Based on Wilson–Frenkel growth law [73] and our nanothermodynamic model [12], we developed a growth kinetic approach to clarify the growth of one-dimensional nanostructures forming inside CNTs upon vapor deposition in detail. Furthermore, the model is employed in the synthesis of Ga nanowires upon chemical vapor deposition by CNTs' template, and the theoretical results are in good agreement with experimental data.

4.1. Growth Kinetic Model

It is well known that CNTs have many potential applications as molecular sieves, membranes, and ''nanopipes'' for precise delivery of gases or liquid [74]. Some studies focused their attention on transport, adsorbed, and condensed phases of gases inside CNTs [75]. For instance, the transport rates of CH4 and H2 gases in CNTs were suggested to be exceptionally high [76]. Therefore, when one end or two ends of CNTs are open, one-dimensional nanostructures could be fabricated inside CNTs upon vapor deposition by nucleating and growing [75]. Based on Wilson–Frenkel growth law of crystals, here we present a simple kinetic model about phase equilibration and phase transition of nuclei to study growth of one-dimensional nanostructures inside CNTs in gases. This theory is formulated based on the assumption: nano-nuclei are perfectly spherical with no structural deformation of the internal structure from the bulk one. The schematic illustration of a nanowire grown inside a CNT upon vapor deposition is shown in Figure 1. When the reaction gases flow along the CNT, the clusters are condensed on the inner wall of the CNT by a series of surface reactions and diffusions. Sequentially, the nucleation of the clusters will occur by phase transition. Eventually, the nuclei form inside the CNT as shown in Figure 11. Naturally, the starting point in the case is the growth of the nucleus inside a CNT in gases. Generally, growth kinetics of crystals in gases follows the classical Wilson–Frenkel growth law [77].

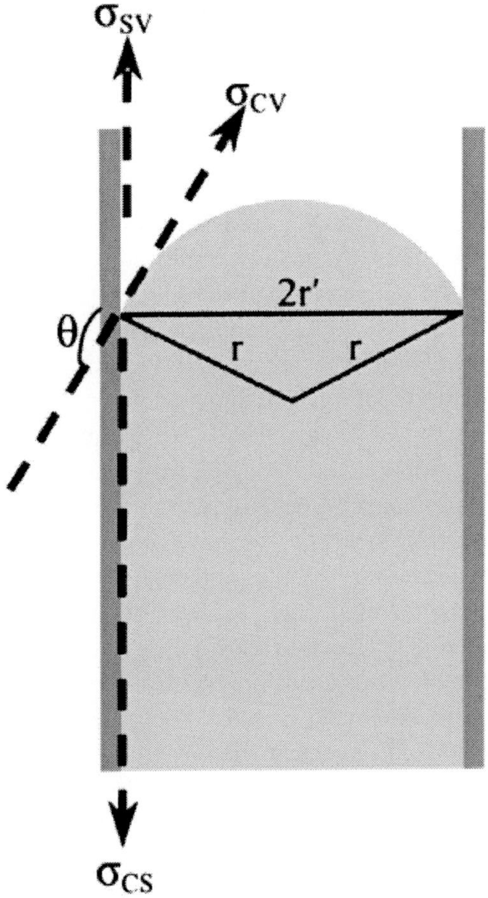

Figure 11. Schematic illustrations of one-dimensional nanostructure grown inside a carbon nanotube.

Accordingly, the growth velocity is driven by the difference in chemical potential between the metastable phase and equilibrium crystal. It is well known that the growth of crystalline nuclei is achieved by trapping their neighbor adatoms that are the mobile surface species and constantly diffuse from a surface site to another site [24]. Adatoms attached at surface sites need the adsorption energy E_a, while those adatoms transporting away from surface sites need to overcome an energy barrier E_b. Thus, the driving force Δg of phase transition from the metastable gas phase to equilibrium crystal is determinate by: $\Delta g = E_a - E_b$. If the adatom amount per unit area and time on the growth surface is N_0, the attachment atom amount N per unit area and time can be expressed by

$$N = N_0 v \exp(-E_a/kT)[1 - \exp(-|\Delta g|/kT)] \tag{4-1}$$

where v is the thermal vibration frequency, k is Boltzmann constant, T is the growth temperature, respectively. Therefore, when $N = N_0$, the crystalline nucleus will grow in the growing direction followed by the layer-by-layer mode. The thickness h of the lay should be equal to the lattice constant of the crystalline nucleus in the growth direction. Accordingly, the growth velocity R of the crystalline nucleus can be expressed by $R = (N/N_0)h$, as follows:

$$R = hv\exp(-E_a/kT)[1-\exp(-|\Delta g|/kT)] \tag{4-2}$$

Actually, when the actual vapor pressure is almost close to the saturation vapor pressure in vapor depositions, $\Delta g \ll kT$. According to the first-degree approximation of Eq. (2) by Thaler formula, we can obtain

$$R = \frac{hv}{kT}\exp(-E_a/kT)\cdot|\Delta g| \tag{4-3}$$

On the other hand, the driving force of phase transition can be expressed by Δg = $-kT \ln(P/P_0)$ where P and P_0 are super-saturation pressure and saturation pressure in vapor depositions, respectively. Based on the nanothermodynamic model [12], from the Laplace–Young equation, the sized-induced additional pressure ΔP of the nano-nucleus as shown in Figure 1 can be expressed by $\Delta P = 2\sigma_{cv}/r$ where σ_{cv} is the interface energy of crystalline nucleus vapor, r is the curvature radius of the nucleus, respectively. Taking account of the effect of nanosized-induced additional pressure on the Gibbs free energy of nucleus inside the CNT, Δg should be expressed by $\Delta g = -2kT\ln(P/P_0)+(2\sigma_{cv}/r)$. Furthermore, considering the Kelvin equation [12,36], the driving force Dg is finally expressed by

$$g_v = -\frac{RT}{V_m}\ln\frac{p}{p_e} - \frac{2\sigma_{cv}}{r'} + \frac{2\sigma_{cv}}{r} \tag{4-4}$$

where M and r are the mole mass and density, respectively. In this way, combining Eq. (3) with Eq. (4), we can describe the growth of the nucleus inside CNTs in Figure 1. Additionally, from Figure 1, one can see that $\cos\theta = (\sigma_{sv} - \sigma_{sc})/\sigma_{cv}$ where θ is the contact angle between the crystalline nucleus and the wall of the CNT, σ_{sv} and σ_{sc} are the interface energies of wall vapor and wall nucleus at a given temperature, respectively. Meanwhile, the curvature radius r of the nucleus relates to the column radius r' (inner radius of CNT) as defined as $r = -r'/\cos\theta$. Therefore, based on Eq.(3), Eq. (4), and the relationship of r and r' above, the growth velocity R of the crystalline nucleus inside CNTs is regarded as the function of the column diameter (inner radius of CNT) in the way.

4.2. Comparison between Theory and Experiment

For comparing our model with the corresponding experimental data, we select a typical growth of the onedimensional nanostructure inside CNTs, i.e., Ga nanowire grown CNTs [56], to check the model. According to the work of Gao and Bando [56] and other references, the thermodynamic parameters of our calculations by Eqs. (3) and (4) are as follows: the thermal vibration frequency v is 1013 (Hz) [27], the activation energy Ea is 2.3 (eV) [78], the temperature T and pressure P are 1073 (K) and 5×10-3 (Pa) [56,79], the saturation pressure P_0 is 4×10-6 (Pa) [36], the contact angle θ is 108° [56], the surface energy σ_{cv} is 0.7 (J/m2) [13], the crystalline spacing h along (001) is 0.766 (nm) [80]. Consequently, the growth rate of Ga nanowires inside CNTs upon chemical vapor deposition is calculated to be about 1.8×10-10 (m/s) based on our model. More importantly, the experimental value of the growth rate of Ga nanowires inside CNTs was in the range of 10-9–10-10 (m/s) [56]. Therefore, our calculations are in agreement with the corresponding experimental data.

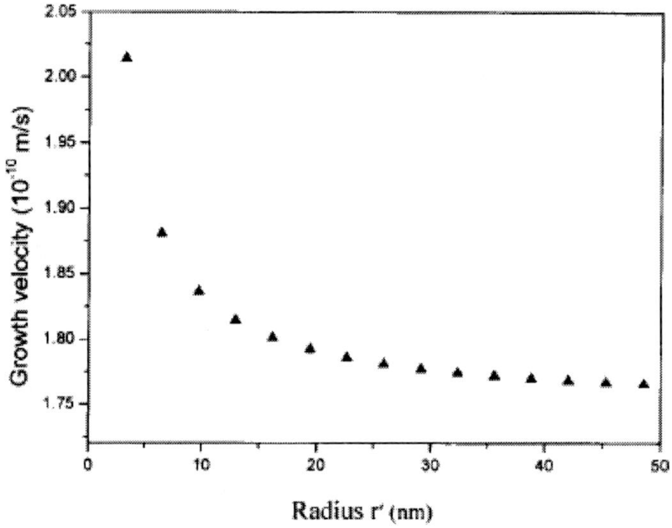

Figure 12. The growth velocity of the metallic Ga nanowire grown inside the CNT as a function of the radius of the CNT.

On the other hand, from Eq. (3), the dependence of the growth velocity R of the nucleus of metallic Ga inside the CNT on the inner radius r' of the CNT was shown in Figure 2, under the conditions of the given thermodynamic parameters such as interface tensions. Obviously, in the Ga nanowire case, the growth of one-dimensional nanostructures inside CNTs is much different from that of two- or three-dimensional structures on the flat of substrates from Figure 12. The diameter of CNTs could effectively influence the growth of one-dimensional nanostructures inside CNTs. Clearly, from Figure 12, we can see that the growth velocity of one-dimensional nanostructures increases with the size of the CNT decreasing at the given deposition conditions such as temperature T and pressure P, and other thermodynamic parameters such as surface and interface tensions. Definitely, when $r' < 10$ nm, the increase of the growth velocity goes much higher with the radius of CNTs continually decreasing. Then, the decreasing of the growth velocity becomes smooth with the radius of CNTs continually increasing, and the decreasing extent of the growth rate also becomes smooth, when $r' > 10$ nm. In fact, the growth is not correlative to the size of CNTs when the radius of CNTs becomes too large. In addition, we calculated the growth rates of GaN and Si3N4 nanowires grown inside CNTs upon vapor deposition. The calculations are consistent with experimental data [70] and the similar dependence of the growth velocity of the nucleus inside the CNT on the inner radius of CNTs is found. Accordingly, the model seems to apply for other nanotubes such as BN or WS2, then, there are the different thermodynamic parameters such as interface energies between nanowires and nanotubes in the different nanotube cases.

In summary, we propose the growth kinetics of one-dimensional nanostructures inside CNTs upon vapor deposition, with respect to the effect of nanosized-induced additional pressure on the growing kinetic parameters, based on the Wilson–Frenkel growth law. Interestingly, these theoretical results are very consistent with the corresponding experimental data. Consequently, we expect the growth kinetics to be a general approach to elucidate growth of one-dimensional nanostructures inside CNTs upon vapor deposition.

5. SIZE-DEPENDENT SHAPE EVOLUTION OF NANOWIRE

Well known, self-assembly processes require atomic or clustering interactions as the thermodynamic driving force to organize those atoms or clusters to form nano-sized domain morphologies [13]. For instance, Bartelt *et al.* recently reported the self-assembly of the regular arrays of two-dimensional (2D) vacancy islands via absorbate-driven dislocation reactions [81]. Therefore, it is essential to pursue the physical and chemical origin causing the nanostructure to control the growth. On the other hand, the surface shape of 1D nanostructures plays a particular important role in their application. For instance, nanowires as wraparound gate transistors or core-shell heterostructures require good control of the surface shape to achieve uniform cross section and minimize carrier scattering at rough interface [82]. Accordingly, a quantitative understanding of the surface shape formation of 1D nanostructures is needed for the application of interest.

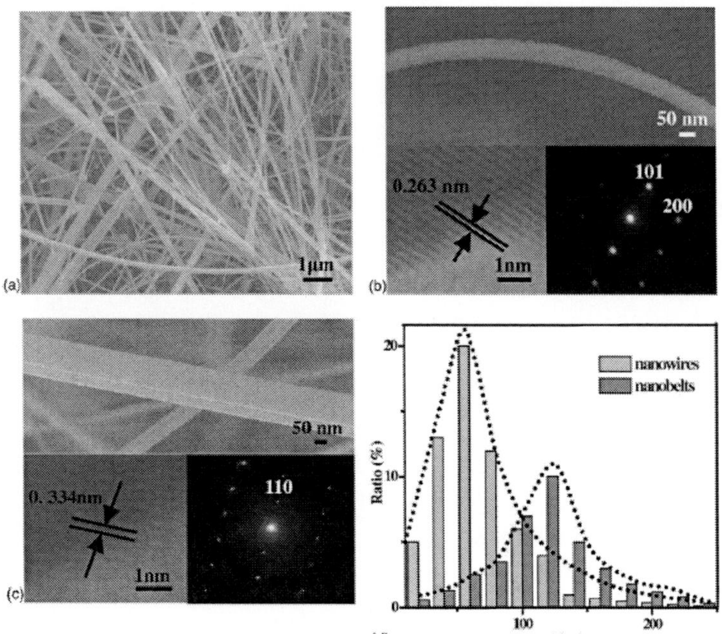

Figure 13. (a) Low-magnifying Field emission scanning electron microscopy (FESEM) image of SnO_2 nanowires and nanobelts. (b) High-magnifying FESEM image of a nanowire, in which a corresponding high-resolution transmission electron microscopy (HRTEM) image and selected area electronic diffraction (SAD) are shown below left right, respectively. (c) High-magnifying FESEM image of a nanobelt (the white line contrast on the nanobelt is another nanobelt whose wider plane of the cross-section inclines extremely towards left inside so as to seem thinner than the practical size), in which a corresponding HRTEM image and SAD are shown below left right, respectively. (d) The SEM statistics result of the radial size distribution of nanowires and nanobelts in the total prepared 1D nanostructures.

Recently, we experimentally observed the size-dependent morphological evolution of the 1D SnO2 nanostructure growth using metal catalysts assistant thermal chemical vapor transport and condensation (VLS process). In detail, we found that the shape formations of nanowires and nanobelts of SnO2 are size dependent, i.e., the wire is favorable when the size is less than 90 nm and the belt is favorable when the size is more than 90 nm (Figure 13). For this issue, we proposed the nucleation thermodynamics, the growth kinetics, and the shape transition thermodynamics to elucidate our findings on the basis of the thermodynamic analysis at the nanometer scale [8, 11, 12, 22, 83, 84]. Note that the size-dependent shape transformation from the initial stage of the nanowires nuclei to the nanobelts nuclei originates from the thermodynamic driving force of nucleation and growth of 1D nanostructures.

5.1. Nucleation Thermodynamic Consideration

Generally, the Gibbs free energy is an adaptable measure of the energy of a state in phase transformation among competing phases. Thermodynamically, the phase transformation is promoted by the difference of the Gibbs free energies [24]. Therefore, we perform the nucleation of SnO2 on the basis of the thermodynamic nucleation at the nanometer scale [22,84]. Considering the nanowire growth originates from the column-shape nuclei, and the formation of nuclei of the nanowire shown in Figure 1 (b) is a process of the reaction precursors extracting from the saturated Au catalyst [30], the Gibbs free energy difference of a column-shape nucleus is expressed

$$\Delta G_1 = -\Delta g_v \pi r_1^2 L_1 + \pi r_1^2 \left(\delta_1 + \delta_1'' \right) + 2\pi r_1 L_1 \delta_1 \tag{5-1}$$

where δ_1 and δ_1'' are the nucleus-vapor and the nucleus-liquid interface energies, and the r_1 and L_1 are the radius and the height of the nucleus. The $\Delta g_v = -RT/V_m ln(P/P_e)$ is the Gibbs free energy difference per unit volume, in which T, R and V_m are the temperature, the gas constant, and the mole volume, and . P and Pe are the partial pressure of the Sn and the Sn vapor phase pressure in thermal equilibrium coexistence with the liquid of composition in a flat surface. From Eq. (1), we deduce that the critical radius of r^*_1, the critical height of L^*_1 and the critical energy of ΔG^*_1 of nuclei are $2\delta_1/\Delta g_v$, $2(\delta_1+\delta_1'')/\Delta g_v$, and $4\pi\delta_1^2(\delta_1+\delta_1'')/\Delta g_v^2$.

Considering the nanobelt growth originates from the rectangle-shape nuclei. Note that, we use square-shape nuclei instead of rectangle-shape nuclei for simplifying the calculation in our case. Thus, The Gibbs free energy difference of a rectangle-shape nucleus is expressed

$$\Delta G_2 = -\Delta g_v r_2^2 L_2 + r_2^2 \left(\delta_2 + \delta_2'' \right) + 4 r_2 L_2 \delta_2 \tag{5-2}$$

where δ_2 and δ_2'' are the nucleus-vapor and nucleus-liquid interface energies, and the r_2 and L_2 are the side-length and the height of nucleus. Similarly, we have that the critical radius of r^*_2, the critical height of L^*_2 and the critical energy of ΔG^*_2 of nuclei are $4\delta_2/\Delta g_v$, $2(\delta_2+\delta_2'')/\Delta g_v$, and $16\delta_2^2(\delta_2+\delta_2'')/\Delta g_v^2$.

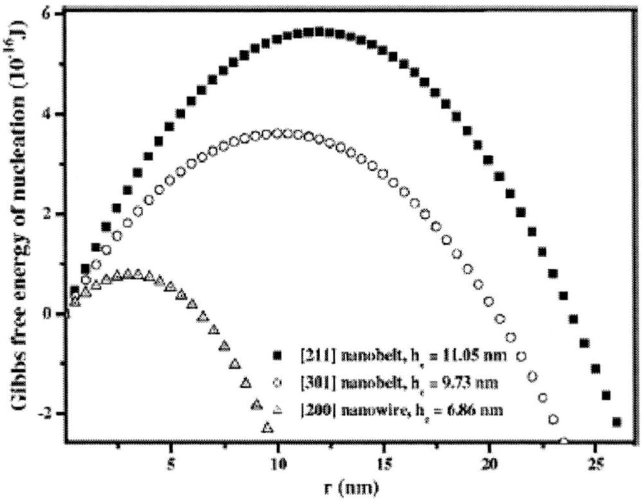

Figure 14. The nucleation energies of a column-shape nucleus of nanowires and a rectangle-shape nucleus of nanobelts growing along the (211) and the (301) directions, respectively.

Considering different growth directions of nanowires (the [200] direction) and nanobelts (the [211] or the [301] direction), Au catalyst, and the VLS nucleation mechanism [11,22], δ_1 and δ_2 are 1.14 and 2.14 (or 1.82) Jm-2 [85,86], δ_1'' and δ_2'' are 1.32 and 1.82 (or 1.66) Jm-2 [22,87], T and V_m are 1223 K, and 16.24 cm3/mole, and P and P_e are 3.29 and 1.047 Pa. Accordingly, r^*_1 and L^*_1 of nanowire nuclei are calculated to be 3.18 and 6.86 nm, r^*_2 and L^*_2 of nanobelt nuclei are calculated to be 11.94 (the [211] direction) or 10.18 nm (the [301] direction) and 11.05 (the [211] direction) or 9.73 nm (the [301] direction), and the size-dependence of the nanowire and nanobelt nucleation is shown in Figure 14. Clearly, the radial size of the critical nucleus of nanobelts is approximately two times larger than the diameter of the critical nucleus of nanowires. Moreover, the nucleation energy of nanowire nuclei is lower than that of nanobelt nuclei that means that the nucleation probabilities of nanowires is higher than that of nanobelts, which are well consistent with our observations that the ratios of wires and belts in the total synthesized nanostructures are 63 % and 37 %, respectively.

5.2. Growth Kinetic Consideration

Once the nucleation takes place, the increases in volume afforded by the continuous supply of the reaction precursors extracting from the saturated Au catalyst contributes to the increases in two feature sizes, radial size, and height, of nanowires and nanobelts on the basis of the VLS process [11,22,87]. The growth mechanism sketch map of the nanowire or the nanobelt is shown in Figure 15 (a). Thus, the radial growth rate of 1D nanostructures determines their final size. Well known, the surface free energy of nuclei and the concentration of the reaction precursors play key roles in the growth of nuclei [57], in which the surface free energy is consisted of the surface energy [88-89] and the strain energy [90,91]. Meanwhile, the supply of the reaction precursors comes from the axial direction reaction precursors extracting from the saturated Au catalyst. The supply mechanism lead to the growth of the radius and height of 1D nanostructures, i.e., (i) axial supply resulting in an augment in the height (P_{aa}), (ii) axial supply resulting in an augment in the radius (P_{ar}) as follows [90]

$$P_{ij} = \frac{\exp\left(-\frac{\Delta E_{ij}}{KT}\right)}{\sum_{i,j}\exp\left(-\frac{\Delta E_{ij}}{KT}\right)}, \Delta E_{ij} = \Delta E_{Sij} + \Delta E_{Tij}, i, j = a, r. \quad (5\text{-}3)$$

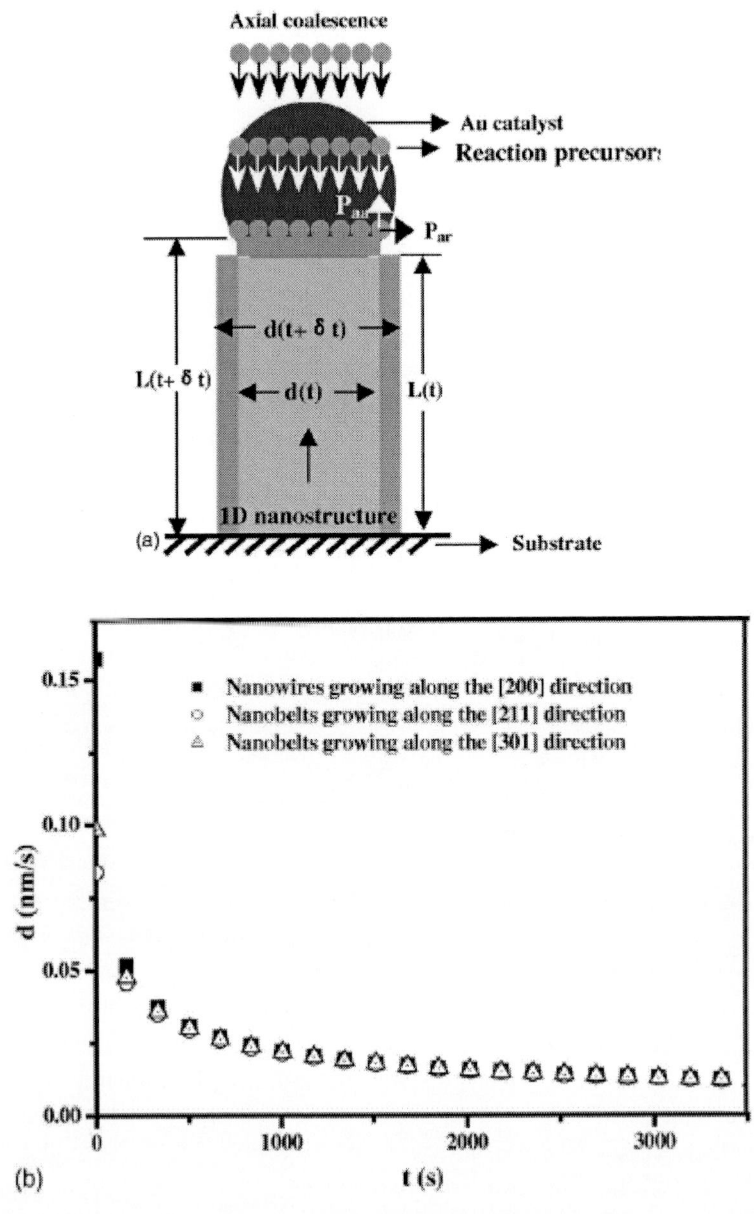

Figure 15. (a) The growth mechanism sketch map of the 1D nanostructure (the radial cross section of the nanowire or nanobelt is shown here). (b) The growth rates of nanowire and nanobelt.

K is Boltzman constant and T is the growth temperature. Thinking of the free energy ΔE_{ij} gaining in each case, we assume that the radial size of nanowires and nanobelts is large enough to make the surface energy ΔE_{Sij} much smaller than the strain energy ΔE_{Tij}. Thus, in each case, ΔE_{Tij} is written by [90]: $\Delta E_{Taa} = E_m(1-\sigma_m)\varepsilon_a^2, \Delta E_{Tar} = E_m\delta_m\varepsilon_a^2$, where E_m is the elastic modulus, δ_m is the Poisson's ratio of nanowires or nanobelts, and ε_a is the strain in the axial. Therefore, δt, the volume additivities of nanowires along the radial and axial are shown as follows [90]:

$$\frac{1}{4}L_1(t)d_1(t)\frac{\partial d_1(t)}{\partial t} = A\frac{1}{4}P_{ar}d_1(t)^2 \tag{5-4}$$

$$\frac{1}{4}d_1(t)^2\frac{\partial L_1(t)}{\partial t} = 2A\frac{1}{4}P_{aa}d_1(t)^2 \tag{5-5}$$

Similarly, in the case of nanobelts, we have

$$\frac{1}{4}L_2(t)d_2(t)\frac{\partial d_2(t)}{\partial t} = A\frac{1}{4}P_{ar}d_2(t)^2 \tag{5-6}$$

$$\frac{1}{4}d_2(t)^2\frac{\partial L_2(t)}{\partial t} = 2A\frac{1}{4}P_{aa}d_2(t)^2 \tag{5-7}$$

where $d_1(t)$ and $d_2(t)$ are the diameter (which is the function of the time t) of nanowires and the side-length (which is the function of the time t) of nanobelts, L1(t) and L2(t) are the height (which is the function of the time t) of the nanowires and the nanobelts, A is the constant that are relative with the growth. In the Eq. 4, the left section of the equation is the radial volume increase of the nanowire and the right section of the equation is the volume increase resulting from the axial supply precursors extracting from the Au catalyst, which leading to the augment of the radius of the nanowire. In the Eq. 5, the left section of the equation is the axial volume increase of the nanowire and the right section of the equation is the volume increase resulting from the axial supply precursors extracting from the Au catalyst, which leading to the augment of the height of the nanowire. Meanwhile, in the Eq. 6, the left section of the equation is the radial volume increase of the nanobelt and the right section of the equation is the volume increase resulting from the axial supply precursors extracting from the Au catalyst, which leading to the augment of the side-length of the nanobelt. In the Eq. 7, the left section of the equation is the axial volume increase of the nanobelt and the right section of the equation is the volume increase resulting from the axial supply precursors extracting from the Au catalyst, which leading to the augment of the height of the nanobelt.

Neglecting the difference in strain energies in every type of growth, we attain approximate conclusions as follows

$$d_1(t) \approx \left[2(t + 2r*_1^{\,2})\right]^{\frac{1}{2}},\ d_2(t) \approx \left[2(t + \frac{r*_2^{\,2}}{2})\right]^{\frac{1}{2}} \quad (5\text{-}8)$$

where t is the growth time of nanostructures after nucleating, and $r*_1$ and $r*_2$ are critical sizes of nanowires and nanobelts. According to Eq. (8), the growth rates of nanowires and nanobelts are shown in Figure 15(b). Distinctly, both growth rates of nanowire and nanobelt are basically the same during the growth of nanostructures. Therefore, the final size of nanobelts is approximately two times larger than that of nanowires, as $r*_2$ is double of $r*_1$. This theoretical result is in well agreement with our findings shown in Figure 13(d).

5.3. Shape Transition Thermodynamic Consideration

During the initial stage of the growth of nanowire nuclei, there are two probabilities of the shape transitions. One is from the wire nuclei to belt nuclei with an increase to the volume of nanowire nuclei, and another one is from the belt nuclei to wire nuclei with and increase to the volume of nanobelt nuclei. In order to clarify which shape transition is preferable in thermodynamical, we need to compare the Gibbs free energies of wire and belt nuclei. Therefore, the Gibbs free energy of the shape transition, i.e., the Gibbs free energy difference between wire nuclei and belt nuclei, is given as $\Delta G = V\Delta P + \delta\Delta S$ [92], where _P is a pressure difference between nanowire nuclei and nanobelt nuclei, which is small enough to ignore. $V = \pi d_1 2L_1/4 = d_2 2L_2$ is the volume of nanowire nuclei and nanobelt nuclei, in which d_1 and L_1 are the diameter and height of nanowire nuclei, and d_2 and L_2 are the side length and height of the nanobelt nuclei, and $L_2 = \pi d_1 2L_1/4d_2 2$. δ is the surface energy difference of the (200) plane of nanowire nuclei and the (211) plane of nanobelt nuclei. ΔS is the surface area difference between nanowires nuclei and nanobelt nuclei. Thus, ΔG can be expressed

$$\Delta G = \delta_2(4d_2L_2 + d_2^2) + \delta_2'' d_2^2 - \delta_1(\pi d_1 L_1 + \pi d_1^2/4) - \pi d_1^2 \delta_1''/4 \quad (5\text{-}9)$$

According to Eq. (9), when d_1 is equal to 150, 175 and 200 nm and L_1 = 12 nm, the relationships between ΔG and d_2 are shown in Figure 16(a). Clearly, three curves intersect at the A point (ΔG = 0 and d_2 = 43 nm), and $\Delta G < 0$ in these regions of AB, AC and AD. In detail, in these regions, $\Delta G < 0$ means the Gibbs free energy of nanowire nuclei is larger than that of nanobelt nuclei, suggesting that the shape transition from wire nuclei to belt nuclei is probable. In instance, the nanowire nuclei with diameters of 150, 175 and 200 nm could transform into the nanobelt nuclei with radial sizes of 60, 70, and 80 nm, respectively. Note that, $\Delta G > 0$, when $d_1 <$ 120 nm, implying that the shape transition from wire nuclei to belt nuclei of nanowire nuclei with the diameter less than 120 nm is not thermodynamically expected. When the nanobelts grow along the [301] direction, according to the Eq. (9), the relationships between ΔG and d_2 are shown in Figure 16(b). Clearly, three curves intersect at the E point (ΔG = 0 and d_2 = 34 nm), and $\Delta G < 0$ in these regions of EF, EG and EH. The nanowire nuclei with diameters of 150, 175 and 200 nm could transform into the nanobelt nuclei with radial sizes of 60, 70, and 80 nm, respectively. Note that, $\Delta G > 0$, when $d_1 <$ 90 nm, implying that the shape transition from the wire nuclei to the belt nuclei of nanowire nuclei with the diameter less than 90 nm is not thermodynamically expected. To sum up, the difference between the calculation results according to two different growth directions of the nanobelts is not so large that the conclusions are basically consistent and acceptable. Therefore, the larger nanowire nuclei sizes are, the easier the shape transition from wire nuclei to belt nuclei is, from the viewpoint of thermodynamic driving force. In fact, these theoretical predictions are not only consistent with experiments shown in Figure 13(d), but also are physical origins of the experimental observations that both the wire formation with small size and the belt formation with large size are favorable during the growth of nanostructures.

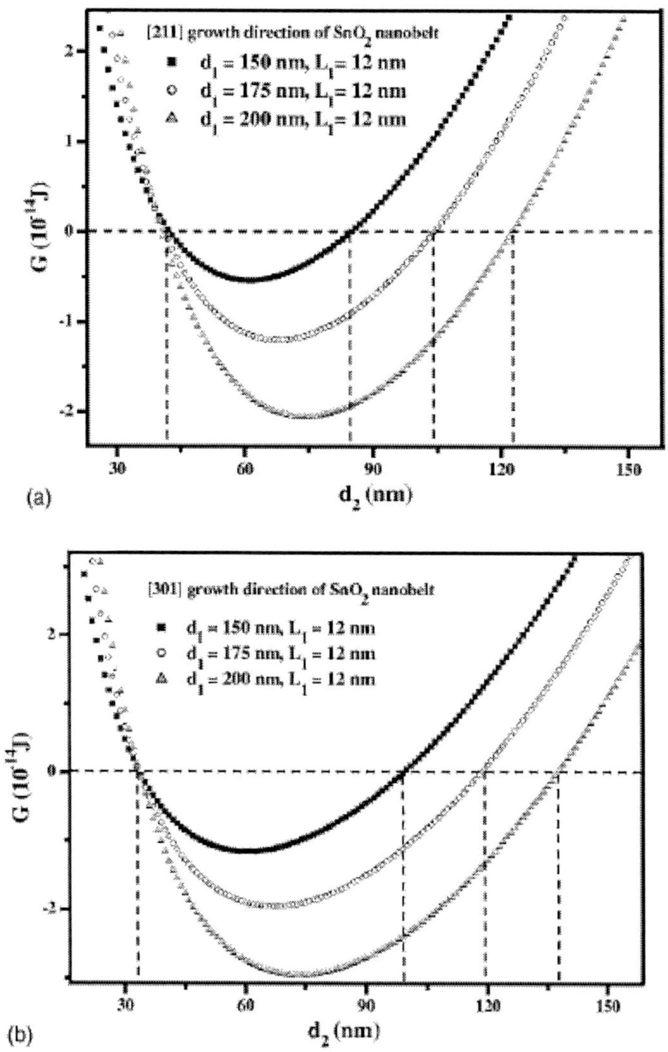

Figure 16. Dependence of the Gibbs free energy of the shape transition during the initial stage of the nuclei growth. (a) The shape transition from the wire nuclei to the belt nuclei that grows along the (211) direction on the radial size of the nanowires nuclei under conditions of d_I =150, 175, and 200 nm, respectively, and L_I = 12 nm. (b) The shape transition from the wire nuclei to the belt nuclei that growing along the (301) direction on the radial size of a nanowires nuclei under conditions of d_I = 150, 175, and 200 nm, respectively, and L_I = 12 nm.

In summary, we have systematically investigated the physical and chemical mechanisms of the shape evolution of 1D SnO2 nanostructures in both experimental and theoretical. Based on experiments and theories, we found that the thermodynamic driving force, the Gibbs free energy difference between two phases, is always responsible for the shape choice in the different growth stages such as nucleation, growth, and structural transition of 1D nanostructures, implying that the thermodynamic driving force is one of the physical origins causing the interesting morphological transformation in the growth of 1D nanostructures. Meanwhile, our studies suggested that the revealed physical mechanisms could be expected to generally apply to control the shape of 1D nanostructures during the growth for purpose of fundamental researches and potential applications.

6. Conclusion

1D nanostructures such as wires, rods, belts, and tubes have become the focus of intensive research owing to their unique applications in mesoscopic physics and fabrication of nanoscaled devices, as they not only provide a good system to study the electrical and thermal transport in 1D confinement, but also are expected to play an important role in both interconnection and functional units in fabricating electronic, optoelectronic, and magnetic storage devices with nanoscale dimension. Naturally, a great challenge for materials scientists is how to flexibly control the self-assembly and synthesis processes of 1D nanostructures for these purposes of fundamental researches and potential applications. Considering the self-assembly and synthesis processes of 1D nanostructures are relatively new material processing methods, it is important to develop new theoretical tools to investigate the physics and chemistry involved in these processes. For this issue, we have systemically established a series of new theoretical tools, i.e., thermodynamic and kinetic approaches, to address the nucleation, growth and phase transition of 1D nanostructures on the basis of the VLS process, which not only elucidate the physical and chemical mechanisms involved in the self-assembly and synthesis processes of 1D nanostructures, but also predict the growth behavior. Importantly, these theoretical tools provide a fundamental understanding to flexibly control the self-assembly and synthesis processes of 1D nanostructures. Therefore, we expected these new theoretical methods to be general approaches applicable to elucidate the formation of 1D nanostructures.

Acknowledgment

The National Science Foundation of China (90306006 and 50525206), the Ministry of Education (106126) and the Natural Science Foundation of Guangdong (01009737) supported this work.

REFERENCES

[1] Xia, Y.; Yang, P.; Sun, Y.; Wu, Y.; Mayers, B.; Gates, B.; Yin, Y.; Kim, F.; Yan, H. *Adv. Mater.* 2003, *15,* 353.
[2] Wang, Z. L. *J. Phys.: Condens. Matt.* 2004, *16*, R829.
[3] Chamberlin, R. V. *Nature* 2000, *408*, 337.
[4] Alivisato, P. *Nano Lett.* 2001, *1*, 109.
[5] Chamberlin, R. V. *Phys. Lett. A* 2003, *315*, 313.
[6] Abe, S.; Rajagopal, A. K. *Phys. Rev. Lett.* 2003, *91,* 120601.
[7] Wang, Z. L.; Kong, X.; Zou, J. M. *Phys. Rev. Lett.* 2003, *91,*185502.
[8] Zhang, C. Y.; Wang, C. X.; Yang, Y. H.; Yang, G. W. *J. Chem. Phys. B* 2004, *108*, 2589.
[9] Wang, C. X.; Yang, Y. H.; Liu, Q. X.; Yang, G. W. *J. Phys. Chem. B* 2004, *108*, 728.
[10] Wang, C. X.; Yang, Y. H.; Xu, N. S.; Yang, G. W. *J. Am. Chem. Soc.* 2004, *126*, 11303.
[11] Wang, C. X.; Chen, J.; Yang, G. W.; Xu, N. S. *Angew. Chem. Int. Edit.* 2005, *44*, 7414.
[12] Wang, C. X.; Yang, G. W. *Mater. Sci. and Eng. -R* 2005, *49*, 157.
[13] Pohl, K.; Bartelt, M. C.; Figuera, J. de la; Bartelt, N. C.; Hrbek, J.; Hwang, R. Q. *Nature* 1999, *397*, 238.
[14] Plass, R.; Last, J. A.; Bartelt, N. C.; Kellogg, G. L. *Nature* 2001, *412*, 875.
[15] Wagner, R. S.; Ellis, W. C. *Appl. Phys. Lett.* 1964, *4*, 89.
[16] Wagner, R. S.; Ellis, W. C.; Jackson, K. A.; Arnold, S. M. *J. Appl. Phys.* 1964, *35*, 2993.
[17] A. M. Morales and C. M. Lieber, *Science* 1998, *279*, 208.
[18] Duan, X. F.; Lieber, C. M. *Adv. Mater.* 2000, *12,* 298.
[19] Chen, C. C.; Yeh, C. C. *Adv. Mater.* 2000, *12,* 738.
[20] Wu, Y.; Yang, P. *Chem. Mater.* 2000, *12,* 605.
[21] Wu, X. C.; Song, W. H.; Wang, K. Y.; Hu, T.; Zhao, B.; Sun, Y. P.; Du, J. J. *Chem. Phys. Lett.* 2001, *336,* 53.
[22] Wang, C. X.; Wang, B.; Yang, Y. H.; Yang, G. W. *J. Phys. Chem. B* 2005, *109,* 9966.
[23] S. H. Tolbert and A. P. Alivisatos, *Annu. Rev. Phys. Chem.* 1995, *46*, 595.
[24] Yang, G. W.; Liu, B. X. *Phys. Rev. B* 2000, *61*, 4500.
[25] Kelton, K. F. *Solid State Phys.* 1991, *45,* 75.
[26] Tan, T. Y.; Li, N.; Gösele, U. *Appl. Phys. Lett.* 2003, *83,* 1199.
[27] Louchev, O. A.; Sato, Y.; Kanda, H. *Appl. Phys. Lett.* 2002, *80,* 2752.
[28] Jiang, Q.; Zhang, S. H.; Li, J. C. *Solid State Commun.* 2004, *130,* 581
[29] Kirkaldy, J. S. *Diffusion in the Condensed State,* The Universities Press, Belfast, 1987.
[30] Feng, S. Q.; Yu, P. D.; Zhang, H. Z.; Bai, Z. G.; Ding, Y. *J. Cryst. Growth* 2000, *209,* 513.
[31] Givargizov, E. I. *Highly Anisotropic Crystals,* Boston, 1986.
[32] Hardy, S. C. *J. Cryst. Growth* 1984, *69,* 456.
[33] Lew, K. K.; Pan, L.; Dicksy, E. C.; Redwing, J. M. *Adv. Mater.* 2003, *15,* 2073.
[34] Cui, Y.; Lauhon, L. J.; Gudiksen, M. S.; Wang, J. *Appl. Phys. Lett.* 2001, *78,* 2214.
[35] Lee, S. T.; Wang, N.; Lee, C. S. *Mater. Sci. Eng. A* 2000, *286,* 16.

[36] Lide, D. R. *CRC Handbook of Chemistry and Physics*, 71st ed. Chemical Rubber, Ohio, 1990–1991.
[37] *Sargent-Welch Scientific Company, Table of Periodic Properties of the lements,* Skokie, IL, 1980, pp. 1–2.
[38] King, H. W. in *Physical Metallurgy*, 3rd ed., edited by R. W. Cahn and P. Haasen, North-Holland Physics, Amsterdam, 1983, pp. 59–63.
[39] *Binary Alloy Phase Diagram*, edited by T. B. Massalsky, ASM, New York, 1986, p. 1108.
[40] The saturation concentration of Si (accounting for 50 at %) is defined by $C^*=NA/(M\text{Si}/2\text{Si})+(M\text{Fe}/2\text{Fe})$, where *NA*, *M*Si, *M*Fe, Si, and Fe, are the Avogadoro constant, mole mass of iron, mole mass of silicon, density of liquid silicon, and density of liquid iron, respectively.
[41] Wang, N.; Tang, Y. H.; Zhang, Y. F.; Yu, D. P.; Lee, C. S.; Bello, I.; Lee, S. T. *Chem. Phys. Lett.* 1998, *283,* 368.
[42] Wang, N.; Tang, Y. H.; Zhang, Lee, C. S.; Bello, I.; Lee, S. T. *Chem. Phys. Lett.* 1999, *299,* 237.
[43] Morales, A. M.; Lieber, C. M. *Science* 1998, *208*, 279.
[44] Sunkara, M. K.; Sharma, Miranda, S.; R.; Lian, G.; Dickey, E. C. *Appl. Phys. Lett.* 2001, *79*, 1546.
[45] Kamins, T. I.; Williams, R. S.; Basile, D. P.; Hesjedal, T.; Harris, J. S. *J. Appl. Phys.* 2001, *89*, 1008.
[46] Westwater, J.; Gosain, D. P.; Tomiya, S.; Usui, S.; Ruda, H. *J. Vac. Sci. Technol. B* 1997, *15*, 554.
[47] Hiruma, K.; Yazama, M.; Katsuyama, T.; Ogawa, K.; Haraguchi, K.; Koguchi, M.; Kakibayashi, H. *J. Appl. Phys.* 1995, *77*, 447.
[48] Duan, X.; Lieber, C. M. *J. Am. Chem. Soc.* 2000, *122*, 188.
[49] Ohlsson, B. J.; Björk, M. T.; Magnusson, M. H.; Deppert, K.; Samuelson, L.; Wallenberg, L. R. *Appl. Phys. Lett.* 2001, *79,* 3335.
[50] Gudiksen, M. S.; Lieber, C. M. *J. Am. Chem. Soc.* 2000, *122*, 8801.
[51] Cui, Y.; Lauhon, L. J.; Gudiksen, M. S.; Wang, J. F.; Lieber, C. M. *Appl. Phys. Lett.* 2001, *78*, 2214.
[52] Davis, S. H. *Proc. Roy. Soc. London. Ser. A* 1969, *310*, 341.
[53] Galdi, G. P.; Straughan, B. *Arch. Rational Mech. Anal.* 1989, *89*, 211.
[54] Hwang, N.M.; Hahn, J.H.; Poon, D.Y. *J. Cryst. Growth* 1996, *160*, 87.
[55] Gao, Y.H.; Bando, Y. *Nature* 2002, *415,* 599.
[56] Gao, Y.H.; Bando, Y. *Appl. Phys. Lett.* 2002, *81*, 3966.
[57] Liu, Q. X.; Wang, C. X.; Yang, Y. H.; Yang, G. W. *Appl. Phys. Lett.* 2004, *84*, 4568.
[58] Fissel, A.; Kaiser, U.; Ducke, E.; Schröter, B.; Richter, W. *J. Crystal Growth* 1995,*154*, 72.
[59] Ho, M. H.; Sun, Y. C. *Surf. Sci.* 2002, *516*, L540.
[60] Liu, Z. Q.; Zhou, W. Y.; Sun, L. F.; Tang, D. S.; Zou, X. P.; Li, Y. B.; Wang, C. Y.; Wang, G.; Xie, S. S. *Chem. Phys. Lett.* 2001, *341*, 523.
[61] Yu, H; Buhro, W. E. *Adv. Mater.* 2003, *15,* 416.
[62] Iijima, S. *Nature* 1991, *354,* 56.
[63] Ajayan, P. M.; Iijima, S. *Nature* 1993, *361*, 333.
[64] Dai, H.; Wong, E. W.; Lu, Y. Z.; Fan, S.; Lieber, C. *Nature*, 1995, *375*, 769.

[65] Tsang, S. C.; Chen, Y. K.; Harris, P. J.; Green, M. L. *Nature*, 1994, *372*, 159.
[66] Hsu, W. K.; Hare, J. P.; Terrones, M.; Harris, P. J.; Kroto, H. W.; Walton, D. R. *Nature* 1995, *377*, 687.
[67] Subramoney, S. *Adv. Mater.* 1998, *10*, 1157.
[68] Pradhan, B. K.; Kyotani, T.; Tomita, A. *Chem. Commun.* 1999, *67*, 595.
[69] Wong, E. W.; Maynor, B. W.; Burns, L. D.; Lieber, C. M. *Chem. Mater.* 1996, *8*, 2041.
[70] Han, W.; Fan, S.; Li, Q.; Hu, Y. *Science* 1997, *277*, 1287.
[71] Terrones, M.; Grobert, N.; Hsu, W. K.; Zhu, Y. Q.; Hu, W. B.; Terrones, H.; Hare, J. P.; Kroto, H. K.; Walton, D. R. *MRS Bull.* 1999, *24*, 42.
[72] See: A special issue on carbon nanotubes, *Acc. Chem. Res.* 2002, *35*, No. 12.
[73] Wilson, A. Philos. Mag. 1900, 50, 609; Frenkel, J. *Phys. Z.* 1932, *1*, 498.
[74] Wang, Q.; Challa, S. R.; Shol, D. S.; Johnson, J. K. *Phys. Rev. Lett.* 1999, *82*, 956.
[75] Calbi, M. M.; Cole, N. W.; Gatica, S. M.; Bojan, M. J.; Stan, G. *Rev. Mod. Phys.* 2001, *73*, 857.
[76] Skoulidas, A. I.; Ackerman, D. M.; Johnson, J. K.; Sholl, D. S. *Phys. Rev. Lett.* 2002, *89*, 185901.
[77] Halpin-Healy, T.; Zhang, Y. C. *Phys. Rep.* 1995, *254*, 215.
[78] Yang, H.; Labella, V. P.; Bullock, D. W.; Ding, Z.; Smathers, J. B.; Thibado, P. M. *J. Cryst. Growth* 1999, *201-202,* 88.
[79] Liu, Z.; Gao, Y.; Bando, Y. *Appl. Phys. Lett.* 2002, *81*, 4844.
[80] Mighell, A. D.; Santoro, A.; Donny, J. D. H. *International tables for x-ray crystallography*, 1969, Vol. I.
[81] Thurmer, K.; Carter, C. B.; Bartelt, N. C.; Hwang, R. Q. *Phys. Rev. Lett.* 2004, *92,* 106101.
[82] Ross, F. M.; Tersoff, J.; Reuter, M. C. *Phys. Rev. Lett.* 2005, *95,* 146104.
[83] Liu, Q. X.; Wang, C. X.; Yang, G. W. *Phys. Rev. B* 2005, *71*, 155422.
[84] Liu, Q. X.; Wang, C. X.; Xu, N. S.; Yang, G. W. *Phys. Rev. B* 2005, *72*, 085417.
[85] Oviedo, J.; Gillan, M. J. *Surf. Sci.* 2000, *463*, 93.
[86] Slater, B.; Catlow, C. R.; Gay, D. H.; Williams, D. E.; Dusastre, V. *J. Phys. Chem. B* 1999, *103*, 10644.
[87] Vitos, L.; Ruban, A. V.; Skriver, H. L.; Kollar, J. *Surf. Sci.* 1998, *411*, 186.
[88] Kwon, S. J.; Park, J. Q. *J. Chem. Phys.* 2005, *122*, 214714.
[89] Witt, F.; Vook, R. W. *J. Appl. Phys.* 1968, *39*, 2773.
[90] Kwon, S. J. *J. Phys. Chem. B* 2006, *110*, 3876.
[91] Pelleg, J.; Zerin, L. Z.; Lungo, S. *Thin Solid Films* 1991, *197*, 117.
[92] Liang, L. H.; Liu, F.; Shi, D. X.; Liu, W. M.; Xie, X. C.; Gao, H. J. *Phys. Rev. B* 2005, *72*, 035453.

In: Nanotechnology Research Developments
Editor: R. Jiménez-Contreras, pp. 173-203
ISBN: 978-1-60021-899-6

Chapter 4

PREPARATION AND CHARACTERIZATION OF NANOPARTICLES AND NANOCRYSTALLINE BULKS OF PURE RARE-EARTH METALS

Xiaoyan Song
College of Materials Science and Engineering,
Key Laboratory of Advanced Functional Materials,
Ministry of Education, Beijing University of Technology,
Beijing 100022, China

ABSTRACT

Because of the extremely high chemical reactivity and hence the considerably rigorous requirements for the equipments to preserve a high purity of the product, the preparation and characterization of nanostructured pure rare-earth metals are still a big challenge in nanoscience and nanotechnology. In this chapter, followed a thermodynamic description of the energy state and the thermal stability of rare-earth nanoparticles and nanocrystalline bulks, the practical preparation and characterization of nanoscale rare-earths are depicted in detail. Different from a number of existing techniques that have been developed to synthesize nanocrystalline bulk materials, a home-designed configuration of the "oxygen-free" in-situ synthesis system, where the inert-gas condensation is combined with the SPS technology into an entirely closed and vacuumized system, is mainly introduced. With the technically designed route of sequentially performed processes, i.e. amorphization of nanoparticles, nucleation and growth of the short-range ordered "clusters" inside the nanoparticle, and the complete nanocrystallization, the ultrafine nanocrystalline bulk can be obtained with the grain size much smaller than the initial nanoparticle size. Microstructures and properties of some typical members of the rare-earth family are presented ordinally. The prominently changed thermal, physical and mechanical properties imply the promising applications of the nanocrystalline rare-earth metals and alloys in the fields of functional and structural materials. Finally, although the chapter is focused on rare-earth nanoparticles and nanocrystalline bulks, the proposed thermodynamic models and preparation techniques are potentially applicable to a large variety of nano metal materials, hence may establish a new way for studies on microstructures and performances of metallic nanomaterials, especially of those highly reactive in the air.

1. INTRODUCTION

Rare-earths are a group of 17 metal elements with atomic numbers 21, 39, and 57–71; the name lanthanides is reserved for the elements 58–71. The unique configuration of unpaired 4f electrons and the rich structures of the energy levels enable the rare-earth metals to possess many particular physical and chemical properties that are of world-wide interests, such as high electrical conductivity, large magnetic moment, and very high complexation reactivity [1,2]. Based on these properties, the rare-earth elements and compounds have been applied in many advanced materials such as high-performance magnets, fluorescent materials [3], chemical sensors, autocatalysts [4], and high-temperature superconductors [5]. For example, Nd2Fe14B and SmCo5/Sm2Co17 are powerful rare-earth permanent magnets, and Sm-Co permanent magnets have the highest resistance to demagnetization of any known magnetic material and very high intrinsic coercive force [6,7]. There is an increasing demand for high purity rare-earth metals and oxides particularly for use in heat storage and dispersal [8], corrosion resistance [9], as well as for environment-favorable applications, such as in pigments for paint and plastics [10] to replace heavy metal (e.g. Pb, Cr, Cd and Hg) pigments; in cement manufacture to reduce the temperature of calcination and help save energy [11]; and in refrigeration components as a potential substitute for chlorofluorocarbon (CFC) [12].

It is known that materials at nano-scale have properties significantly different from those at larger scales. There are two principal reasons for this: first, nanomaterials have a highly increased relative surface or boundary area, this can make materials more chemically reactive and affect their strength or electrical properties; second, quantum effects can begin to dominate the behavior of matter at the nano-scale, which affect the optical, electrical and magnetic behavior of materials.

The three-dimensional nanoscale materials, i.e. nanoparticles, classified as particles less than 100nm in diameter, and nanocrystalline metals, made up of nanometer-sized grains, are in the category that is discussed in this chapter. For the nanoscale rare-earth particles, due to a much greater surface area per unit mass hence a higher proportion of atoms at the surface compared with those of larger particles, some new or enhanced size-dependent properties such as chemical reactivity and optical behavior may exhibit. For the nanocrystalline rare-earth metals, as the grain size decreases, the interface area within the material greatly increases; this can significantly affect mechanical, electrical and magnetic properties of rare-earths. For example, it has been shown that magnets made of nanocrystalline Sm-Co grains possess unusual magnetic properties [13], as high coercivity can be obtained because magnetization flips cannot easily propagate past the extremely increased grain boundaries. This could lead to applications in motors, analytical instruments like magnetic resonance imaging (MRI), used widely in hospitals, and microsensors.

Since remarkably improved functional and structural properties can be expected for the nanoscale rare-earth materials, the related investigations have attracted growing interests in multiple interdisciplinary fields. However, because of the extremely high chemical reactivity and hence the considerably rigorous requirements for the equipments to preserve a high purity of the product, the preparation and characterization of nanoscale pure rare-earth metals are still a big challenge in nanoscience and nanotechnology. To synthesize nanoscale rare-earths is not simply another step in the miniaturization of materials with some existing 'top-down' and 'bottom-up' techniques, the chemical activity and physical properties of rare-earths, as well as the impact of those on the preparation route, have to be given careful considerations. Consequently, due to the confinement of the preparation technique, many important features of the nanoscale rare-earth metals such as the physical, the chemical, the thermal, and the mechanical characteristics have been rarely reported so far. Whereas, the corresponding research topics are of great importance both for the development of nanoscience and nanotechnology and for extending the applications of the rare-earth metals.

Although a number of reviews and books have been published on the nanoscience and nanotechnology, especially, there have been numerous reports on preparation and characterization of novel nanostructured materials, very few of them comprehensively deal with the features of nanoparticles and nanocrystalline bulks of the rare-earth family. In this consideration, this chapter is an effort to depict to the maximum extent possible the features of rare-earth nanoparticles and nanocrystalline bulks, presenting more detailedly the progress of our research specialization. The author tried to include as many as possible details in the following sections, however, it should be understood that the content of the chapter is still far from an exhaustive description of all the characteristic aspects of nanoscale rare-earths.

In this chapter, focused on the nanoscience and nanotechnology of nanoscale rare-earths, the content is outlined as follows: in Section 2, the energy states and the thermodynamic features of nanoparticles and nanocrystalline bulks are described by the developed thermodynamic models, with calculations performed on nanoscale rare-earth metals; then preparation and microstructure analysis of rare-earth nanoparticles and nanocrystalline bulks are presented in Section 3 and Section 4, respectively; characterization of thermal, physical and mechanical properties of nanocrystalline rare-earth bulks and comparison with those of conventional polycrystalline rare-earth metals are expounded in Section 5.

2. Thermodynamic Analysis of Nanoparticles and Nanocrystalline Bulks

It has been demonstrated in literature [14,15] that for nanoscale materials, owing to the significantly increased total surface area (e.g. nanoparticles) or the grain boundary area (e.g. nanocrystalline bulks), many new features in the crystal structure, interface, thermodynamics and phase transitions were along with introduced. Since the effects of the surface and the grain boundary on the total thermodynamic properties of materials were not emphasized or even neglected in the conventional thermodynamic models, the thermodynamic studies on nanomaterials, which are aimed to comprehend the thermal features of a material system including its surface or interface as well as the bulk, are needed extraordinarily to be carried out. In this section, the dependence of the fundamental thermodynamic functions on the nanoparticle size or the nano-grain size will be developed, based on which the thermal and physical properties, as well as the phase-transition features and thermal stability of the nanomaterials, may be quantitatively described.

2.1. Gibbs Energy of Nanoparticles

The thermodynamic equilibrium between solid and liquid phases in a small particle system was firstly considered nearly a century ago by Pawlow [16]. In 1954, Takagi [17] observed that the melting temperatures of pure Pb and Bi decreased with decreasing the size of the particles. From then on, several studies have been carried out to explain the dependence of the melting points of metals and alloys on the metallic particle size [18-23]. In recent years, Tanaka and his group [24-26] have evaluated the surface properties as well as phase diagrams of a few binary alloys in the small particle system on the basis of thermodynamic databases. The present thermodynamic model of the energy state of metal nanoparticles has been developed based on the Tanaka model of Gibbs energy in small particle systems.

For a single solid nanoparticle with its radius r, the total Gibbs energy of the particle is the sum of the surface energy and the Gibbs energy in the internal bulk of the particle, i.e.

$$G^{Total,Sol} = G^{Surface,Sol} + G^{Bulk,Sol} \tag{1}$$

The surface energy of the particle can be calculated as

$$G^{Surface,Sol} = (2\sigma^{Sol} V_{mol} / r) \cdot V / V_{mol} \tag{2}$$

where σSol is the surface tension of the solid particle; Vmol and V are the mole volume and the volume of the solid particle, respectively. For the surface tension of pure solid metals at temperature T, the estimation in the Tanaka model [24,25] is used, i.e.

$$\sigma^{Sol} = 1.25 T_{mp} / 1000 - 0.0001 \cdot (T - T_{mp}) \tag{3}$$

where T_{mp} is the melting temperature of the material in the conventional bulk form.

The Gibbs energy of the internal bulk of the particle can be calculated by the classic thermodynamic model of the metal bulk material with a general function as

$$G^{Bulk,Sol} = G_{Sol}(T) \cdot V / V_{mol} \tag{4}$$

GSol(T) is the Gibbs free energy of the material in the conventional bulk form, which is a function of the temperature and can be obtained from the thermodynamic databases, e.g. SGTE. As the nanoparticle with a spherical shape is assumed, the volume of the particle is $V=4/3\pi r^3$. From eqs. (1)-(4), the total Gibbs energy of the single-phased solid nanoparticle is expressed as

$$G^{Total,Sol} = 4/3\pi r^3 / V_{mol} \cdot G_{Sol}(T) + 4/3\pi r^2 \cdot (0.0027 T_{mp} - 0.0002 T) \tag{5}$$

Thus, the Gibbs energy of the nanoparticles can be evaluated by the temperature and the particle size, and the intrinsic material parameters such as its bulk melting temperature and the specific heat capacity (from which GSol(T) can be deduced).

The Gibbs energy of the rare-earth nanoparticles were calculated by eq. (5). Figure 1 shows as examples the Gibbs energy of nanoparticles in the rare-earth family such as Sm, Nd, Gd and La with given crystal structures. It can be seen that the Gibbs energy of nanoparticles changes with both the temperature and the particle size. For a certain size class of the particles, the Gibbs energy reduces with increasing the temperature. At a given temperature, the Gibbs energy reduces with the increase of the nanoparticle size. The above changing features of the Gibbs energy of particles result in the spontaneous tendencies of segregation of fine nanoparticles and rapid coarsening of nanoparticles with increasing the temperature.

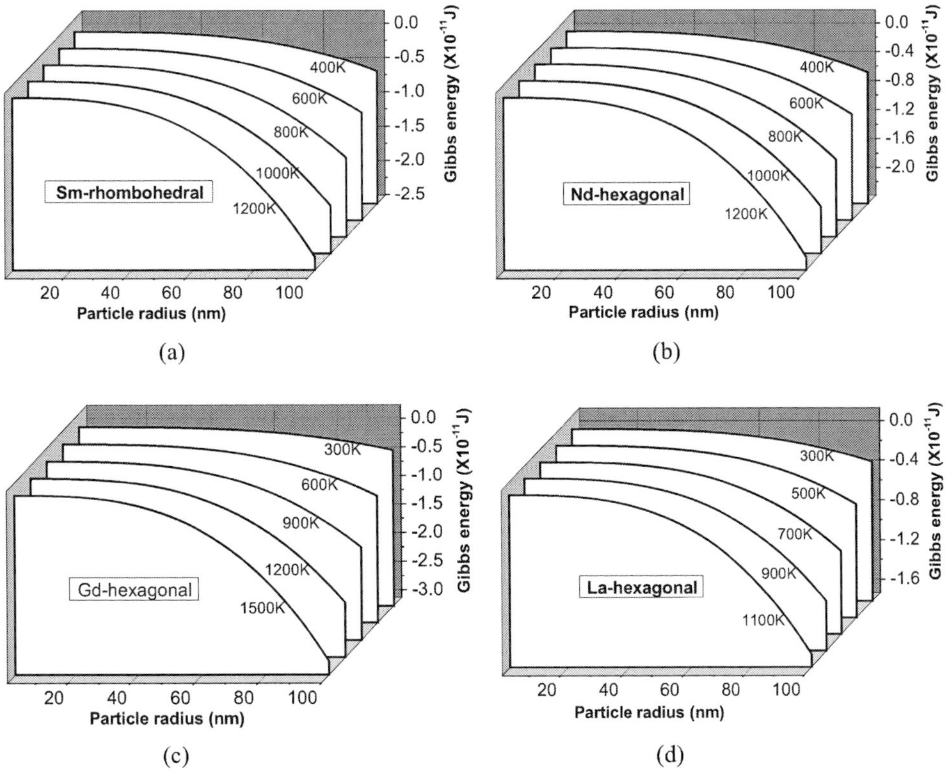

Figure 1. The Gibbs energy of nanoparticles as a function of temperature and particle size of rare-earth metals of (a) Sm, (b) Nd, (c) Gd and (d) La.

2.2. Melting Temperatures of Rare-Earth Nanoparticles

In the condition of phase equilibrium, the melting point of the particle is the temperature at which the Gibbs energy of the solid particle is equal to that of the particle liquid, i.e. $GLiq=G^{Total, Sol.}$ The Gibbs energy of the liquid state of the particle is estimated as

$$G^{Liq} = G_{Liq}(T) \cdot V / V_{mol} \tag{6}$$

where GLiq(T) is the Gibbs free energy at the liquid state of the material evaluated by the conventional thermodynamic model, it is a function of the temperature and can also be obtained from the thermodynamic databases. Vmol and V are the mole volume and the volume of the liquid phase, respectively, here the possible tiny change of the volume while the particle melts is neglected. Equalize eqs. (5) and (6), one obtains

$$T = 13.5T_{mp} - 5 \cdot 10^3 (G_{Liq}(T) - G_{Sol}(T)) \cdot r / V_{mol} \tag{7}$$

which is an implicit function equation between the melting temperature of the nanoparticle and the particle size, containing the Gibbs free energy functions of the solid and liquid phases, respectively.

Figure 2 shows as examples the size-dependence of the melting temperatures of rare-earth nanoparticles of Sm, Nd, Gd and La with given crystal structures, which are estimated by eq. (7).

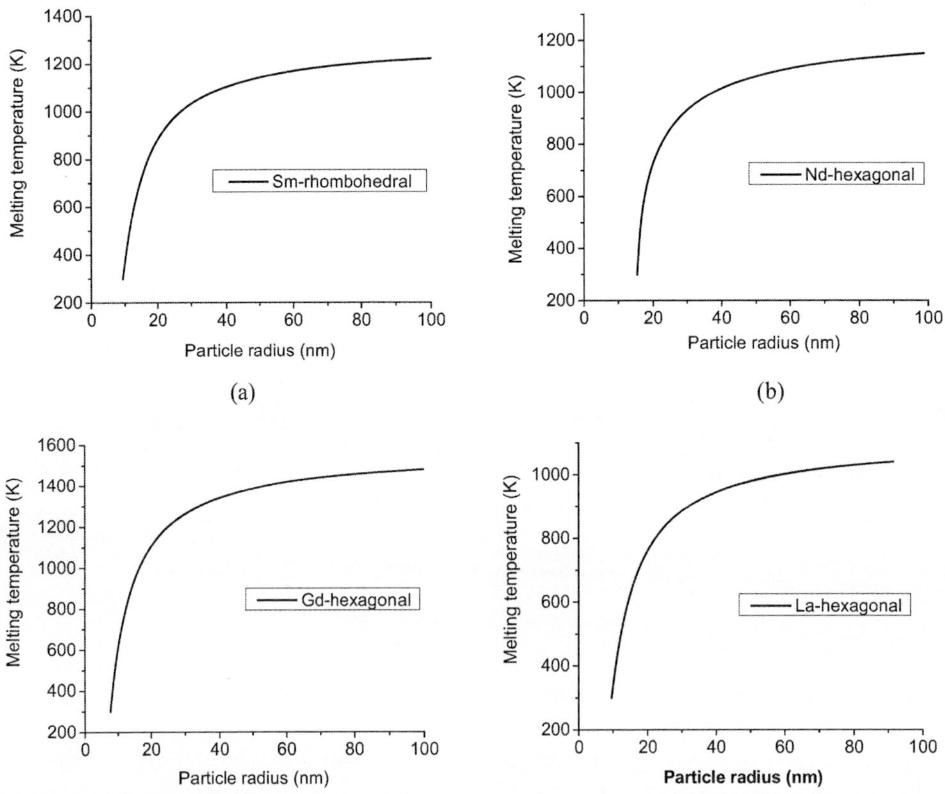

Figure 2. Model prediction on the size-dependence of the melting temperatures of rare-earth nanoparticles of (a) Sm, (b) Nd, (c) Gd and (d) La.

It can be seen that the rare-earth particles have the similar reduction tendency of the melting points with the decrease of the particle size. Generally, when the particle size is reduced to below 30 nm, there is a sharp decrease of the melting temperature of the nanoparticle. As the particle size decreases, the number of atoms at the surface increases, e.g. a particle of size 10 nm has 20% of its atoms on its surface, and at 3 nm 50% of its atoms. Thus nanoparticles have a much greater surface area per unit mass compared with larger particles. Due to the vibration relaxation of atoms at the surface, as the amplitude of vibration reaches 10-20% lattice constants, the melting takes place starting from the surface.

As is known, manufactured nanoparticles are typically not products in their own right, but generally serve as raw materials, ingredients or additives in other materials. For most applications, nanoparticles will be fixed, for example, consolidated into a bulk of the same material or within a composite. Based on the model prediction of the melting temperatures of the rare-earth nanoparticles, the processing parameters of consolidation can be optimized for the methods of powder metallurgy. Since the melting points of nanoparticles are greatly reduced as compared with those of the conventional bulk material and the coarse particles, the consolidation of nanoparticles can be performed at lower sintering temperatures.

2.3. Thermodynamic Model of Nanocrystalline Metal Bulk

As soon as the nanocrystalline materials are produced, the stability of nanostructures becomes most importance, namely, the nanoscaled grain structure should stay sufficiently stable over a useful period of temperature and/or time to retain the attractive properties of the nanocrystalline material. Rapid grain growth is the principal process of the destabilization of nanocrystalline structures. Nano-grain growth behaviors have been studied by various experimental techniques such as TEM [27-29], DSC [30], X-ray diffraction [31,32], and Raman [14] and Mössbauer spectroscopy [33], as well as Monte-Carlo [34] and molecular dynamics simulations [35]. Very recently, a few studies have attempted to investigate the grain growth kinetics in nanocrystalline materials with a thermodynamic point of view [36-38].

With reducing the grain size, the volume fraction of grain boundaries is significantly increased in the nanocrystalline materials. Structurally, a nanocrystalline material consists of two components: nanometer-sized crystallines composing of atoms that locate in the regular crystal lattice, and boundary regions composing of atoms that are partially ordered or disordered with a lower density [39]. In other words, a nanocrystalline material is inherently heterogeneous, or thermodynamically non-equilibrium on a nanometer scale. As compared to the conventional coarse-grained materials, the thermal stability of nanocrystalline materials is much more strongly affected by the grain boundaries. From the thermodynamic point of view the atomically disordered nano-grain boundaries influence the configurational and vibrational entropy as well as the enthalphy of the system [40]. Due to the fact that nanocrystalline materials contain a large volume fraction of grain boundaries, the thermal properties cannot be sophisticatedly described by the classic thermodynamic functions that have been derived for the polycrystalline materials where the contribution of the grain boundaries to the thermal features of the material is generally neglected.

Based on a "dilated crystal model", Fecht [41] and Wagner [42] established the thermodynamic basis for describing the thermal properties of interfaces in nanocrystalline metals. Fecht calculated some thermodynamic parameters of nano-grain boundaries of Cr, Pd and Hf in terms of a universal equation of state (EOS) [41]. Whereas, Wagner calculated the thermal properties of nano-grain boundaries of Cu, Pd and Ir using the quasiharmonic Debye approximation (QDA) [42]. Later Lu [43] and Meng et al [44] applied QDA to analyze the nano-crystallization of NiP and Ni from the amorphous state and the nanocrystalline γ-Fe→α-Fe phase transformation, respectively. Both the EOS and QDA approaches showed good agreement between the theoretical calculations and the experiments [41-44]. However, these models concentrated attention on interpreting the anomalies in the phase transformation behaviors of nanocrystalline materials as compared with those of polycrystalline specimens, but were not applied to analyze the characteristics of nano-grain growth. Considering that the thermal stability of the nano-grain boundary is remarkably influenced by its inherent non-equilibrium properties, the grain boundary migration is probably controlled by its thermodynamic energy state. Thus, the conclusions concerning the nano-grain growth kinetics without correlating with the thermodynamic features of nano-grain boundaries, are likely to be incomplete or unsubstantiated. Therefore, a model that has been recently developed in our group [37] which correlated the thermodynamics of nano-grain boundaries and the grain growth kinetics in nanocrystalline materials will be depicted in this section. The model can be used to describe the thermodynamic properties, as well as to predict the thermal stability of nanocrystalline metal bulks.

2.3.1. Thermodynamic Functions of Nanocrystalline Interfaces

To describe the structure state of the interfaces in nanocrystalline materials, Fecht [41] and Wagner [42] proposed a dilated crystal model, where a nano-grain boundary is equivalent to a dilated crystal with same crystallographic structure as the grain interior, but it has an expanded unit cell in which the nearest-neighbor separation between atoms is larger than the equilibrium interatomic distance as in a perfect crystal. Thus, a concept of 'excess volume', when compared to that of the perfect crystal, was introduced to characterize the features of the nano-grain boundary structure, which was defined as [41-44]:

$$\Delta V = \frac{V_b}{V_0} - 1 \tag{8}$$

where Vb is the atomic volume in the 'dilated crystal' at the nano-grain boundary, V0 is the atomic volume of perfect crystal in the equilibrium state. Distinctly, ΔV characterizes the deviation extent of the structure of nano-grain boundary with respect to the crystal. The excess volume ΔV, which is directly related to the nano-grain size (there is no conclusive relationship between the grain size and ΔV so far, though, it is recognized that ΔV increases with the decrease of grain size), is the most significant intrinsic parameter that affects the thermal properties of nano-grain boundary, as have been shown theoretically in Ref. [41-44]. Furthermore, it is a key parameter that correlates the thermodynamics of nano-grain boundary with the behavior of nano-grain growth, as will be shown in the following content.

Based on the dilated crystal model, the EOS and QDA approaches need to be combined in order to derive the fundamental thermodynamic functions for the nano-grain boundaries, where the nano-scale effects can be elaborated quantitatively. Assuming that the thermal features of the nano-grain boundary is equivalent to those of the dilated crystal with the same excess volume, the fundamental thermodynamic functions of the nano-grain boundary, including the excess enthalpy ΔH, the excess entropy ΔS, and the excess free energy ΔG, which are generated if compared with the grain interior, are given by [41]

$$\Delta H = E + PV_b \tag{9}$$

$$\Delta S = C_V \gamma \ln(V_b / V_o) \tag{10}$$

$$\Delta G = \Delta H + C_V (T - T_R) - T[\Delta S + C_V \ln(T - T_R)] \tag{11}$$

where E is the binding energy of atoms at the nano-grain boundary [45,46], and can be calculated using the following equation:

$$E = E_0 E^*(a^*) \tag{12}$$

E0 is the binding energy of atoms at the equilibrium state, and can be estimated [47] by

$$E_0 = \frac{(3.45K_B)^2}{12\pi r_0^3 \lambda_0^2 B_0} \tag{13}$$

where λ_0 and B_0 are the zero-pressure linear expansion coefficient and the modulus at the equilibrium state, respectively. K_B is the Boltzman constant. r_0 is the interatomic distance at the equilibrium state (the equilibrium Weigner-Seitz cell radius [48]). $E^*(a^*)$ is a universal function [45] with the expression

$$E^*(a^*) = (-1 - a^* - 0.05a^{*3})e^{-a^*} \tag{14}$$

where

$$a^* = \frac{r_b - r_0}{l} \tag{15}$$

Here rb is the interatomic distance in the dilated crystal at the nano-grain boundary, l is a scaling length characteristic of the metal [49].

P in eq. (9) is a negative pressure developed at the grain boundary as a result of the reduction of the grain size, and can be approximated as a function of the atomic volume at the grain boundary and the absolute temperature [49],

$$P(V_b, T) = \frac{3B_O}{(V_b / V_0)^{2/3}}(1 - (V_b / V_0)^{1/3}) \exp[\eta_O (1 - (V_b / V_0)^{1/3})] + \alpha_O B_O (T - T_R) \tag{16}$$

α_0 is the zero-pressure bulk expansion coefficient at the reference temperature T_R. η_0 is a function of the derivative of the bulk modulus with respect to the pressure at the zero point, and is expressed as]1)/[(2/3₀₀−∂∂==PPBη.

γ in eq. (10) is the Grüneisen parameter of the dilated crystal at the nano-grain boundary, and is expressed by eq. (17) as suggested by Wagner [42], CV in eqs. (10)-(11) is the specific heat capacity at constant volume. In EOS, CV was a constant value of 3kB for each atom in the harmonic approximation at high temperature and is independent of ΔV [41]. To include the effect of ΔV, the Debye temperature Θ and Debye function D(Θ/T) [42],

$$\gamma = -1 - \frac{V_b}{2}\left[\frac{\partial^2 P/\partial V_b^2 - 10P/9V_b^2}{\partial P/\partial V_b + 2P/3V_b}\right] \quad (17)$$

which are both influenced by ΔV, are introduced to describe C_V:

$$C_V = 3R\left[12\left(\frac{T}{\Theta}\right)^3 \int_0^{\Theta/T} \frac{x^3}{e^x - 1}dx - \frac{3(\frac{\Theta}{T})}{e^{\Theta/T} - 1}\right] \quad (18)$$

where R is the gas constant. The specific heat capacity at constant pressure C_P can be obtained from C_V by the relationship:

$$C_P - C_V = -T\frac{\left(\frac{\partial P}{\partial T}\right)_{V_b}^2}{\left(\frac{\partial P}{\partial V_b}\right)_T} = \frac{\alpha_0^2 B_0^2 V_b T}{B(V_b)} \quad (19)$$

where . $B(V_b) = -V_b(\partial P/\partial V_b)_T$.

The Debye temperature depends on the interatomic distance rb at the nano-grain boundary and is written as [42]

$$\Theta(r_b) = \left[\frac{\varphi'(r_b)}{\varphi''(r_0)}\right]^{1/2} \Theta_0 \quad (20)$$

with $\varphi'(r_b) = \partial\varphi/\partial r_b$, $\varphi''(r_0) = (\partial^2\varphi/\partial r^2)_{r=r_0}$. Θ_0 is the Debye temperature of the crystal in the grain interior at the equilibrium state, ϕ is the Morse function [50] and expressed as

$$\varphi(r_b) = D\{\exp[-4\gamma_0(r_b/r_0 - 1)] - 2\exp[-2\gamma_0(r_b/r_0 - 1)]\} \quad (21)$$

D is a fitting parameter, γ_0 is the Grüneisen parameter of the crystal at the equilibrium state. Thus, the dependence of the Debye temperature on the excess volume is derived as

$$\Theta(\Delta V) = \Theta_0 \left\{ 2\exp[-4\gamma_0((1+\Delta V)^{1/3} - 1)] - \exp[-2\gamma_0((1+\Delta V)^{1/3} - 1)] \right\} \tag{22}$$

Now, the thermodynamic parameters, including specific heat capacities C_V and C_P, the excess enthalpy ΔH, the excess entropy ΔS, and the excess free energy ΔG, are obtained as the functions of the excess volume ΔV and the temperature T from the equations (9)-(19).

2.3.2. Thermodynamic Properties of Nano-Grain Boundaries

To investigate the thermodynamic properties of nanocrystalline metals, the α-Sm (Sm-rhombohedral) bulk with nanocrystalline structure was selected from the rare-earth family as an example. The initial parameters that are needed for the thermodynamic calculations include: r0 = 1.9930Å [46], α0 = 31.2· 10-6/K [51], B0 = 100.5 GPa at TR =300K [52], and Θ0 ≈ 200K as estimated from experimental measurements in our group.

The thermodynamic parameters of nanocrystalline Sm-rhombohedral are shown in Figs. 3-4 as a function of the excess volume at different temperatures. CV approaches to a constant value of 3NkB (N, the Avogadro's number) and is independent of ΔV at high temperatures (Figure 3a). Whereas, there is an abrupt increase of CP above a certain excess volume (Figure 3b). For the thermodynamic functions of nano-grain boundaries, as shown in Figure 4, the excess enthalpy ΔH increases with ΔV (Figure 4a), indicating an increase of the heat storage capacity of the nano-grain boundaries with decreasing the grain size. With the increase of ΔV and hence that of the fraction of atoms at the grain boundaries, the excess entropy ΔS increases (Figure 4b) as a result of expanded volume of the disordered array of atoms. For the excess free energy ΔG as a function of ΔV, it is found that at relatively high temperatures there exists a maximum corresponding to a critical volume, and this critical value decreases with increasing the temperature (as indicated by arrows in Figure 4c). This implies that there will be a distinct change of the thermal stability of the nano-grain structure at the critical volume (or the critical nano-grain size) for a given temperature.

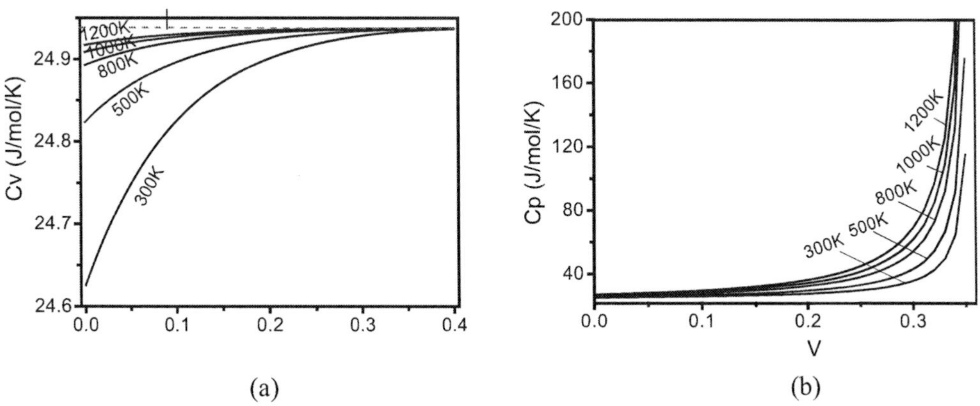

Figure 3. Calculated dependence of (a) the specific heat capacity at constant volume CV and (b) the specific heat capacity at constant pressure CP on the excess volume ΔV at different temperatures. Nanocrystalline Sm-rhombohedral is used as an example.

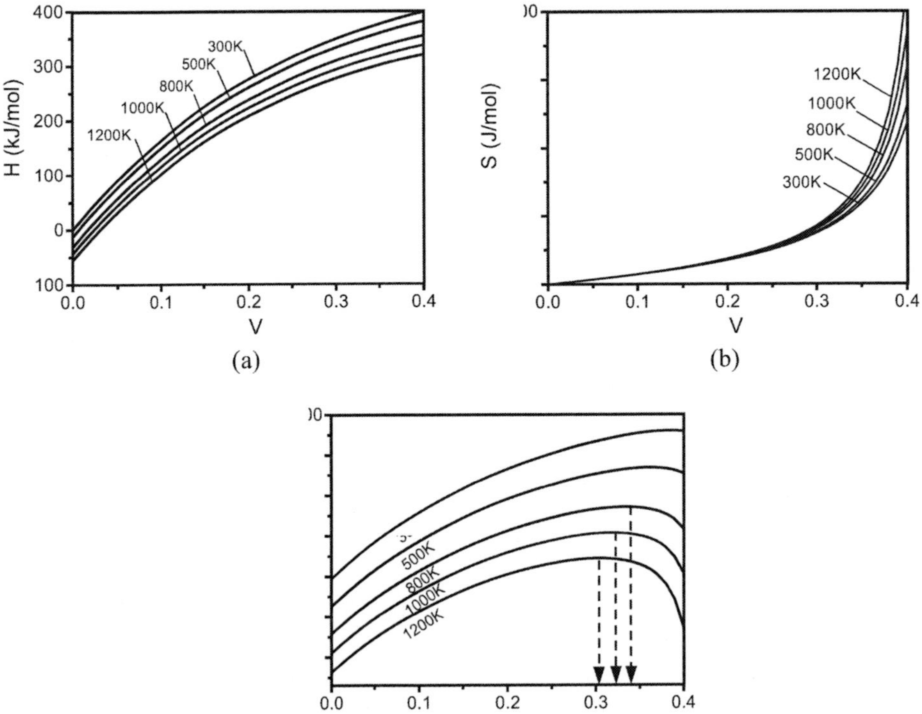

Figure 4. Calculated changes of (a) the excess enthalpy ΔH, (b) the excess entropy ΔS and (c) the excess free energy ΔG of nano-grain boundaries with the excess volume ΔV at different temperatures. Nanocrystalline Sm-rhombohedral is used as an example.

2.4. Thermal Stability of Nanocrystalline Metal Bulk

The grain growth behavior in nanocrystalline metals and alloys has been investigated rather extensively in experiments. It has been found that grains grow very slowly at low homologous temperatures in the nanocrystalline sample while rapidly above a certain temperature [28,29,53,54], and that grain growth becomes negligible for longer annealing times [29,55]. The empirical time exponent n, which is usually used to describe the isothermal grain growth, has been found to have variations of 2, 3 or 4 with the annealing time in nanocrystalline Cu [27], indicating deviation from the conventional polycrystalline metals, where n is close to 2 and is basically independent of time. According to the characteristics of nano-grain growth in the available reports, it is conjectured that grain growth in nanocrystalline materials occurs in a manner different from that in polycrystalline materials, thus is hard to be reasonably described by the conventional theories for the normal coarse-grain growth.

The thermodynamic model that describes the metastable energy state of nano-grain boundaries is more appropriate to explain the grain growth behavior in nanocrystalline metals from the viewpoint of the driving force. It is predicted by the thermodynamic model (Section 2.3) that at relatively high temperatures there exists a maximum in the excess free energy ΔG corresponding to a critical volume ΔV_C.

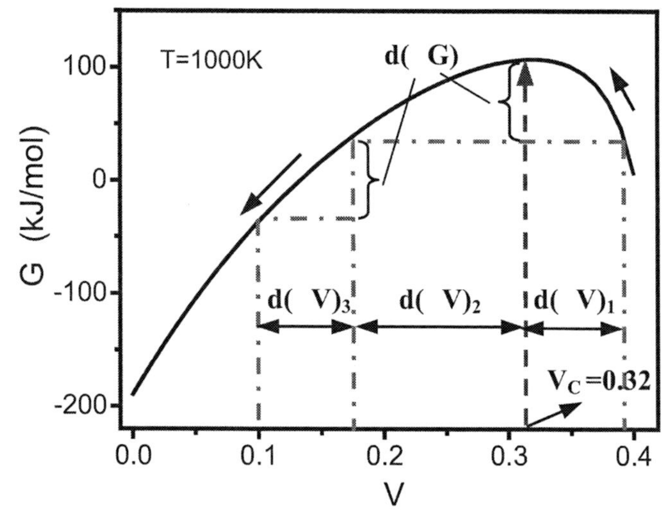

Figure 5. The excess free energy of nano-grain boundaries as a function of the excess volume, calculated for the nanocrystalline Sm-rhombohedral at T=1000K, showing the destabilization possibility in certain conditions.

As an example shown in Figure 5, at a temperature of T=1000K, when ΔV is comparatively smaller or larger than ΔV_C, e.g. $\Delta V = \Delta V_C - d(\Delta V)_2$, or $\Delta V = \Delta V_C + d(\Delta V)_1$, ΔG is lower than the maximum, implying that the system has a relatively more metastable energy state in both coarser- and finer-grain structures than in that with a critical nano-grain size. In the case of $\Delta V > \Delta V_C$, due to the fact that the excess free energy is relatively lower than those in the nano-grain structures with coarser grain sizes, i.e.

$$\Delta G_{(\Delta V = \Delta V_C + d(\Delta V)_1)} \leq \Delta G_{((\Delta V_C - d(\Delta V)_2) \leq \Delta V \leq \Delta V_C)},$$

the nanostructure with small grain size below the critical value may have higher thermal stability than the coarser nano-grain structures. This means that an energy barrier has to be overcome, e.g. by thermal activation, in order to initiate the nano-grain growth, namely, the evolution of the metastable nano-grain structure from very fine grain size towards the more stable coarser-grain structures. Although experimental data on the growth behavior of small grains in a range of a few nanometers are very scarce, a fairly high stability of the nanocrystalline metals with a mean grain size of 5-15 nm has been found in a very recent study [56].

2.4.1. Temperature-Varying Nano-Grain Growth

As indicated in Figure 5, the excess free energy of the nano-grain boundaries deviates from the progressive increase above a critical excess volume ΔVC, which implies that the thermal stability of the nanocrystalline material will change abruptly at a certain critical nano-grain size that is corresponding to ΔVC. This means that for a given nano-grain structure with an excess volume ΔV>ΔVC, the thermodynamically metastable state will be destabilized when external energy is supplied, for example by thermal activation, resulting in a critical temperature TC, at which the nanocrystalline structure will lose the thermal stability due to the high energy state. Except the thermal activation, the energy stored in the nanocrystalline structure, e.g. from the preparation history of the nanocrystalline material, may have the similar effect as supplying energy to trigger the thermodynamic instability in the nanostructure with grain size below the critical value.

It has been reported systematically [54] that for the pure nanocrystalline Co, in a series of one-hour annealing experiments performed in a wide range from room temperature to 1173K, rapid discontinuous grain growth occurred in a narrow intermediate temperature range between 773 and 873K with the mean grain size increased by approximately eight times. Similar behaviors of discontinuous nano-grain growth, on which the effects of solute segregation [28,57,58] and/or pinning of the precipitates [53,55,58] are eliminated, have been observed in nanocrystalline Al [59], the electrodeposited nanocrystalline Ni-Fe alloys with a fcc single phase [29], and the single-phased nanocrystalline Fe–Si and Fe–Si–Al alloys [60], respectively. According to the conditions in which the rapid grain growth occurred [29,54,59,60], it is confirmed that the discontinuous nano-grain growth is driven by the excess free energy of nano-grain boundaries in combination with the thermal energy or the stored energy from the preparation process of the nanocrystalline material.

2.4.2. Isothermal Nano-Grain Growth

As shown in Figure 4(c) and Figure 5, if $\Delta V<\Delta V_C$, i.e. the nano-grain size is larger than the critical value, the excess free energy of nano-grain boundaries reduces monotonically with the increase of grain size, implying that the grains will grow continuously from the initial size scale. According to the dependence of ΔG on ΔV when $\Delta V<\Delta V_C$, for the same reduction amount of ΔG, the release of the excess volume will decrease with the grain growth process, i.e. $d(\Delta V)_2>d(\Delta V)_3$ (see Figure 5), indicating that with the grain growth the increase in grain size becomes smaller. More specifically, for the same $d(\Delta G)$ at T=1000K, as marked in Figure 5, in the nano-grain structure having $\Delta V=0.32$ as a starting value, the nano-grains will grow into a grain size corresponding to $\Delta V=0.18$ (with $d(\Delta V)_2=0.14$), whereas in a coarser nano-grain structure with $\Delta V=0.18$, the grain growth under the equivalent driving force will lead to a grain size corresponding to $\Delta V=0.10$ (with $d(\Delta V)_3=0.08$). Therefore, it is predicted thermodynamically that at a constant temperature, once the nano-grain growth is initiated, the nano-grain grows faster in the stage with smaller grain size than in later stages, thus the rate of nano-grain growth is decreasing monotonically with the increase of grain size, i.e. the isothermal nano-grain growth has a slowing-down kinetics.

The experimental results of isothermal nano-grain growth in the pure nanocrystalline Co at T=803K was demonstrated in reference [37]. It was observed that the finer grains have a higher growing rate than the coarser ones. The grain growth exponent, n, which was estimated by the measurement of grain sizes at different times, was found to vary approximately in a range of 1.1~ 4.4. As exhibited in experiments, for the isothermal nano-grain growth, the growth rate decreases monotonically with time (or grain size), indicating that at the critical temperature the rapid nano-grain growth takes place at an early stage of anneal holding. The experimental finding of the slowing-down kinetics of isothermal nano-grain growth in the nanocrystalline metal bulk agrees well with the thermodynamic predications.

To summarize, the high energy states of the nanoparticles and the nanocrystalline bulks can be described by the thermodynamic models developed by aiming at features of nanostructures. The derived dependence relationships of the free energy and the melting temperature on the particle size provide a guide for the safekeeping and characterization of metal nanoparticles, as well as the design of the consolidation processing for the preparation of nanocrystalline bulks. The predicted thermodynamic properties and the thermal stability of the nanocrystalline metals, help to control the nano-grain growth and to investigate the physical and mechanical properties of nanocrystalline metal bulks as well as their nano-scale dependence. In the following sections, the preparation and characterization of nanoparticles and nanocrystalline bulks of rare-earth metals will be described.

3. Preparation and Characterization of Nanoparticles of Rare-Earth Metals

Studies on synthesis technologies and the related nanoparticle-based metal nanostructures have attracted great interests with their increasingly wide applications in the fields of information, energy, environmental and medical technologies. Unagglomerated nanoparticles with narrow size distribution are preferred for practical applications and technologies starting from nanoparticles. A number of techniques for the preparation of nanoparticles that satisfy this requirement have been developed mainly including liquid and gas phase processes. There are various liquid phase methods developed for preparing metal nanoparticles [61], such as the chemical reduction, sol-gel, reverse micelle process, pyrolysis, and spray pyrolysis methods. For the gas phase methods, condensation of the corresponding metal vapors is one of the predominant technologies for synthesis of metal nanoparticles [14,15,61]. At present characterization of nanoparticles is concentrated on measurements of particle diameter and particle size distribution, observations on morphology/shape and surface characteristics.

For the particularity of the extremely high chemical activity of pure rare-earth particles, hence the requirement of high purity for the safekeeping and the further usage of rare-earth nanoparticles, a technique of inert-gas condensation that applies processes of solid $\leftrightarrow$ gas phases transformation in an enclosed system will be introduced in this section for the preparation of rare-earth nanoparticles.

3.1. Preparation of Rare-Earth Nanoparticles

The inert-gas consolidation method was first generated by Gleiter and coworkers for preparing the materials that demonstrated the exciting properties of nanostructured materials [14]. In their method, a metal was typically vaporized into a low density gas by electrical heating, then metal vapors migrate from the hot source into a cooler inert gas. The decreasing temperature leads to a rapid reduction in the equilibrium vapor pressure and correspondingly a high supersaturation state. Subsequently the vapors rapidly nucleate, forming a large number of nanoparticles.

The inert-gas condensation for preparing rare-earth nanoparticles was carried out with a home-made instrument in the author's group, which was appended with forced inert-gas flow and surrounding cooling of the evaporation chamber with cyclic cold-water, as compared with some previously reported inert-gas condensation instruments [62], in order to enhance the deposition efficiency and the dispersing degree of the produced nanoparticles. It was found that the inert-gas pressure, the applied voltage, the input current and the distance between the cathode and the vaporizing metal, as well as the vapor pressure of the metal, all influence to different contents the resultant particle size and the production rate of the nanoparticles. In the highly purified inert (Ar or He) gas at a pressure of ~10-1 Torr, the mean size of the condensed particles could be controlled in the range of 20-80 nm when the applied voltage was adjusted from 10 to 20V, the input current from 80 to 200A, and the evaporation time from 0.5 to 3 hours. When the other conditions are fixed, as the inert-gas pressure increases, both the mean size and the production rate of the rare-earth particles increase as a result of the higher density of plasma generated between the two electrodes, hence the higher collision frequency of extremely small particles that leads to coagulation and growth of particles. Increasing the voltage and/or the current intensity result in a higher production rate and probably coarser particle size as a reason of the temperature increase in the electrode zones. With reducing the distance between the cathode and the vaporizing metal, the evaporation process becomes faster, and the production rate is increased.

A series of experiments showed that the production rate depends strongly on the vapor pressures of rare-earth metals. Generally, the larger the difference between the melting and the boiling temperatures of the rare-earth metal, the higher is the transition temperature for the sharp increase of the vapor pressure of the rare-earth, and the lower is the production rate of the rare-earth nanoparticles by the inert-gas condensation method. A figure demonstrating this rule is shown in Figure 6 including most rare-earths. In principle, any metals whose transition temperatures for the sharp increase of the vapor pressures are not very high can be possibly used to prepare the nanoparticles by the inert-gas condensation method. Those rare-earth metals that nanoparticles have been made in our work are shown with bold lines in Figure 6. Table 1 shows correspondingly the production rate of the rare-earth nanoparticles made by inert-gas condensation.

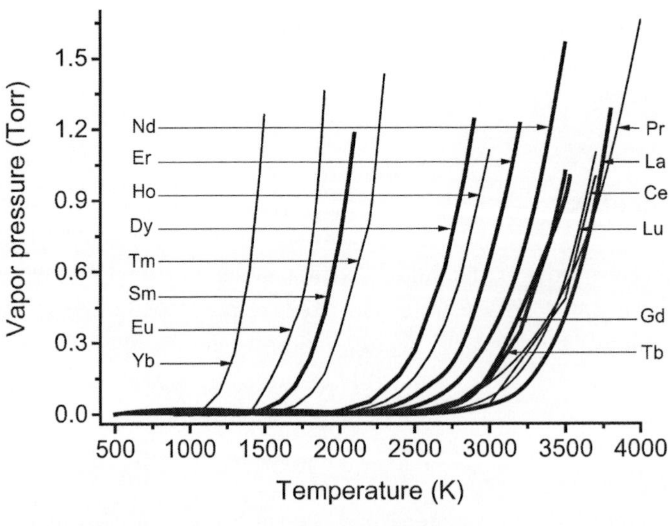

Figure 6. The vapor pressures of rare-earth metals via temperature.

Table 1. The production rates of rare-earth nanoparticles prepared by inert-gas condensation, together with the melting and boiling points of the rare-earths

Rare-earth metals	Sm	Dy	Er	Nd	Tb	Gd	La
Melting point (K)	1345	1682	1795	1289	1630	1585	1193
Boiling point (K)	2064	2835	3136	3341	3496	3539	3730
Production rate (g/h)	300	200	150	80	50	50	5

3.2. Microstructure of Rare-Earth Nanoparticles

Since the pure rare-earth nanoparticles are so active that become burning immediately when they are exposed in the air, the prepared rare-earth nanoparticles were preserved in the glove box filled with highly purified argon gas. Also because of their extremely high chemical activity, at present the rare-earth nanoparticles cannot be detected by X-ray diffraction (XRD) method yet. The transmission electron microscopy (TEM) was chosen as a master characterization method for the shape, size and size distribution, and crystal structure of rare-earth nanoparticles. The nanoparticles were put into the acetone while in the glove box and subsequently fetched onto the copper-grid film for the TEM observation.

The TEM images of the prepared rare-earth nanoparticles are shown in Figure 7, using Sm, Nd, Dy and Tb as examples, together with the high-resolution TEM (HRTEM) images of single rare-earth particles, selected area electron diffraction (SAED) patterns and its indexing of the particles, and the particle size (diameter) distributions. It is observed from Figure 7 that the inert-gas condensed rare-earth nanoparticles have a comparatively homogeneous size distribution with mean diameters of 25-40nm. The nanoparticles keep basically the same crystal structure of the raw polycrystalline rare-earth metal, as indicated by the indexing of the SAED patterns of the particles. In an overview, each nanoparticle is a single-crystal.

4. Preparation of Nanocrystalline Bulks of Rare-Earth Metals

4.1. Preparation Route of Nanocrystalline Bulks

In the past two decades, a number of techniques have been developed to synthesize nanocrystalline bulk materials, such as inert-gas condensation and consolidation [63], electrodeposition [64], severe plastic deformation [65], crystallization of amorphous solid [66], surface mechanical attrition [67], and powder metallurgy [68-70]. However, it is hard to produce nanocrystalline bulk materials with controllable grain sizes in a wide range below 100 nm. Particularly, in conventional powder metallurgy processes, the rapid coarsening of the particles occurs very often, even with the nanoparticles as the starting powder, grains in the scale of submicro- or even micrometer may form in the resultant bulk. Thus, in the preparation of bulk materials by the conventional powder-metallurgical methods, the grain size in the synthesized bulk is generally larger than the initial particle size [68-70].

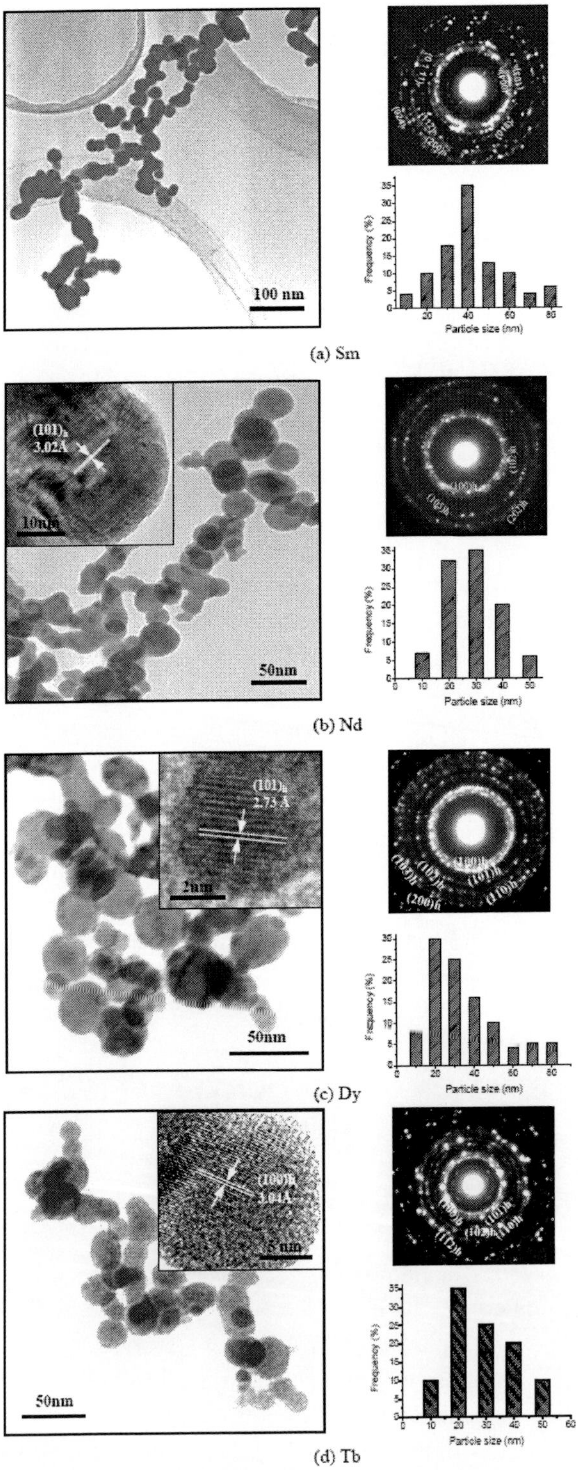

Figure 7. TEM images of some of the prepared rare-earth nanoparticles, together with the HRTEM images of single rare-earth particles (inset in each TEM image), SAED patterns and its indexing (upper right in each partial figure), and size distributions (lower right in each partial figure) of the nanoparticles of (a) Sm, (b) Nd, (c) Dy and (d) Tb, respectively.

An advanced powder metallurgical technology, so-called as spark plasma sintering (SPS), has been fast developed in recent years. It is a solid consolidation process that has the key advantage of the very high degree of densification obtained at low sintering temperatures and within a short time, compared with the conventional sintering methods. The SPS process applies simultaneously high-intensity pulsed direct current and uniaxial pressure, which offers the possibilities of using rapid heating rates (e.g. achieving 1000°C/min) and very short holding times (e.g. 3-5min) to obtain highly dense samples [71-73]. As a reason, grain growth in the sintering samples can be effectively restrained. Up to now the SPS technology has been applied to fabricate various materials including metal and alloys, compounds, ceramics, composites, especially, some new-type materials such as bulk amorphous and nanocrystalline materials [74], multi-scaled structure and function graded materials [75] and etc.

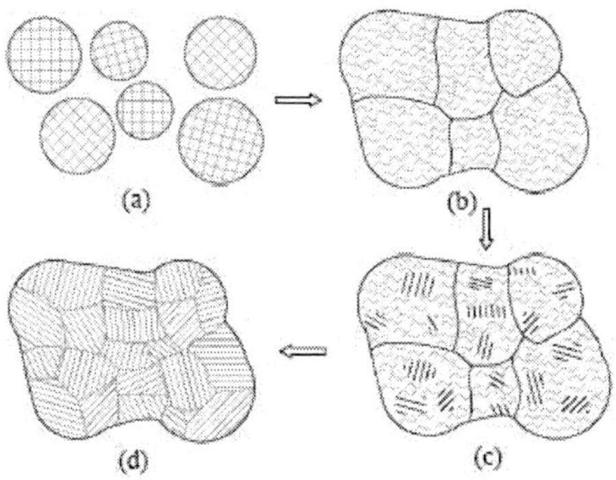

Figure 8. Schematic diagram of preparing metal bulk with the ultrafine nanocrystalline structure with (a) single-crystal nanoparticles, (b) amorphization of nanoparticles in SPS consolidation process, (c) nucleation and growth of the short-range ordered "clusters" inside the nanoparticle, and (d) complete nanocrystallization.

A novel technique combining inert-gas condensation with the SPS technology has been recently established in our group [52,56]. Taking into account the extremely high chemical activity of rare-earth nanoparticles, we integrated the configuration of inert-gas condensation with SPS equipment into an entirely closed and vacuumized system, named as "oxygen-free" in-situ synthesis system where the oxygen concentration is controlled below 0.5ppm. According to the special mechanisms of SPS consolidation, which were proposed [71,72] and developed [76,77] in the literature, when the pulsed current passes through the powder of rare-earth metals that has a high electrical conductivity while a relatively low heat conductivity, a high temperature can be reached rapidly in the powder particles. Based on the model calculations [76], at local sites of the sintering particles, the temperature induced by the Joule-heating can be much higher than that measured by the thermocouple placed in the wall of the graphite die, and there is a sharp increase of the temperature from the particle body to the particle-to-particle interface. Since the temperatures of phase transition of the inert-gas condensed nanoparticles are reduced due to the nano-scale effect as referred to Section 2.1, the high temperature induced by the Joule-heating at the local site of each nanoparticle can exceed its glass transition temperature, and can even reach the melting point at the interfaces between the nanoparticles. As a result, the nanoparticles are likely to have a metallic-glass (amorphous) body and a melted boundary surface. Enlightened by the experimental finding that a metastable amorphous state exists in the compact powder sample in SPS process under certain conditions of sintering parameters and nanoparticle size classes, a novel preparation scheme for metal bulks with ultrafine nanocrystalline structures was proposed for the first time by the author's group. The procedure is composed of sequentially arranged processes of: amorphization of nanoparticles, nucleation and growth of the short-range ordered "clusters" inside the nanoparticle, and the complete nanocrystallization. The diagram of the above route is shown in Figure 8.

4.2. Rare-Earth Bulks with Ultrafine Nanocrystalline Structure

Using the route introduced in Section 4.1, pure metal bulks of a series of typical members in the rare-earth family, e.g. Sm, Gd, Nd, Dy, Tb and Er, have been prepared with the ultrafine nanocrystalline structures.

For different rare-earth metals, the processing parameters in the preparation procedure were modified depending on the particle size and the melting temperature of the rare-earth nanoparticles. For the modification, the prediction on the size dependence of the melting temperatures of rare-earth nanoparticles, which was proposed in Section 2.2, was used as the foremost reference data. In a glove box connected with the powder reservoir of the inert-gas condensation apparatus, the prepared rare-earth powder was fed into the high-strength graphite die, which was subsequently sent to the SPS equipment (SPS-5.40-IV/ET, Sumitomo Coal Mining Co., Ltd) by a sliding rail. An external pressure was then applied to the powder through the upper and lower punches. In the sintering process, the external pressure was kept constant, and the temperature, the heating rate and the holding time were adjusted in order to achieve the consolidated bulks with nearly full relative density and desired nanocrystalline microstructures. The details in the step sintering processes to obtain the amorphous matrix and the final nanocrystallized structure can be found in references [52] and [56] by using the preparation of nanocrystalline Sm as an example. The microstructures of the prepared nanocrystalline rare-earth bulks, using some members of the rare-earth family as examples, i.e. Sm, Nd, Dy and Tb, are shown in Figure 9.

It is observed that compact ultrafine nano-grains formed in the rare-earth bulks with the grain sizes of 5-15 nm, an example displaying by arrows the different orientations of ultrafine nano-grains is given in Figure 9(a). The crystal structures of the prepared nanocrystalline rare-earth bulks were kept the same as those of the raw polycrystalline metals at the room temperature, as analyzed by the SAED patterns. Especially, the grain sizes in the resultant rare-earth bulks (ca. 10-15nm on average) are distinctly smaller than the initial nanoparticle sizes (ca. 40-60nm on average), which may be the first demonstration that the traditionally accepted relationship, that the grain size in the sintered bulk is larger than the initial nanoparticle size, can be changed. Applying the process of nucleation and growth of the short-range ordered "clusters" in the amorphous matrix inside the nanoparticle is the most significant advantage of the preparation route proposed in Section 4.1. In virtue of this technique, nanocrystalline bulk materials with controllable grain sizes can be obtained, which will improve the material properties in various aspects by applying the scale effects; moreover, the prepared ultra-fine nanocrystalline bulks can be used to study some special features of nanostructured materials.

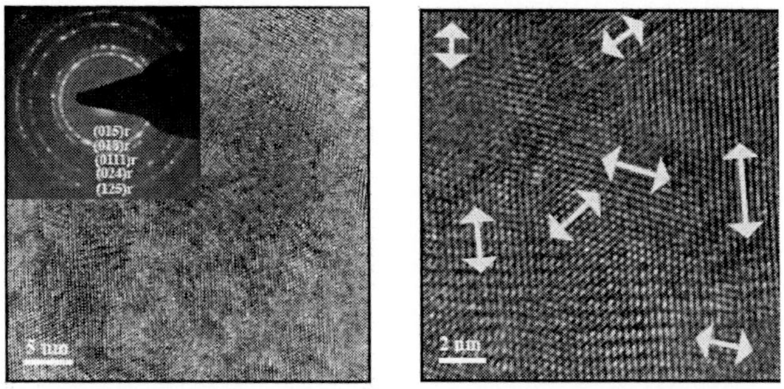

(a) Sm

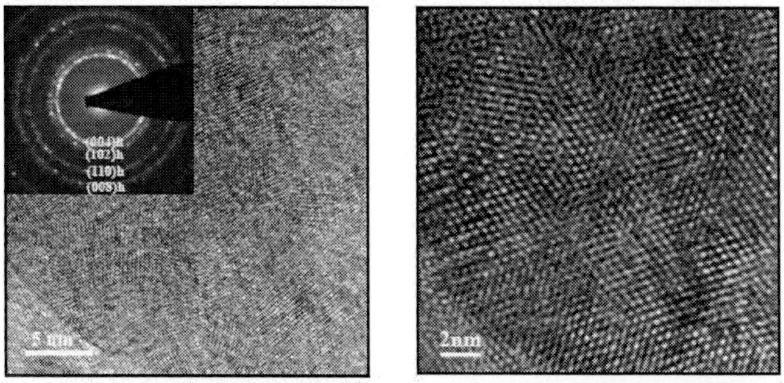

(b) Nd

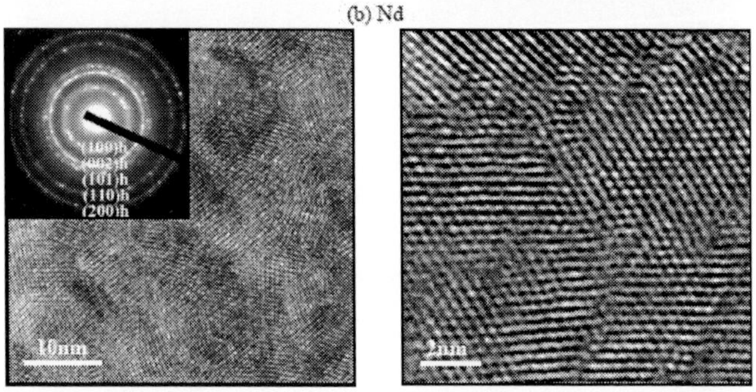

(c) Dy

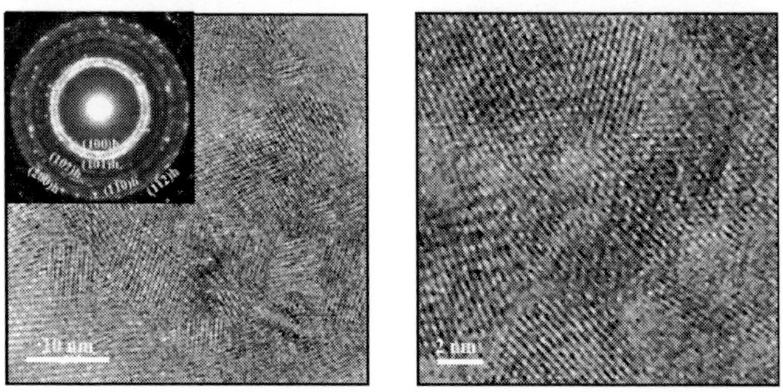

(d) Tb

Figure 9. The nanocrystalline microstructures of rare-earth metals prepared by combining inert-gas condensation and SPS, shown with HRTEM images (left in each partial figure), SAED patterns and its indexing (upper left in each partial figure), and locally enlarged areas revealing ultrafine nano-grains (right in each partial figure) of (a) Sm, the arrows indicating the orientations of nano-grains, (b) Nd, (c) Dy and (d) Tb, respectively.

5. Properties Characterization of Nanocrystalline Rare-Earth Metals

It is considered that as the nanocrystalline microstructures of rare-earth metals are obtained, especially when the ultrafine nano-grain structures are achieved, the rare-earth metals would present different properties from those of the polycrystalline rare-earths. In order to characterize the properties of nanocrystalline rare-earth metals, the thermal, physical and mechanical properties of the prepared nanocrystalline bulks were measured for the typical members in the rare-earth family. For the sake of comparison, the measurements of properties on the polycrystalline rare-earth metals were performed correspondingly. To keep the purity of the samples, all these measurements were performed under highly pure argon or helium gas to protect from oxidization or reactions.

The remarkably changed thermal properties (e.g. specific heat capacity), physical properties (e.g. electrical resistivity and thermal conductivity) and mechanical properties (e.g. micro-hardness and bulk modulus of elasticity) of nanocrystalline Sm, and magnetic properties (e.g. Curie transition temperature and magnetic moment) of nanocrystalline Gd, have been reported in literature [56,78]. Here a detailed demonstration of the properties of nanocrystalline rare-earth metals is given by using Nd and Dy as examples.

5.1. Thermal Properties

5.1.1. Specific Heat Capacity

The isobaric heat capacities (Cp) were measured on both the prepared nanocrystalline and the raw polycrystalline rare-earth bulks by two apparatuses of DSC 404C, PEJASUS and DSC Q1000, TA Instruments Inc. respectively for each sample, in order to verify the data repeatability.

The curves of the specific heat capacities of Nd and Dy bulks are shown in Figure 10 for the nanostructured samples and the raw polycrystalline metals, respectively. It is noted that the specific heat capacity of the nanocrystalline bulk has a stronger increasing tendency with the temperature than that of the polycrystalline metal. Furthermore, compared with the conventional polycrystalline metals, the specific heat capacities of the ultrafine nanocrystalline rare-earth bulks are apparently larger in the whole range of the testing temperature. For example, compared with the Cp values of the raw polycrystalline bulks, which are ca. 32.3 J/mol/K for Nd and 28.5 J/mol/K for Dy at 350oC, the Cp values of the ultrafine nanocrystalline Nd and Dy bulks are ca. 47.0 and 35.4 J/mol/K at the same temperature, which have been increased approximately by 50% and 20%, respectively. As is known that the heat capacity of a material is closely related to its vibrational and configurational entropies of atoms [79,80], the increase of the heat capacity in the ultrafine nanocrystalline bulks reflects the substantial contribution of the highly dense nano-grain boundaries to the thermodynamic features of the material. of (a) Nd and (b) Dy, together with the comparison with those of the corresponding raw polycrystalline metals.

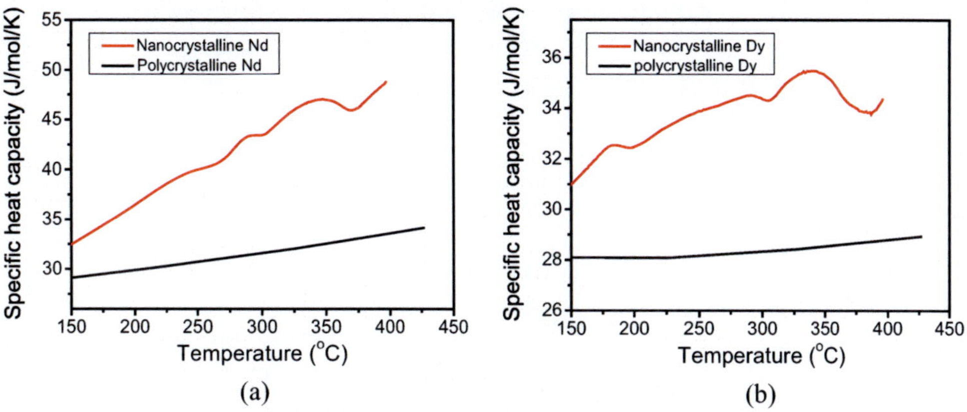

Figure 10. The specific heat capacities of the prepared ultrafine nanocrystalline bulks of (a) Nd and (b) Dy, together with the comparison with those of the corresponding raw polycrystalline metals.

5.1.2. Thermal Expansion Coefficient

The thermal expansion coefficients (the linear expansion, αL) were measured for the prepared nanocrystalline and the raw polycrystalline rare-earth bulks with the NETZSCH DIL 402EP apparatus, respectively.

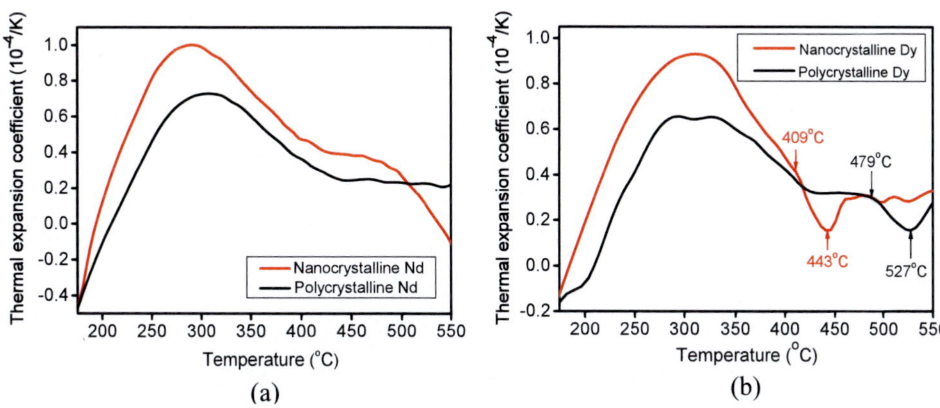

Figure 11. The thermal expansion coefficients of the prepared ultrafine nanocrystalline bulks of (a) Nd and (b) Dy, together with the comparison with those of the corresponding raw polycrystalline metals.

The variations of the thermal expansion coefficients of Nd and Dy bulks with the temperature are shown in Figure 11 for both the nanocrystalline samples and the raw polycrystalline metals, respectively. In the range of the testing temperature, the αL value of the ultrafine nanocrystalline bulk is generally larger than that of the polycrystalline metal, except some points corresponding to phase transformations. The appearance of sharp wave in the αL curve reveals phase transformation at the corresponding temperature. It is observed in Figure 11(b) that, the starting temperature of phase transformation from hexagonal to rhombohedral crystal structure is about 479°C in the polycrystalline Dy, and the temperature corresponding to the wave hollow is about 527°C (which is consistent with the value in thermodynamic database). However, these two temperatures are about 409°C and 443°C in the ultrafine nanocrystalline Dy bulk, respectively. Therefore, the starting temperature for phase transformation in the ultrafine nanocrystalline Dy bulk is reduced by 70°C from that of the polycrystalline metal. Similarly, the starting temperature for hexagonal → rhombohedral phase transformation in the ultrafine nanocrystalline Tb bulk is found reduced by about 65°C from that of the polycrystalline Tb.

5.2. Physical Properties

5.2.1. Electrical Resistivity

The electrical resistivities were measured for the prepared nanocrystalline and the raw polycrystalline rare-earth bulks with the Model ZEM-2, ULVAC apparatus, respectively.

Figure 12 shows the changes of the electrical resistivities of Nd and Dy bulks with the temperature for the nanocrystalline samples and the raw polycrystalline metals, respectively. As shown in the figure, the electrical resistivities of all the rare-earth samples increase with temperature as expected in normal electrical property of metal materials. However, at the same temperature, the electrical resistivities of the prepared ultrafine nanocrystalline rare-earth bulks are higher than those of the raw polycrystalline metals, which can be explained by the "quantum confinement" [14,15], which is essentially due to the changes in the atomic structure as a result of the direct influence of the ultra-small length scale on the energy band structure. The characteristic electronic properties of the nanocrystalline structures are a result of tunneling currents, purely ballistic transport and the coulomb blockade effects [61].

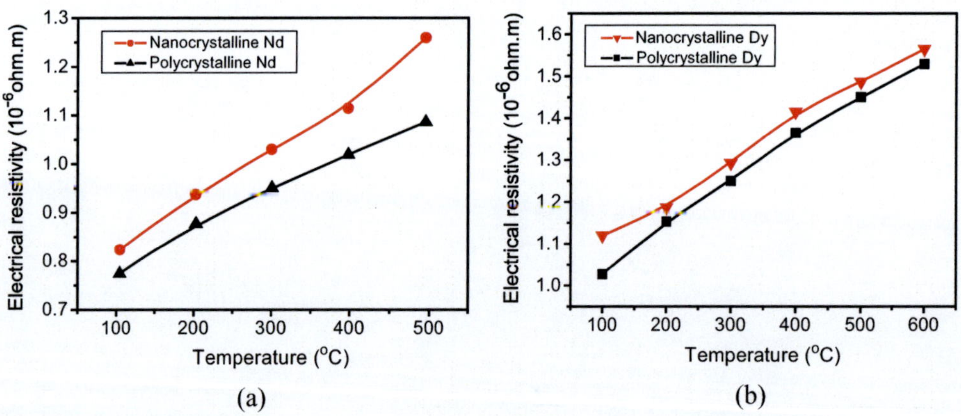

Figure 12. The electrical resistivities of the prepared ultrafine nanocrystalline bulks of (a) Nd and (b) Dy, together with the comparison with those of the corresponding raw polycrystalline metals.

5.2.2. Thermal Conductivity

The thermal conductivities were measured for the prepared nanocrystalline and the raw polycrystalline rare-earth bulks with the TC-7000H apparatus, respectively.

Figure 13 shows the changes of the thermal conductivities of Nd and Dy bulks with the temperature for the nanocrystalline samples and the raw polycrystalline metals, respectively. It is noted that the thermal conductivities of Nd samples vary little with the temperature, while those of Dy samples increase with the temperature. For the same rare-earth, the thermal conductivity of the prepared ultrafine nanocrystalline bulk is apparently lower than that of the polycrystalline metal, which is contrary to the changing tendency of the electrical resistivity as the microstructure transforms from polycrystalline to nanocrystalline. As compared with the raw polycrystalline metal, the thermal conductivity of the ultrafine nanocrystalline bulk is reduced by ca. 78% on average for Nd and by 27-31% for Dy.

In general, the thermal conductivity of a material contains two portions of contributions [2,81], one is generated from the phonon scattering, another is from the current carrier. In the ultrafine nanocrystalline bulk, the larger fraction of the nano-grain boundaries increases apparently the probability of phonon scattering, hence reduces the thermal conductivity from the phonon scattering effect. On the other hand, the high density of the nano-grain boundaries increases the resistance to the motion of the current carrier and reduces the migration rate, thus increases the thermal conductivity from the contribution of the current carrier. However, as the portion of the thermal conductivity from the phonon scattering effect has a much stronger impact on the total thermal conductivity of the material, the thermal conductivity of the ultrafine nanocrystalline bulk is lower than that of the raw polycrystalline metal.

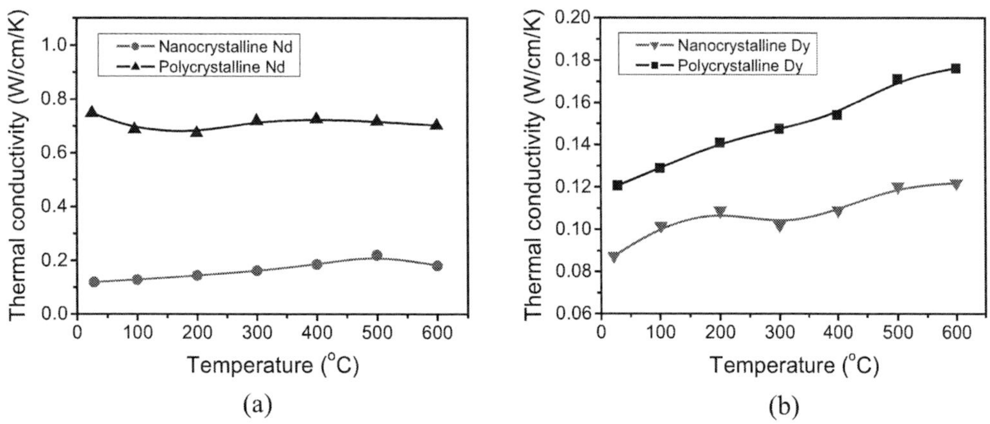

Figure 13. The thermal conductivities of the prepared ultrafine nanocrystalline bulks of (a) Nd and (b) Dy, together with the comparison with those of the corresponding raw polycrystalline metals.

5.3. Mechanical Properties

The mechanical properties of the nanocrystalline rare-earth bulks are characterized by the micro-hardness and the modulus of elasticity, measured by the nano-indentation method [82] with the Hysitron Trobolndenter apparatus.

5.3.1. Micro-Hardness

The measurements of the micro-hardness on the nanocrystalline and the raw polycrystalline Nd and Dy bulks, respectively, are shown in Figure 14. With the loads in a range of 500-4000 μN, the values of the micro-hardness become stable when the load is increased to above 2500μN. It can be seen that the micro-hardness of the prepared ultrafine nanocrystalline rare-earth bulks are apparently higher than those of the raw polycrystalline metals. For example, the micro-hardness are 3.1 GPa for ultra-fine nanocrystalline and 1.3 GPa for polycrystalline Nd bulks, respectively, which is approximately 1.4 times increased.

For Dy, the micro-hardness of the ultra-fine nanocrystalline bulk is approximately 2.5 times as high as that of the raw polycrystalline metal.

5.3.2. Modulus of Elasticity

Figure 15 shows the measured modulus of elasticity of the nanocrystalline and the raw polycrystalline Nd and Dy bulks, respectively. The similar feature as that of micro-hardness is found, i.e. the modulus of elasticity of the prepared ultrafine nanocrystalline rare-earth bulks are much higher than those of the raw polycrystalline metals. For example, the modulus of elasticity are 64.5 GPa for ultra-fine nanocrystalline and 28.0 GPa for polycrystalline Nd bulks, respectively, which is increased approximately as 2.3 times high. For Dy, the modulus of elasticity of the ultra-fine nanocrystalline bulk is approximately 2.0 times as high as that of the raw polycrystalline metal.

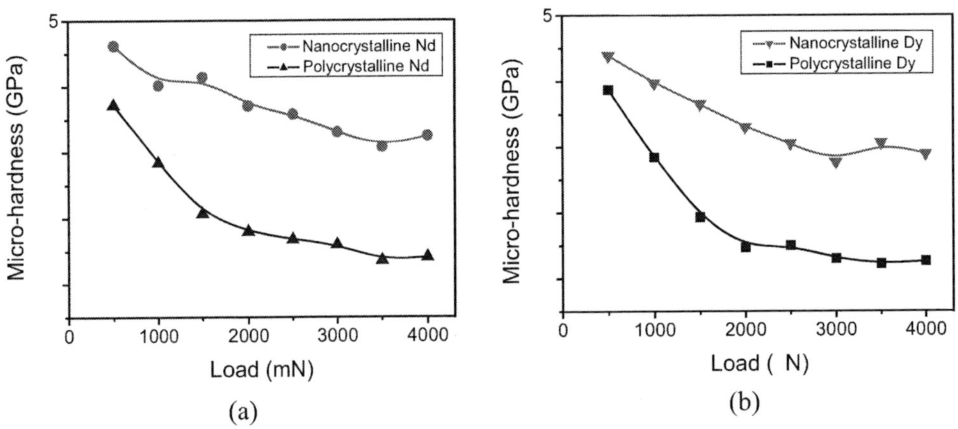

Figure 14. The micro-hardness of the prepared ultrafine nanocrystalline bulks of (a) Nd and (b) Dy, together with the comparison with those of the corresponding raw polycrystalline metals.

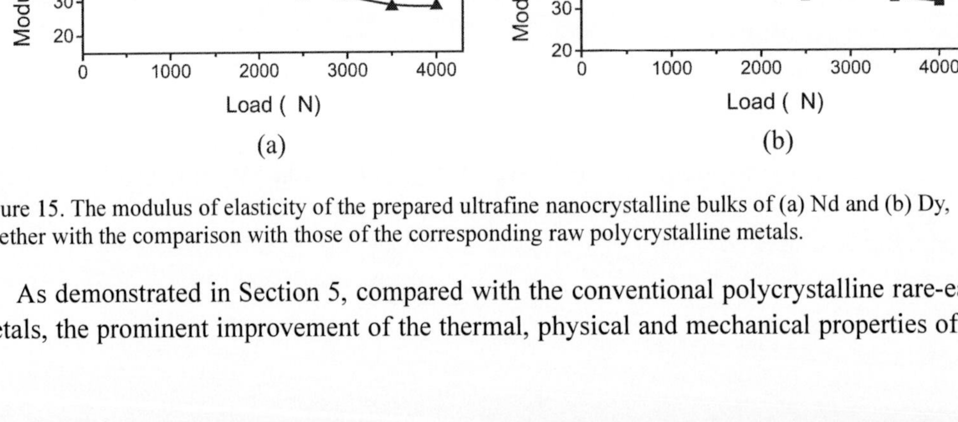

Figure 15. The modulus of elasticity of the prepared ultrafine nanocrystalline bulks of (a) Nd and (b) Dy, together with the comparison with those of the corresponding raw polycrystalline metals.

As demonstrated in Section 5, compared with the conventional polycrystalline rare-earth metals, the prominent improvement of the thermal, physical and mechanical properties of the prepared ultrafine nanocrystalline rare-earth bulks have been found, which indicates promising applications of the nanocrystalline rare-earth metals and alloys in functional materials, such as heat-storage and thermal-electric materials, as well as in structural materials, such as rare-earth-light metal alloys, nanocrystallized rare-earth-based metallic glass, and rare-earth doped composites.

6. CONCLUSION

Since the nanostructured rare-earth metals are significant in fundamental research and technological applications, whereas there are very few journal reviews or text books that comprehensively deal with the features of nanoparticles and nanocrystalline bulks of the rare-earth family, an effort has been made in this chapter to depict the energy state, thermal stability, microstructure and property of rare-earth nanoparticles and nanocrystalline bulks, respectively, which are research fields of our specialization.

The fundamental thermodynamic functions of metal nanoparticles and nanocrystalline bulks deduced by our group were introduced firstly, which describe the high energy state and the dependence of thermodynamic properties on both the temperature and the nanoparticle size or the nano-grain size. The predicted melting temperatures of the rare-earth nanoparticles as a function of the particle size help to design the consolidation processing for preparing the nanocrystalline rare-earth bulks. Based on calculations of the excess free energy, the thermal stability of nanocrystalline bulks, including the temperature-varying and the isothermal nano-grain growth behavior, can be quantitatively described. The possible occurrence of discontinuous rapid grain growth in the intermediate temperature range for the temperature-varying case, and the monotonically decreasing nano-grain growth rate with the annealing time for the isothermal case, were predicted for rare-earth nanocrystalline bulks. The thermodynamics-based description of the thermal stability provides a guide for the control of nano-grain growth in rare-earth nanocrystalline bulks in the situation of hot-working treatment on them after preparation.

The preparation of nanoparticles and nanocrystalline bulks of pure rare-earth metals was then introduced. A home-made configuration of combining the inert-gas condensation with the SPS technology in an entirely closed and vacuumized system, where the oxygen concentration is controlled below 0.5ppm, has been established in our group by taking into account the extremely high chemical activity of nanoscale rare-earths. The structures, as well as the nano-grain sizes in the consolidated rare-earth bulks, can be tailored either by changing the nanoparticle size or by modifying the sintering parameters. Ultrafine nanocrystalline bulk can be obtained with the grain size much smaller than the initial nanoparticle size using sequentially performed processes, i.e. amorphization of nanoparticles, nucleation and growth of the short-range ordered "clusters" inside the nanoparticle, and the complete nanocrystallization. This technique has a primary advantage of an "oxygen-free" in-situ synthesis of the pure rare-earth bulks with controllable nanostructures over the conventional power metallurgy techniques for nanomaterials preparation, it can be potentially used to prepare any nanocrystalline bulk materials if the metal precursors can be thermally evaporated without much difficulty.

The properties of the prepared nanocrystalline rare-earth bulks were introduced subsequently. Particularly, the thermal properties (e.g. specific heat capacity), physical properties (e.g. electrical resistivity and thermal conductivity) and mechanical properties (e.g. micro-hardness and modulus of elasticity) of the ultrafine nanocrystalline rare-earth bulks were characterized and compared with those of the corresponding raw polycrystalline rare-earth metals, respectively. The prominently changed thermal, physical and mechanical properties imply the promising applications of the nanocrystalline rare-earth metals and alloys in the fields of functional and structural materials. Accordingly, the advanced properties and important applications of nanoscale rare-earth metals will stimulate increasingly great research efforts in the future, and the challenge may lie in the integration of nanostructured rare-earths into the devices and components.

Finally, although the focus of this chapter are rare-earth nanoparticles and nanocrystalline bulks, the proposed thermodynamic models and preparation techniques are applicable to a large variety of nano metal materials, hence establish a new route for the studies on microstructures and performances of metallic nanomaterials, especially of those highly reactive in the air.

ACKNOWLEDGEMENTS

Our nanoscience and nanotechnology research is supported by the National Natural Science Foundation of China (50401001, 50671001), the Program for New Century Excellent Talents in University (NCET 2006), the Key Project of Science and Technology Innovation Engineering, Chinese Ministry of Education (705004), and the Program of Beijing New Star of Science and Technology (2004B04). The author sincerely acknowledges Prof. Jiuxing Zhang, from Beijing University of Technology, and Markus Rettenmayr, from Friedrich-Schiller-University Jena, for their valuable comments and suggestions. The author would also like to acknowledge her collaborators and co-researchers.

REFERENCES

[1] Gschneider, K. A. In Book *The Rare Earth*; Spedding, F. H.; Daane, A. H. Ed.; 1st Ed.; J Wiley Press: New York, 1961, pp 5-20.

[2] Gschneider Jr., K. A.; Eyring, L. *Handbook on the Physics and Chemistry of Rare Earths;* ISBN 0 444 85022 8; North-Holland Publishing: Amsterdam, 1978; Vol. 1, pp 1-15; Vol. 2, pp 164-170; Vol. 16, pp 107-120.

[3] Uda, T.; Jacob, K. T.; Hirasawa, M. *Science* 2000, 289, 2326-2331.

[4] Harrison, P. G.; Bailey, C.; Azelee, W. J. *Catalysis*. 1999, 186, 147-159.

[5] Laurent, C.; Zoltán, A.; Gál, T. P.; Braun, F. J.; DiSalvo, B. B.; Jürgen, H. M. *J. Sol. St. Ch.* 2001, 162, 90-95.

[6] Legrand, B. A.; Perrier de la Bathie, R.; Tournier, R.; Chateigner, D. *J. Magn. Magn.* 1997, 173, 20-28.

[7] Kerschl, P.; Handstein, A.; Khlopkov, K.; Gutfleisch, O.; Eckert, D.; Nenkov, K.; Téllez-Blanco, J.-C.; Grössinger, R.; Müller K.-H.; Schultz, L. *J. Magn. Magn.* 2005, 290-291, 420-423.
[8] Yamaguchi, K.; Itagaki, K.; Kim, D.-Y.; Ohtsuka, M. *J. Alloy. Com.* 1995, 221, 161-168.
[9] Amadeh, A.; Pahlevani, B.; Heshmati-Manesh, S. *Corros. Sci.* 2002, 44, 2321-2331.
[10] Fray, D. J. *Science* 2000, 289, 2295-2300.
[11] Worrell, E.; Martin, N.; Price, L. *Energy* 2000, 25, 1189-1214.
[12] Dai, W.; Shen, B. G.; Li, D. X.; Gao, Z. X. *J. Alloy. Com.* 2000, 311, 22-25.
[13] Sellmyer, D. J.; Skomski, R. *Scrip. Mater.* 2002, 47, 531-535.
[14] Gleiter, H. *Acta Mater.* 2000, 48, 1-29.
[15] Tjong, S.C.; Chen, H. *Mater. Sci. Eng. R.* 2004, 45, 1-88.
[16] Pawlow, P. Z. *Phys. Chem.* 1909, 65, 1-8.
[17] Takagi, M. *J. Phys. Soc. Jpn.*1954, 9, 359-365.
[18] Wronski, C. R. M.; *Br. J. Appl. Phys.* 1967, 18, 1731-1737.
[19] Couchman, P. R.; Jesser, W. A. *Nature*, 1977, 269, 481-486.
[20] Borel, J. P. *Surf. Sci.* 1981, 106, 1-7.
[21] Saka, H.; Nishikawa, Y.; Imura, T. *Phil. Mag.* A. 1988, 57, 895-906.
[22] Sasaki, K.; Saka, H. *Phil. Mag.* A. 1991, 63, 1207-1221.
[23] Lee, J.G.; Mori, H.; Yasuda, H. *Phys. Rev.* B. 2002, 66, 012105-1-4.
[24] Tanaka, T.; Hara, S. Z. *Metallkd.* 2001, 92, 467-472.
[25] Tanaka, T.; Hara, S. Z. *Metallkd.* 2001, 92, 1236-1241.
[26] Tanaka, T.; Lee, J.; Hirai, N. In Book *Chemical Thermodynamics for Industry*; Letcher, T. Ed.; ISBN 0 85404 591 0; Thomas Graham House: Cambridge, 2004, 209-218.
[27] Ganapathi, S. K.; Owen, D. M.; Chokshi, A. H. *Scrip. Metall. Mater.* 1991, 25, 2699-2704.
[28] Malow, T. R.; Koch, C. C. *Acta Mater.* 1997, 45, 2177-2185.
[29] Ebrahimi, F.; Li, H. *Scrip. Mater.* 2006, 55, 263-266.
[30] Ivanisenko, Y.; Wunderlich, R. K.; Valiev, R. Z.; Fecht, H. *J. Scrip. Mater.* 2003, 49, 947-952.
[31] Lu, K. *Scrip. Metall. Mater.* 1991, 25, 2047-2052.
[32] Winning, M.; Rollett, A. D. *Acta Mater.* 2005, 53, 2901-2907.
[33] Duhaj, P.; Šveca, P.; Sitek, J.; Janickovic, D. *Mater. Sci. Eng.* A. 2001, 304, 178-186.
[34] Liu, Z. J.; Shen, Y. G. *Acta Mater.* 2004, 52, 729-736.
[35] Millett, P. C.; Selvam, R. P.; Molecular, A. S. *Acta Mater.* 2006, 54, 297-303.
[36] Millett, P. C.; Selvam, R. P.; Saxena, A. *Acta Mater.* 2007, 55, 2329-2336.
[37] Song, X. Y.; Zhang, J. X.; Li, L. M.; Yang, K. Y.; Liu, G. Q. *Acta Mater.* 2006, 54, 5541-5550.
[38] Liu, F.; Chen, Z.; Yang, W.; Yang, C. L.; Wang, H. F.; Yang, G.C. *Mater. Sci. Eng. A.* 2007, 457, 13–17.
[39] Song, X. Y.; Gao, J. P.; Zhang, J. X. *Acta Phys. Sinca.* 2005, 54, 1313-1319.
[40] Gartner, F.; Bormann, R.; Birringer, R.; Tschope, A. *Scrip. Mater.* 1996, 35, 805-810.
[41] Fecht, H. J. *Phys. Rev. Lett.* 1990, 65, 610-613.
[42] Wagner, M. *Phys. Rev. B.* 1992, 45, 635-639.
[43] Lu, K. *Phys. Rev. B.* 1995, 51, 18-27.
[44] Meng, Q.; Zhou, N.; Rong, Y.; Chen, S.; Hsu, T. Y. *Acta Mater.* 2002, 50, 4563-4570.

[45] Rose, J. H.; Ferrante, J.; Smith, J. R. *Phys. Rev. B*. 1983, 28, 1835-1840.
[46] Rose, J. H.; Smith, J. R.; Guinea, F.; Ferrante, *J. Phys. Rev. B*. 1984, 29, 2963-2969.
[47] Guinea, F.; Rose, J. H.; Smith, J. R. *Appl. Phys. Lett*. 1984, 44, 53-55.
[48] Vinet, P.; Ferrante, J.; Smith, J. R.; Rose, J. H. *J. Phys. C*. 1986, 19, L467-L473.
[49] Vinet, P.; Smith, J. R.; Ferrante, J.; Rose, J. R. *Phys. Rev. B*. 1987, 35, 1945-1953.
[50] Gerifalco, L. A.; Weizer, V. G. *Phys. Rev*. 1959, 114, 687-692.
[51] Villars, P.; Calvert, L. D. Pearson's *Handbook of Crystallographic Data for Intermetallic Phases;* ISBN 0 87170 219 3; American Society for Metal: Metals Park, Oh, 1985, Vol. 2, pp 5277-5278.
[52] Song, X. Y.; Zhang, J. X.; Li, E. D.; Lu, N. D.; Yin, F. X. *Nanotechnology*. 2006, 17, 5584-5589.
[53] Driver, J. H. *Scrip. Mater*. 2004, 51, 819-823.
[54] Song, X. Y.; Yang, K. Y.; Zhang, J. X. *J. Nanosci. Nanotechnol*. 2005, 5, 2155-2160.
[55] Xun, Y.; Lavernia, E. J.; Mohamed, F. A. *Mater. Sci. Eng. A*. 2004, 371, 135-140.
[56] Song, X. Y.; Zhang, J. X.; Yue, M.; Li, E. D.; Zeng, H.; Lu, N. D.; Zhou, M. L.; Zuo, T. Y. *Adv. Mater*. 2006, 18, 1210-1215.
[57] Kirchheim, R. *Acta Mater*. 2002, 50, 413-419.
[58] Choi, P.; Silva, M.; Klement, U.; Al-Kassab, T.; Kirchheim, R. *Acta Mater*. 2005, 53, 4473-4481.
[59] Gianola, D. S.; Petegem, S. V.; Legros, M.; Brandstetter, S.; Swygenhoven, H. V.; Hemker, K. J. *Acta Mater.* 2006, 54, 2253-2263.
[60] Zuo, B.; Saraswati, N.; Sritharan, T.; Hng, H. H. *Mater. Sci. Eng. A*. 2004, 371, 210-216.
[61] Kuchibhatla, S.; Karakoti, A. S.; Bera, D.; Seal, S. *Prog. Mat. Sci.* 2007, 52, 699-913.
[62] Hai, N. H.; Lemoine, R.; Remboldt, S.; Strand, M.; Shield, J. E.; Schmitter, D.; Kraus, R. H.; Espy, M.; Leslie-Pelecky, D. L. *J. Magn. Magn*. 2005, 293, 75-79.
[63] Gleiter, H. *Prog. Mater. Sci.* 1989, 33, 223-315.
[64] Erb, U.; El-Sherik, A. M.; Palumbo, G.; Aust, K. T. *Nanostruct. Mater*. 1993, 2, 383-390.
[65] Valiev, R. Z.; Korznikor, A. V.; Mulyukov, R. R. *Mater. Sci. Eng. A*. 1993, 168, 141-148.
[66] Yoshizawa, Y.; Oguma, S.; Yamauchi, K. *J. Appl. Phys*. 1988, 64, 6044-6046.
[67] Zhang, H. W.; Hei, Z. K.; Liu, G.; Lu, J.; Lu, K. *Acta Mater*. 2003, 51, 1871-1881.
[68] Dogan, C.; Rawers, J.; Govier, D.; Korth, G. *Nanostruct. Mater*. 1994, 4, 631-644.
[69] Lillo, T. M.; Korth, G. *Nanostruct. Mater*. 1998, 10, 35-43.
[70] Zhang, H. W.; Gopalan, R.; Mukai, T.; Hono, K. *Scrip. Mater*. 2005, 53, 863-868.
[71] Tokita, M. In *Proceedings of the 2000 Powder Metallurgy World Congress*; Kosuge, K.; Nagai, H. Ed.; Japanese Society of Powder and Powder Metallurgy: Kyoto, 2001, pp 729-732.
[72] Mamedov, V. *Powd. Metall*. 2002, 45, 322-328.
[73] Shen, Z. J.; Johnson, M.; Zhao, Z.; Nygren, M. *J. Am. Ceram. Soc*. 2002, 85, 1921-1927.
[74] Shearwood, C.; Fu, Y. Q.; Yu, L.; Khor, K. A. *Scrip. Mater*. 2005, 52, 455-460.
[75] Tokita, M. in *Proceedings of the Seventh International Symposium on Functionally Graded Materials*; Beijing, 2002, pp 1-8.
[76] Song, X. Y.; Liu, X. M.; Zhang, J. X. *J. Am. Ceram. Soc*. 2006, 89, 494-500.

[77] Vanmeensel, K.; Laptev, A.; Hennicke, J.; Vleugels, J.; Biest, O. *Acta Mater.* 2005, 53, 4379-4388.
[78] Yue, M.; Zhang, J. X.; Zeng, H.; Wang, K. *J. Appl. Phys. Lett.* 2006, 89, 232504-232508.
[79] Sun, N. X.; Lu, K. *Phys. Rev. B.* 1996, 9, 54-57.
[80] Rupp, J.; Birringer, R. *Phys. Rev. B.* 1987, 15, 36-40.
[81] Tritt, T. M.; Nolas, G. S.; Slack, G. A.; Cohn, J. S. *J. Appl. Phys.* 1996, 11, 8412-8415.
[82] Mirshams, R. A.; Parakala, P. *Mater. Sci. Eng. A*, 2004, 372, 252-260.

In: Nanotechnology Research Developments
Editor: R. Jiménez-Contreras, pp. 205-245
ISBN: 978-1-60021-899-6

Chapter 5

Novel Manufacturing and Processing Technologies of Nanoporous Silicon

Jia-Chuan Lin and Wei-Chih Tsai
Department of Electronics Engineering, St. John's University,
Taipei 25135, Taiwan, R.O.C
Institute of Microelectronics, Department of Electrical Engineering,
National Cheng Kung University, Tainan 70101, Taiwan, R.O.C.

Abstract

Porous silicon (PS) films consisting of many pores and pillars are widely used to yield efficient visible photoluminescence (PL) and electroluminescence (EL) at room temperature. Such light-emission behaviors are primarily attributed to electron confinement in the nanocrystals that constitute the PS film. Also, the PS film shows a unique electrical property of negative difference conductance (NDC) for the carrier mobility difference in different sizes of pores and pillars.

The techniques of PS formation have been developed by many different methods. Among them, electrochemical anodization is the most commonly used. The anodization is performed in a hydrogen fluoride (HF) solution with an anodic current on the sample. However, it is very difficult to control the manufacturing parameters precisely in the manufacturing processes. Especially in nano-scale, the anisotropic etching or selective etching cannot be easily achieved in the wet-etching method. However, it is well known that nano-scaled PS (NPS) shows light emission more readily in the visible range than macro-PS does. Therefore, precise control of the NPS formation is very important.

In this chapter, the authors introduce three novel manufacturing technologies. In the first method, the Hall-effect is applied in the electrochemical anodization. By Lorentz force, the carriers can be well swept to the allocated areas. In the second method, a forward biased pn-junction is used for bottom-hole assistance. The NPS films with nano-scaled pores and high-aspect-ratio pillars can be formed. In the third method, the authors proposed a simple and mask-free method for the fabrication of patterned PS using a localized electric field. The electric field is applied by patterned electrodes (anode and cathode) which are placed horizontally underneath the sample. No masking-layer or related photo-lithography processes are needed in this method. Strong visible photoluminescence emissions in PS can be obtained by these novel methods.

In addition, the surface roughness of NPS leads to a high contact resistance to the metal layer of the applications on EL, and hence a waste of an electrical power. To enhance the EL efficiency and stabilize performance, a good metal contact to the light-emitting layer (NPS layer) is required. In the authors' study, a supercritical fluid (SCF) technique is explored to overcome the processing problem on metal contact. The SCF is a liquid or gas material used in a state above the critical temperature and critical pressure where gases and liquids can coexist. The use of a CO2 SCF and silver (Ag) nanoparticles is shown to improve the electrical contact between the metal and NPS. Therefore, the contact resistance can be largely reduced, and the power consumption of EL on the PS structure can be greatly reduced by the method.

The morphology, porosity, and photoluminescence of the NPS prepared by the proposed method are investigated. This novel technology will have much benefit in the applications of nano-electronic and opto-electronic devices.

1. Introduction

1.1. Overview

Porous silicon (PS) consisting of many pores and silicon (Si) pillars is discovered in the 1950s [1]. As we know, PS structures are widely used to yield efficient visible photoluminescence (PL) at room temperature [1, 2]. Such light-emission behaviors are primarily attributed to electron confinement in the nano-crystals that constitute the porous structure [3-5]. The PS structure is prepared by electrochemical etching in a hydrogen fluoride (HF) solution. It is possible to control the size of the pores by adjusting the etching-process parameters including the formation conditions of PS strcutre. The PS structure with different scaled Si pillars/pores lead to different material properties. Besides, it is also known that the nano-scaled PS (NPS) layers are easily to show stronger emitting light in the visible range at room temperature.

In this chapter, some novel manufacturing and processing technologies of NPS would be discussed. Some new ideas for NPS manufacturing technologies are introduced. Among them, the Hall-effect is applied in the electrochemical anodization [6]. By Lorentz force, the carriers can be well swept to the allocated areas. In addition, a forward biased pn-junction is also used for bottom-hole assistance [7]. NPS films with nano-scaled pores and high-aspect-ratio pillars can be formed by this method. Moreover, a simple and mask-free method for the fabrication of patterned PS using a localized electric field is proposed [8]. The electric field is applied by patterned electrodes (anode and cathode) which are placed horizontally underneath the sample. No masking-layer or related photo-lithography processes are needed in this method. These ideas contain original information, theoretical and experimental in the field of nanoscaled electrochemistry. The principle, physical mechanism and experimental results by these methods are demonstrated.

In our experiments, the electrical properties of NPS samples on *current-voltage (I-V)* characteristics are measured by Kiethley 236 source-measure unit. The optical properties on photoluminescence (PL) feature are measured on the PL setup with a He-Cd laser (excitation wavelength of 325 nm) and a Hamamatsu R928 photomultiplier detector. All scanning electron microscope (SEM) imagines are performed by JEOL JSM-6335FNT.

1.2. Future Developments

The NPS is a promising material in the areas of optoelectronics and microelectronics due to their unique electrical and optical properties. The NPS shows room temperature visible luminescence characteristics which are not observed in bulk or single-crystalline Si. Such unique properties and low cost make the NPS become a good candidate for the realization of NPS-based optical devices. The electroluminescence (EL) of NPS is discovered in the 1990s [2]. Then, some optioelectronic devices based on the NPS materials such as light emitting diodes (LEDs), flat-panel displays, cathode-ray displays, nonvolatile light emission memories, photodetectors, solar cells, and sensors [9-13] are also being developed. Conventional bulk Si-based material which is an indirect bandgap has the disadvantages in a poor optical radiative efficiency and shows luminescence only outside visible range. However, NPS material overcomes those disadvantages. In addition, the surface area density of NPS is very large because of considerable quantities of porosities. It is about 600 times more than the surface area density of the normal polished Si surface [14]. It implies that the surface of NPS can lead to a better efficiency of solar cells [15].

Furthermore, some applications on the tunable active microcavity resonators, optical waveguides, laser mirrors, photonic crystals, silicon-on-insulator (SOI) structures, gas sensors, membranes, and molecular sieves also have been proposed for the future developments [16-21].

2. Forming Mechanisms

The NPS forming mechanism is plotted in figure 2.1 schematicly. The illustrative equation of the overall process during PS formation can be expressed as below:

$$Si + 2HF + 2h^{+} \rightarrow SiF_2 + 2H^{+}$$
$$SiF_2 + 4HF \rightarrow H_2 + H_2SiF_6 \quad (for\ divalent\ anodic\ reaction)$$
$$Si + 2HF + 4h^{+} \rightarrow SiF_4 + 4H^{+}$$
$$SiF_4 + 2HF \rightarrow H_2SiF_6 \quad (for\ tetravalent\ anodic\ reaction)$$

The etching rate is determined by the hole (h^{+}) accumulation in the adjacent regions of the HF electrolyte and Si atoms. It implies that the etching rate and etching area can be controlled by the hole accumulation quantity and area. It is an important idea in this chapter's discussion.

As to the NPS formation mechanism, at least five mechanisms are proposed. (a) The Beale model (figure 2.2) [22-25] (b) The diffusion-limited model (figure 2.3) [26-29] (c) The quantum model (figure 2.4) [30-32] (d) The rate model [33] (e) The two-component resistor network model [34].

A brief summary for the first three common models are described below: Beale *et al.* pointed out that space charge regions exist between the PS layers and Si bulk. The electric field lines concentrate at surface irregularities, and focus current flow at the PS pore tips and locally enhance dissolution there.

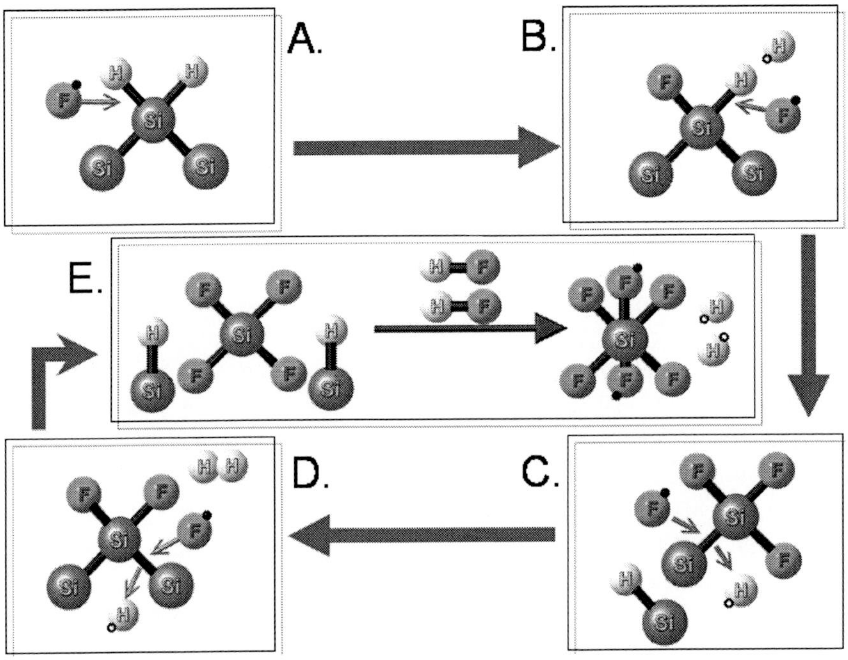

Figure 2.1. The schematic diagram of NPS forming mechanism.

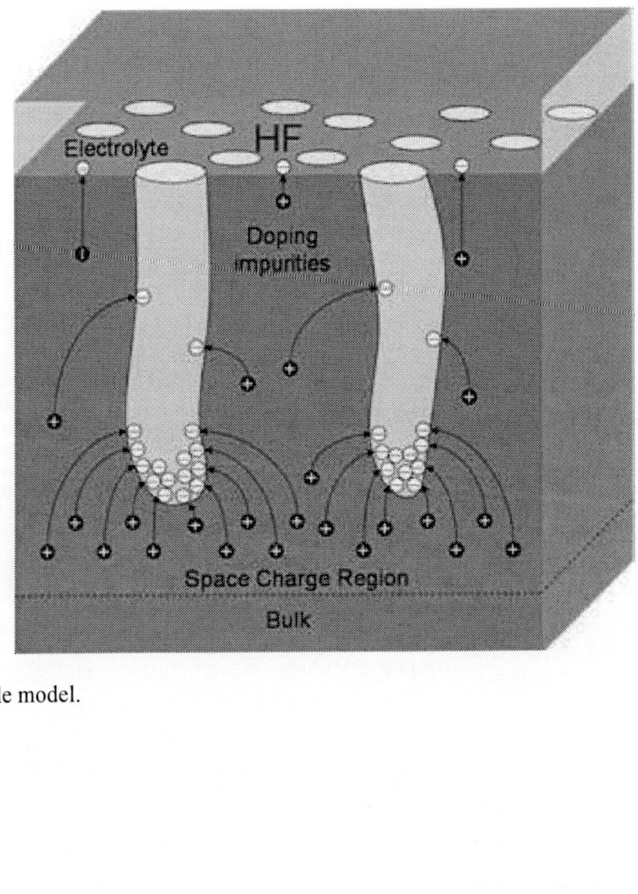

Figure 2.2. The Beale model.

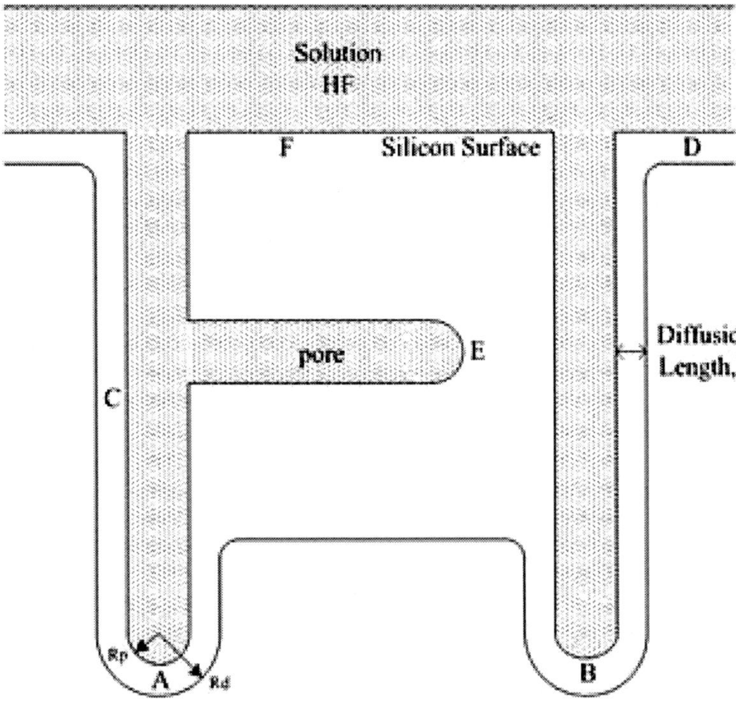

Figure 2.3. The diffusion-limited model.

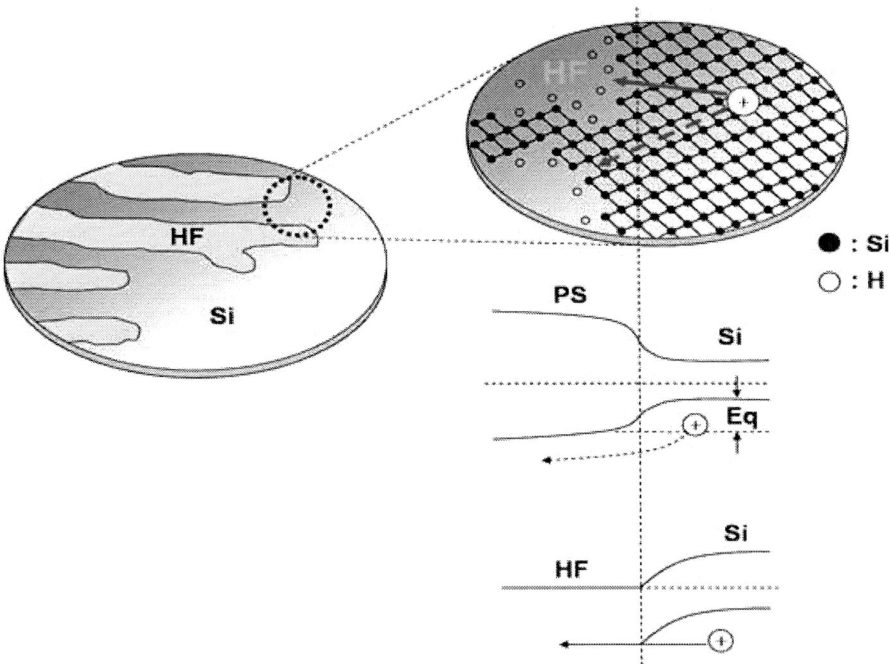

Figure 2.4. The quantum model.

Anodic current therefore anisotropicly directed to the pore tips. The diffusion-limited model proposed by Smith and Collins describes the formation of PS in terms of a random walk. The model is based on the the mechanism of diffusion-limited aggregation by Witten and Sander. It is the first computer simulated theoretical model. The holes diffuse to the Si surface and react with Si surface atoms. As to the quantum model proposed by Lehamann, it assumes the PS formation the possibility of a quantum confinement for charge carriers in the pore walls due to their small thickness.

3. Perpendicular-Electric-Field-Assisted Method

The NPS formation techniques have been developed by many methods, including electrochemical anodization [4, 35, 36], stain etching [37-39], spark-erosion [40] and vapor etching technique [41-42]. Among them, electrochemical anodization is the most commonly used.

The electrochemical etching is usually performed under an electrical field which is perpendicular to the sample surface. And the sample is dipped into a HF solution. The etching rates are controlled by adjusting the etching current densities and electrolyte compositions. It is well known that the etching current is governed by the hole generation at the Si surface to assist anodic oxidation during the electrochemical etching. It is possible to control the size of the pores and the material properties by adjusting process parameters including the formation conditions. The schematic diagram of the experimental setup for preparing NPS samples is given in figure 3.1. The main body of the HF electrolyte container is made of Teflon materials.

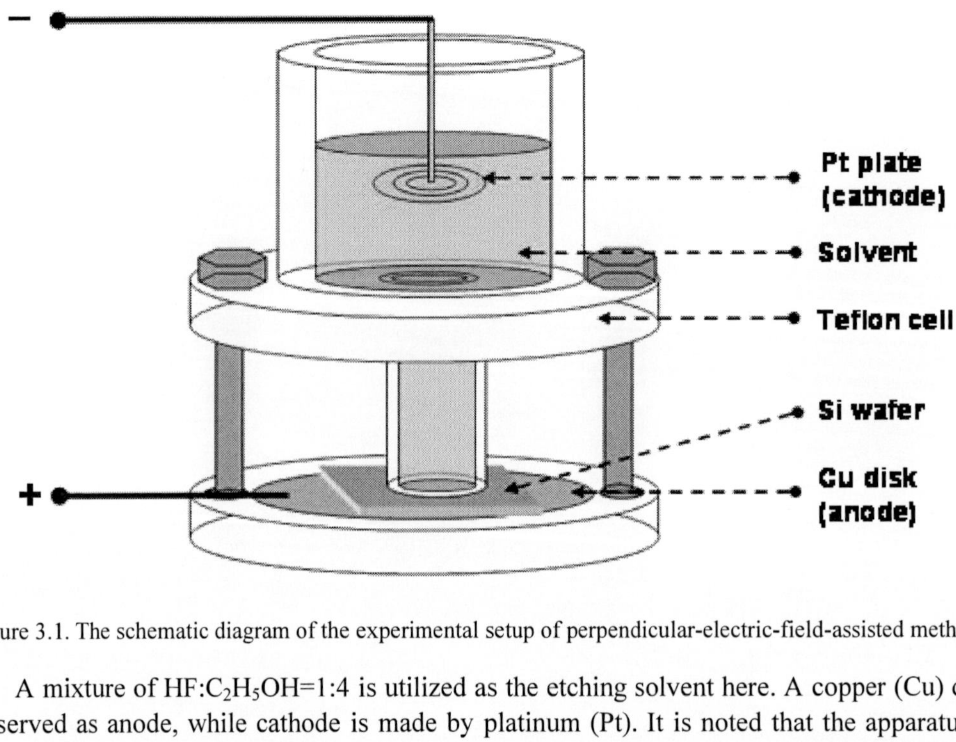

Figure 3.1. The schematic diagram of the experimental setup of perpendicular-electric-field-assisted method.

A mixture of $HF:C_2H_5OH=1:4$ is utilized as the etching solvent here. A copper (Cu) disk is served as anode, while cathode is made by platinum (Pt). It is noted that the apparatus is designed with a vertical-standing arrangement so as to easily remove the hydrogen bubbles from the NPS surface and to improve the uniformity of the NPS samples. In our study, the NPS samples are prepared on Si (100). Prior to the etching process, a sintered aluminum (Al) film is used to form an intimate backside ohmic contact to allow a homogeneous anodization current flow. The anodization etching current is supplied by an external constant current source. The schematic mechanism is plotted in figure 3.2. The typical top view and cross-section SEM images of the NPS sample are shown in figures 3.3.

For p-type Si, the holes are the majority charge carriers, and the p-type NPS (p-NPS) layers are easy to produce. Conversely, n-type NPS (n-NPS) is very difficult to form, as a result of the lack of holes. Previously, most of the research done on NPS has been made on hole-rich p-type Si. However, the control of n-NPS is necessary since an application based on n-type substrate is possible in light emitting diode (LED) technology and microelectronic technology. [43, 44] Therefore, some assistance methods are used for solving the problems and described as belows.

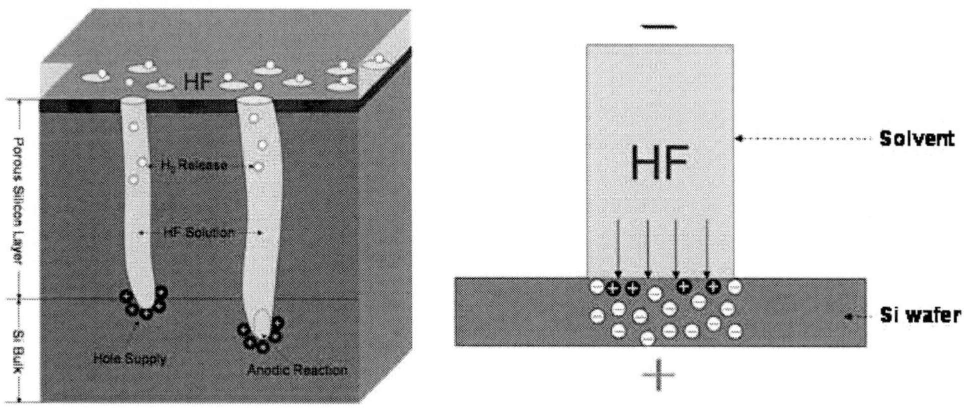

Figure 3.2. The schematic mechanism of perpendicular-electric-field-assisted method.

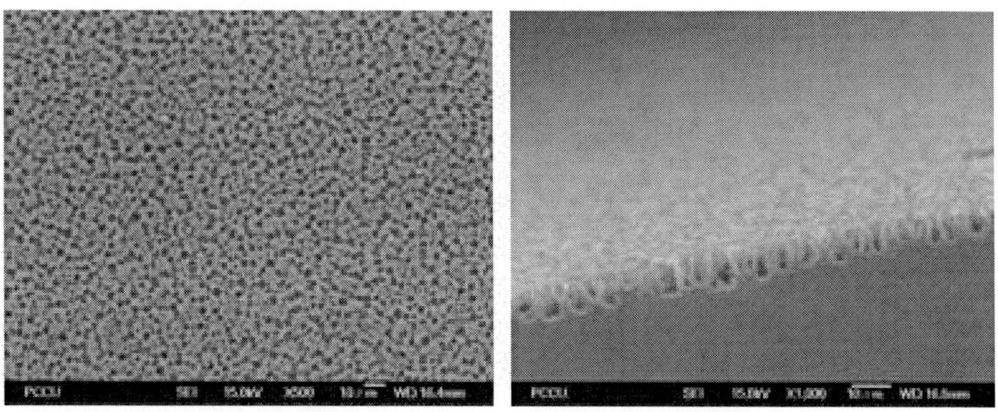

Figure 3.3. The SEM images of the NPS sample of perpendicular-electric-field-assisted method.

4. Illumination-Assisted Method

To get an n-NPS, illumination-assisted method is the most popular way to generate the holes required in the electrochemical etching process on the hole-poor n-type samples (as shown in figure 4.1.) [45, 46]. However, the actual photo-energy absorbing by atoms is very sensitive to the illumination-source intensity and electrolyte environment in which the conditions are not constant in the process. The top view and cross-section SEM images of the n-NPS sample are shown in figures 4.2. Since the illumination intensity is related to the distance, besides the absorption coefficient and the wavelength, only top-layer atoms are photo-excited and only the surface layers under illumination generate the electron-hole pairs. The actual photo-energy absorbing by Si atoms is very sensitive to the illumination-source intensity and electrolyte environment, and the conditions are not constant in the process. The etching rate will gradually decrease because the illumination is very difficult to reach the deeper layers. In other words, the only-illumination-assisted etching approach is depth-limited. The illumination related factors and limitation may affect the NPS morphology and formation. Therefore, cone-shape pores in NPS structures are often formed in this method [47]. Besides, it is also known that the n-PS obtained under illumination usually consists of a nano-PS layer and macro-PS layer. [48]. As we know, macro-PS is difficult to show room temperature light emitting in the visible range [4]. Therefore, in next sections, some ideas are proposed.

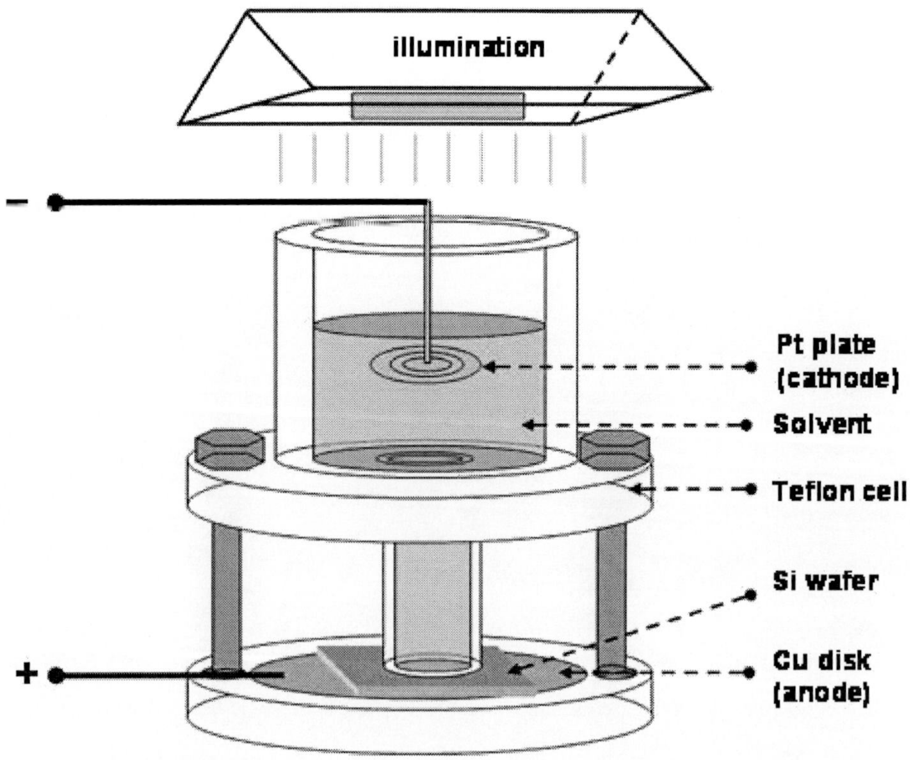

Figure 4.1. The schematic diagram of the experimental setup of illumination-assisted method.

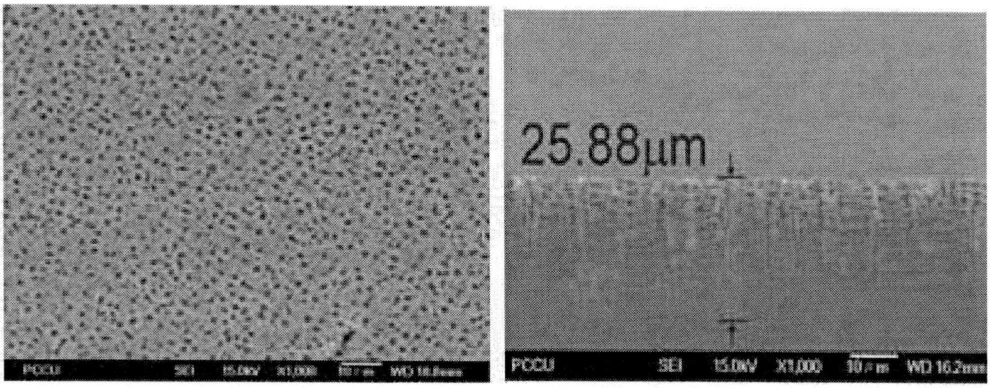

Figure 4.2. The SEM images of the NPS sample of illumination-assisted method.

5. Hall-Effect-Assisted Method

In this section, an alternative method based on Hall-effect is applied on n-NPS fabrication (as shown in figure 5.1). The majority carriers in n-type Si (electrons) are swept down by Lorentz force. Enough holes continuously appear on the surface layer to participate in chemical reaction during the etching process.

The schematic diagram of the experimental setup for preparing n-NPS is given in figure 5.1. A mixture of HF: C_2H_5OH=1:4 is also utilized as the etching solvent in the section 3. The NPS samples are prepared on n-type Si (100). The sample is exposed to the Hall-effect environment. An electrical field (a voltage V_x) is biased on the samples in x-direction and makes a current I_x. In y-direction, a magnetic field (B_y) is placed perpendicular to the current. Charge carriers flowing in the samples will experience a Lorentz force ($F_z = qE_z = qV_x \times B_y$), as indicated in the figure 5.1. The majority carriers (electrons) are swept down by $\vec{F}_z$. A Hall voltage (V_H) across the semiconductor is built in z-direction. By applying a sufficiently large bias-voltage and a magnetic field density, the upper layers of the semiconductor are depleted and then inverted from an n-type to a p-type semiconductor. Enough holes can reveal on the upper layers to participate in chemical reaction during the etching process. The schematic mechanism is plotted in figure 5.2. The etching rate is determined by the hole (h^+) accumulation in the adjacent regions of the HF electrolyte and Si atoms. It can be found that the hole accumulation increases with the magnetic field.

There are two special points in this method. First, the illumination sources are not necessary. Therefore, no illumination-limit has to be concerned in the formation of deep NPS layer. In the second, the hole supporting is owing to the inversion layer (from n-type to p-type) creating on the upper layers. So, no vertical metal electrodes are needed to be soaked in the electrolyte to serve as a cathode. By the way, the illumination-free and metal-electrode-free etching container can be well designed as a hermetic container with a strong cover for the safety of handling corrosive HF electrolyte. Also, the metal-electrode contamination can be prevented in HF electrolyte.

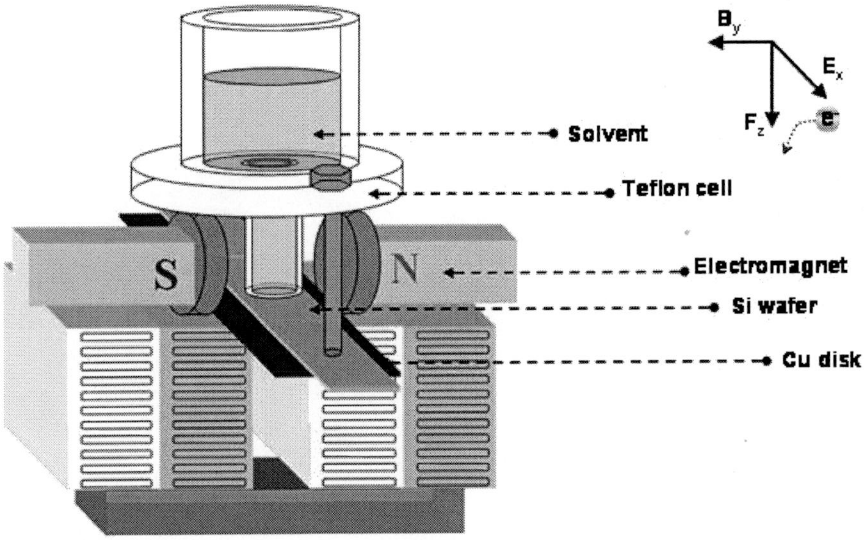

Figure 5.1. The schematic diagram of the experimental setup of Hall-effect-assisted method.

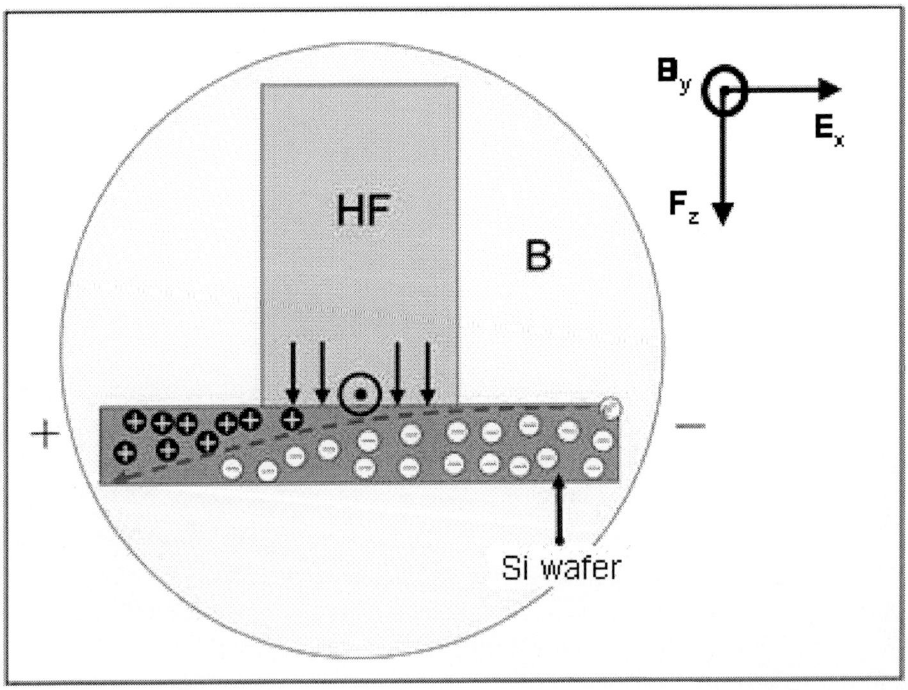

Figure 5.2. The schematic mechanism of Hall-effect-assisted Method.

Under a magnetic field (B_y) of about 20 milliTesla and a voltage (V_x) of about 60 Volts, NPS layers are successfully fabricated by the Hall-effect etching method without any illumination sources. The top view and cross-section SEM images of the NPS sample are shown in figure 5.3 and 5.4, respectively.

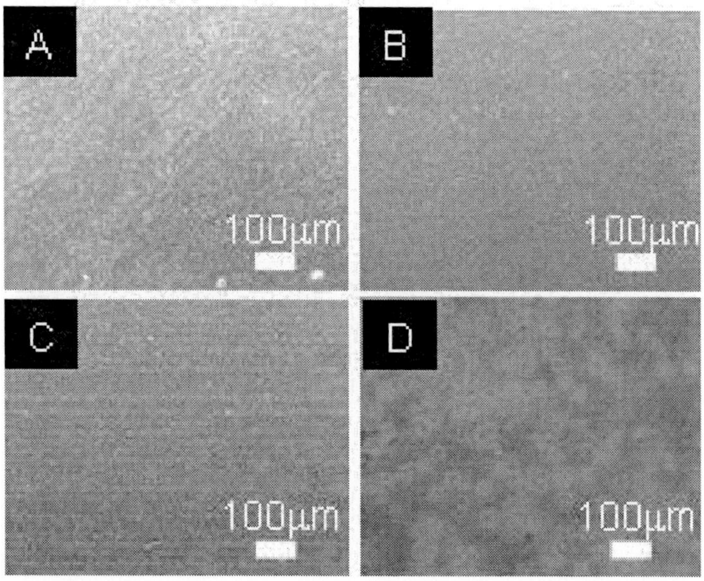

Figure 5.3. The top view SEM images of the NPS sample of Hall-effect-assisted method.

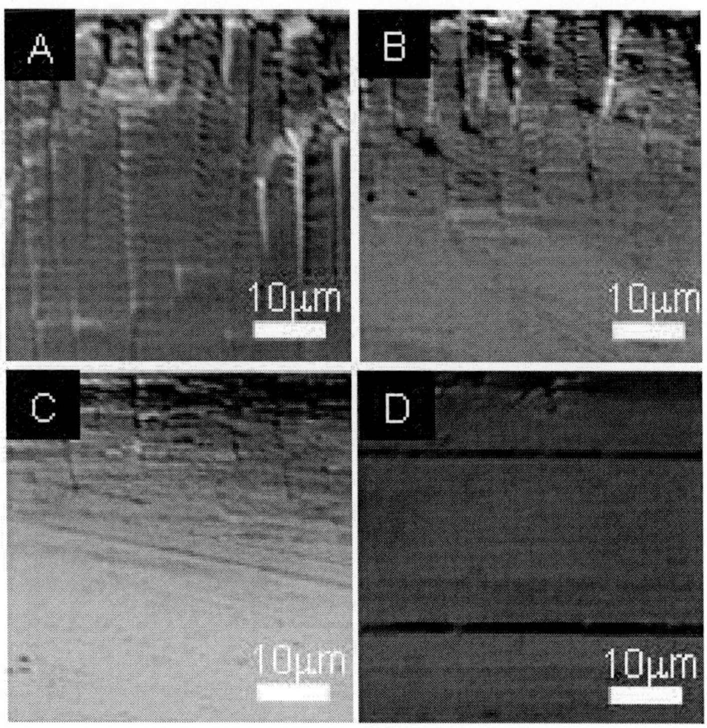

Figure 5.4. The cross section view SEM images of the NPS sample of Hall-effect-assisted method.

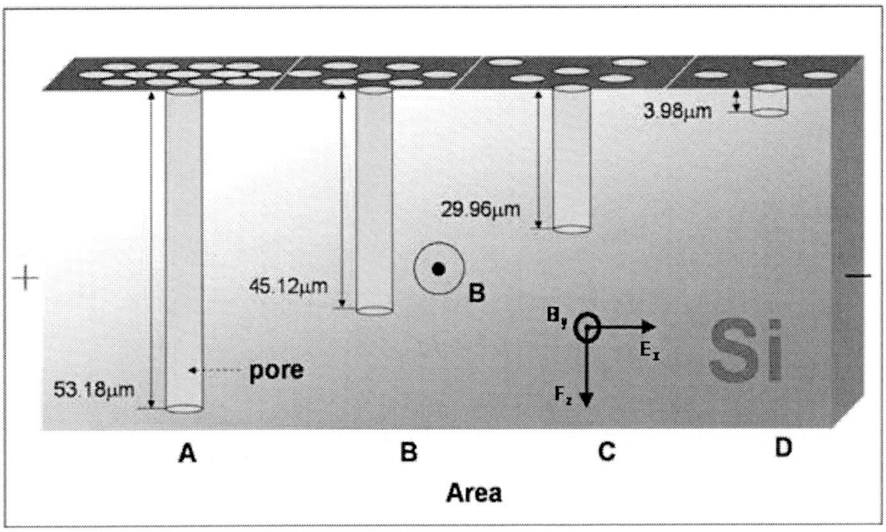

Figure 5.5. The schematic of the pore depths and distribution.

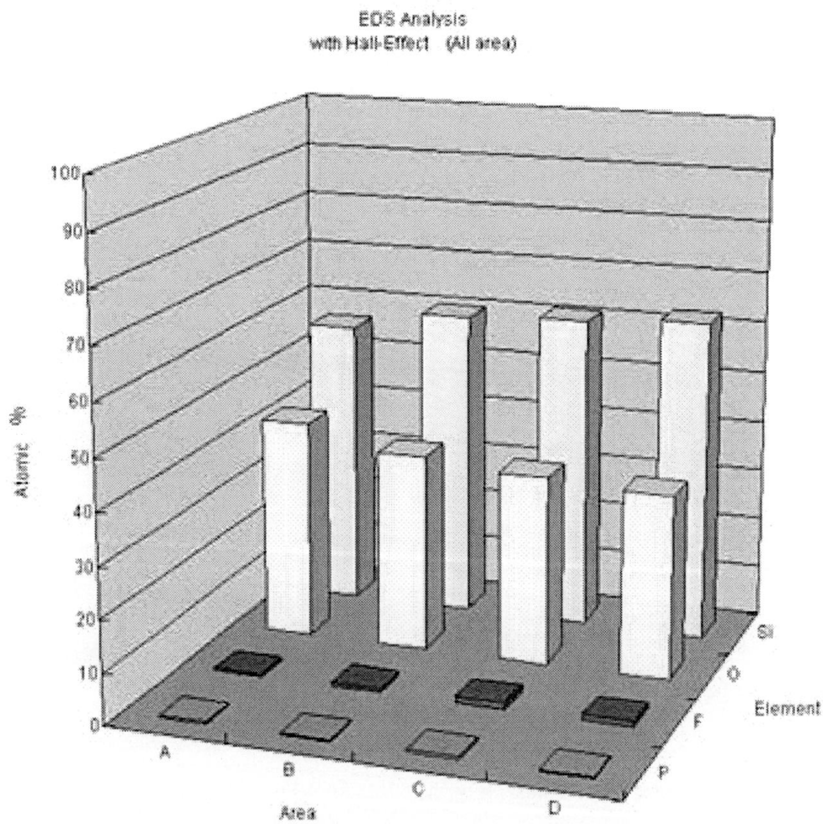

Figure 5.6. The EDS analysis of the NPS sample of Hall-effect-assisted method.

A larger voltage drop across the x-direction will cause a larger gradient-situation of etching extent of NPS. To easily show the gradient-situation for further investigation, a larger

V_x is adopted in this study. On the other hand, a small V_x is suggested for the film uniformity. Four regions (A, B, C and D) are defined on the sample for article describing, as shown in the figure 5.3 and 5.4. The pores are narrow and straight with a high aspect ratio because the hole-supplying mechanism is bottom-up. The mechanism is different from the only-illumination-assisted method in which it only relies on the illuminated surface atoms. The pore depths of the NPS layers are about 53 μm (region A), 45 μm (region B), 29 μm (region C) and 3 μm (region D), respectively. Figure 5.5 shows the schematic of the pore depths and distribution for the four regions. And figure 5.6 demonstrates the energy dispersive spectrometer (EDS) analysis. The experiment results are helpful to understand the variation of NPS morphology at different positions under Hall-effect. The variation of morphology in A~D regions implies that the areas closer to the anode can accumulate more holes to participate in etching reaction. Such plenty of holes following the bottom-up paths lead to high-aspect-ratio pores with a deeper and straighter shape. The hole concentration and originating source determine the pore shapes and NPS morphology.

In only-illumination-assisted n-NPS, after surface pores (of few μm) are formed, photo-energy is not completely received by the deeper layers. Therefore, the deeper layers are difficult to form the required NPS in the darkness. Only a thin part of NPS layer contributes in the PL. Meanwhile, a lateral etching keeps going on and cone-shape pores are produced. Thus, the n-NPS visible luminescence with a depth-limitation is usually less than p-NPS. In the Hall-effect-assisted method, the visible luminescence of the n-NPS is very similar to the hole-rich p-NPS. An excitation wavelength of 325 nm from a He-Cd laser is used for the PL measurements. Figure 5.7 are the PL spectra of A~D regions under the conditions of $B_y = 20$ milliTesla and $V_x = 60$ Volts. From regions A to D, the PL intensity, etching extent and the porosity decrease as shown in figure 5.6 and 5.7.

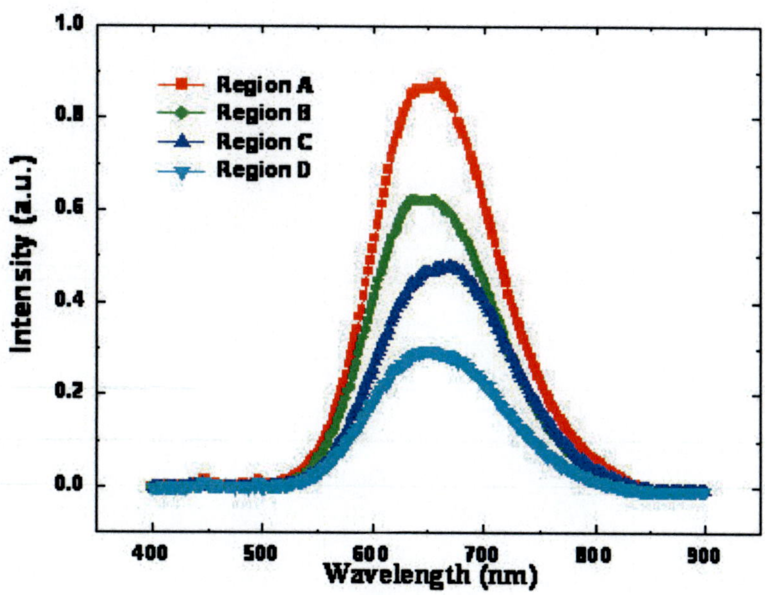

Figure 5.7. The PL spectra from A to D regions of n-NPS sample.

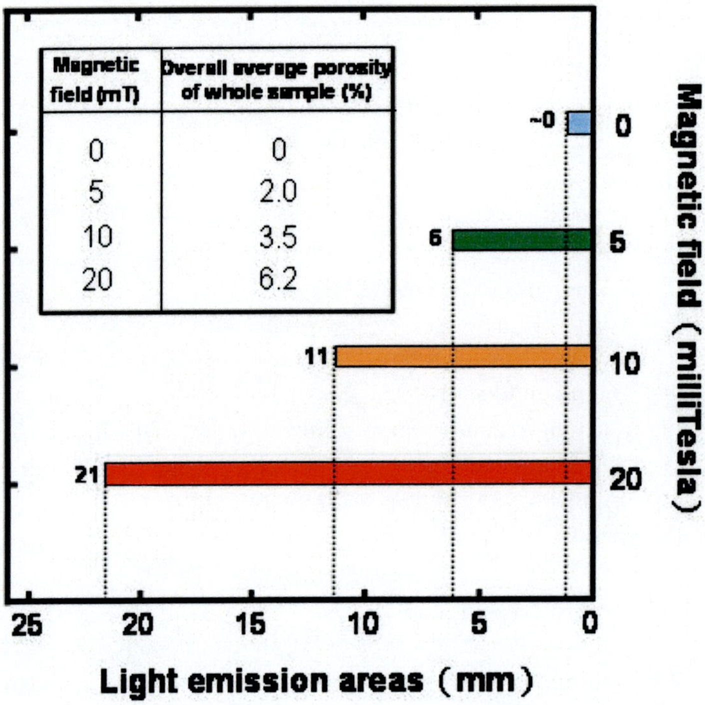

Figure 5.8. The influence of magnetic field (B_y) on the area size of light emission. The overall average porosities of the samples are also shown in the inset.

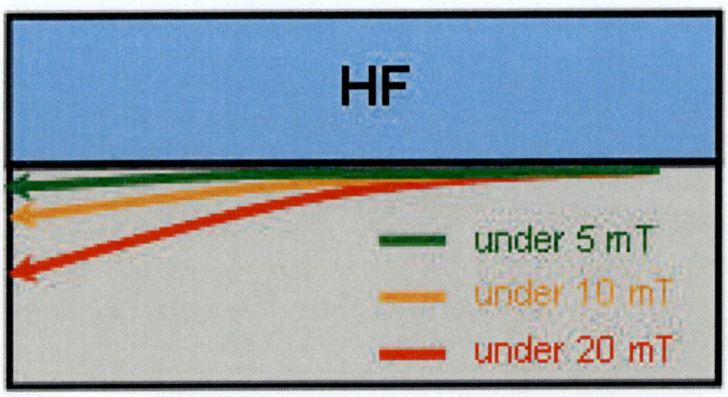

Figure 5.9. The schematic mechanism of the influences of magnetic field for Hall-effect-assisted method.

The high PL intensity in region A is attributed to the high hole-concentration, and hence the high porosity. Obviously, the PS sample produces a strong emission of orange-red luminance under ultraviolet (UV) illumination. The PL intensity is proportional to the hole concentration that is appearing on A~D regions gradually. This implies that the selective-etching and the gradient light-emission on the identical sample by this method is possible.

In the section, the Hall-effect-assisted method can provide an anisotropic and illumination-free technology for n-NPS manufacturing. A strong visible light is obtained on n-NPS at room temperature. Also, a gradient etching can be formed on the identical sample and that is controlled by a electric field and magnetic field.

6. Buried-P-Layer-Assisted Method

In this section, another method (named as buried-p-layer-assisted method) is proposed for manufacturing n-NPS layers. A mount of holes are supported by a forward biased pn-junction in which the buried p-layer plays the role as a hole-source underneath the n-layer. For the hole-assisting direction, it is bottom-up and straight-upward. Therefore, no depth-limitation should be taken into consideration in this approach. The schematic mechanism of the biased pn structures for bottom-hole-assistance is plotted in figure 6.1. The pillars and pores are narrower, straighter, deeper, and more condensed than those prepared by the conventional only-illumination-assisted approach. The proposed buried-p-layer-assisted method also can be superimposed on the illumination-assisted method in the electrochemical anodization (named as superimposition-assisted approach). The superimposition-assisted approach will be described in the section 7.

The schematic diagram of the experimental setup for preparing n-NPS layer is given in figure 6.2 (a). Figure 6.2 (b) shows the case of difficulty in electrochemical etching in n-Si sample without any hole-assistance (illumination-assistance and bottom-hole-assistance). When the illumination-assistance is applied, as shown in figure 6.2 (c), the electron-hole pairs are generated and accumulating only on the top surface layers in which some tiny lateral etching accompanies with the vertical etching. On the other hand, in the buried-p-layer-assisted method as figure 6.2 (d), an electrical field across the anode and cathode forward biases the np junction, and holes flow straight-upward from p- to n-region. Sufficient holes can reach to the pore tips to participate in chemical reaction during the etching process, even in the dark.

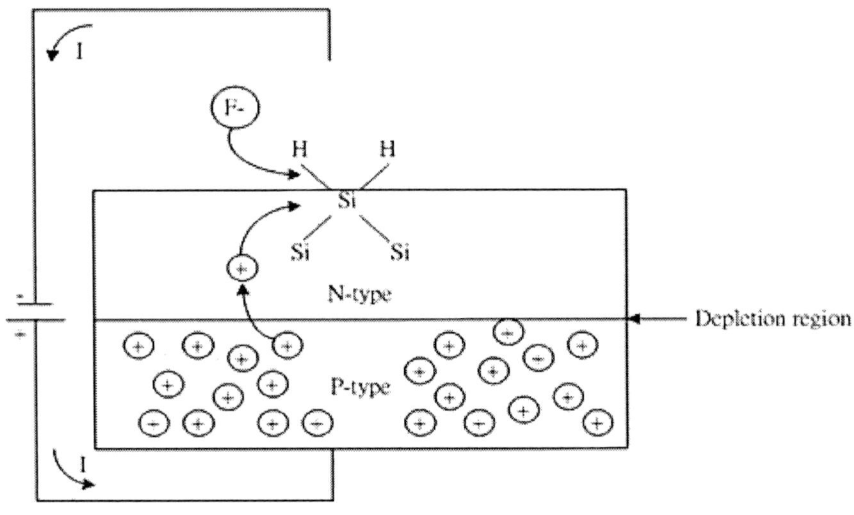

Figure 6.1. The schematic mechanism of the biased pn structures for buried-p-layer-assisted method.

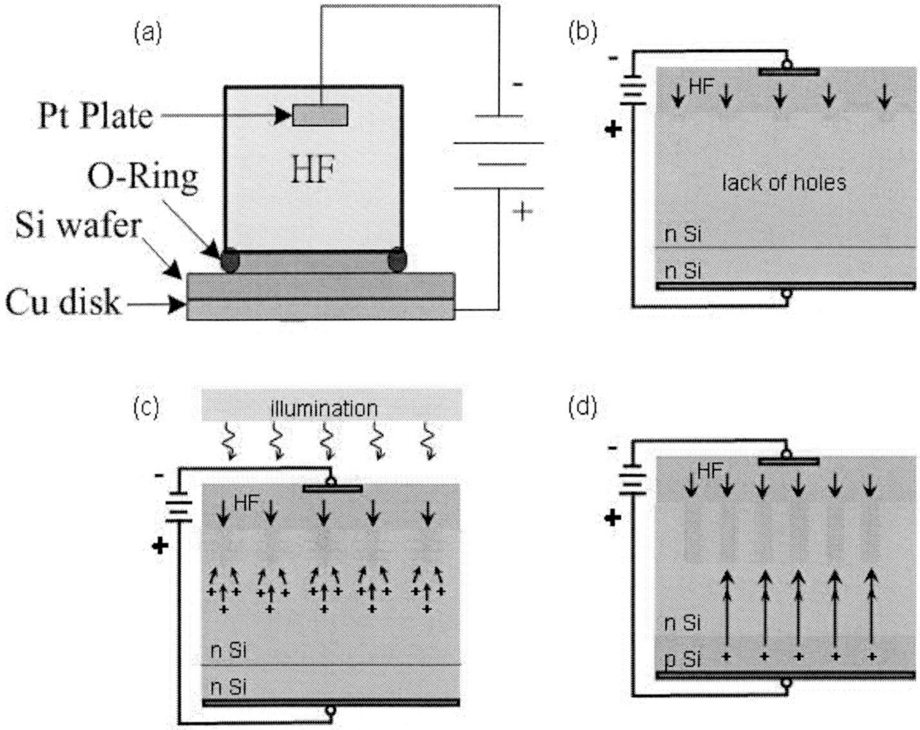

Figure 6.2. (a) The schematic diagram of experimental setup of buried-p-layer-assisted method, (b) the sketch of electrochemical etching in n-on-n Si without any hole-assistance, (c) the mechanism sketch with the illumination-assistance approach, and (d) the mechanism sketch with the bottom-hole-assisted approach.

Figure 6.3. The SEM images of the NPS samples of buried-p-layer-assisted method.

The top view and cross section view of NPS films are successfully fabricated by the buried-p-layer-assisted method with or without illumination-assistance. The top view and cross-section SEM images of the NPS samples are shown in figure 6.3. Two NPS samples with n-epilayers on n-substrates (n-on-n) and one NPS samples with n-epilayers on p-substrates (n-on-p) are prepared, and named as samples A, B, and C, respectively.

In sample A (n-on-n), not any hole-assistance is applied on the sample, and then the electrochemical reaction can not successfully proceed. The NPS layer fails to be formed on sample A. For the conventional case based on the only-illumination-assistance, it is applied on sample B (n-on-n). In sample C (n-on-p), which is prepared by buried-p-layer-assisted method, the hole-supplying relies on the p-substrate without any illumination. Furthermore, etching solution ratio of C_2H_5OH: HF =2:1 is used in all samples.

In the sample B (with illumination-assistance), the macropores clearly appear on the surface. The surface pores are cone-shape (V-groove), especially even clearer in sample B. The electron-hole pairs only rely on the top-surface layers by illumination. After surface pores (of several ten um) are formed, photo-energy is not completely received by the deeper layers. Therefore the deeper layers are illumination-limited and the NPS pores are difficult to be formed. A lateral etching keeps on while the vertical etching is proceeding, hence crosswise-branched pores and tissue-like branch structures are formed.

Conversely, in sample C which is based on the buried-p-layer-assisted method, the hole-supplying path is bottom-up and no depth-limitation. The pore shapes are narrower, deeper and straighter. Also, high-aspect-ratio nano-scaled pores (aspect ratio = 300) can be obtained, and the pore structures are channel-like instead of tissue-like that always occurs in illumination-assisted approach. As shown in figure 6.3 (c), the pore depths of the PS layers can be up to 100 μm and the pore width can less than 300 nm. It is useful to serve as thick-layer light emitting applications.

An excitation wavelength of 325 nm from a He-Cd laser is used for the PL measurements. Figure 6.4 shows the PL spectra of samples A~C. Clearly, the buried-p-layer-assisted method can effectively improve the light-emitting density.

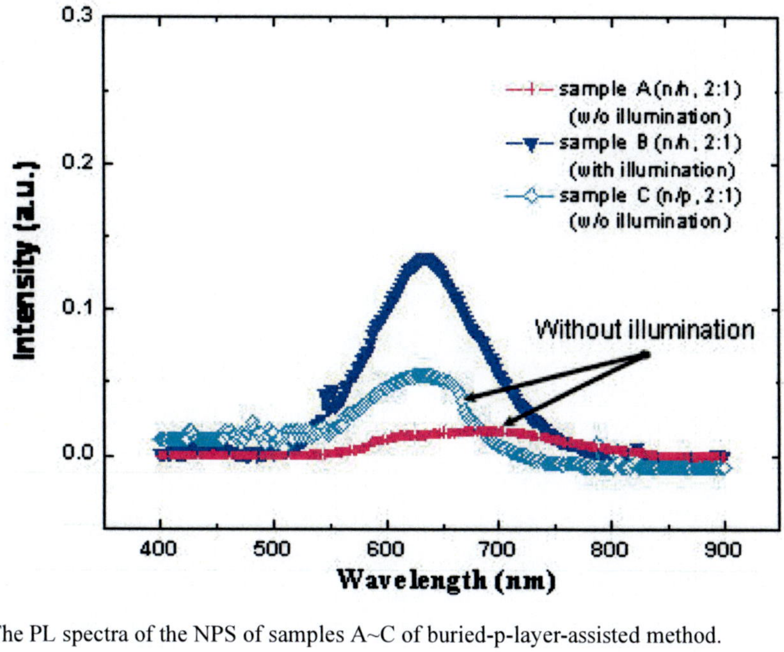

Figure 6.4. The PL spectra of the NPS of samples A~C of buried-p-layer-assisted method.

The measured PL that is usually attributed to quantum size effects is a combination result of light emissions from upper layer (macroporous) and lower layer (nanoporous). Thus no significant size dependence of the PL peak energy can be easily observed and concluded. Obviously, the PS samples produce a strong emission of orange-red luminance under UV illumination.

The buried-p-layer-assisted method based on a forward biased np-junction for manufacturing n-PS layer is proposed in this section. The buried-p-layer-assisted method can overcome the illumination-limitation and depth-limitation problems in only-illumination-assisted method. With the buried-p-layer-assistance, the anodization etching is almost anisotropic. This feature has great potential in fabricating processes.

In the section, clear visible PL emissions at room-temperature are obtained on the hole-poor n-type samples even in the dark. Strong visible photoluminescence emissions are demonstrated on the hole-poor n-type porous layer at about 650 nm.

7. Lateral-Electric-Field Method

In this section, a mask-free method is proposed for the fabrication of p-type NPS using a localized electric field. The electric field is applied by the patterned electrodes (anode and cathode) which are horizontally placed underneath the sample. No masking-layer and related photo-lithography process are needed in this method. Besides, no metal electrodes and hence no metal-pollution in electrolyte have to be concerned in the formation of NPS. The morphology of the NPS prepared by this method is investigated. Strong visible photoluminescence emissions in the selected areas of NPS are demonstrated on NPS at about 650 nm.

As we know in the previous section, the etching current is governed by the hole accumulation at the Si surface to assist anodic oxidation during the electrochemical etching. Most of the researches done on NPS have been made on the whole Si samples.

However, the patterning of NPS is necessary for the applications on microelectronic process, microelectromechanical system and optoelectronic integration. Some patterning method are done by masking layer (etch-resistant film) [49, 50]. As the patterned NPS is defined by masks, masking-design and related photo-lithography processes are required. Since no masking-layer is required, and hence no anodization-time limitation should be concerned in this method.

In the study, the schematic diagram of experimental setup for preparing NPS is given in figure 7.1. The samples are prepared on p-type Si (100). A mixture of HF:C_2H_5OH=1:4 is also utilized as the etching solvent here. The etching rate is determined by the hole (h^+) accumulation, so the hole accumulation strongly depends on the distribution and intensity of electric field in this method.

The patterned NPS is formed by a localized electric field during the manufacturing process as illustrated in figure 7.2. The hole distribution in Si is controlled by the electric field applied across the planar metal electrodes (anode and cathode) which are attached beneath the Si sample.

The geometry patterns of metal electrodes are designed on a reusable printed circuit board (PC board) as shown in figure 7.3.

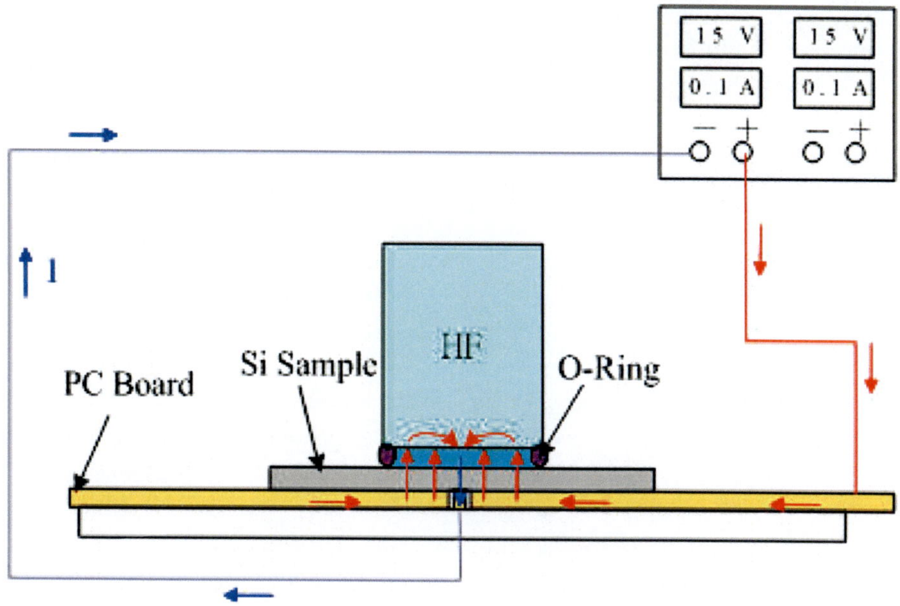

Figure 7.1. The schematic diagram of the experimental setup of lateral-electric-field method.

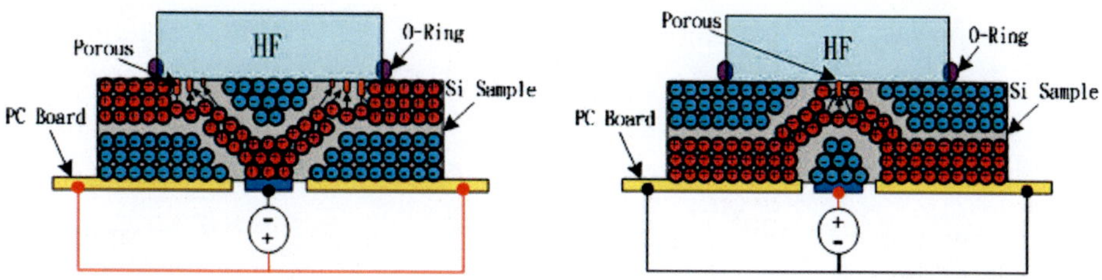

Figure 7.2. The schematic illustration of mechanism and the influence of a localized electric field.

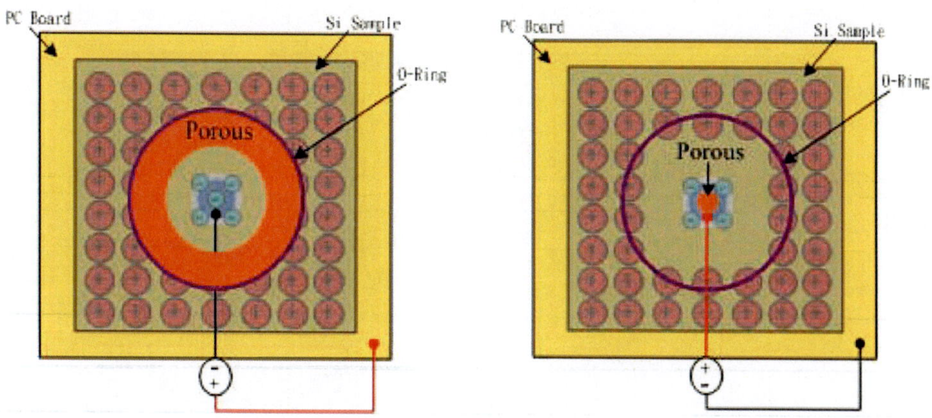

Figure 7.3. The geometry patterns of metal electrodes are designed on a reusable PC board.

For the simplification of fabrication procedure, the back-side surface of Si sample directly and closely touches on the PC board, and the procedure of back-side Ohmic contact can be omitted in our sample preparing. Though extra electric power may consume on the metal-semiconductor contact barrier without the Ohmic contact procedure, however, the experimental results do not change. In the cathode-related region (above cathode), holes are drifted downward the cathode-electrode and the top layers are depleted of holes. On the contrary, the holes in the anode-related region (above anode) drift upward the Si/HF interface layer to participate in chemical reaction during the etching process. Such a mechanism forms a patterned NPS. No else perpendicular electrodes are needed to be soaked in the electrolyte to serve as a cathode in this method.

Two samples (sample No.1 and sample No.2) are dealt with the etching processes under localized electric fields of 15 Volts for 30 minutes. On sample No.1, the inner-circle region is attached on the cathode electrode and negatively charged; on the contrary, the outer-circle region is attached on the anode electrode and positively charged. Meanwhile, the sample No.2 is biased in the same way but exchanging the electric polarity.

Patterned NPS layers are successfully formed on the two samples, and the naked-eye photos are shown in figure 7.4 (a) and (b), respectively. A strong emission of orange-red luminance appears in the selected areas (outer-circle region in sample No.1 and inner-circle region in sample No.2) under UV illumination without any mask.

The top view SEM images and the distribution of the four regions are shown in figure 7.5. For further studying and article describing, four regions (A, B, C and D) are defined on the sample No.1. A voltage drop across electric terminals would cause a non-uniform distribution of holes. To easy show the relationship between the etching degree and hole distribution for further investigation, a typical circle-shape light-emitting pattern is designed in this study. It is believed that other shape patterns (e.g. four-spot pattern in sample No.3) as shown in figure 7.4 (c) could be possible by proper designs of electric field distribution which are under our taken.

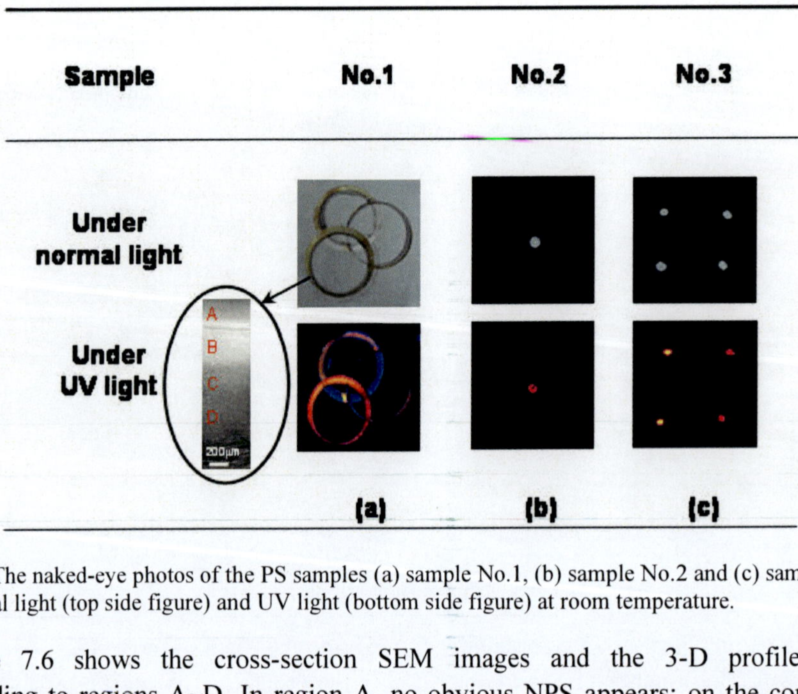

Figure 7.4. The naked-eye photos of the PS samples (a) sample No.1, (b) sample No.2 and (c) sample No.3 under normal light (top side figure) and UV light (bottom side figure) at room temperature.

Figure 7.6 shows the cross-section SEM images and the 3-D profile as well corresponding to regions A~D. In region A, no obvious NPS appears; on the contrary, the pores are drastically produced in region B, C, D for the sufficient hole-supplying. The pore depths of the NPS layers are about 0 μm (region A), 2.6 μm (region B), 8.4 μm (region C), 11.3 μm (region D), respectively. The experiment results are helpful to understand the variation of morphology under different etching conduction and extent.

In the method, the visible luminescence of the selected area is similar to the conventional method; however, the luminescence can not be found in non-selected areas. This indicates that the selective-etching and the gradient light-emission on the identical sample in one step is possible. With different geometry electrode, different NPS pattern can be produced.

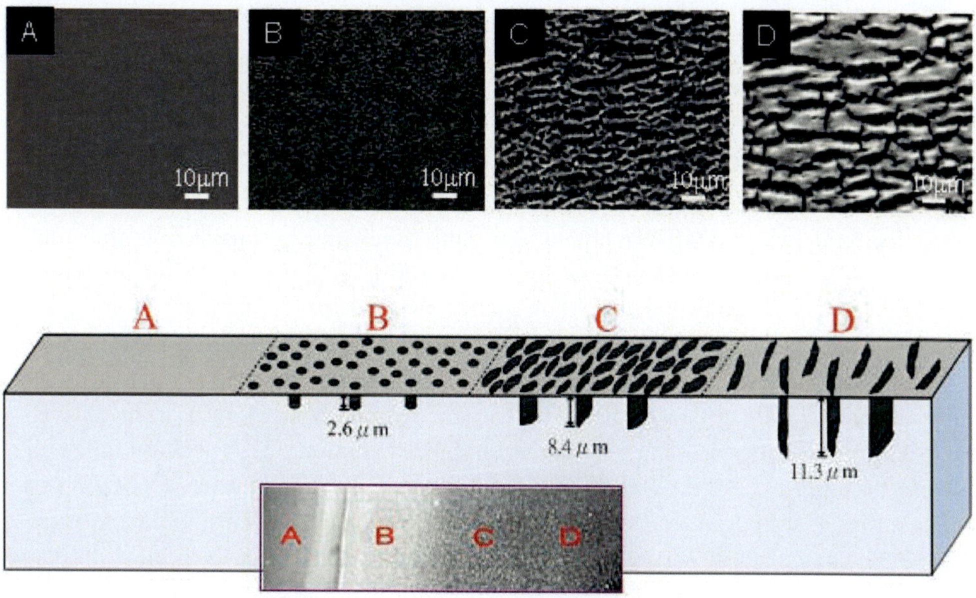

Figure 7.5. The top view SEM images and the distribution of the NPS sample of lateral-electric-field method.

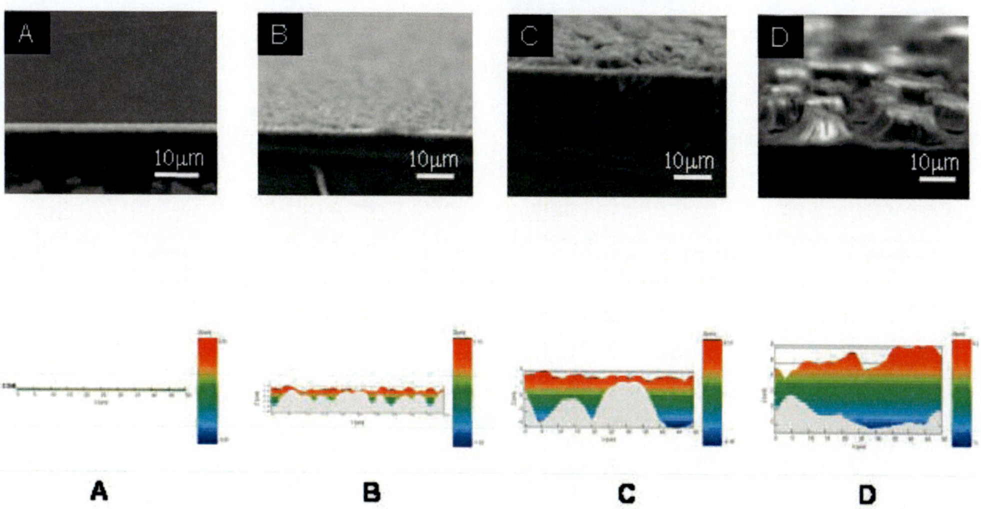

Figure 7.6. The cross-section SEM images and the 3-D profile of the NPS sample of lateral-electric-field method.

8. Superimposition-Assisted Method

As we mentioned above, perpendicular electric filed, lateral electric filed, magnetic field, illumination, buried p-layer are all important impact factors on the NPS formation. All factors are optional hole-source in the fabrication of NPS and can be super-imposed on the manufacturing processes. In this section, some examples are demonstrated for the superimposition-assisted method.

8.1. Illumination-Assisted Method with Buried-P-Layer-Assisted Method

The light-emission properties of n-NPS with an illumination-assistance and a buried-p-layer-assistance are explored in sections 4 and 6, respectively. In this section, illumination-assisted method and buried-p-layer-assisted method are superimposed on n-Si. The n-NPS film that has nano-scaled pores and high-aspect-ratio Si pillar arrays are stably fabricated. It is known that NPS is easier to show room temperature light emission in the visible range than macro-PS. The influences of anodic current density on the formation, morphology and properties of n-NPS are measured. The *I-V* characteristics and PL features of n-NPS films are investigated for different anodic-current cases. Since the anisotype-junction is forward biased during the anodization process, many holes can drift straight-upward from p-layer and participate in the electrochemical reaction. At room temperature, high peak-to-valley current ratios of about 117.3 can be obtained in negative difference conductance (NDC) region. In addition, strong visible light emissions are clearly observed under UV excitation.

The schematic diagram of the experimental setup for preparing n-NPS layers is given in figure 8.1. In figure 8.2, the left-side shows the case by the only-illumination-assisted method. The generation of top-photo-excited holes is almost isotropic and only occurs on the top layers. Therefore macro-pores with some tiny lateral branch etchings (tissue-like) are often observed in this way. And the pore has a depth limitation. On the other hand, as sketch in the right-side of figure 8.2, when the buried p-layer holes are superimposed by a forward biased pn-structure, some extra part of holes will drift straight-upward from p-layer to n-layer, and reach the pore tips to participate in electrochemical reaction during the anodization etching process. The supplying paths of buried p-layer holes are bottom-up and anisotropic, so the problems about depth limitation and lateral etching can be solved.

To analyze the impact of anodic-current densities on the etching process, three n-NPS formation cases which have different etching current densities are compared. They are specified as case-A (10 mA cm^{-2}), case-B (20 mA cm^{-2}) and case-C (40 mA cm^{-2}). An identical anodic time of 30 mins is used in all cases.

The top views and cross section views of SEM images on n-NPS samples are shown in figure 8.3 and 8.4, respectively. When the anodization current is 10 mA cm^{-2} (case-A), as shown in figures 8.3 (a) and 8.4 (a), the sample surface is shallowly etched, and some small pores with about 8 μm length are formed. In this case, the pores are formed on the top layer with illumination, so the width is wider than those in the deep layers of cases-B and C. In case-B, a clear macro-porous boundary reveals on the top view, as shown in figures 8.3 (b). Besides, in figures 8.4 (b), deep and straight pores and pillars appear in deep layers in which the anodization reaction relies on the holes from buried p-layer.

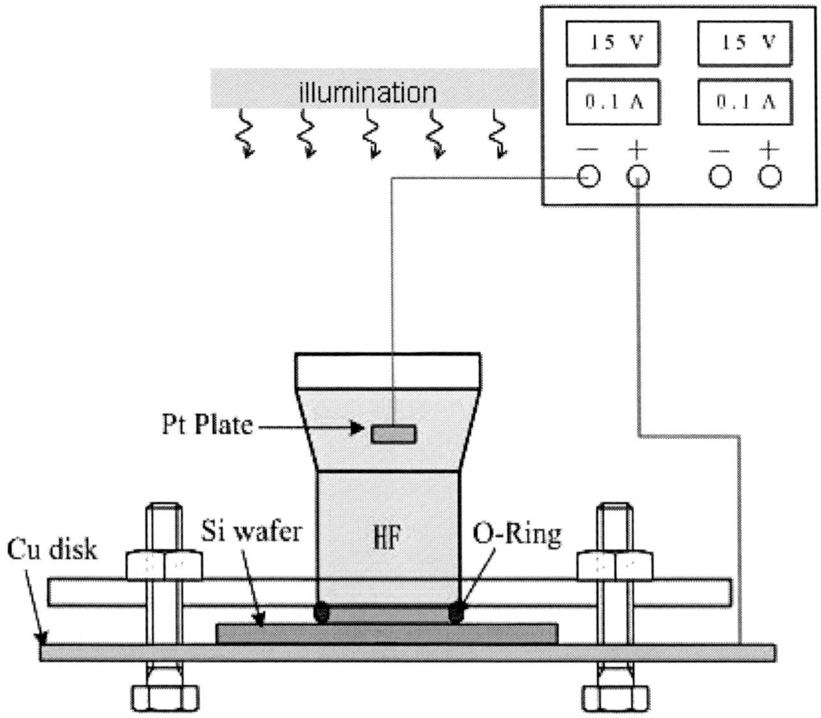

Figure 8.1. The schematic diagram of the experimental setup of illumination-assisted method with buried-p-layer-assisted method.

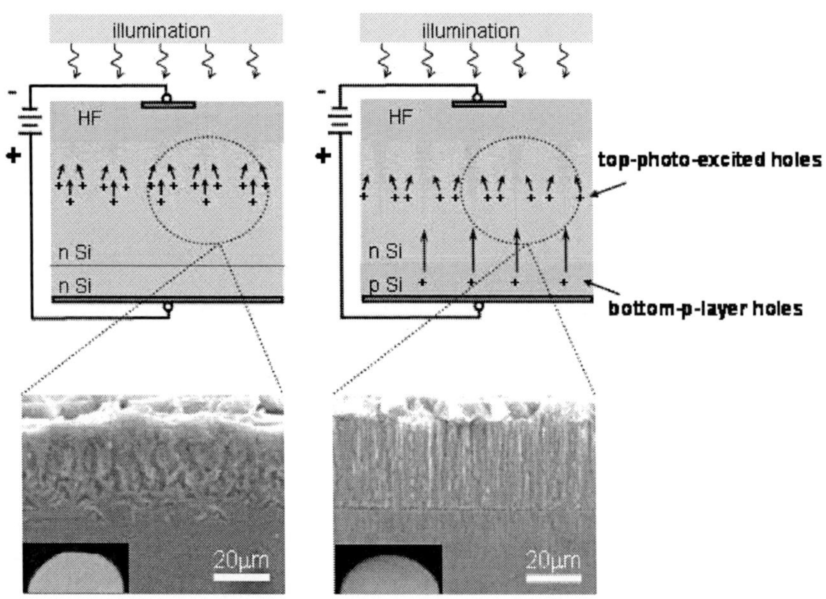

Figure 8.2. The illustrative mechanism of conventional method for n-NPS formation (left side figure) and the illustrative mechanism of n-NPS formation with the assistance of bottom-p-layer holes based on a pn structure (right side figure).

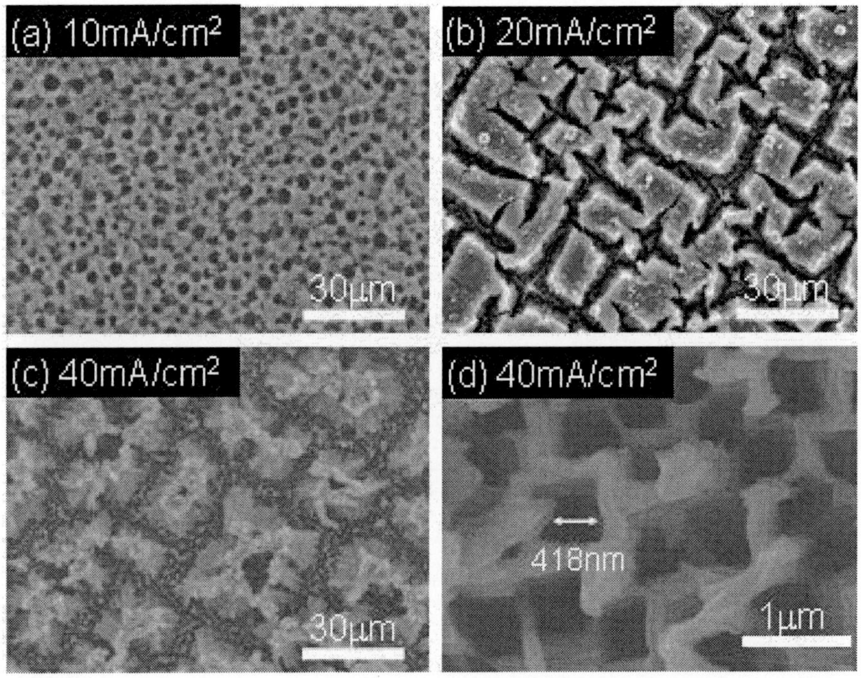

Figure 8.3. The top view SEM images of the NPS sample of illumination-assisted method with buried-p-layer-assisted method.

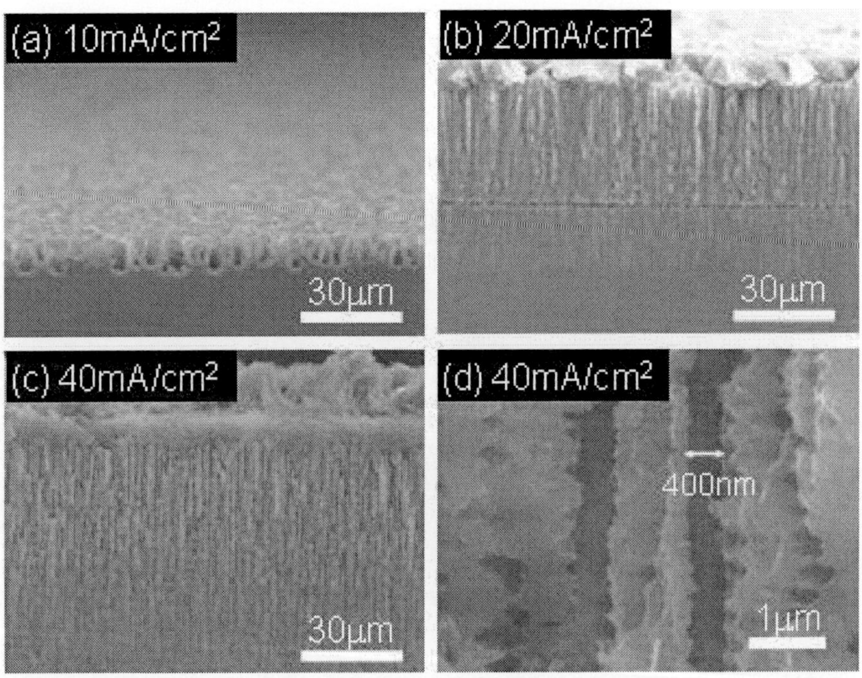

Figure 8.4. The cross section view SEM images of the NPS sample of illumination-assisted method with buried-p-layer-assisted method.

In case-C, with the highest anodic current density and drastic reaction, the vertical etchings on pore-tips proceed more rapidly. As shown in figures 8.3 (c) and 8.4 (c), the surface macro-pores expand larger than case-B. The SEM pictures of case-C at high magnification are shown in figures 8.3 (d) and 8.4 (d). It is clear that the samples formed with the superimposition of buried p-layer holes, the pore shapes are deep, straight and narrow to nano-scale, which profiles are very different from conventional method only with top-photo-excited holes. In this method, a high-aspect-ratio of 181 can be obtained. Besides, the n-NPS structures are channel-like instead of tissue-like that always occurs in conventional illumination method. The pore depths of the n-NPS layers can be up to 72 μm and the pore width can less than 400 nm. It is useful for the applications on deep-trench etching and thick-layer high-density light emission.

Figure 8.5 shows the *I-V* characteristic of cases A, B and C. All curves reveal the rectification properties due to the built-in potential in pn junction. With the smallest anodic current density in case-A, the NPS pores are the shallowest and hence a highest conductance than the other two cases is measured. It is well known that the resistivity of NPS layer is higher than bulk-Si layer [51]. In cases B and C, with higher anodic current densities than case-A, the PS layers are easily formed and that causes them to have higher resistances. Clear NDC phenomena are observed in case-B and case-C at room temperature. Their PVCRs are as high as 117.3 (case-B) and 9.5 (case-C), respectively. The NDC in NPS film is caused by the carrier transfer from high-mobility pillars to low-mobility pillars [52]. Since the widths of all pillars are not totally identical in NPS films, different (higher/lower) carrier mobilities appear in different (wider/narrower) pillars with corresponding lower/higher energy bandgaps. Among these different size pillars, the space confinement and surface scattering effect cause the carrier mobility in narrower pillars to be lower than wider pillars. Initially, when the applied voltage is turned on, the carriers prefer conducting by the wider pillars rather than narrower pillars. So the carriers are mainly concentrated in wider pillars during the first time stage. However, when the applied voltage (electric field) increases to a critical point, a drastic scattering occurs. In wider pillars, the carriers have sufficient energy to overcome the band discontinuity and to jump over the nanoscale pores into the narrower pillars. Many carriers are redistributed to all pillars (including narrower and wider ones) during the second time stage, and hence the average mobility as well as conductance reduces. The sudden switching from a higher conductance to a lower conductance produces the NDC phenomenon. The excellent NDC performance may have potential applications on microwave, multi-level switching, and memory devices [53-57].

However, NDC does not appear in case-A. Like real-space transfer diodes [24], the carriers in wider pillars must be accelerated and "heated" by the electric field so much that they are able to transfer to a "cooler" pillar. In case-A (the NPS prepared at 10 mA cm^{-2}), the porous etching depth is shallow, and the pillar length is not long enough for carrier acceleration. Therefore, no NDC phenomena are observed. However, in cases-B and case-C with higher anodic current densities than case-A, long Si pillars are easily formed by our method, and NDC phenomena are clearly observed at room temperature. In our study, the NDC can not be observed in every sample. In cases with pillars that are too short (shallow etching depths) or those with extremely uniform sizes of pillars, the NDC is difficult to be obtained.

Figure 8.6 shows the PL spectra of cases A, B and C. Such orange-red emissions are primarily attributed to the quantum confinement effect in the nano-scaled Si crystals that constitute the NPS film [2, 58, 59].

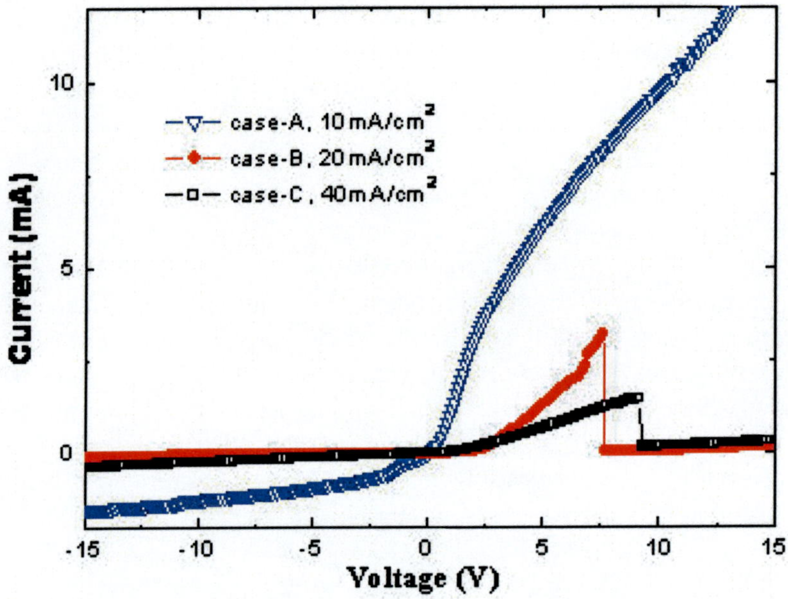

Figure 8.5. The I-V characteristics of cases A, B and C.

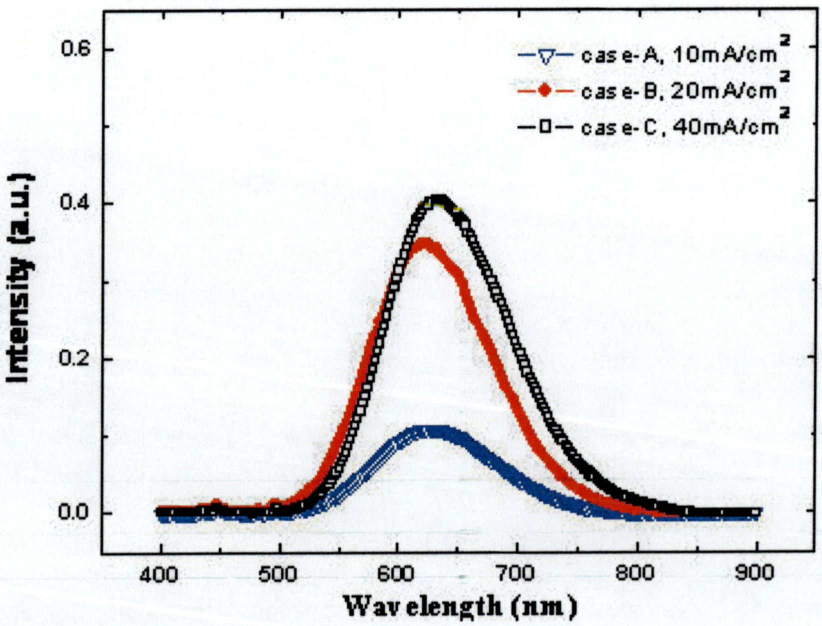

Figure 8.6. The PL spectra of cases A, B and C.

The quantum confinement effect for explaining the luminescence on NPS has ever been proposed by Canham [2, 60]. Radiative recombination in a confined nanoscale Si crystallite is proved to be more efficient than that in bulk Si. It is well known that the effective bandgap will increase when a Si structure is confined into a quantum scale. In this section, the energy bandgap is increased to about 1.9~2.0 eV by quantum confinement, and then the light with a wavelength of 650~620 nm emits. In addition, the case with higher current density that results in deep-size pores (pillars) produces a higher PL intensity. It also means that the assistance of bottom-p-layer holes can effectively improve the light emission intensity.

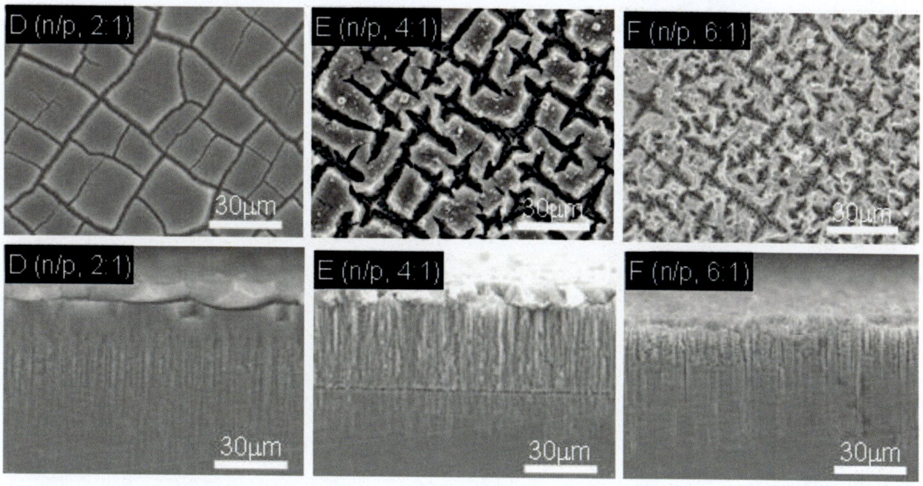

Figure 8.7. The top view SEM images of cases D, E and F.

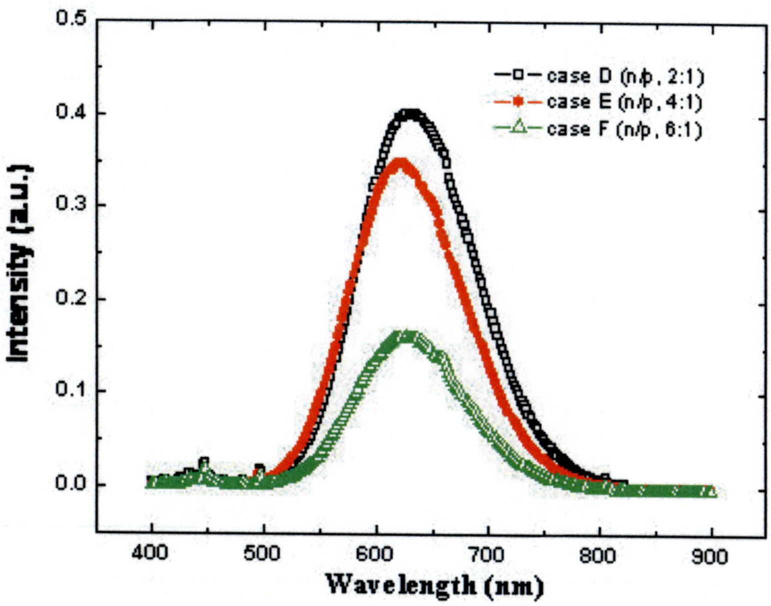

Figure 8.8. The PL spectra of cases D, E and F.

In addition, superimposition-assisted approach is applied on cases D, E and F. The top views and cross section views of SEM imagesof the samples are shown in figure 8.7. In cases D, E and F (n-on-p), the illumination-assistance and buried-p-layer-assisted are also simultaneously superimposed (superimposition-assisted approach). Furthermore, etching solution ratio of C_2H_5OH: HF =2:1 is used in cases D. In samples E and F, the ratios of 4:1 and 6:1 are used, respectively. To enhance the etching rate and etching depth, superimposition-assisted approach is useful to serve as thick-layer light emitting applications. The visible luminescence of the n-NPS can be drastically enhanced by the superimposition of buried-p-layer-assisted and illumination-assistance. This method can be used to improve the performance of photovoltaic devices that convert higher energy into visible wavelength light.Clear visible PL emissions at room-temperature are obtained. Strong visible photoluminescence emissions are demonstrated on the hole-poor n-type porous layer (as shown in figure 8.8).

8.2. Hall-Effect-Assisted Method with Perpendicular-Electrical-Field Method

The perpendicular-electric-filed-assisted method and Hall-effect-assisted method is desciribed in section 3 and 5, respectively. In this section, these two methods are superimposed to enhance the PL on n-NPS (without any illumination).

The schematic diagram of the experimental setup for preparing n-NPS is given in figure 8.9. A mixture of HF: C_2H_5OH=1:1 is utilized as the etching solvent here. No illumination equipment is needed in our process design. The NPS samples are prepared on n-type Si (100). A Si wafer with copper (Cu) contact is served as anode, while cathode is made by platinum (Pt). A sintered aluminum (Al) film is used to form an intimate backside ohmic contact to Si sample prior to the etching process. By adjusting the magnitude of the correspondent magnetic field and voltage, sharpened Si-tips can be fabricated.

Since $F_z = qV_x \times B_y$, different NPS structures reflect differences in the magnitude of the correspondent magnetic field and voltage. Electropolishing occurs for high magnetic field and voltage. At lower magnetic field and voltage, the dissolution process is changing and NPS layer is formed. By adjusting the magnitude of the correspondent magnetic field and voltage, sharpened Si-tips can be fabricated. Under a vertical anodization voltage of 30 Volts, Hall-effect assistance with a magnetic field (By) of 10 milliTesla and a voltage (Vx) of 30 Volts, which are the optimum conditions successfully support the etching process without any illumination sources. The top view and cross-section SEM images of the NPS sample are shown in figures figure 8.10 and 8.11, respectively. A larger voltage drop across the x-direction will cause a larger gradient-situation of etching extent of NPS. To clearly show the gradient-situation for further investigation, a larger Vx is adopted in this study. If a uniformed NPS film is needed, a small bias voltage (Vx), a small device area, or rotating Hall-effect apparatus are advised to solve this non-uniformity problem. Four regions (A, B, C and D) are defined on the sample for article describing, as shown in figure 8.12. The NPS films have high-aspect-ratio Si-tips because the hole-supplying mechanism is bottom-up.

All the Si-tips are vertical align to the substrate surface. The heights of Si-tips are about 78 μm (region A), 50 μm (region B), 31 μm (region C) and 16 μm (region D), respectively. Their diameters are 22 μm (region A), 18 μm (region B), 16 μm (region C) and 11 μm (region D), respectively. Figure 8.12 shows the schematic of the tip height and distribution for the four regions. And figure 8.13 shows the EDS analysis. The experiment results are helpful to understand the variation of morphology under different etching conduction and extent.

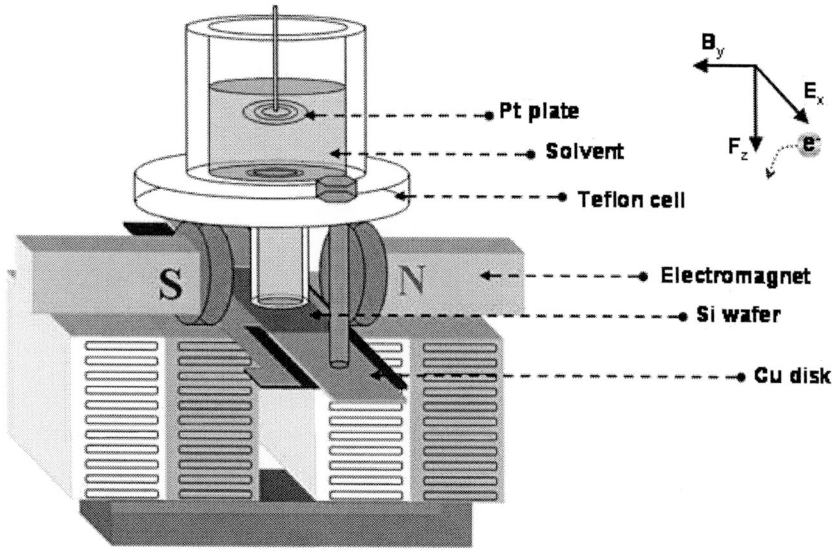

Figure 8.9. The schematic diagram of the experimental setup of Hall-effect-assisted method with buried-p-layer-assisted method.

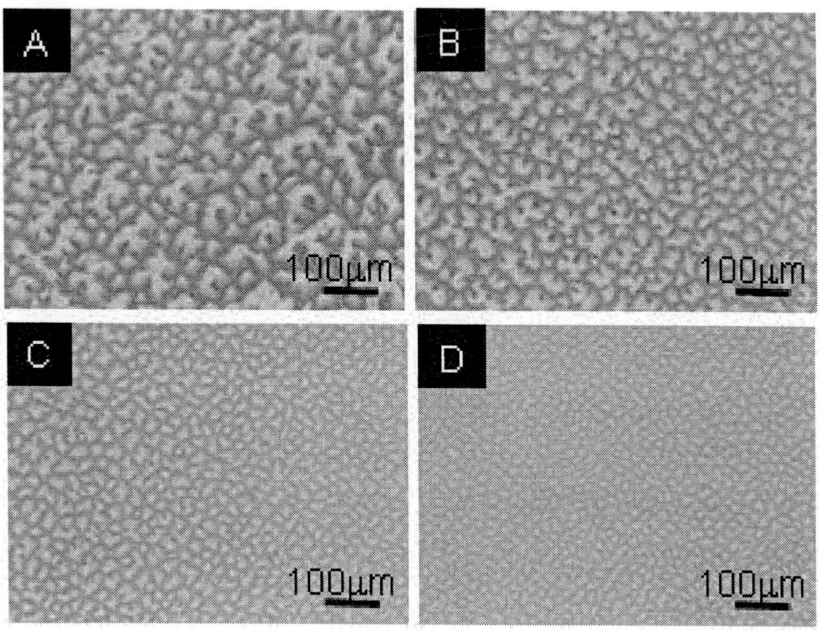

Figure 8.10. The top view SEM images of the NPS sample of Hall-effect-assisted method with buried-p-layer-assisted method.

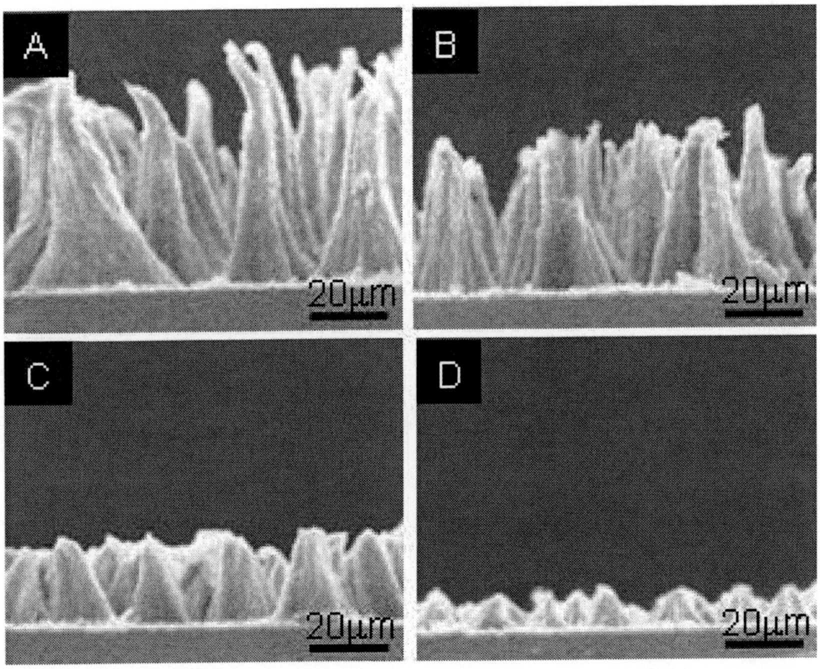

Figure 8.11. The cross section view SEM images of the NPS sample o of Hall-effect-assisted method with buried-p-layer-assisted method.

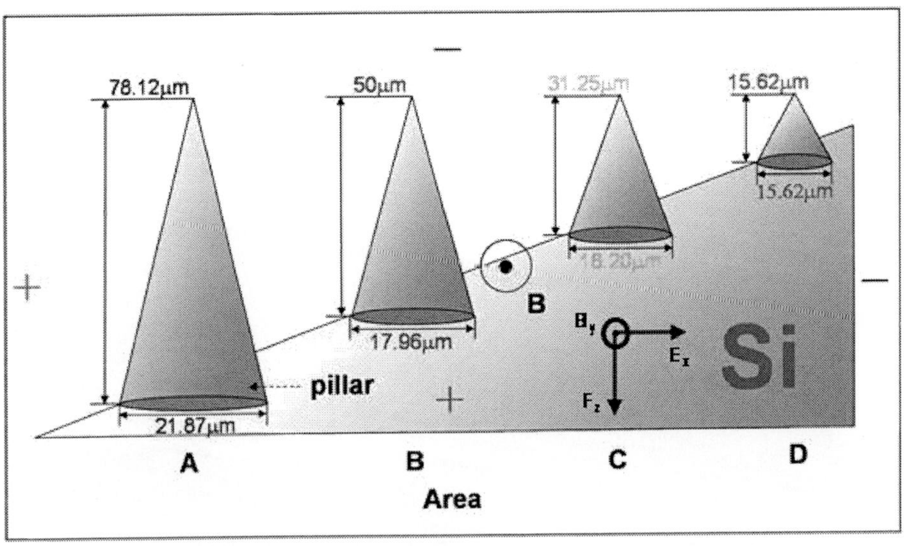

Figure 8.12. The schematic of the Si-tips and distribution.

Figure 8.14 shows the variation in time of anodization current under a constant voltage of 30 Volts. With the hole-effect-assistance on n-Si (case-1), the curve tendency of anodization current is very similar to that on p-Si (case-2) that is fairly different to conventional n-Si without any hole-assistance (case-3).

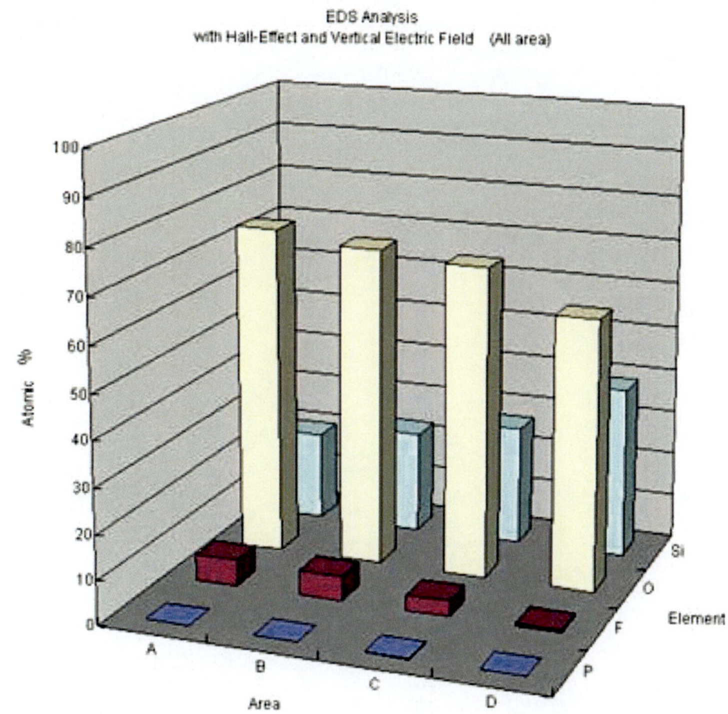

Figure 8.13. The EDS analysis of the NPS sample of Hall-effect-assisted method with buried-p-layer-assisted method.

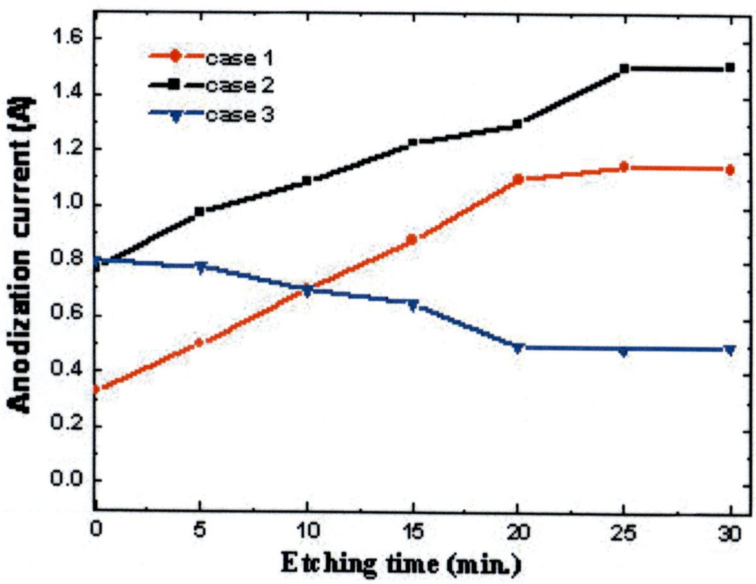

Figure 8.14. The variation in time of anodization current: the hole-effect-assistance on n-Si (case-1), conventional p-Si (case-2) and conventional n-Si without any hole-assistance (case-3).

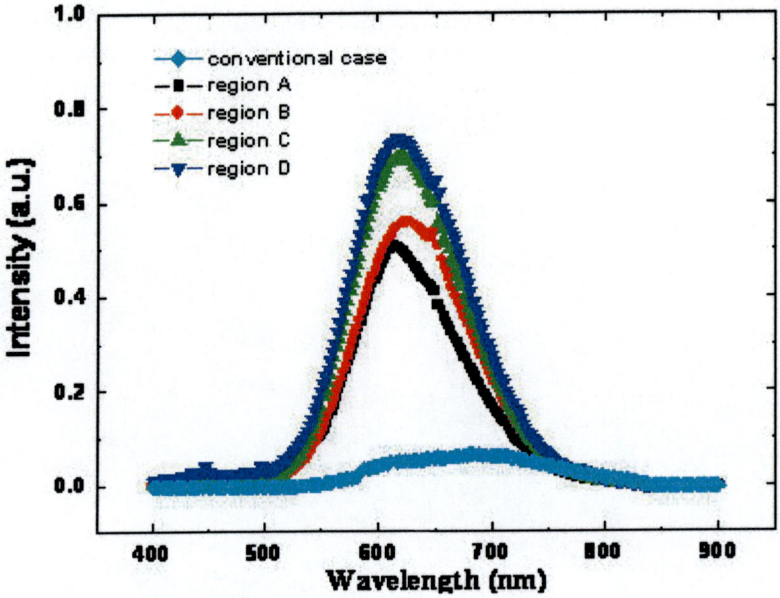

Figure 8.15. The PL spectra of A~D regions.

The increasing tendencies with time in case-1 and case-2 are resulted from the successful processing of chemical reaction which produces the hole drift current as well as the thinning down of bulk-Si layers which decrease the total resistances. The total resistance of the galvano-circuit can be express as follows:

$$\begin{aligned} R_{total} &= R_{electrolyte} + R_{measurement-system} + R_{etched-layer} + R_{bulk-Si-layer} \\ &\cong R_{etched-layer} + R_{bulk-Si-layer} \end{aligned}$$

In case-1 and case-2, the etched layers that are formed with slim Si-tips let the electrolyte among the tips serve as the major conducting path. Therefore the dominant variation of total resistance (R_{total}) is the term of bulk-Si layer ($R_{bulk-Si-layer}$) and hence the current will increase with the decreasing of $R_{bulk-Si-layer}$. However, in case-3, the chemical reaction cannot keep going for the lack of holes. Slight etching forms some narrow pore and Si-pillars. The resistances of Si-pillars are larger than that of bulk-Si-layer ($R_{etched-layer} > R_{bulk-Si-layer}$), so the dominant variation of R_{total} is the changing from $R_{bulk-Si-layer}$ to $R_{etched-layer}$. Such situation is different from case-1 and case-2. As to the initial-state points (time=0) of current, a drop from 0.80 A (case-3) to 0.33 A (case-1) results from the electron depletion by Hall-effect on the surface layer in n-Si and hence the increasing of $R_{bulk-Si-layer}$.

In conventional illuminated n-NPS, after surface pores (of few um) are formed, photo-energy is not completely received by the deeper layers. Therefore, the deeper layers are difficult to form the required NPS in the darkness [27]. Only a thin part of NPS layer contributes in the PL. In our method, the visible luminescence of the n-NPS is similar to the hole-rich p-NPS.

An excitation wavelength of 325 nm from a He-Cd laser is used for the PL measurements. Figure 8.15 are the PL spectra of A~D regions under the Hall-effect assistance (By = 10 milliTesla and Vx = 30 Volts). The PL intensity is inversely proportional to the diameter of Si-tip that appears on A~D regions gradually. The highest PL intensity in region D can be attributed to its small size of tip-width. The results consist with the quantum confinement theory. In the quantum confinement model, the visible PL phenomenon is explained by the changes of the Si band structure when dimensions are reduced to the nano-scale size.

9. Supercritical-Fluid Method for Metal-Semiconductor Contact

The EL of NPS is discovered in the 1990s. In the applications on EL, the electrical consumption power is an important issue. However, the surface roughness of NPS leads to a high contact resistance to the metal layer and, hence, an electrical power waste. To enhance the EL efficiency and the stable performance, a good metal contact to the light-emitting layer (NPS layer) is required. Some efforts have been made to improve the metallic contact between the metal and NPS. Herino *et al.* [61] and Dhar *et al.* [62] use the electroless Ni plating technique to make electrical contacts with NPS. Gelloz *et al.* propose a low porosity superficial layer between NPS and the top contact to achieve low-resistance contacts [63, 64]. Metal films are always deposited on NPS layers by conventional physical vapor deposition (PVD) methods in most other experiments [65, 66]. However, the roughness surface of NPS layer consisting of nano scale Si pillars and pores is very difficult to attach with metal atoms by conventional metal deposition techniques. Conventional approaches always result in high contact resistances in metal-NPS contacts. Therefore, a supercritical fluid (SCF) technique is explored to overcome this problem. The results are now presented in this section.

In the section, supercritical carbon dioxide (CO2) is used as a solvent. Silver (Ag) nanoparticles are produced and successfully fill the NPS pores on the surface layer by CO2 SCF equipment which is made by the High Pressure Equipment Company (SFT-150 DLOW SHEET) (figure 9.1). The *I-V* characteristics of the samples are measured and analyzed at room temperature. The curve is also compared to that is formed by the conventional metal deposition approach (without SCF processes). Good metal contacts to the NPS layers are obtained by the new proposed approach. In addition, a NDC curve is clearly observed that is usually suppressed due to the high contact resistance between metal and NPS form by conventional deposition methods. The occurrence of NDC curve in NPS is also ever been observed by Kwei *et al.* In their presentation, they attribute the NDC phenomenon to the carrier mobility difference of thick and thin wires in the NPS layers [55].

The schematic diagram of SCF equipment setup is shown on the left side in figure 9.2. The CO2 pump can rapidly produce the high pressures which are required for SCF processing. The main part of the system is an interflow oven. The interflow oven includes both the sample holder and the SCF collector. As the chemical reaction is carried out, the valve of CO2 cylinder is closed and the decompression valve is opened quickly. Then, CO2 decompress in a transient time. Meanwhile, the Ag nanoparticles are produced and filled the pores in the NPS.

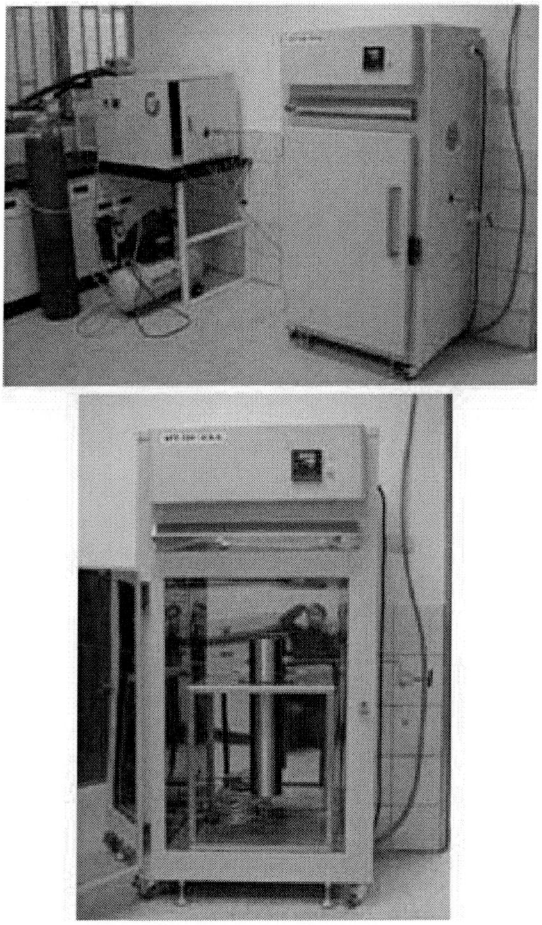

Figure 9.1. The CO_2 SCF equipment (SFT-150 DLOW SHEET).

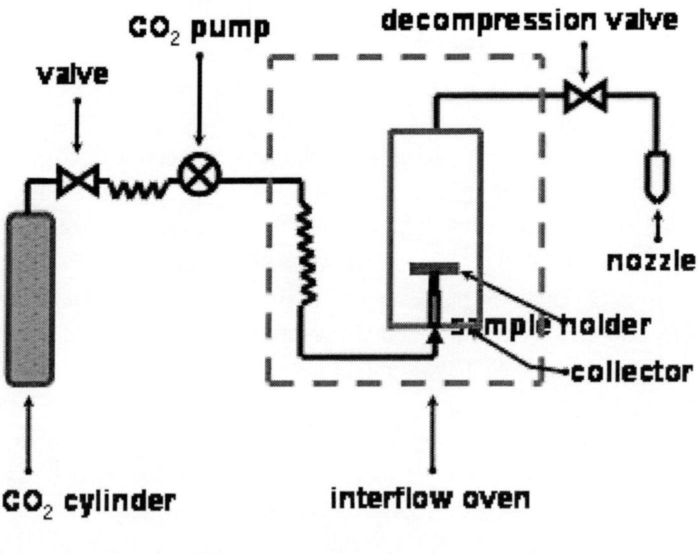

Figure 9.2. The schematic diagram of SCF equipment setup.

The SCF processing involves the chemical reduction of soluble compounds in supercritical CO2 to yield nano scaled Ag extraction. The CO2 solvent is the most commonly used in SCF systems because of its characteristics of nontoxic, nonflammable, low cost, low power consumption, room temperature operation, and simple production. In the study, silver nitrate (AgNO3) (solubility of 10%, 20% and 30%) are utilized as precursors of Ag nanoparticles extraction. Under a high pressure (200 bars) and a high temperature (100°C), Ag nanoparticles are extracted from AgNO3 by the chemistry reduction of CO2 solutions during the SCF process. The illustrative equation of the overall process during Ag nanoparticles extraction can be expressed as below:

$$8\ AgNO_3 + NaBH_4 + 7\ OH^- \rightarrow 8\ Ag + 8\ NO_3^- + 4\ H_2O + H_3BO_3 + Na^+$$

By contrast with the SCF process, when this reaction is carried out in an ultrasonic generator, many large-sized and randomly-distributed Ag particles are produced and accumulates on the top surface of NPS pillars.

The top view and cross-section scanning electron microscope (SEM) images of the NPS sample without SCF process are shown in figure 9.3 (a) and 9.3 (b), respectively. It shows that the pore diameters are about 800~2000 nm and the pore depths are about 10~20 μm. Narrow- and deep-trench NPS structures (with high aspect ratios of 1:25) are prepared and focused in our studies, whose surfaces are extremely rough, and it is always difficult to produce good electrical contacts between them and metal films by conventional methods.

Figure 9.4 (a)-(f) show the top view and cross-section SEM images of the NPS samples with SCF processes in which three kinds of AgNO3 concentration (10%, 20% and 30%) are used, respectively. Figure 9.4 (a) and (b) show the Ag nanoparticles with sub-100 nm fill the most part of the deep-trench NPS pores when using 10% AgNO3 solution.

The other cases with different solutions of AgNO3 (20% and 30%) are shown in figures (c)-(f). The results show that the sizes of these Ag nanoparticles are bigger than those in figures 9.4 (a) and 2 (b) when the concentration of AgNO3 increases. In figures 9.4 (e) and 2 (f), a mass of large-scaled Ag particles are extracted and congregated upon the NPS surface but cannot fill the trench due to the large size of Ag particles. Such a coating style is similar to a conventional deposition method.

It is concluded that the size and the morphology of metal particles can be controlled by changing the solution of reaction, and 10% AgNO3 is the best solution of Ag nanoparticles extraction in our study.

As illustrated in figure 9.4 (g)-(i), EDS mappings show the distribution analysis of Ag nanoparticles (shown in several white spot areas that are dependent on the resolution). The figures show the Ag nanoparticles (sub-100 nm in width) are widely distributed over the NPS surface homogenously when AgNO3 solution is 10%. A mass of large-scaled Ag particles are produced and assembled upon the surface of NPS layer when the concentration of AgNO3 increases. The above analyses can reach the same conclusions as figure 9.2 that the different AgNO3 solutions will lead to different grain sizes of Ag particles.

Table 1 demonstrates the EDS analysis of solids formed using solutions with different concentrations of AgNO3. The atomic percentage of Ag element decreases when the concentration of AgNO3 increases. When the concentration of the AgNO3 solution is higher than 10%, the filling of the NPS pores becomes more difficult.

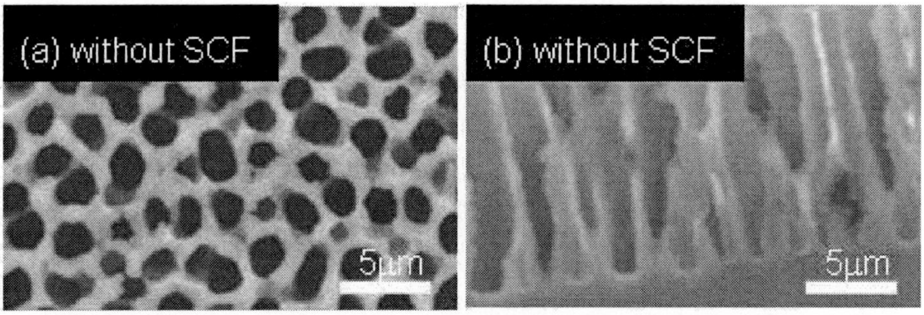

Figure 9.3. The SEM images of the NPS sample without SCF process.

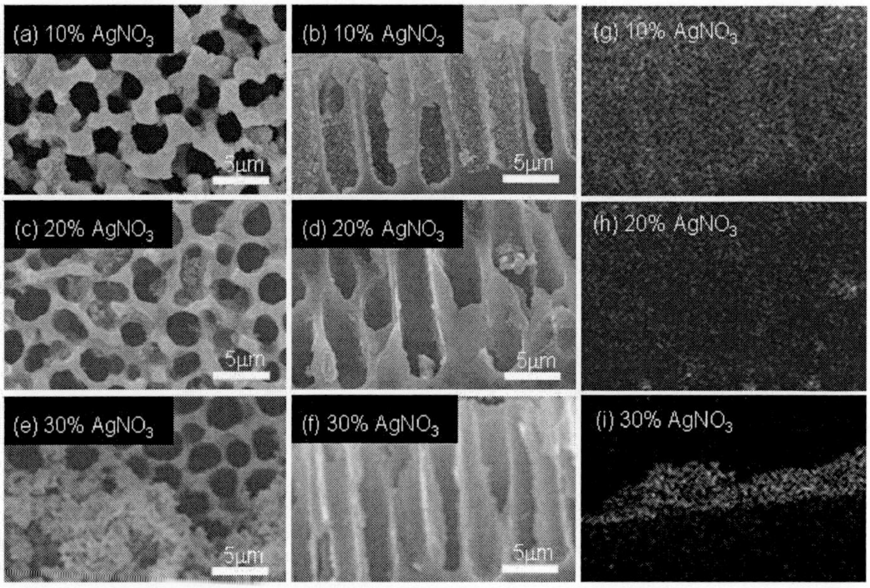

Figure 9.4. (a)-(f) The SEM images of the NPS sample with SCF process, and (g)-(i) EDS mappings of the NPS samples with different solutions of $AgNO_3$.

Table 9.1. The The EDS analysis of solids with different solutions of AgNO3

$AgNO_3$ (concentr ation)	10%	20%	30%
Element	Atomic(%)	Atomic(%)	Atomic(%)
O	52.49	26.41	57.35
Si	39.96	68.75	41.34
Ag	7.55	4.84	1.31
Total	100.00	100.00	100.00

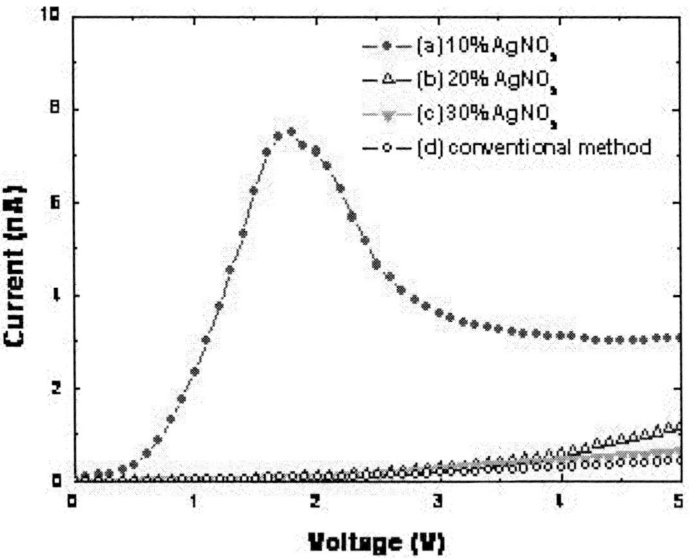

Figure 9.5. The *I-V* characteristics of NPS samples by SCF method.

The *I-V* measurement results of the NPS samples are shown in figure 9.5. The curves *a-c* show the *I-V* characteristics of the NPS samples that are treated with SCF processes. The curve *d* is the *I-V* characteristic of the NPS sample using conventional method for comparison in which the metal film is deposited on the NPS surface by PVD. It is clearly shown that the electrical contact is effectively improved with 10% AgNO3 solution (curve a). A nonlinear *I-V* characteristic with a peak current point at 1.8 V is clearly shown that is usually suppressed in poor metal contact. However, the SCF processes with 20% (curve b) and 30% (curve c) AgNO3 concentrations do not work in force because the Ag particles only accumulate on the top NPS pillars and cannot fill the NPS pores successfully. The results shown in curves *b* and *c* are very similar to curve *d* in which large-sized Ag particles are deposited and the metal film is formed on the top. Identically, good electrical contacts cannot be obtained in NPS pores and, hence, high contact resistances produce. The high contact resistance causes the suppression of NDC phenomenon. Apparently, the contact resistance is successfully reduced (about 250:1) due to the pore filling with Ag nanoparticles by SCF method (10% AgNO3). A good improvement of metal contact to the interfaces of NPS layers is achieved.

The method can be used to improve the metal contact to the NPS surface by the SCF technique. During the SCF process, Ag nanoparticles with sub-100 nm diameter are formed at 100oC under a pressure of 200 bars and fill the NPS pores. The size and the morphology of particles are controlled by changing the solutions. The best AgNO3 concentration of 10% is strongly advised in the study. In addition, the *I-V* characteristics are measured and discussed. It is demonstrated that the conventional metal deposition methods produce high contact resistances and, however, the new approach by SCF technique improves the performance substantially. Also, the NDC phenomenon of NPS structures can be clearly revealed resulting from the reduction of contact resistance. It is believed that the power consumption of EL on the NPS structure can be greatly reduced by this new approach. The study results will benefit the applications of electroluminescence, microelectronic and optoelectric devices.

10. CONCLUSIONS

In conclusion, several new ideas for NPS formations are introduced, and some related experimental results are presented. The conventional perpendicular-electric-field-assisted method for anodization process is described. Also, the conventional illumination-assisted method for n-NPS formation is mentioned. In addition, Hall-effect is applied to assist the manufacturing of n-NPS. The etching process is anisotropic and illumination-free. An uniform hole distribution and gradient etching are produced. Different etching strength in four regions on the same sample is explored. The electric field and magnetic field control the hole accumulation, and hence NPS formation and light-emission areas. A strong visible light is obtained on n-NPS at room temperature.

Also, a new buried-p-layer-assisted method based on a forward biased np-junction for manufacturing n-PS layer is studied. The method can overcome the illumination-limitation and depth-limitation problems in conventional only-illumination-assisted method. With the bottom-hole-assistance from the buried p-layer, the anodization etching is almost anisotropic. This feature has great potential in semiconductor manufacturing. The clear visible PL emissions at room-temperature are obtained on the hole-poor n-type samples, even in the dark. Strong visible PL emits at about 650 nm. Of course, the visible luminescence of the n-NPS can be drastically enhanced by the superimposition of buried-p-layer-assisted method and illumination-assisted method.

Remarkly, the buried-p-layer-assisted method can produce an n-NPS film with deep, straight and high-aspect-ratio Si pillar arrays. It is interesting that, at room temperature, both the properties of high PVCR in NDC and strong light emission in PL can exist on such n-NPS films together. The key factor for the n-NPS formation is the special design of anisotype pn structure that supports orderly holes from bottom p-layers.

Furthermore, with the superimposition of Hall-effect assistance and perpendicular-electric-field assistance, a film with the sharpened Si-tips can be obtained. A strong visible light is also obtained on the Si-tips at room temperature.

Alternatively, a new method for the fabrication of p-type patterned NPS only by electric field is proposed. The manufacturing process is anisotropic and illumination-free. No masking-layer and related photo-lithography processes are needed in this method.

On the other hand, the method is put forward to improve the metal contact to the NPS surface by the SCF technique. During the SCF process, Ag nanoparticles are formed at 100oC under a pressure of 200 bars and fill the NPS pores. The size and the morphology of particles are controlled by changing the solutions. In addition, by current-voltage characteristic measurements, it is proved that the SCF method improves the performance substantially. Also, the NDC phenomenon can be clearly revealed resulting from the reduction of contact resistance. It is believed that the power consumption of EL on the NPS structure can be greatly reduced.

In this chapter, these studies on NPS provide alternative methods of fabricating electronic and optoelectronic devices.

REFERENCES

[1] L.T. Canham. (1990). *Appl. Phys. Lett.*, 57, 1046.
[2] P. McCord, S. L. Yau and A. J. Bard. (1992). *Science*, 257, 68.
[3] Bisi, Stefano Ossicini, and L. Pavesi. (2000). *Surf. Sci. Rep.,* 38, 1.
[4] G. Cullis, L. T. Canham and P. D. J. Calcott. (1997). *J. Appl. Phys.*, 82, 909.
[5] P. M. Fauchet. (1996). *J. Lumin.,* 70, 294.
[6] J. C. Lin, P. W. Lee and W. C. Tsai. (2006). *Appl. Phys. Lett.*, 89, 121119.
[7] J. C. Lin, W. L. Chen and W. C. Tsai. (2006). *Opt. Express*, 14, 9764.
[8] J. C. Lin, H. T. Hou and W. C. Tsai. (2007). *Microelectron. Eng.*, 84, 336.
[9] G. Smestad and H. Ries. (1992). *Solar Energy Materials Solar Cells*, 25, 51.
[10] G. Willeke, H. Nussbaumer, H. Bender and E. Bucher. (1992). *Solar Cells*, 25, 51.
[11] T. Taliercio, M. Dilhan, E. Massone, A. M. Gue, B. Fraisse and A. Foucaran. (1992). *Thin Solid Films*, 255, 310.
[12] L. Z. Yu and C. R. Wie. (1992). *Electron. Lett.*, 28, 911.
[13] Y. Watanabe, Y. Arita, T. Yokoyama and Y. Igarashi. (1975). *J. Electrochem. Soc.*, 122, 1351.
[14] H. J. Moon, K. H. Jung, H. P. Hyung, J. K. Joong, H. H. Sang and Y. C. Se. (1997). *Thin Solid Films*, 309, 490.
[15] H. Art. (2004) *Alternative Approaches for DMFC Design:Silicon-Based Systems 6th Annual SMALL FUEL CELLS.*
[16] S. Chan and P. M. Fauchet. (1999). *Appl. Phys. Lett.*, 75, 274.
[17] W. H. Zheng, P. Reece, B. Q. Sun, and M. Gal. (2004). *Appl. Phys. Lett.*, 84, 3519.
[18] V. Lehmann. (1993). *J. Electrochem. Soc.*, 140, 2836.
[19] J. D. Benjamin, J. M. Keen, A. G. Cullis, B. Innes, and N. G. Chew. (1986). *Appl. Phys. Lett.*, 49, 716.
[20] Z. Gaburro, P. Bettotti, M. Saiani, L. Pavesi, L. Pancheri, C. J. Oton, and N. Capuj. (2004). *Appl. Phys. Lett.*, 85, 555.
[21] Z. Gaburro, C. J. Oton, L. Pavesi, and L. Pancheri. (2004). *Appl. Phys. Lett.*, 84, 4388.
[22] C.Pickering, M.J. Beale, D.J. Robbins, P.J. Pearson, and R. Greef. (1984). *J. Phys. C: Solid State Phys.*, 17, 5535.
[23] M.J. Beale, J.D. Benjamin, M.J. Uren, N.G. Uren, N.G. Chew, and A.G. Cullis. (1985). *J. Cryst. Growth*, 73, 622.
[24] M.J. Beale, J.D. Benjamin, M.J. Uren, N.G. Uren, N.G. Chew, and A.G. Cullis. (1985). *Appl. Phys. Lett.*, 46, 86.
[25] M. Young, M. I. Beale and J.D. Benjamin. (1985). *Appl. Phys. Lett.*, 46, 1133.
[26] R.L. Smith, S.F. Chuang, and S.D. Collins. (1988). *J. Electron. Mater.*, 17, 533.
[27] R.L. Smith and S.D. Collins. (1989). *Phys. Rev. A*, 39, 5409.
[28] R.L. Smith and S.D. Collins. (1992). *Phys. J. Appl.Phys.*, 71, R1.
[29] T.A. Witten and L.M. Sander. (1983). *Phys. Rev. B*, 27, 5686.
[30] A.J. Read, R.J. Needs, K.J. Naish, L.T. Canham, P.D.J. Calcott, and A. Qteish. (1992). *Phys, Rev. Lett.*, 69, 1232.
[31] G. D. Sanders and Y. C. Chang. (1992). *Phys. Rev. B*, 45, 856.
[32] V. Lemann, U. Gösele. (1991). *Appl. Phys. Lett.*, 58, 856.

[33] V. P. Parkhutik, J. M. Albella, J. M. Martinez-Duart, J. M. Gomez-Rodriguez, A. M. Baro, and V.I. Shershulsky. (1993). *ibid.*, 2, 366.

[34] H. Yan and X. Hu. (1993). *J. Appl. Phys*., 73, 4324.

[35] L. T. Canham, W. Y. Leong, T. I. Cox and L. Taylor. (1992). *Appl. Phys. Letts.*, 61, 2563.

[36] H. D. Fuchs, M. Stutzmann, M. S. Brandt, M. Rosenbauer, J. Weber, A. Breitschwerdt, P. Deak and M. Cardona. (1993). *Phys. Rev. B*, 48, 8172.

[37] Sarathy, S. Shih, K. H. Jung, C. Tsai, K. H. Li, D. L. Kwong and J. C. Campbell. (1992). *Appl. Phys. Lett.*, 60, 1532.

[38] Ksendzov, R. W. Fathauer, T. George, W. T. Pike and R. P. Vasquez. (1993). *Appl. Phys. Lett.*, 63, 200.

[39] S. Kalem and M. Rosenbauer. (1995). *Appl. Phys. Lett.*, 67, 2551.

[40] M. Saadoun, H. Ezzaouia, B. Bessais, M. F. Boujmil and R. Bennaceur. (1999). *Solar Energy Mater. Solar Cells*, 59, 377.

[41] S. Kalem and O. Yavuzcetin. (2000). *Opt. Express*, 6, 7.

[42] M. Saadoun, N. Mliki, H. Kaabi, K. Daoudi, B. Bessaïs, H. Ezzaouia and R. Bennaceur. (2002). *Thin Solid films*, 405, 29.

[43] N. Lalic and J. Linnros. (1996). *Thin Solid Films*, 276, 155.

[44] Richter, P. Steiner, F. Kozlowski and W. Lang. (1991). *IEEE EDL.*, 12, 691.

[45] M. J. Eddowes. (1990). *J. Electroanal. Chem*., 280, 297.

[46] Stumper and L. M. Peter. (1991). *J. Electroanal. Chem*., 309, 325.

[47] A. Yaron, S. Bastide, J. L. Maurice, C. Levy-Clement. (1993). *J. Lumim.*, 57, 67.

[48] E. Galun, A. Lagoubi, R. Tenne, C. Levy-Clement. (1993). *J. Lumim*. 57, 125.

[49] G. Nassiopoulos, S. Grigoropoulos, L. Canham, A. Halimaoui, I. Berbezier, E. Gogolides and D. Papadimitriou. (1995). *Thin Solid films*, 255, 329.

[50] Krüger, R. Arens-Fischer, M. Thönissen, H. Münder, M. G. Berger, H. Lüth, S. Hilbrich and W. Theiss. (1996). *Thin Solid films*, 276, 257.

[51] D. Mao, K. J. Kim, Y. S. Tsuo, and Arthur J. Frank. (1995). *J. Phys. Chem.*, 99, 3643.

[52] K. Lee, C. H. Chu, Y. C. Tseng, J. M. Shyr, and C. H. Kao. (2000). *IEEE EDL.*, 21, 587.

[53] Sibille, J. F. Palmier, H. Wang, J. C. Esnault and F. Mollot. (1990). *Appl. phys. Lett.* 56, 256.

[54] Reklaitis, A. Krotkus, A. Geizutis and M. Asche. (1999). *Semicond. Sci. Techno.*, 14, 341.

[55] S. J. Koester, K. Ismail, K. Y. Lee, and J. O. Chu. (1997). *IEEE EDL*., 18, 432.

[56] Saitoh, H. Harata, and T. Hiramoto. (2004). *Appl. Phys. Lett.*, 85, 6233.

[57] H. Lee and Y. H. Jeong. (2004). *IEEE Trans. Nanotechnol.*, 3, 377.

[58] N. Saeta and A. C. Gallagher. (1995). *J. Appt. Phys.*, 77, 4639.

[59] G. D. Sanders and Y. C. Chang. (1992). *Phys. Rev. B*, 45, 9202.

[60] G. Cullis and L. T. Canham. (1991). *Nature*, 353, 335.

[61] R. Herino, P. Jan and G. Bomchil. (1980). *J. Electrochem. Soc.*, 127, 1339.

[62] S. Dhar and S. Chakrabarti. (1996) *Semicond. Sci. Technol*., 11, 1231.

[63] Gelloza and N. Koshida. (1998). *Appl. Phys. Lett.,* 73, 2021.

[64] B. Gelloz and T. Nakagawa. (2000). *J. Appl. Phys.,* 88, 4319.

[65] Andsager, J. Hillard, J. M. Hetrick, L. H. AbuHassan, M. Plisch and M. H. Nayfeh. (1993). *J. Appl. Phys.,* 74, 4783.

[66] W. Lang, P. Steiner and F. Kozlowski. (1993). *J. Lumines,* 57, 341.

In: Nanotechnology Research Developments
Editor: R. Jiménez-Contreras, pp. 247-275
ISBN: 978-1-60021-899-6

Chapter 6

MOVPE GROWTH OF HIGH DENSITY InAs/InGaAs/InP QUANTUM DOTS FOR MID-INFRARED EMISSION

Tang Xiaohong and Yin Zongyou
Photonics Research Center, School of Electrical and Electronic Engineering
Nanyang Technological University, Singapore 639798

ABSTRACT

Emission in mid-infrared range, 2 – 5μm, has very important applications in variety of areas including range-finding, laser surgery, remote trace-gas sensing, secure free-space communications, etc. Owing to reduction of the dimensionality carriers' confinement, quantum dots (QDs) based optoelectronic devices, e.g. semiconductor lasers, show novel properties when compared with their counterpart quantum well based devices.

In this article, self-assembly InAs quantum dots grown on InGaAs/InP matrix for mid-infrared emission using metal-organic vapor phase epitaxy is introduced. Effects of the growth conditions on the dot density and size uniformity of the quantum dots have been investigated in detail. A two-step growth method for growing high density and uniform quantum dots has been introduced. Improvement in QDs density and size uniformity in the two-step growth has been discussed in details. By changing the indium content of the InxGa1-xAs matrix layer of the InAs/InxGa1-xAs/InP quantum dot structure, emission wavelength of the InAs QDs structure has been extended to the longest >2.35 μm measured at 77K. Full-width at half-maximum (HWHM) of the photoluminescence emission spectrum of the InAs quantum dot structure was measured as narrow as 25.5 meV.

INTRODUCTION

Attempts to use InAs nanostructures based on InP substrates to extended the emission wavelength into mid-infrared (mid-IR) region of 2 –3 μm are attracting much attention, where mid-IR lasers have very important military, biomedical, environmental and industrial applications, including range finding, lidar atmospheric detection, laser surgery, molecular spectroscopy and remote sensing of atmospheric and planetary gases, etc. It has been shown that three-dimensional carrier confinement of quantum dot (QD) structures promise ultra-low threshold current and temperature independent semiconductor lasers with very high differential gain [1,2]. Also, using InP as the substrate, the device fabrication processes are mature and cheaper when compared with that of conventional antimonide substrate based quantum well (QW) mid-IR lasers. Stranski-Krastanov (S-K) self-assembled InAs quantum dots (QDs) and quantum-dash lasers have been demonstrated at various wavelengths from 1.60 to 2.04 μm [3–6]. To extend the emission wavelength of InAs/InP QDs longer than 2 μm, InGaAs has been inserted into the InAs/InP QD structure as the confinement layers of the InAs QDs structures [7-10].

However, InAs QDs self-assembled on an InGaAs/InP matrix generally have large size dispersion. Large size variation of the InAs QDs is due to the indium adatoms' migration [11,12] and the anisotropic surface diffusion [13,14] on top surface of the InGaAs/In matrix during the InAs QDs growth. In order to reduce the size dispersion of InAs QDs formed on an InGaAs/InP matrix, different techniques have been investigated [11,12,15]. But, the photoluminescence (PL) emission spectra from the InAs QD structures are still not narrow enough. Further improvement in the QD size uniformity and dot density is essential in order to make them suitable for mid-IR optoelectronic device applications.

In this research, InAs QDs formed on InP substrate by metal-organic vapor phase epitaxy (MOVPE) for mid-IR emission has been studied. Effects of the growth conditions on formation of the InAs QDs on InGaAs/InP matrix have been investigated in details. To achieve high dot density and narrow size dispersion, a two-step growth method has been developed and investigated. With the optimized growth conditions, the size uniformity and dot density of the InAs QDs formed by using the two-step growth has been improved by 50% and >80%, respectively, when compared to that of the InAs QDs grown on InP by using conventional S-K growth method with the similar MOVPE growth conditions. To extend emission wavelength of the InAs QD structure, indium graded content InxGa1-xAs layers are used as the matrix/barrier layers of the InAs QD structure. The indium content of the InxGa1-xAs matrix/barrier layers is linearly increased from x = 0.53 (lattice matched to InP) to x = 0.65 ~ 0.80. It has been received that, when compared to that of InAs QDs with the lattice matched In0.53Ga0.47As barriers, emission wavelength of the InAs QDs embedded in graded In0.53→(0.53+y)Ga0.47→(0.47-y)As barriers redshifts by 13.5 nm when the indium content gradation, y, of the In0.53→(0.53+y)Ga0.47→(0.47-y)As matrix/barrier layers is increased by every 0.01. The longest emission wavelength of the InAs/InxGa1-xAs/InP QD structure reaches > 2.35 μm when measured at 77K, which is the longest emission wavelength of InAs QD structure so far reported. The full width at half maximum (FWHM) of the PL spectrum of the InAs QD structure reaches 25.5 meV. These results would be promising for development of mid-IR QD optoelectronic devices on InP or GaAs substrates.

EXPERIMENT

All the samples were grown in a horizontal low-pressure MOVPE reactor (Aixtron, AIX200) with gas foil rotation of the susceptor. The metal-organic (MO) sources used in the growths were tertiarybutylarsine (TBA), tertiarybutylphosphine (TBP), trimethylgallium (TMGa) and trimethylindium (TMIn). To improve the safety of MOVPE growth process, high purity (99.999%) nitrogen gas further purified by a MonoTorr phase II Getter column with the dew point below –100oC was used as the carrier gas instead of using explosive hydrogen gas. The reactor pressure was set at 20 mbar for growing the InAs QDs, and it was set at 100 mbar for depositing InP buffer and InxGa1-xAs barrier layers. A 200 nm InP buffer layer was grown on InP (001) substrate with the temperature increased from 600 oC to 630 oC before growing the QD structure. Then, the reactor temperature was lowed down for growing the InAs QDs. As shown in Figure 1, the InAs QDs are grown on a 30 nm un-relaxed InxGa1-xAs/InP matrix. Atomic Force Microscopy (AFM) was used to take the surface morphology of the QDs. For photoluminescence measurement, the InAs QDs are covered by a symmetric InxGa1-xAs upper barrier and then capped with an InP cap layer.

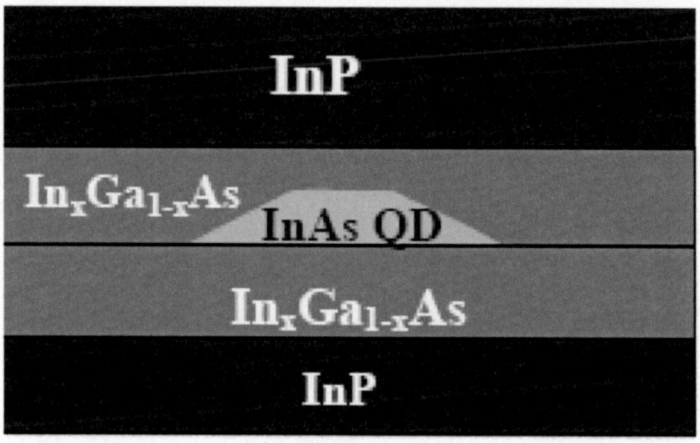

Figure 1. Schematic layer structure of the InAs/InGaAs/InP quantum dot structure.

A. S-K Self-Assembly Growth of InAs/ InxGa1-xAs/InP QDs

In S-K self-assembly growth of InAs QDs, the reactor pressure was set at 20 mbar and the V/III ratio was set at 10. The total gas flow rate in the reactor was Q_{total} = 3.1 slm. To grow the InA QDs, a very thin InAs layer, few monolayers (MLs), was grown as normal; and then the TMIn source was closed for the post-growth interruption before cooling down the system and the InAs QDs were formed self-assembly because of large lattice mismatch. Formation of the QDs is affected by changing the growth conditions including the growth time, post-growth interruption, substrate temperature and source flux, etc, which are discussed in details below.

1) Effect of the Growth Time

To study the effect of the growth time (or layer thickness) of InAs on the formation of InAs QDs, five samples, samples (a) to (e), were grown by changing the InAs layer growth time. The substrate temperature and InAs growth rate was kept at 500 oC and 0.15 ML/sec, respectively. All the other growth conditions were kept same for growing the five samples. Figure 2 (a) to (e) compares the surface morphology of the five samples with different InAs layer growth time. It is observed that when the growth time was 10 seconds or less, no QD was formed. When the InAs layer growth time was increased to 12 seconds, QDs started to be formed as shown in Figure 2(c), but with very low density, ~1×109cm-2. Average dot height, h_g, of the samples was calculated by using the following formula [17].

$$\log_{10} h_g = \frac{\Sigma(n \log_{10} h_n)}{N} \quad (1)$$

where N is the total number of the dots counted on the sample, n is the number of the QDs with the same dot height of h_n. The average dot diameter was calculated using the similar equation. The calculated average dot height and diameter of sample (c) is 10.0 nm and 35.5 nm, respectively. Based on the dot size and number of the QDs counted, volume of the InAs QDs can be calculated by taking the QDs with pyramid shape. The total volume of InAs QDs grown of the sample was equivalent to an InAs nominal layer. The InAs norminal layer thickness of the InAs QDs is calculated by using the AFM scanned area to divide the calculated InAs QDs volume in the scanned area of the sample. We have taken that 1 ML = 0.3 nm in the calculations.

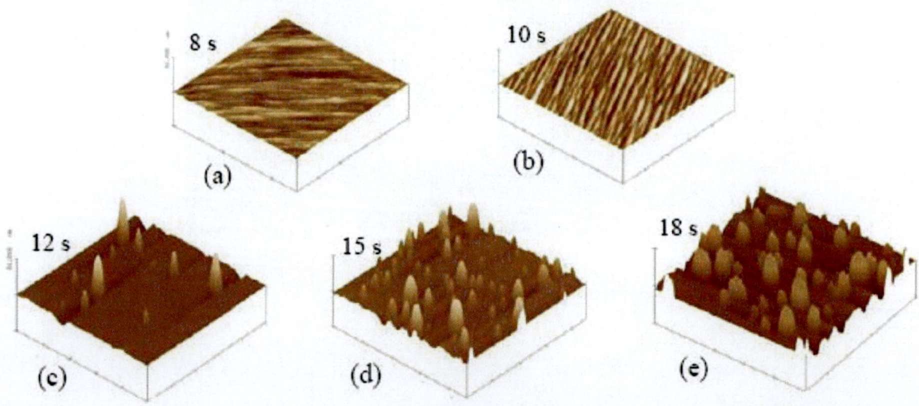

Figure 2. 1×1 μm^2 AFM images of the InAs samples with different growth times. The growth time for sample (a), (b), (c), (d) and (e) was 8 s (1.2 ML), 10 s (1.5 ML), 12 s (1.8 ML), 15 s (2.25 ML) and 18 s (2.7 ML), respectively.

The InAs wetting layer thickness of a sample was determined by deducting the QDs' nominal layer thickness from the total grown InAs layer thickness. Figure 3 shows the evolution of the grown InAs materials consumed for growing the wetting layer and the QDs as a function of the InAs growth time. It shows that if the total deposited InAs layer thickness is less than 1.5 ML, as in the cases of sample (a) and (b) in Figure 2, no QD is formed, only InAs wetting layer was grown. The QDs started to be formed when the total deposited InAs layer was thicker than 1.8 ML, as shown in samples (c), (d) and (e).

For sample (d), the total deposited InAs layer thickness is 2.25 ML, which is increased by 0.45 ML when compared with that of sample (c). But, the QDs' nominal layer thickness of sample (d) is 0.7 ML, which is 0.55 ML thicker than that of sample (c) whose QDs nominal layer thickness is only 0.15 ML. The wetting layer thickness of sample (c) and (d) is calculated of 1.65 ML and 1.55 ML, respectively, as shown in Figure 3.

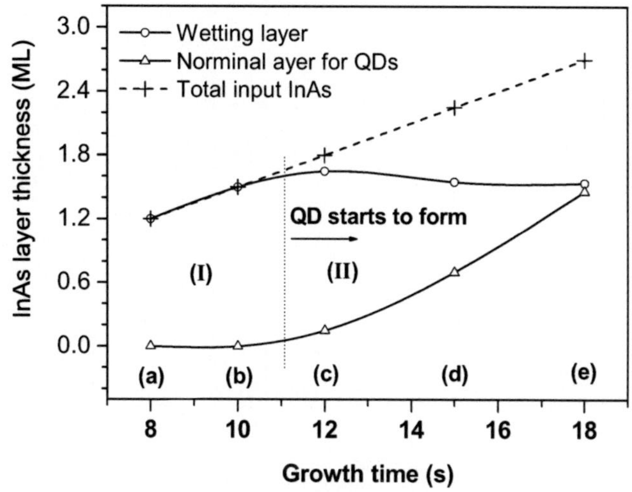

Figure 3. InAs material consumption for the wetting layer growth and the QD formation under different InAs deposition time. The two solid curves show the materials for growing wetting layer and QDs, and the dot line shows the total input InAs materials.

Reduction of the wetting layer thickness, or increase of the QDs nominal layer thickness from sample (c) to (d) shows that some InAs material was decomposed from the wetting layer and transferred to the formation of the QDs. Similar decomposition of wetting layer was observed in the InAs/GaAs QDs growth [17]. However, when the InAs growth time was increased to 18 seconds, the total grown InAs layer was very thick. The InAs islands no longer accommodated the strain, misfit dislocations were formed between the island-substrate interfaces. Incoherent islands were formed and grew into large clusters by consuming the InAs dots nearby, i.e. the ripening [18]. Therefore, the QDs of sample (e) are obviously larger than those of sample (d). An optimized growth time of 15 seconds has been received under these growth conditions.

2) Effect of the Post-Growth Interruption

To study the effects of the post-growth interruption time on the QDs formation, four samples, samples (f), (g), (h) and (i), were grown at 500 oC with the same grown InAs layer thickness of 2.25 ML. All the growth conditions of the four samples were the same except the post-growth interruption time. The post-growth interruption time for the four samples was changed from 0 to 20 seconds. For sample (f) with 0 second post-growth interruption, the TMIn source was closed at the same time when the heater of the reactor was switched off when the growth reached the target layer thickness. While for growing sample (g) to sample (i), after the TMIn source had been witched off to stop the InAs layer growth, the reactor was kept at the growth temperature for 5 seconds, 10 seconds and 20 seconds, respectively, before switched off the heater. Figure 4 shows the AFM measurement results of the four samples. The QDs' size distribution, dot density, dot size standard deviation and surface depth of the four samples are summarized in Table 1. The four samples have similar dot density when the post-growth interruption time was changed from 0 to 20 seconds, but the sample's dot size dispersion is sensitive to the post-growth interruption time.

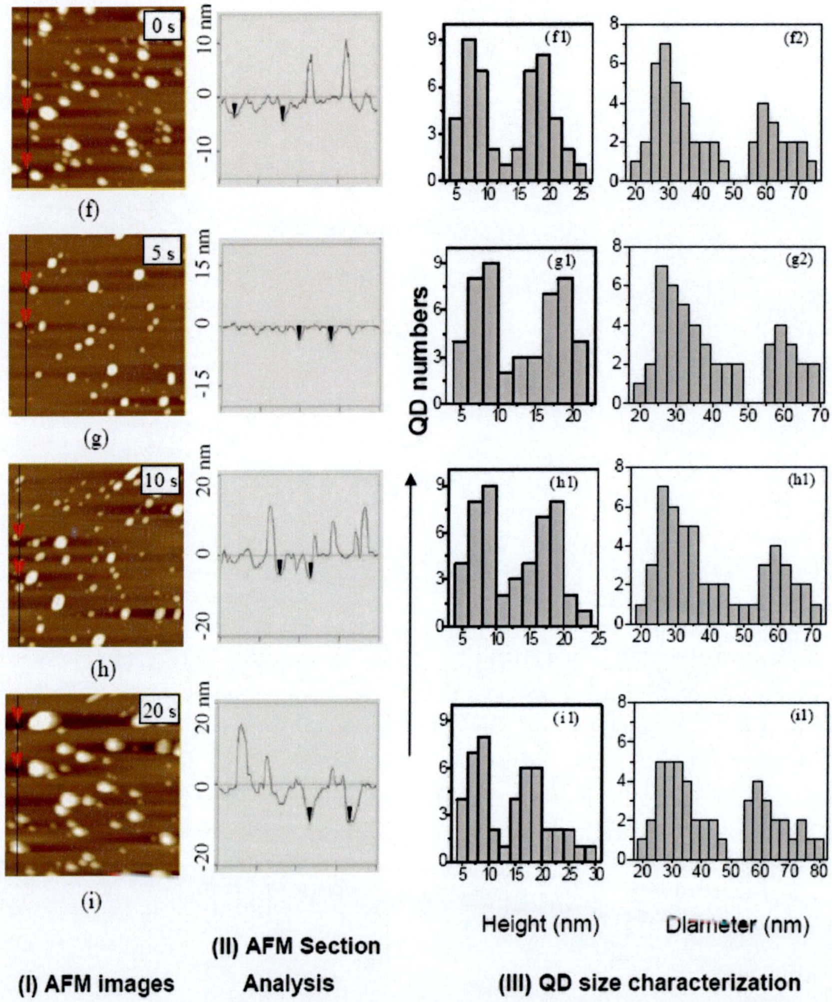

Figure 4. (I) Top-view 1×1 μm^2 AFM images, (II) AFM in-situ Section Analysis of the surface depth along a cross-sectional line across the image, and (III) The InAs QDs size characterization of samples (f), (g), (h) and (i).

Table 1. Comparison of QDs density, QDs' height & diameter standard deviations and wetting layer surface depth for the samples (f), (g), (h) and (i)

Compared parameters	Sample (f)	Sample (g)	Sample (h)	Sample (i)
QD density ($10^{10}/cm^2$)	0.63	0.61	0.62	0.6
QD's height standard deviation (nm)	6.2	5.2	6.0	7.8
QD's diameter standard deviation (nm)	15.2	14.3	15.1	16.9
Surface depth (nm)	3.7	3.5	4.8	8.7

With zero second post-growth interruption, standard deviation of the QDs' height of sample (f) is 6.2 nm. When the interruption time was 5 seconds, standard deviation of sample (g)'s QD height reduces to 5.2 nm. However, standard deviation of the QDs' height increases to 7.8 nm of sample (i) when the post-growth interruption time was increased to 20 seconds. The post-growth interruption time of 5 seconds gives sample (g) the minimum QDs' height standard-deviation, i.e. the narrowest QDs height distribution. Similar trend of the QDs' diameter standard deviation has been received with different post-growth interruption time. The results show that there exists a post-growth interruption time window for forming the uniform QDs. Out of this window, too long or too short of the post-growth interruption will cause large dot size distribution of the QDs formed. This can be explained by the surface kinetics of the QD epitaxy growth. On epitaxy surface, a 3-D nucleation center acts as a low energy site which attracts surrounding adatoms. Adatoms accumulate at the nucleation center to make it grows bigger [19]. For sample (f), the heater was turned off immediately when the InAs reached the target layer thickness, the surface adatoms had no time to diffuse across the surface to be efficiently absorbed by the 3-D nucleation center. Small islands or nuclei had no time to grow bigger. For sample (g), the reactor was maintained at the growth temperature for 5 seconds when the InAs growth was stopped, the small nuclei kept growing by absorbing the surface adatoms during the post-growth interruption. Since smaller dots have larger aspect ratio, they grow faster [20], while the bigger dots grow slower. So that, the QDs finally formed are more uniform in size. However, if the post-growth interruption time is too long, as for sample (i), when the surface adatoms are consumed out, some islands grow further into large clusters to relax the lattice mismatch strain between the QDs and the matrix layer underneath by consuming the material of the wetting layer. This process results in rough surface of the wetting layer underneath and a very broad QDs size distribution of sample (i). Dislocations will be generated in the large clusters formed. They act as non-radiative recombination centers and degrade the QDs' PL performance [21]. Using H2 as carrier gas in MOVPE growth of InAs QDs with similar growth conditions, high-quality and uniform QDs were formed with much longer post-growth interruption time, from 20 seconds to 30 seconds [22,.23]. The shorter post-growth interruption needed when the InAs QDs grown in N2 ambient may be due to the lower thermal conductivity and higher viscosity of nitrogen gas used in this study than that of hydrogen gas [24]. Because of the lower thermal conductivity and higher viscosity of nitrogen gas, the growth surface temperature cools down slower when the reactor heater is turned off and the unstable surface adatoms stay on the surface longer than adatoms in H2 ambient during the interruption before they are desorbed. Shorter interruption is needed for the atoms being finally absorbed by the nucleation center before the surface temperature becomes too low after switch off the heater.

In general, with current growth conditions, QDs formed with different post-growth interruption times in sample (f), (g), (h) and (i) are still with low dot density and broad size dispersion. In order to increase the dot density and improve the QDs size dispersion, the growth temperature, inlet source flux and growth mode are further investigated.

3) Effect of the Growth Temperature

To study the effects of growth temperature on the formation of InAs QDs, six samples, sample (j) to (o), were grown at different temperature in the range from 420 oC to 520 oC. The total InAs layer thickness was still target at 2.25 ML and the post-growth interruption time was set at 5 seconds. Figure 5 shows the top-view AFM images of the six samples grown at different temperatures.

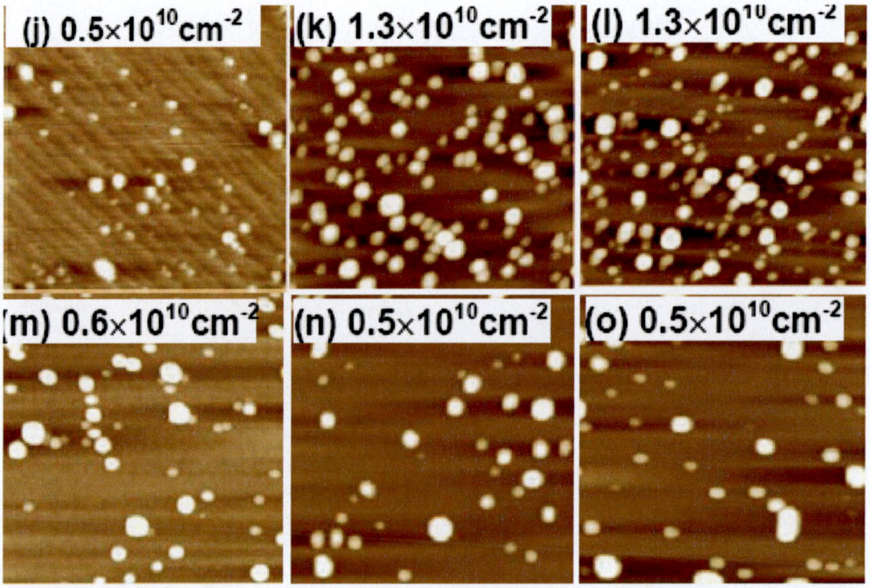

Figure 5. Top-view 1×1□m^2 AFM images of the InAs QD samples grown at different temperatures. Sample (j), (k), (l), (m), (n) and (o) was grown at 420°C, 440°C, 460°C, 480 °C, 500 °C and 520 °C, respectively. The measured dot density of each sample is inserted.

High QD density of around 1.3×1010 cm-2 is obtained when the samples were grown at the temperature range from 440 oC to 460 oC. Out of this temperature range, when the sample was grown at temperature either higher or lower by 20 oC, the samples' QD density reduces dramatically to around 0.5×1010 cm-2. When the growth temperature was above 480 oC, large InAs islands with low density were obtained. This is due to the longer diffusion length of the adatoms at higher growth temperature [25]. During the post-growth interruption, larger islands grew further by consuming small islands during the coarsening process [26]. This can be seen from the AFM images of samples (m), (n) and (o), where some large dots are surrounded with sparse distribution of very small dots. Low density QDs were formed at low growth temperature, ≤ 420 oC. This is because: (i), the critical thickness of the wetting layer for starting the QD formation increases when the growth temperature is lower due to the reduction of adatoms' diffusion length at lower temperature [26]; (ii), the diffusion coefficients of the source adatoms are smaller at lower the growth temperature, which reduces the growth efficiency [27]. Therefore, the materials available for growing the InAs QDs were reduced when the growth temperature was below 420 oC, which resulted in the low density of InAs QDs formed.

The distribution and standard deviation of the QDs' diameter and height versus the growth temperature are plotted in Figure 6. Figures 6(i) and (ii) show bimodal distributions of the QDs' diameter and height when the growth substrate temperature was increased. Sample (k), which was grown at 440 oC, has the smallest standard deviation of QDs' diameter and height, as shown in Figure 6(iii). The results show that more uniform QDs with high dot density have obtained when the sample was grown at 440 oC with the TMIn flux set at 30 sccm. But, the growth temperature of 440 oC is too low for growing the InP cap layer of the InAs QD structures. In the following investigations, the growth temperature was set at 500 oC.

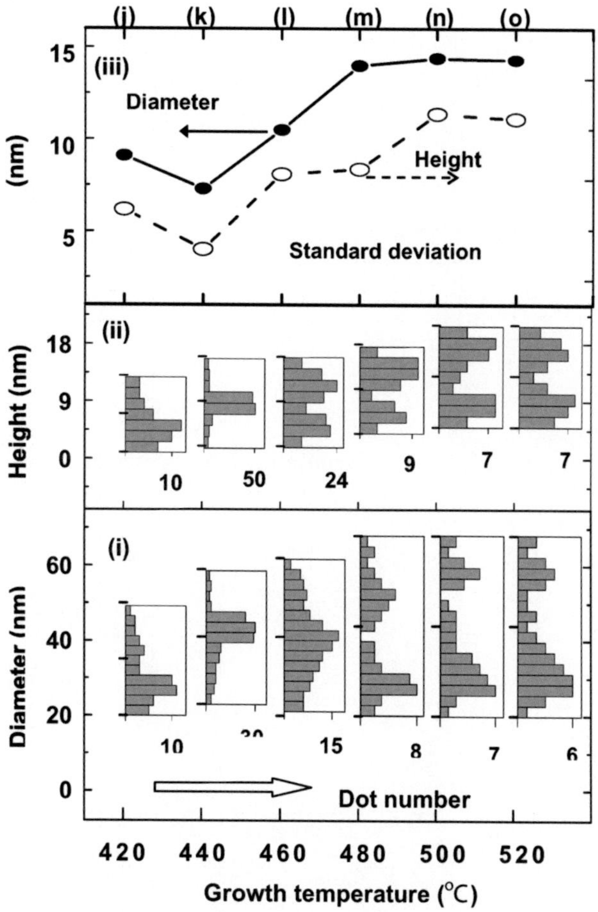

Figure 6. InAs QDs' size distribution and standard deviation of samples (j), (k), (l), (m), (n) and (o) which were grown at different temperatures: (i) Histograms of QDs' diameter distribution, (ii) histograms of QDs' height distribution and (iii) standard deviation of QDs' diameter and height as a function of deposition temperature.

4) Effect of the Source Flux

In MOVPE growths of QD structures, the molar ratio of the total input group-V sources versus the group-III sources, or V/III ratio, is generally larger than 1. The growth is with rich group-V source elements ambient. In this study, the V/III ratio was set at 10. The effects of input source flux on the InAs QDs formation were investigated by changing the input group-III (TMIn) and group V (TBA) source fluxes while the V/III ratio was kept same during the InAs QDs growth. Samples (p) and (q) were grown by setting the input TMIn flux at 30 sccm and 75 sccm, respectively. The substrate temperature was set at 500 oC and the post-growth interruption was kept 5 seconds for growing the two samples. To keep the same deposited InAs layer thickness, the InAs growth time was changed accordingly when the input TMIn source flux was changed. As shown in Table 2, higher density InAs QDs are formed in sample (q) than that of sample (p). This is because at temperature of 500 oC the surface reaction dominates the epitaxy growth, the growth rate is in mass flow control. The InAs growth rate is proportional to the inlet TMIn flux. When the TMIn flow flux was increased in growing sample (q), the growth rate was higher and more nuclei were formed. QD density of sample (q) is higher.

When the substrate temperature was lowered to 440 oC, as for growing sample (r) and (s), the epitaxy growth was in surface reaction limited growth. The InAs growth rate did not increase accordingly when the TMIn input flux was increased. The InAs layer deposited in sample (s) was too thin to form the QDs when the growth time was reduced to 6 seconds.

Table 2. Top-view 1×1□m2 AFM images of InAs QDs grown with different TMIn input flux & inflow time at 440 oC and 500 oC, respectively

		Growth methods	
		Slow growth	Fast growth
TMIn flux & inflow time		30 sccm &	75 sccm &
Growth temperature (°C)	500	(p) $0.5\times10^{10}cm^{-2}$ 	(q) $1.3\times10^{10}cm^{-2}$ 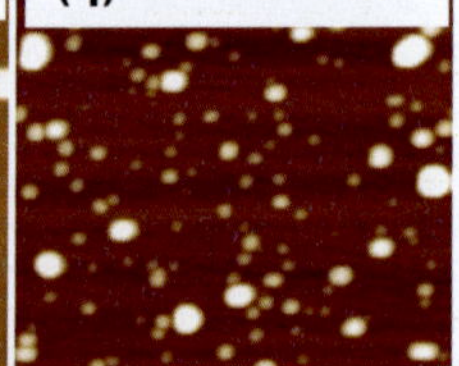
	440	(r) $1.3\times10^{10}cm^{-2}$ 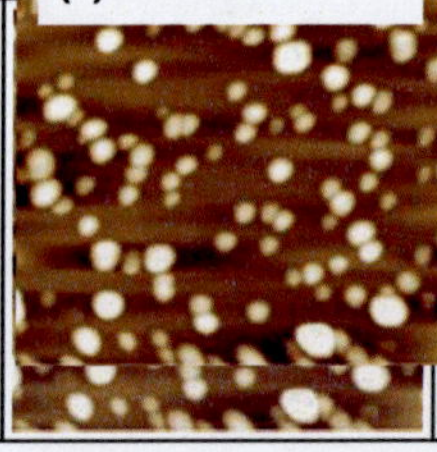	(s) $0cm^{-2}$ 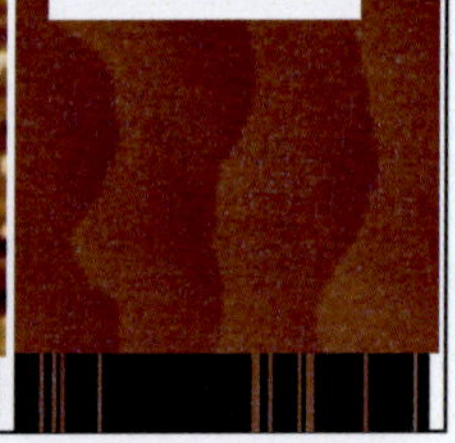

B. A New Two-Step Growth of QDs

To improve the dot density and uniformity, we proposed a two-step growth method for growing the InAs QDs [28]. In this QD growth method, InAs QD growth is divided into two steps. Figure 7 compares the inlet TMIn flux flow in conventional S-K growth (A) and in the two-step growth (B).

In step 1 growth of two-step growth QDs, a thin InAs layer is deposited with high TMIn source flux, and then the growth is paused for an internal interruption by closing the TMIn source. InAs nuclei are formed during the internal interruption. And then, the TMIn source is opened again for the step 2 growth. In the step 2 growth, the InAs layer is grown with lower growth rate. When the target layer thickness is reached, the TMIn source is closed for post-growth interruption and the InAs QDs are finally formed self-assembly. For comparison in this study, sample (t) and (u) was grown by the two-step growth method at 500 oC and 440 oC, respectively. The measured AFM images of the two samples are shown in Table 3. For comparison, AFM images of samples (q) and (s) which were grown by the conventional S-K self-assembly mentioned above are also shown in Table 3. Dot density of sample (u) is 0.2×1010 cm-2 which is higher than that of the sample (s) which was grown at the same temperature but by using the conventional S-K growth method. Increased the growth temperature to 500 oC, the QDs density of sample (t) which was grown by two-step growth

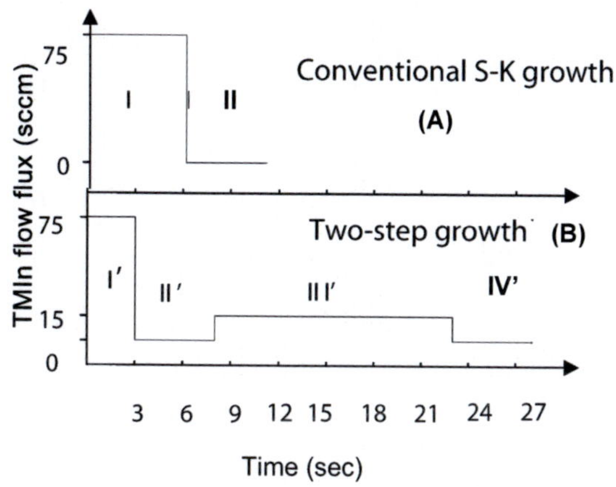

Figure 7. Schematic diagram of TMIn source flow settings during the InAs QDs growth. In conventional S-K growth (A), I: InAs deposition; II: post-growth interruption; in two-step growth (B): I′: step-1 InAs deposition; II′: internal interruption; III′: step-2 InAs growth; IV′: post-growth interruption.

Table 3. Top-view 1×1μm2 AFM images of InAs QDs grown with fast growth and two-step growth at 440 oC and 500 oC, respectively

		Growth methods	
		Fast growth	Two-step growth
TMIn flux & inflow time		75 sccm & 6 sec	75+15 sccm & 3+15 sec
Growth temperature (°C)	500	(q) $1.3\times10^{10}cm^{-2}$ 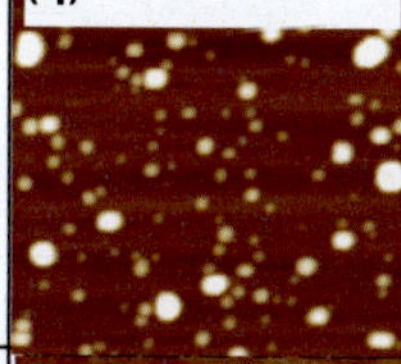	(t) $1.7\times10^{10}cm^{-2}$ 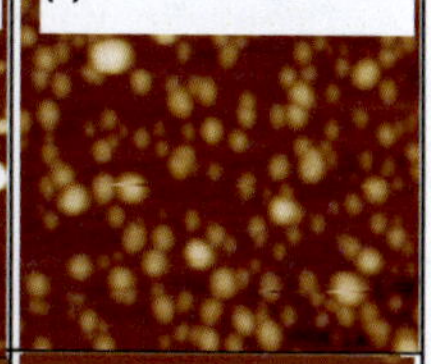
	440	(s) $0cm^{-2}$ 	(u) $0.2\times10^{10}cm^{-2}$ 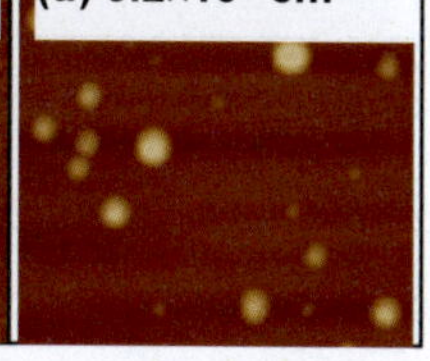

method is about 30% higher than that of sample (q) which was grown at same temperature but by using the conventional S-K mode. The QDs' diameter and height distributions of samples (q) and (t) are shown in Figure 8. The dot size dispersion of sample (t) is much narrower than that of sample (q). The standard deviation of the QDs' diameter and height of sample (q) is 12.9 nm and 7.6 nm, respectively, while that of sample (t) is 4.2 nm and 2.3 nm, respectively. These results show that the QDs' size uniformity grown by the two-step growth method has been improved.

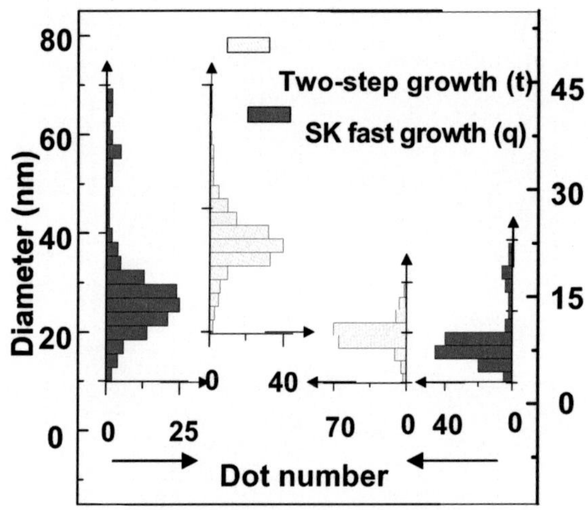

Figure 8. Histograms of the QDs' diameter (left) and height (right) of the samples grown by conventional S-K fast growth, sample (q), and by two-step growth, sample (t). The sampele were grown at 500 °C.

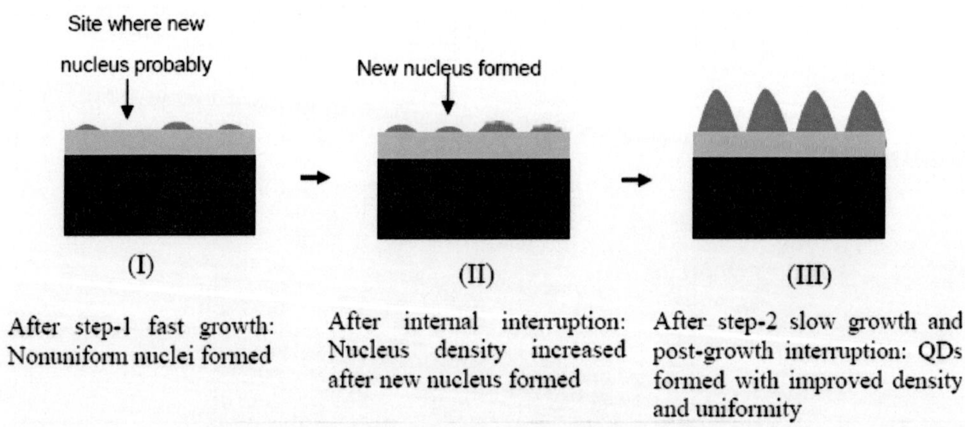

Figure 9. Schematic diagrams show the formation of InAs QDs in the two-step growth.

A schematic diagram of Figure 9 shows the QDs formation in the two-step growth: un-uniform nuclei are formed in the first step fast growth; during the internal interruption, these nuclei continue grow by absorbing the surface adatoms. At the same time, some of the adatoms self-nucleate between the formed dots, more QDs are formed in this way, which increases the QD density. All the formed dots and nuclei continue growing during the step-2 growth. In step-2 growth, the TMIn source is set at low flux so that the dots grow slowly into

QDs. At the same time, more new nuclei can be formed and grow also in step-2 growth and they all grow into QDs during the post-growth interruption. The higher edge energy barrier of the larger dots prevents the adatoms approaching the large dots and making them bigger[29,30]. As a result, the dot density is increased much by using the two-step growth and because the larger islands/dots grow much slower than the small islands/dots in setp 2 growth [30], the size uniformity of the QDs is also improved in two-step growth. To optimize the growth conditions of the two-step growth, effects of growth parameters of two-step growth on the QDs formations have been investigated.

1) Effect of the Internal Interruption Time

Different from conventional S-K self assembly growth of QDs, in two-step growth of QDs, an internal interruption is introduced between the two step growths. The internal interruption time is important in forming uniform QDs in the two-step growth. To study the effect of the internal interruption time on the QD formations in two-step growth, four samples, sample (v) to (y), were grown at 500 oC and the total InAs grown material was still kept at it's nominal layer thickness of 2.25 ML. The TMIn source flow flux and the growth time for each step of the MOVPE growths of the four samples are listed in Table 4. Figure 10 shows the 1×1 μm2 AFM surface images of the four InAs QDs samples grown by the tow-step growth under different internal interruption times. For comparison, a reference sample was grown using conventional S-K one-step growth under the similar growth conditions. Figure 10 shows that the dot density of sample (v), (w), (x) and (y) is 1.5×1010 cm-2, 1.7×1010cm-2, 1.7×1010cm-2 and 0.7×1010cm-2, respectively. It has been reported that in conventional S-K growth of QDs, higher QD density can be achieved if the InAs is grown with a faster growth rate [25,31].

Table 4. The TMIn source flow flux settings and the growth time for each growth step in MOCVD growth of the InAs QD samples

Compared samples	Two-step growth			
	Step 1	**Internal interruption**	Step 2	Post-growth interruption
(v)	**0.5**	**s**		
(w)	**2**5 sccm	**s**	15 sccm	5 s
(x)	**5** & 3 s	**s**	& 15 s	
(y)		**60 s**		

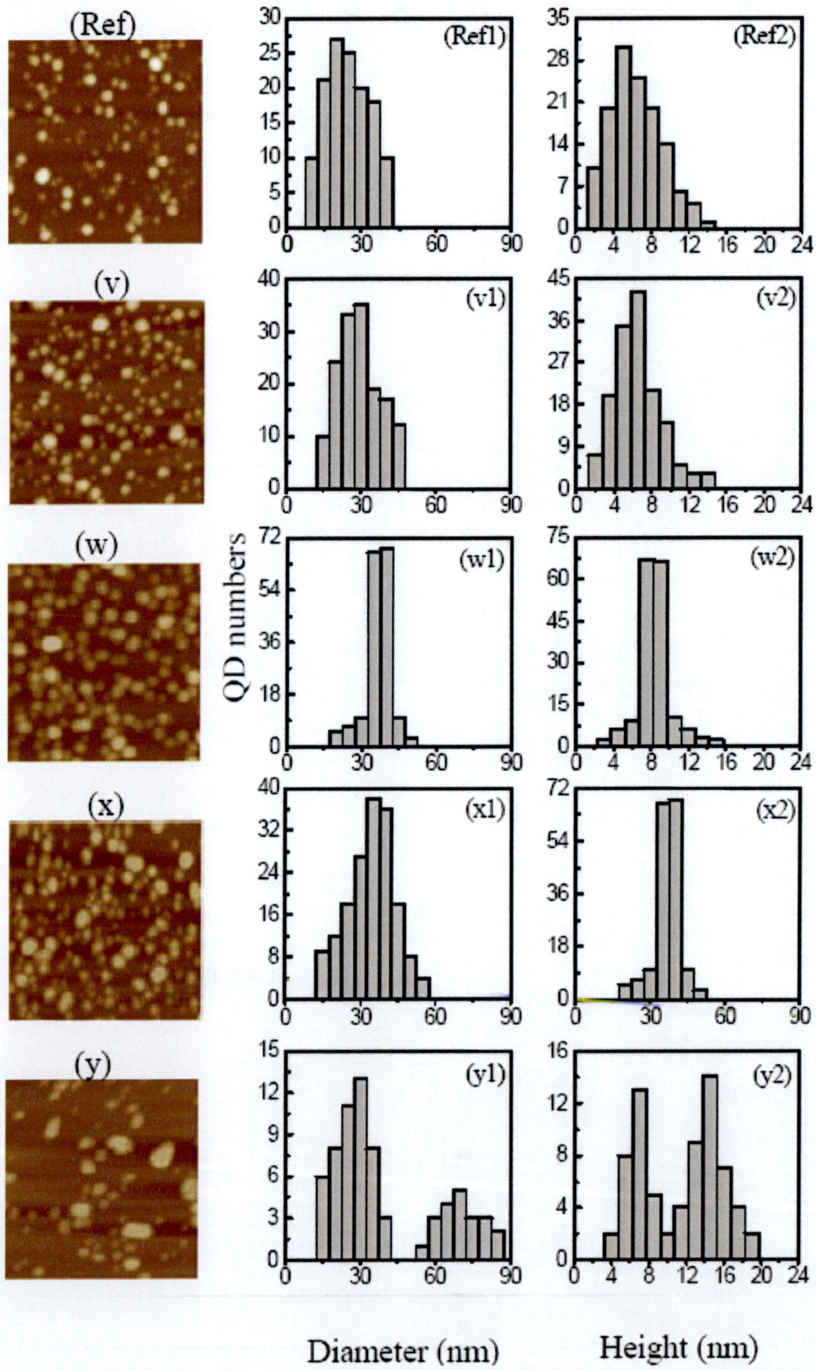

Figure 10. 1×1 μm2 AFM surface images of the Ref. InAs QDs sample which was grown by conventional S-K fast growth and the samples (v) to (y) which were grown by two-step growth with different internal interruption time (t_{int}). The dot diameter and height histograms associated to each sample are shown on right.

In this study, the Ref. sample was grown with the fast InAs growth rate and the dot density received is measured of 1.3×1010cm-2. However, for samples (v) and (x) ,which were grown by using the two-step growth method, they have even higher dot density than that of the Ref. sample, although the TMIn flow flux in step two growth of the two-step growth was only set at 15 sccm which was much smaller than that set from growing the Ref sample, where the TMIn was set at 75 sccm. As shown in the histograms in Figure 10, the average diameter of the InAs QDs grown by two-step growth is ~ 36 nm, while for the Ref sample, the average QDs diameter is ~ 25 nm. But, the QDs' diameter dispersion of sample (w) grown by using the two-step method with the internal interruption time *t*int = 2 seconds is much narrower than that of the QDs grown by the conventional S-K method. More uniform QDs are grown using the two-step growth. As shown in Figure 10, the internal interruption time, *t*int, strongly affects the size uniformity and dot density in two-step growth of QDs. When *t*int is too short, as for sample (v), there is not enough time to form the new nuclei during the internal interruption. In this case, the growth is similar to that of conventional S-K growth. If *t*int is too long, as in the case of sample (y), very large dots/islands or clusters are formed during the internal interruption by consuming the materials from the wetting layer or adjacent smaller islands/dots. These large dots/islands are ripened further by merging neighbor small dots/islands in the step 2 growth, resulting in very large dot diameter dispersion and very low dot density [32].

2) Effect of the Source Flux in Step-2 Growth

To investigate the effects of the InAs growth rate in step 2 growth on the formation of QDs in the two-step growth, four samples, samples (w)–(w‴), were grown with different TMIn source flow fluxes in step 2 growth of two-step growth of InAs QDs, while the other conditions were kept same. As shown in Table 5, the TMIn flux in step 2 growth of InAs was set at 15, 30, 56.3, 112.5 sccm and the InAs was grown for 15, 7.5, 4 and 2 seconds, respectively, for growing these four samples to keep the InAs thickness in step 2 growth the same. Figure 11 shows the 1×1μm2 AFM surface images of the four samples. When the TMIn flux in step 2 was increased from 15 sccm (sample (w)) to 112.5 sccm (sample (w‴)), the dot density increased by 47%, reaches 2.5x1010cm-2 of sample (w'''. The average dot diameter increases from 36 to 43 nm, but the average height of the QDs decreased from 8.4 nm for sample (w) to 5.6 nm for sample (w‴). The dot diameter dispersion was also observed to be increased when the growth rate in the step 2 growth was increased.

Table 5 The TMIn source flow flux settings and the growth time for each growth step in MOVPE growth of the InAs QD samples

Compared samples	Two-step growth			
	Step 1	Internal interruption	**Step 2**	Post-growth interruption
(w)	75 sccm & 3 s	2 s	**15 sccm & 15 s**	5 s
(w')			**30 sccm & 7.5 s**	
(w'')			**56.3 sccm & 4 s**	
(w''')			**112.5 sccm & 2 s**	

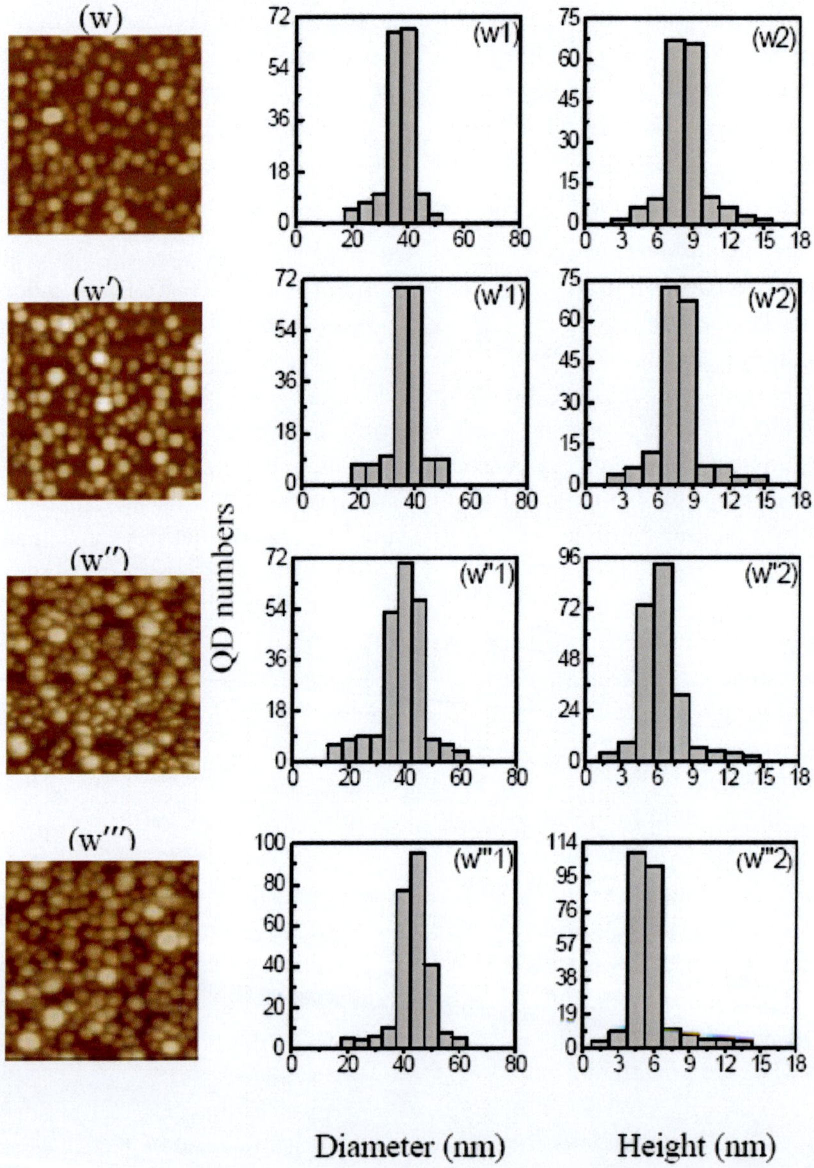

Figure 11. 1×1 μm2 AFM surface images of InAs QD samples grown by two-step growth with different TMIn source flux in step 2 growth. The total InAs layer thickness of all the sample are kept same. The dot diameter and height histograms associated to each image are also shown.

With the faster growth rate, the growth time for the step 2 growth of InAs was shorter; the dots formed in the step 2 growth do not have enough time to release all their strain energy by increasing the aspect ratio. Thus, QDs formed under such conditions are large in diameter but with low height [20].

So far the average diameter and height of the InAs QDs formed is around 40 nm and 7 nm, respectively, which is much larger than that of the InAs QDs grown by MOVPE using TBA and H2 carrier gas. It has been reported that the average dot diameter and height of InAs QDs grown by MOVPE in H2 ambient was around 20 nm and 5 nm, respectively, [23] for which the deposited InAs material thickness was of 2.6 MLs which is even thicker than the

InAs layer deposited of the samples in this study. The reason for larger dots formed in this study grown in N2 ambient than those formed in H2 ambient may be because of the lower thermal conductivity and larger viscosity of N2 as mentioned previously [22]. The adatoms stayed around the formed islands for longer time in N2 ambient, which increases the probability for them to be absorbed by the islands before they are desorbed or brought away to the exhaust. As a result, total efficiency of the material consumed in forming QDs in N2 ambient was increased and hence the larger QDs were formed.

3) Effects of Step-1 Nucleation in Two-Step Growth of QDs

In two-step growth of QDs, 3D island nucleation starts in step-1 growth and then followed by kinetic self-limiting growth of uniform QDs in step-2 atomic layer epitaxy (ALE) growth. Optimization of step-2 growth in the two-step growth of QDs has been investigated above. However, the first step nucleation has important impact in the morphology and crystal-quality of the QDs finally formed in the two-step growths.

In this section, effects of nucleation in the step-1 growth on the morphology and crystal quality of the InAs/InGaAs/InP QDs formed by the two-step growths have been investigated in detail. In the two-step growths, the optimized ALE growth conditions were used in the step 2 growth for all the samples. The InAs QDs are grown on a strained In0.53-0.72Ga0.47-0.28As/InP matrix. The detailed growth conditions for InAs QDs layer growths were listed in Table 6.

Table 6. Comparison of growth rate (samples (e)-(h)), growth temperature (samples (i)-(m)) and V/III flux ratio (samples (n)-(u)) for growing InAs QDs on In0.53-0.72Ga0.47-.28As/InP matrix. The other common growth conditions are also listed

Two-step growth / QDs samples	Step 1	Internal interruption	Step 2 (ALE)	Post-growth interruption
(e) – (h)	**0.42 1.68 ML/sec** (550 °C & V/III = 6.5)	2 sec	4-cycle monolayer growth:	5 sec
(i) - (m)	**Temp: 460 600 °C** (0.84 ML/sec & V/III = 6.5)			
(n) – (u)	**V/III ratio = 0 11** (0.84 ML/sec & 550 °C)			

(i) Effect of the Growth Rate

Four samples, sample (e), (f), (g) and (h) were grown by setting the growth rate in step-1 at 0.42, 0.84, 1.26 and 1.68 ML/sec, respectively. The other growth conditions for the four samples were kept the same. Top-view AFM images and the corresponding low temperature PL spectra of these samples are shown in Figure 12. The QDs' size standard deviation, dot density, PL peak intensity and linewidth are summarized in Figure 13.

The results show that when the growth rate is around 0.84 ML/sec in step-1 growth, sample (f) has the highest dot density, strongest PL emission intensity and narrowest PL spectrum, which indicates the narrowest dot size dispersion and high crystal-quality of the QDs of sample (f). The kinetic theory of stress driven QD growth shows that the 3D island surface density rises with increasing the growth rate, *V*, approximately as *V*3/2.[33].

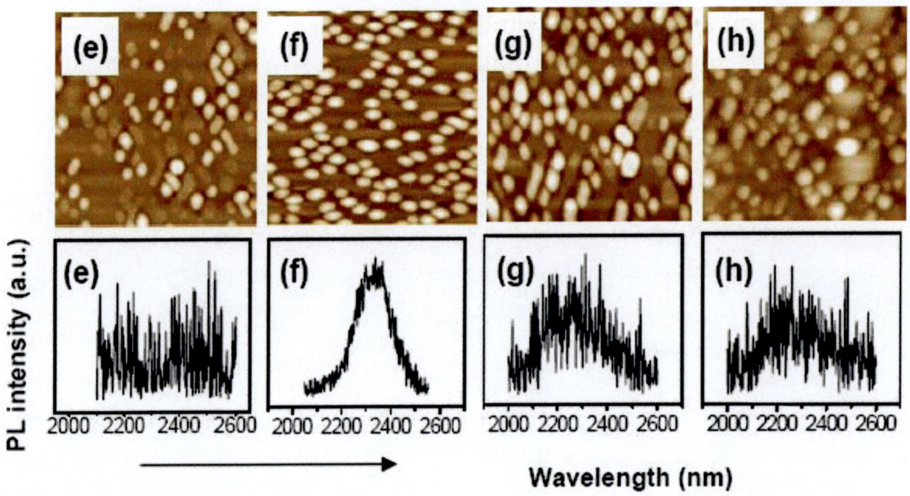

Figure 12. AFM (1×1μm2) top-view images and 20K-PL spectra for samples (e) - (h) with different growth rates in step-1 growth.

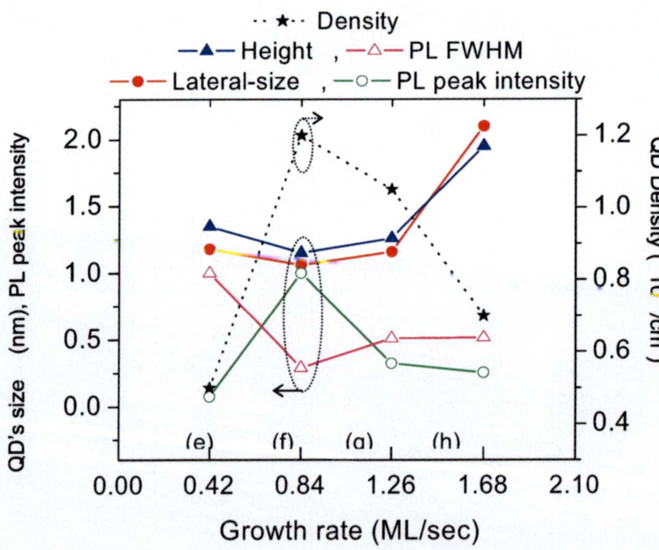

Figure 13. Comparison of the QDs density, dot size standard deviation (σ), PL peak intensity and linewidth of samples (e) – (h) grown with different growth rate.

In the two-step growth, QDs growth is divided into two steps. In step-1 growth, high-density nuclei are formed. Then the formed nuclei further develop into the QDs in step-2 ALE growth. As predicted in Ref. [33], the characteristic time of the nucleation during the QD formation is much shorter than the time for the nuclei being developed into the QDs. In two-step growth of QDs, if the growth rate, thus the nucleation rate, in step-1 growth is too fast, ≥ 1.26 ML/sec, a large amount of strain energy must be dissipated with the insufficient time for single-crystalline material formation. This causes the formation of non-crystalline defects in the QDs. These defects are with lower energy barrier and grow faster than the crystallographically-selective slow-growing single-crystals [22]. These defects will become the seeds for forming the incoherent islands or even the large clusters. As a result, the crystal quality, dot density and size uniformity of the QDs will be affected if the growth rate is too high in step-1 growth. This is shown from the measured results of samples (e), (f), (g) and (h) in Figure 12 and Figure 13. When the growth rate in step1 growth was increased properly from 0.42 ML/s to 0.84 ML/s, the dot density, size uniformity and PL emission of the sample has been improved much. But when the growth rate is increased to 1.26 ML/s and above with less than 2.7-second duration in step-1 growth, the dot density decreases due to the small QDs being consumed for the ripening growth of the incoherent islands growing into the large clusters. This obviously degrades the dot size-uniformity and PL emission intensity of the samples.

(ii) Effect of the Growth Temperature

To investigate the effects of the substrate temperature on the InAs/InGaAs/InP QDs formation, five samples, samples (i), (j), (k), (l) (= (f)) and (m), were grown at the temperature of 460, 500, 520, 550 and 600 oC, respectively, while the growth rate in step 1 was set at the optimized growth rate of 0.84 ML/s and all other growth conditions were kept the same for growing the five samples. Figure 14 shows the AFM images and the low temperature PL spectra of the five samples. The dot size standard deviation, dot density, PL peak intensity and PL spectrum linewidth of the samples are summarized in Figure 15. It has been observed: (I) when grown at low temperature, 460 oC, bi-modal sized QDs were formed in sample (i); when the temperature was increased to 550 oC, the dot size dispersion becomes narrower, but it becomes broader again when the temperature was increased to 600 oC; (II) The highest dot density of sample (k) was grown at 520 oC; (III) Sample (l) with the best PL performance was grown at 550 oC.

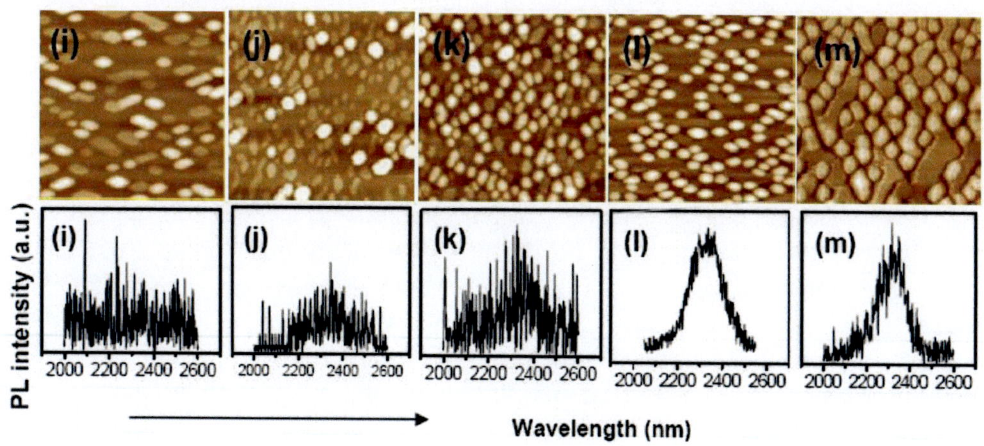

Figure 14. 1×1μm2 AFM top-view images and 20K-PL spectra for samples (i)–(m) grown at different temperatures.

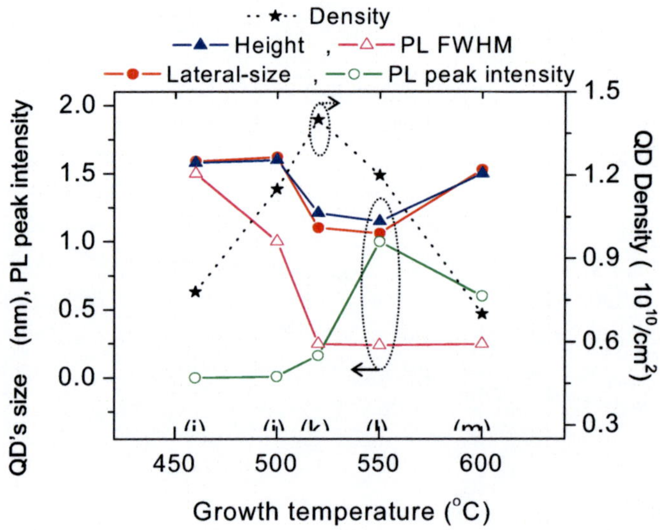

Figure 15. The samples' QD density, dot-size standard deviation (σ), PL peak intensity and linewidth changes with the growth tempertures.

At low growth temperature, ≤ 500 oC, the growth rate is low, there was not enough time to grow enough material for forming the QD nuclei in the step-1 growth, and only wetting layer was grown. The QDs were mainly grown in step-2 ALE growth. Therefore, the AFM image of the QDs of sample (i) which was grown at 460 oC looks similar to that of the QDs grown using normal ALE growth mode. When the temperature was increased to 520 oC, nuclei formed in step-1 were all developed into the QDs during step-2 growth. So, high density QDs of 1.4×10^{10}/cm^2 have been received in sample (k). However, the dot density of the samples decreases when the growth temperature was increased above 520 oC and the dot size becomes bigger, which is caused by the increased diffusion length of adatoms at higher temperature [34]. The better PL performance of sample (l) which was grown at 550 oC, rather than that of sample (k) with higher dot density, is because crystal quality of the formed QDs is better when grown at higher temperature. Material grown at higher temperature normally has better crystal quality due to the reduction of the inner grown defects, such as non-/poly-crystalline structures. Non-crystallines are generated at the QDs/matrix interface due to the large lattice mismatch strain between the InAs dots and the InGaAs/InP matrix. At higher growth temperature these non-crystalline structures are broken, and hence the crystal quality of the QDs has been improved. But, if the growth temperature is too high, ≥ 600 oC, as in the case of sample (m), the sample's dot-density is reduced because the larger size dots are formed at higher temperature. So that, sample (m)'s PL intensity is lower than that from sample (l). The results show that an optimized growth temperature of 550 oC which gives high dot density, uniform dot size and best PL performance of the QDs has been received under the current conditions.

(iii) Effect of the V/III Flux Ratio

Above investigations show that the dot density of the samples under optimized growth conditions received is around 1.2×1010/cm2. It can be calculated that the corresponding QDs' surface coverage of the sample is only ~23%. In order to enhance the sample's PL performance further, it is very important to grow the sample with higher dot density. It has shown that the dot density does not change much by optimizing the growth rate and substrate temperature during the growth. In this investigation, we have tried to increase the dot density by optimizing the inlet source flux of group-V source TBA flux in step-1 growth.

In this study, the TMIn group-III source flux and all the other growth conditions were kept unchanged for growing the samples. Only group-V source flux of TBA for step-1 growth was changed from 0 to 4.52×10-4 mol/min, with the V/III ratio correspondingly changed from 0 to 11. Samples (n), (o), (p), (q), (r), (s), (t) and (u) were grown by using the two-step growth with the input source flux V/III ratio in step-1 growth set at 0, 0.2, 0.5, 1.1, 2, 4, 6.5 and 11, respectively, by adjusting the TBA flux. Figure 16 shows the AFM top surface images of these samples and their low-temperature PL spectra. The measured dot density of the samples grown with different input V/III ratio in step 1 growth is summarized Figure 17.

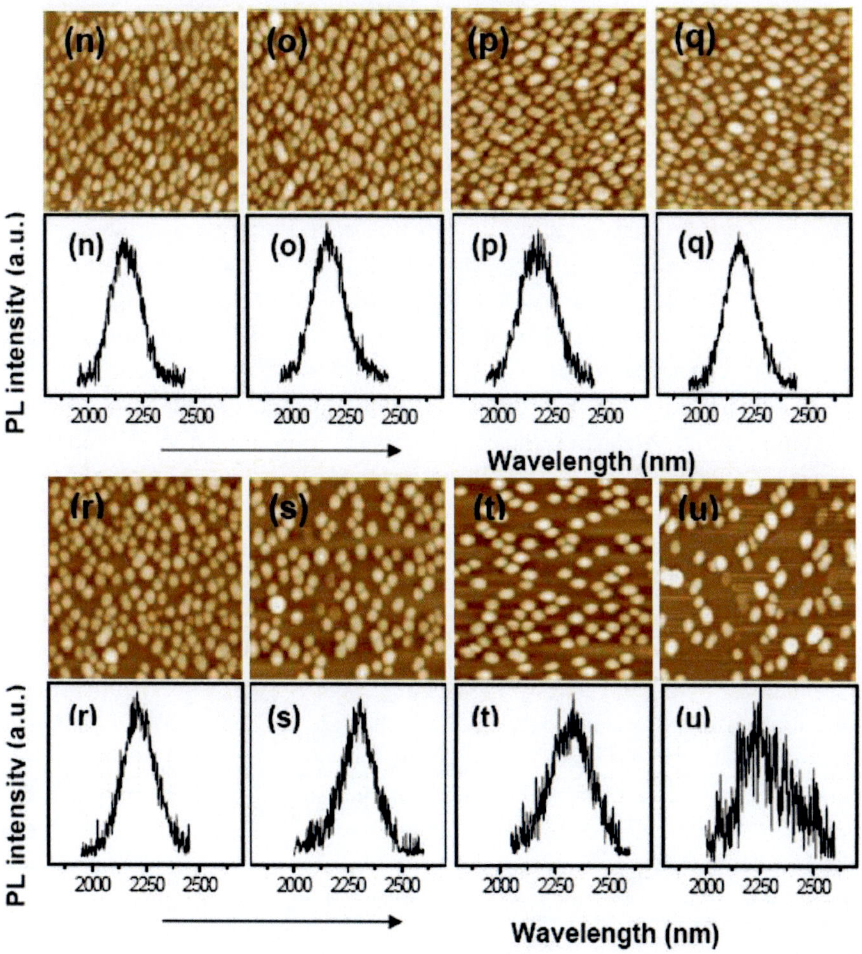

Figure 16. 1×1μm2 AFM top-view images and low temperature PL spectra of the two-step grown InAs QD samples (n), (o), (p), (q), (r), (s), (t) and (u) with different V/III source flux ratio in step-1 growth.

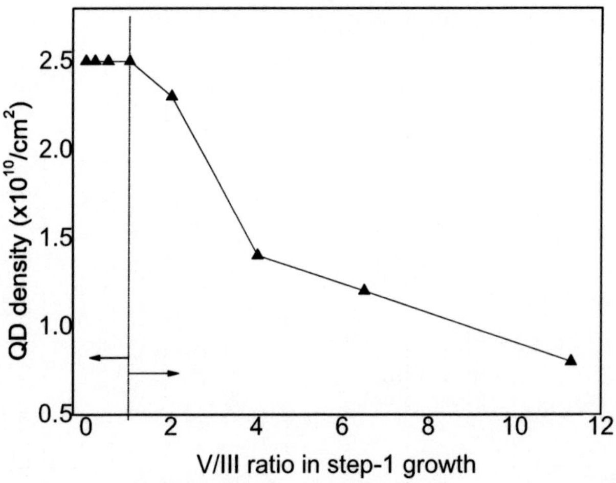

Figure 17. Dot density of the sample changes with the V/III ratio used in step-1 growth of samples (n), (o), (p), (q), (r), (s), (t) and (u).

Very high dot density, ~2.5×1010 cm-2, has been achieved with very low V/III ratio during the first step growth and it slightly changes when the V/III ratio changed from 0 to 1. For the V/III ratio is >1, the formed QD density drops sharply when the V/III ratio was increased. This observation is different from what has been reported where the dot density increased with the V/III source flow ratio [35, 36]. The drop of the dot density in the range of V/III >1 can be explained as the surface-diffusion length of indium adatom is increased when the group V atom concentration on the epitaxy growth surface is reduced by lowering the V/III ratio [37, 38]. With the longer surface diffucion length, the effective coverage of the indium adatoms on the growth surface is enlarged. The formed nuclei density is thus increased with the reduction of V/III ratio during the step-1 growth in two-step growth of InAs QDs, which results in higher dot density. Besides, it has also been shown that a thinner critical thickness is obtained when the group V source, TBAs, inlet flux is decreased in S-K growth of QDs [39]. Or in other wards, the low group V source flow can accelerate the onset of the 3D islands formation in S-K QDs growths, thus resulting in higher density nuclei/QDs with lower inlet V/III ratio.

When the input V/III ratio during the step 1 growth is very low, V/III<1, input indium adatoms are more than the arsenic adatoms on the reaction surface. The extra indium adatoms act as surface surfactant to enforce roughening the growth surface [40]. A roughened growth surface has much lower energy sites for 3D growth and at the same time reduces the critical thickness for the onset of 2D to 3D growth transformation [41]. As a result, the formed QDs' dot-density in the two-step growth becomes higher when the input V/III ratio in step-1 growth reduces to less than 1. It is calculated that the surface coverage of the QDs of sample (q) with setting the V/III input flow ratio in step-1 growth at V/III = 1.1 is ~ 60 %, which is very high. There are not many rooms to further increase the dot density of the sample by further reducing the input flow V/III ratio in step-1 growth. The dot-density of samples (n), (o) and (p) did not increase further when the V/III ratio in step-1 growth was reduced from 1 to zero.

Optical and crystal properties of the InAs QDs samples grown with different input V/III ratio in step-1 growths has been studied using PL measurement. The measured PL spectra of the samples are also shown in Figure 16. Figure 18 plots the PL peak position, peak emission intensity, integrated emission intensity and PL linewidth of the samples grown with different input V/III ratio in step-1 growth.

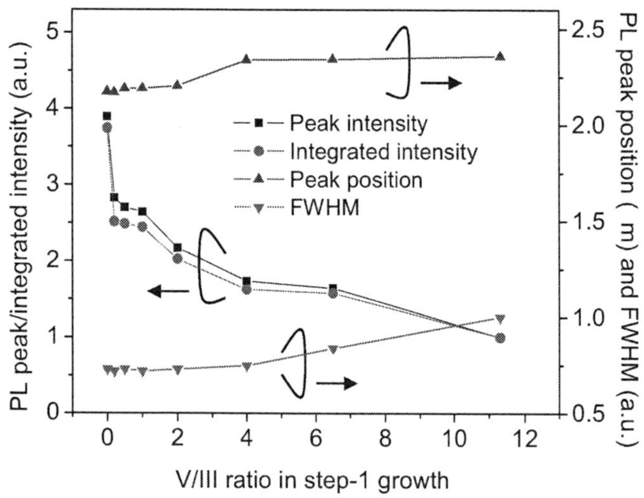

Figure 18. PL peak position, peak intensity, integrated emission intensity and FWHM of the QDs samples as a function of the input V/III ratio in step 1 growth.

It is observed that the PL peak position of the QDs samples slightly blueshifts when the input flow V/III ratio in step-1 growth was reduced in the twp-step growth of QDs. The blue shift of the PL emission is because the dot's height grown with lower input V/III ratio is lower. The peak emission intensity and the integrated emission intensity of the samples are also obersved increase with the input flow V/III ratio in step-1 growth decreasing. These results show that the crystal quality of the QDs grown is not degraded whem the input V/III ratio in step-1 growth was reduced in the two-step growth of InAs QDs. The large increase in the PL emission intensity of the sample grown with the input V/III ratio <1 in step-1 growth is mainly because of the much higher dot density of the sample received. The FWHM of the PL spectra of the samples grown with the input V/III ratio ≤ 2 are similar, which shows the uniformity of dot size and crystal quality of the samples are similar, even though the input V/III ratio was reduced. It is noticeable that for sample (n) with the input V/III ratio in step-1 growth was V/III = 0, the PL intensity is much higher than that of the other samples. The real mechanism of this sharp increasing in the PL emission intensity of the QDs grown with zero input V/III ratio in step-1 growth of the two-step growth InAs QDs needs to be further studied.

High density, uniform and high quality InAs QDs have been grown on the InGaAs/InP matrix with the optimized the growth conditions and by using a two-step growth method. In the following sessions, extending the emission wavelength of the InAs QDs into mid-infrared have been discussed.

C. Extend Emission Wavelength of the QDs Structures

To extend the emission wavelength of the InAs QD structures, InxGa1-xAs has been used as the InAs QDs' matrix layer and their upper barrier. To change the QD structure's emission wavelength, indium content of the InxGa1-xAs matrix layer and upper barrier layer of the QD structures was linearly increased from x = 0.53, which lattice matched to InP, to x = 0.65 ~ 0.80. Figure 19 shows schematically the InAs QD structure of the samples.

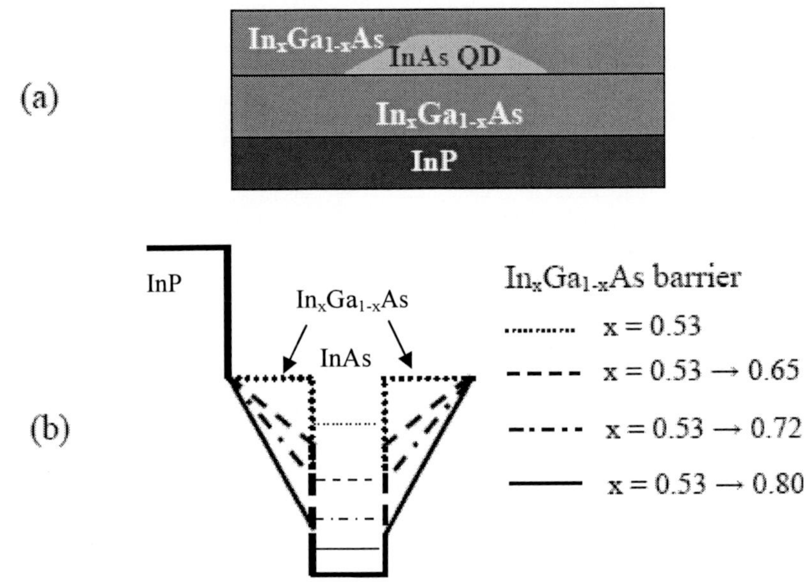

Figure 19. (a) Schematic structure of the InAs/$In_xGa_{1-x}As$ QD samples and (b) the conduction band diagrams of the structures with different $In_xGa_{1-x}As$ barriers.

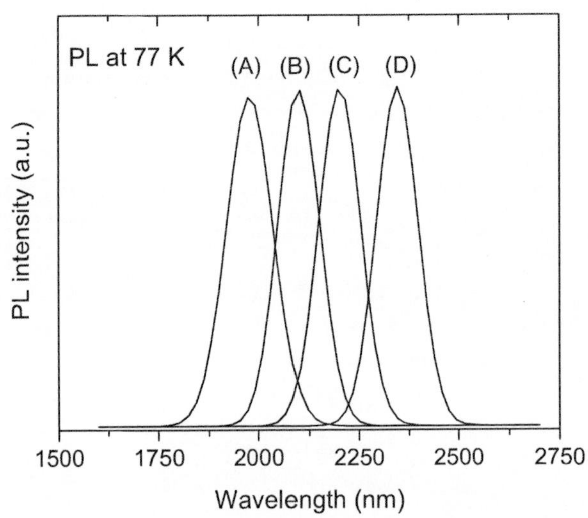

Figure 20. The normalized 77K photoluminescence spectra of the InAs QDs samples grown on different $In_xGa_{1-x}As$/InP matrixes.

Figure 20 shows the normalized 77 K PL spectra of the InAs QDs sandwished between InxGa1-xAs matrixes and barriers with different In content. For sample (A), the InAs QDs are embedded in lattice matched In0.53Ga0.47As/InP matrix and barrier layer. Its peak emission wavelength is at 1.98 μm. To extend the emission wavelength, graded InxGa1-xAs layers are used as the matrix and barrier layer in samples (B), (C) and (D) with the indium content of the graded InxGa1-xAs layer increased linearly from 0.53 to 0.65, to 0.72 and to 0.8, respectively. As shown in Figure 21, the emission peak wavelength of the InAs QDs sandwished between In0.53→(0.53+y)Ga0.47→(0.47-y)As graded matrix and barriers red-shift from that of InAs QDs embedded in lattice matched In0.53Ga0.47As barriers of sample (A). Increasing the indium content gradation, y, of the In0.53→(0.53+y)Ga0.47→(0.47-y)As graded barriers of sample (B), (C) and (D) from 0.12 to 0.27, the InAs QD structure's emission wavelength redshifts by 120 nm, 220 nm and 370 nm, respectively. When the indium content gradation of the In0.53→(0.53+y)Ga0.47→(0.47-y)As graded barrier layer, y, increases by every 0.01, the peak emission wavelength of the InAs QD structure redshifts about 13.5 nm. When the indium gradation, y, increases to 0.27, emission wavelength of the In0.53→0.8Ga0.47→0.2As /InAs/ In0.53→0.8Ga0.47→0.2As/InP QD structure reaches 2.35 μm measured at 77 K, which is, to our best knowledge, the longest inter-band transition emission wavelength of InAs QD structures reported so far. FWHM of the emission spectra of the four QD samples are also plotted in Figure 21. The spectrum becomes narrower when the indium content gradation y of the In0.53→(0.53+y)Ga0.47→(0.47-y)As matrix/barrier increased. Reasons for the narrower emission spectrum of the In0.53→(0.53+y)Ga0.47→(0.47-y)As/InAs/In0.53→(0.53+y)Ga0.47→(0.47-y)As/InP QD structure with the higher indium content gradation of its matrix/barrier are because (i) less lattice mismatch or strain between InAs QDs and the upper InxGa1-xAs barrier layer and (ii) the larger QDs. With the higher indium content of the InxGa1-xAs upper barrier, lattice mismatch between the InxGa1-xAs barrier layer and the InAs QD layer is less, which suppresses the indium mixing or segregation between the InAs QDs and the $In_xGa_{1-x}As$ upper barrier layer [42]. The uniformity of the InAs QDs is not affected by covering the high indium gradation InxGa1-xAs upper barrier layer. The smaller lattic mismatch between the high indium content gradation In0.53→(0.53+y)Ga0.47→(0.47-y)As matrix and InAs makes larger InAs QDs being formed. Optical transition energies of larger QDs will be less affected by the dot size fluctuation [43, 44]. These result in narrower emission spectrum of the In0.53→(0.53+y)Ga0.47→(0.47-y)As/InAs/ In0.53→(0.53+y)Ga0.47→(0.47-y)As/InP QD structure with higher indium content gradation In0.53→(0.53+y)Ga0.47→(0.47-y)As matrix and upper barrier.

Emission wavelength of the InxGa1-xAs/InAs/ InxGa1-xAs/InP QD structure is mainly affected by:

(I) Height of the InAs QDs formed. Since the aspect ratio, which is defined as the QD height versus its diameter, of the InAs QDs formed in the four samples is very small, ~0.18, quantum confinement effect of the QDs is mainly determined by their QD height [45, 46]. Table 1 shows that the aspect ratio of the four samples A, B, C and D, is almost same, 0.167~0.171. The mean dot height of these four samples is also similar, 8.9 nm ~ 9.1 nm. Therefore, the measured red shift of PL peak of samples

(B), (C) and (D) versus that of sample (A) is not because of the sample's QD height variation.

(II) The second factor causes the sample's peak emission redshift is because its $In_xGa_{1-x}As$ barrier layers. Figure 19 (b) compares the conduction band profiles of the four samples under quantum well (QW) approximation [42, 46]. Quantum confinement of carriers in the InAs QDs structure with an $In_xGa_{1-x}As$ graded barriers is reduced, which lowers the transition energy of the QD structure. Thus, its emission wavelength redshifts. The lowest electron-hole transition energies of the four QD structures are calculated by using eight-band $k{\cdot}p$ calculation [47]. Figure 21 compares the calculated first electron-hole transition energies of sample (A), (B), (C), (D) and the measured PL emission peak positions of these four samples. The measured results agree with the calculated transition energies well. It has been noted that the calculated peak emission wavelength of the samples is systematically slightly longer than the experimental results, and the difference between the calculated transition energies and experimental results becomes smaller when the indium composition of the $In_xGa_{1-x}As$ barriers of the QD structure is increased. We believe that this systematically longer calculated emission wavelengths are because of the QW approximation used in the calculation. In the calculation, mean height of the sample's InAs QDs is used as the well width of the approximated QW of the sample. Since the QDs are generally in truncated pyramid shape, the effective height of the QDs is lower than their mean height. This causes the calculated transition energies smaller or the correcponding emission wavelength longer. Further more, we used the mean height of the un-capped InAs QDs of the sample in the calculation, while the actual QD height of the QD structure is expected to be higher because of the indium decomposition from the $In_xGa_{1-x}As$ barrier layer during the growth. This indium decomposition is depended on the $In_xGa_{1-x}As$ barrier layer's indium composition, which is described below. This explains the reduction of the difference between the calculated and measured peak positions of the sample when the indium content of the $In_xGa_{1-x}As$ barrier layer is increased as shown in Figure 21.

(III) Partially decomposition of the $In_xGa_{1-x}As$ barrier layer during the MOVPE growth of the $In_xGa_{1-x}As$ /InAs/$In_xGa_{1-x}As$/InP QD structure effectively increases the height of the InAs QDs [48, 49], which results in a redshift of the QDs' PL emission peaks. This indium decomposition from the $In_xGa_{1-x}As$ upper barrier layer depends on its indium composition. Higher the indium content of the $In_xGa_{1-x}As$ barrier layer, more indium decomposition during the QD structure growth, which makes the InAs QDs higher. This increasement of the QDs' height with the higher indium content of the $In_xGa_{1-x}As$ barrier layer effectively results in a larger the redshift of the QD structrue's emission wavelength.

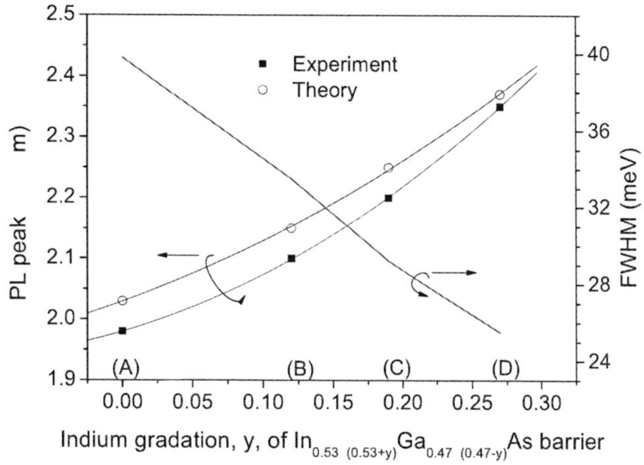

Figure 21. Experimental measured and the theoretical calculated PL peak positions and the measured FWHM of the PL spectra of the InAs QDs embedded in different $In_xGa_{1-x}As/InP$ matrixes.

SUMMARY

In summary, MOVPE growths of mid-infrared emission InAs QDs on InxGa1-xAs /InP matrix have been investigated in detail. Effects of the MOVPE growth conditions on formation of the InAs QDs have been studied. A new two-step growth of QDs based on self-assembly growth mode has been developed and optimized. Under the optimized growth conditions, very high dot density (above 2.5×1010 cm-2) of InAs QDs emitting at >2.3μm wavelength has been received. The surface coverage of the mid-infrared emission InAs QDs reaches ~60% and the FWHM of the QDs' PL spectrum is only 25.5meV. By embedding the InAs QDs in graded In0.53→(0.53+y)Ga0.47→(0.47-y)As barriers, we have demonstrated the InAs QDs emitting in mid-IR range at the longest wavelength of 2.35 μm measured at 77 K. The InAs QDs structures' emission wavelength has been extended by increasing the indium content of the graded InxGa1-xAs barrier layers of the InAs QD structure. Compared with that of the InAs QDs embedded in lattice matched In0.53Ga0.47As barriers, the emission wavelength of the InAs QDs embedded in In0.53→(0.53+y)Ga0.47→(0.47-y)As graded barriers achieves much longer. Redshift of emission wavelength of the InAs QDs structures depends on the indium content gradation, y, of the In0.53→(0.53+y)Ga0.47→(0.47-y)As graded barrier layers. The results show guidance for further development of mid-IR QD lasers based on InP substrates.

ACKNOWLEDGEMENT

This research was sponsored by the Science and Engineering Research Council (SERC) of Agency for Science, Technology & Research (A*STAR), Singapore (SERC Grant No. 052 101 0105) and the Academic Research Fund (Grant No. RGM 23/06) of Nanyang Technological University, Singapore.

REFERENCE

[1] Arakawa and S.Sakaki, *Appl. Phys. Lett.*, 40, (1982) 939.
[2] M. Asada, Y. Miyamoto and Y. Suematsu, *IEEE J. Quantum Elec.*, QE-22, (1986)1915.
[3] R. H. Wang, A. Stintz, P. M. Varangis, T. C. Newell, H. Li, K. J. Malloy, and L. F. Lester, *IEEE Photonics Technol. Lett.* 13, (2001) 767.
[4] R. Schwertberger, D. Gold, J. P. Reithmaier, and A. Forchel, *IEEE Photonics Technol. Lett.* 14, (2002) 735.
[5] Y. Qiu, D. Uhl, R. Chacon, and R. Q. Yang, *Appl. Phys. Lett.* 83, (2003)1074.
[6] Bradshaw J L, Yang R Q, Bruno J D, Pham J T, Wortman D E and Tober R L, *Appl. Phys. Lett.* 81 (2002) 397.
[7] Sun Z Z, Wu J, Liu F Q, Xu HZ, Chen YH, Ye X L, Jiang W H, Xu B and Wang Z G, *J. Appl. Phys.* 88 (2000) 533
[8] Zongyou Yin, Xiaohong Tang, Deny Sentosa and Jinghua Zhao, *Nanotechnology,* 17, (2006) 1646
[9] Qiu Y M and Uhl D, *Appl. Phys. Lett.* 84 (2004)1510.
[10] Landi S M, Pires M P, Tribuzy C V-B, Souza P L, Marega E Jr, Silva A G and Guimaraes P S S, *Phys. Status Solidi* 8 (2005) 317.
[11] Qiu Y M and Uhl D, *Appl. Phys. Lett.,* 84 (2004) 263.
[12] Qiu Y M and Uhl D, *J. Cryst. Growth,* 257 (2003) 225.
[13] Zhang Z H, Pickrell G W, Chang K L, Lin H C, Hsieh K C and Cheng K Y, *Appl. Phys. Lett.* 82 (2003) 4555.
[14] Li H X, Daniels-Race T and Hasan M A., *Appl. Phys. Lett.* 80 (2002) 1367.
[15] Z. H. Zhang, and K. Y. Cheng, *Appl. Phys. Lett.* 83,(2003) 3183.
[16] Y. Maeda, *Phys. Rev. B,* 51, (1995)1658.
[17] J. Johansson, and W. Seifert, *J. Cryst. Growth,* 234, (2002) 132.
[18] F. K. LeGoues, M. C. Reuter, J. Tersoff, M. Hammar, and R. M. Tromp, *Phys. Rev. Lett.*, 73, (1994) 300.
[19] T. T. Ngo, P. M. Petroff, H. Sakaki, and J. L. Merz, *Phys. Rev. B,* 53, (1996) 9618.
[20] F. M. Ross, J. Tersoff, and R. M. Tromp, *Phys. Rev. Lett.,* 80, (1998) 984.
[21] J. G. Cederberg, F. H. Kaats, and R. M. Biefeld, *J. Cryst. Growth,* 261 (2004) 197.
[22] R. L. Sellin, I. Kaiander, D. Ouyang, T. Kettler, U. W. Pohl, D. Bimberg, N. D. Zakharov, and P. Werner, *Appl. Phys. Lett.,* 82,(2003) 841.
[23] J. Tatebayashi, M. Nishioka, and Y. Arakawa, *Appl. Phys. Lett.,* 78,(2001) 3469.
[24] S. C. Saxena, and V. K. Saxena, *J. Phys. A: Gen. Phys.,* 3,(1970)309.
[25] A. A. El-Emawy, S. Birudavolu, P. S. Wong, Y.-B. Jiang, H. Xu, S. Huang, and D. L. Huffaker, *J. Appl. Phys.,* 93 (2003)3529.
[26] C. W. Snyder, J. F. Mansfield, and B. G. Orr, *Phys. Rev. B,* 46 (1992) 9551.
[27] C. C. Hsu, Y. F. Yang, H. J. Ou, E. S. Yang, and H. B. Lo, *Appl. Phys. Lett.*, 71 (1997)3248.
[28] Xiaohong Tang , Zongyou Yin, Jinghua Zhao and Sentosa Deny, *Nanotechnology*, 17,(2006) 295.
[29] M. Kastner, and G. Voigtlander, *Phys. Rev. Lett.*, 82,(1999) 2745.
[30] Y. Chen, and J. Washburn, *Phys. Rev. Lett.*, 77,(1996) 4046.

[31] A. A. El-Emawy, S. Birudavolu, S. Huang, H. Xu, and D. L. Huffakera. *J. Cryst. Growth*, 255, (2003) 213.

[32] A. V. Osipov, F. Schmitt, S. A. Kukushkin, and P. Hess, *Appl. Surf. Sci.*, 188,(2002) 156.

[33] V. G. Dubrovskii, G. E. Cirlin, and V. M. Ustinov, *Phys. Rev. B.* 68, (2003) 75409.

[34] Yin, Z. Y.; Tang, X. H.; Zhao, J. H.; Deny, S. *J. Appl. Phys.* 99 (2006)124306.

[35] M. C. Chen, H. H. Lin and C. W. Shie, *J. Appl. Phys.,* 83,(1998) 3061.

[36] R. Leon, C. Lobo, A. Clark, R. Bozek, A. Wysmolek, A. Kurpiewski and M. Kaminska, *J. Appl. Phys.,* 84,(1998) 248.

[37] Y. Horikoshi, M. Kawashima, and Hiroshi Yamaguchi, *Jpn. J. Appl. Phys*., 25,(1986) 868.

[38] G. S. Solomon, J. A. Trezza and J. S. Harrris, Jr., *Appl. Phys. Lett.,* 66,(1995) 3161.

[39] R. Leon, C. Lobo, J. Zou, T. Romeo, and D. J. H. Cockayne, *Phys. Rev. Lett*., 81, (1998)2486.

[40] C. S. Peng, Q. Huang, W. Q. Cheng, J. M. Zhou, Y. H. Zhang, T. T. Sheng and C. H. Tung, *Appl. Phys. Lett.,* 72,(1998) 2541.

[41] D. J. Eaglesham, F. C. Unterwald and D. C. Jacobson, *Phys. Rev. Lett*., 70, (1993)966.

[42] K. Nishi, H. Saito, S. Sugou, and J. S. Lee, *Appl. Phys. Lett*., 74, (1999) 1111.

[43] A E Zhukov, V M Ustinov, A R Kovsh, A Yu Egorov, N N Ledentsov, N A Maleev, A F Tsatsul'ikov, M V Maximov, Yu G Musikhin, N A Bert, P S Kop'v, D Bimberg, and Zh I Alferov, *Semicond. Sci. Technol*., 14,(1999) 575.

[44] R. Heitz, I. Mukhametzhanov, J. Zeng, P. Chen, A. Madhukar, and D. Bimberg, *Superlattices Microstruct*., 25,(1999) 97.

[45] F. Q. Liu, Z. G. Wang, J. Wu, B. Xu, Q. Gong, and J. B. Liang, *J. Appl. Phys*., 85,(1999) 619.

[46] K. H. Schmidt, G. Medeiros-Ribeiro, J. Garcia, and P. M. Petroff, *Appl. Phys. Lett*., 70,(1997) 727.

[47] Zongyou Yin, Xiaohong Tang, Wei Liu, Daohua Zhang, *J. of Appl. Phys*., 100,(2006) 033109.

[48] M. V. Maximov, A. F. Tsatsul'nikov, B. V. Volovik, D. S. Sizov, Yu. M. Shernyakov, I. N. Kaiander, A. E. Zhukov, A. R. Kovsh, S. S. Mikhrin, V. M. Ustinov, Zh. I. Alferov, R. Heitz, V. A. Shchukin, N. N. Ledentsov, D. Bimberg, Yu. G. Musikhin, and W. Neumann, *Phys. Rev. B*, 62,(2000) 16 671.

[49] W. H. Chang, H. Y. Chen, H. S. Chang, W. Y. Chen, T. M. Hsu, T. P. Hsieh, J. I. Chyi, and N. T. Yeh, *Appl. Phys. Lett*., 86,(2005) 131917-1-3.

In: Nanotechnology Research Developments
Editor: R. Jiménez-Contreras, pp. 277-294
ISBN: 978-1-60021-899-6

Chapter 7

NANOTECHNOLOGY BASED GAS SENSORS

John T.W. Yeow* and Yun Wang
Systems Design Engineering, University of Waterloo
200 University Ave West, Waterloo, Ontario, Canada, N2L 3G1

ABSTRACT

This chapter focuses on novel gas/chemical sensing systems that are based on nanotechnology. Recently, there has been an extensive interest in nanomaterials such as carbon nanotubes, nanowires, nanofibers, and nanoparticles. Their application on gas sensing systems has also been widely studied. Due to their unique micro/nano porous structures and ultra high surface-to-volume ratio, nanostructured materials have a high potential to enhance conventional gas sensors' performance, including the sensitivity, the operation temperature and the recovery time. In this chapter, various nanomaterials-based gas/chemical sensors are presented, including their structure, fabrication, and operation principles.

Gas sensing systems, or chemical sensors, are attracting tremendous interest because of their widespread applications in industry, environmental monitoring, biomedicine, pharmaceutics, and space exploration. Gas sensors with high sensitivity and selectivity are required for leakage detection of explosive gases such as hydrogen, and for real-time detection of toxic or pathogenic gases in industry. There is also a strong demand for the ability to monitor and control of our ambient environment, especially with the increasing concern of the globe warming. NASA researchers can also use similar high performance gas/chemical sensors for identification of atmospheric components of various planets. In addition, nerve agent sensing for homeland security is also at the centre of public concern [1].

Generally, there are several basic criteria for good and efficient gas sensing system: i) high sensitivity and selectivity; ii) fast response time and recovery time; iii) low analyte consumption; iv) low operating temperature and temperature independence; and v) stability in performances. Commonly used gas sensing systems include polymer gas sensors, semiconductor metal oxide gas sensors and surface acoustic wave sensors. Usually polymer

* E-mail: jyeow@engmail.uwaterloo.ca

gas sensors have limited sensitivity, while semiconductor metal oxide gas sensors require high temperatures (approximately 200-500oC). Since the most common gas sensing principle is the adsorption and desorption of gas molecules on sensing materials, it is quite understandable that by increasing the contact interfaces between the analyte and sensing material, the sensitivity can be significantly enhanced.

Recent development of nanotechnology has created huge potential to build highly sensitive, low cost, portable sensors with low power consumption. The extremely high surface-to-volume ratio and hollow structure of nanomaterials is ideal for gas/chemical molecules adsorption and storage. Particularly, the advent of carbon nanotubes (CNTs), which is, arguably, the most widely investigated nanomaterial, has fuelled the inventions of gas sensors that exploit CNTs' unique geometry, morphology and material properties. Upon exposure to certain gases, the change in CNTs' properties or in the sensing medium which is induced by the morphology of CNTs, can be detected by various methods. As a result, CNT-based gas sensing systems and the theoretical analyses of gas adsorption and collision effects on the nanotubes have been subjects of intense research. In addition, nanomaterials other than CNTs such as nanowires, nanofibers, and nanoparticles have also been investigated widely as potential materials for gas sensing applications. This chapter includes a broad but yet in-depth survey of nanomaterials that are used for gas sensing applications. The underlying sensing mechanisms and their corresponding advantages will also be discussed.

1. Carbon Nanotubes Based Gas Sensors

Since being discovered in 1991, CNTs have captured researchers' attention worldwidely. Considerable research have been conducted for new synthesis and preparation techniques, as well as to investigate the properties and potential applications of CNTs. In this section, a brief introduction of CNTs is presented. Illustration of the mechanism of gas molecules adsorption on nanotubes will be discussed and some CNT-based gas sensing systems that have been developed recently will be introduced.

1.1. Brief Introduction of CNTs

CNTs belong to the family of fullerene structures. There are two types of nanotubes: (i) single-walled carbon nanotubes (SWCNTs); and (ii) multi-walled carbon nanotubes (MWCNTs). A SWCNT can be considered as a one-atom-thick layer of graphite rolled up into a seamless cylinder with a diameter of several nanometers, and length on the order of 1-100 microns [2]. Similarly, MWCNTs consist of multiple layers of graphite wrapped up together to form a tube shape, sharing the same central axis. MWCNTs were first observed in 1991, by Iijima of NEC Laboratory in Japan when he saw fine threads in carbon-soot grown by arc-discharge [3]. In 1992, Ebbesen and Ajayan managed to grow MWCNTs at the gram level by the arc-discharge method [4]. In the following year, SWCNTs were first synthesized in small amount using the same method by Bethune et al. [5]. Since then, tremendous CNT related research, including both experimental and theoretical aspects have been conducted to better understand the growth of CNTs and their unique properties

1.1.1. Preparation of Carbon Nanotubes

There are three main methods to prepare CNTs: i) arc-discharge technique; ii) laser ablation technique; and iii) chemical vapor deposition (CVD) technique.

The carbon arc-discharge method involves two carbon electrodes as carbon source. When high DC voltage is applied between the carbon anode and cathode in a vacuum chamber, plasma of the insert gas (typically helium) is generated to evaporate the carbon atoms. The ejected carbon atoms are then deposited on the negative electrode to form CNTs. It is the principal method to produce high quality CNTs, both SWCNTs and MWCNTs, with nearly perfect structures.

The laser ablation technique requires a carbon target to be ablated by intense laser pulses in a furnace. In the presence of an inert gas and a catalyst, CNTs are formed and collected on a cold substrate. Both the arc-discharge and laser-ablation methods require high growth temperature, which is about 3000-4000oC for the evaporation of carbon atoms from solid carbon source [6].

In a CVD system, a gas hydrocarbon source (usually methane, acetylene or ethylene) flows into the reaction chamber and molecules are broken into reactive species at the temperature range of 550-1000oC. The reactive species react with the catalysts (metal nanoparticles) that are coated on the substrate, leading to the formation of CNTs.

Compared with the first two techniques, CNTs can be synthesized at relative low temperature using CVD method. Therefore, this technique is more efficient and allows for scale-up growth of SWCNTs. By modification and calculated control of the growth parameters, vertically aligned MWCNTs growth can be achieved by CVD technique. This may enhance CNTs electronic properties in different applications. High quality SWCNTs can also be obtained by optimization of the catalysts. However, one of the main disadvantages of CVD technique is the relative high defect densities in MWCNTs, which can be attributed to the lack of sufficient thermal energy [6].

In all of these three growth methods, CNTs come with a number of impurities, which may have negative effects on CNTs' inherent properties. In order to make CNT-based devices more efficient and consistent, purification is an important process to obtain high purity nanotubes. Generally, the most commonly observed impurities are carbonaceous impurities and metallic impurities. The carbonaceous components are byproducts of the growth process and the metallic impurities are residual catalysts. To eliminate the carbonaceous impurities, the main method is oxidation. Two approaches are developed: i) gas phase purification, which involves high temperature and has low yield production; and ii) liquid phase purification, where nanotubes are washed in acidic solutions. Nitric acid, sulphuric acid, and their mixture are all commonly used for liquid purification [7]. Metallic impurities can be removed by heating the samples up to the evaporation temperature of the impurity. With the development and modification of different purification procedure, so far CNTs samples with a purity of up to 99.6% can be obtained [8]. The quality of purification can be inspected by microscopic techniques such as scanning electron microscopy or spectroscopic techniques such as Raman spectroscopy [9].

1.1.2. Properties of Carbon Nanotubes

The structure of carbon nanotubes provides them with inherently unique electrical, physical and chemical properties. In terms of electrical properties, CNTs can be either metallic or semiconducting, depending upon the tube diameter and the chirality (the direction in which the graphite sheet is rolled to form the tube) [10]. The chirality is usually represented by an integer pair (*n*, *m*). Nanotubes with $n - m = 3j$ (j being a nonzero integer) are metallic while all the others are semiconducting. The dielectric property of nanotubes is highly anisotropic due to their nearly one dimensional structures, which may enable nanotubes to carry high currents with negligible heating effect[11].

For mechanical and thermal properties, CNTs are the strongest and stiffest fibers that are known currently, because the C-C bond in graphite is one of the strongest in nature. Due to the limitation of measure techniques and the nonuniformity between CNTs prepared by different method, it is very challenging to obtain the accurate value of the Young's modulus and tensile strength. Lourie and Wagner reported the Young's modulus of 2800–3600 GPa for SWCNTs and 1700–2400 GPa for MWCNTs [12]. Yu et al. found the tensile strength of SWCNT bundles in the range of 13 to 52 GPa, while the value for MWCNTs was in the range of 11 to 63 GPa [13]. CNTs have high thermal stability in both vacuum and air. The thermal conductivity is dependent on the chirality. Theoretically, they are expected to be good thermal conductors along the tube direction and good insulators laterally to the tube axis.

1.2. Gas Adsorption on Carbon Nanotubes

The extremely high surface-to-volume ratio and hollow structures of CNTs create extraordinarily large adsorption capacity for gas molecules. Therefore, CNTs are considered to be very good candidates for the application of gas storage and gas sensing. However, the potential application of CNTs for gas/chemical sensing did not gain serious attention until 2000. Kong et al. demonstrated that upon exposure to a very small concentration of certain gas molecules (NO2 and NH3), the electrical resistance of a single semiconducting SWCNT can be dramatically increased or decreased [14]. Inspired by this finding, the effect of O2, Ar, H2 and other gases on the field emission or electrical property of SWCNTs and MWCNTs are widely studied in the following years [15]. Simultaneously, there were also significant developments on theoretical and simulating studies on the interaction between gas molecules and CNTs [15-21].

In a bundle of SWCNTs, there are generally four distinct sites for gas molecules to be adsorbed onto: i) the external *surface* of the bundle; ii) the *groove* formed at the contact between adjacent tubes on the outside of the bundle; iii) the interior *pore* of individual tubes; and iv) interstitial *channel* formed between three adjacent tubes within the bundle [16]. The gas adsorption on these sites is decided by the binding energy of the gas molecule as well as the site availability. Some of these sites may not be available for certain gases because of the gas molecules' dimension and the site diameter. The interior *pore* is only accessible when the SWCNT is uncapped or has defects on the tube walls. Zhao et al. studied the adsorption of various gas molecules (NO2, O2, NH3, N2, CO2, CH4, H2O, H2, Ar) on single SWCNT and SWCNT bundles using first principles method. Calculations of the binding energy (E_B) for hydrogen adsorption have showed that E_B for channels and grooves are larger than that for pores and surfaces; however, pores and surface have much larger available adsorption area than the other two (as shown in Figure 1). Similar results were obtained for other gas molecules that they studied as well. The larger adsorption energy on groove and channels can be explained by the fact that there are more carbon molecules within a specific volume along these two sites [17]. It is also demonstrated by theoretical studies that gases such as NO2, O2 and NH3 have larger binding energies with SWCNT than other gases, which is coincident with experimental results that CNTs are more sensitive to these three gases.

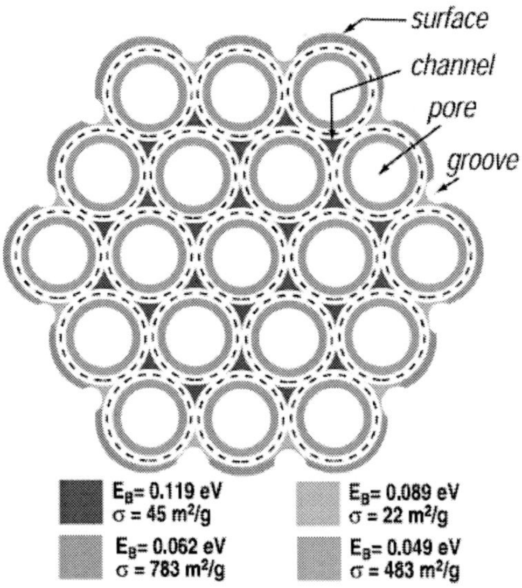

Figure 1. Schematic structure of a SWCNT bundle showing different sites for gas adsorption [16], Copyright Elsevier.

Gas molecule adsorption has effects on both thermoelectric power (S0) and conductance (or resistance) of CNTs. In general, the change in conductance is easier to detecte, so it is more commonly used as the sensing parameter. It is believed that changes in the electronic property of semiconducting SWCNTs are caused by the charge transfer between nanotubes and the adsorbates, which can act as dopants and consequently shift the Fermi energy of the nanotubes. As a result, the nanotubes can be either n-type or p-type doped, depending on whether the adsorbed gas is a charge-donator or a charge-acceptor. O2 and NO2, for example, are charge acceptors with negative charge transfer (-0.06 to -0.14 electron per molecule) and relatively large binding energy compared to other gases [9]. This is consistent with experimental results indicating that the conductance of semiconducting SWCNTs increases significantly upon exposure to O2 and NO2 [18]. Gas sensing properties of MWCNTs and MWCNT films are also reported but much less theoretical work have been done due to the complexity in their tube structures. However, the gas adsorption and charge transfer mechanism may also work for MWCNTs, but metallic nanotubes may play a more important role in MWCNT bundles and films.

1.3. Gas Sensing Devices Based on Carbon Nanotubes

1.3.1. Semiconducting SWCNT (S-SWCNT) FET for Gas Sensing

As addressed in former section, SWCNTs can be either semiconducting or metallic. If two metal contacts are connected to each end of an individual S-SWCNT, the metal/S-SWCNT/metal device exhibits *p*-type transistor characteristics. Based on this, Kong and his coworkers [14] have built an S-SWCNT transistor for chemical sensing. The device was fabricated by controlled CVD growth of individual SWCNTs on SiO2/Si substrates with patterned catalyst islands. Polysilicon in the back side acted as gate of the transistor. The schematic structure is shown in Figure 2. Figure 3 shows the conductance versus time curves of the transistor upon exposure to NO2 and NH3 at room temperature. A sharp increase in conductance of the S-SWCNT FET by about three orders of magnitude was observed after introducing 200 ppm of NO2 into the testing chamber (Figure3a). The response time was about 2-10 second, and the sensitivity (the ratio between resistance after and before gas exposure) was about 100 to 1000. When the same sample was exposed to a 1% NH3 flow after recovery, the conductance of the device decreased dramatically within 1 to 2 minute, and the sensitivity was about 10 to 100 (Figure 3b).

Takao Someya et al. characterized the influence of alcohols adsorption on the resistance of SWCNTs using the same transistor structure [22]. The measurements were carried out under flow of dry nitrogen with and without addition of alcoholic vapor. Figure 4 shows the drain current versus time curves of the device upon exposure to different saturated alcoholic vapor at room temperature (except for tertiary-butanol) with Vg = -20V. For the vapor of methanol, tertiary-butanol, 1-propanol, 2-propanol, and ethanol, the response time were within 5-15 min and the current reduction were in the range of 50%-90%. Short recovery time (less than few seconds) and excellent repeatability can be obtained if the gate voltage is released briefly and then restored before the next cycle, and no degradation of the sensor performance was observed over 1 hour's operation. Moreover, the different response characteristic of the device to different alcoholic vapor also suggests its potential application on the identification of gas sources.

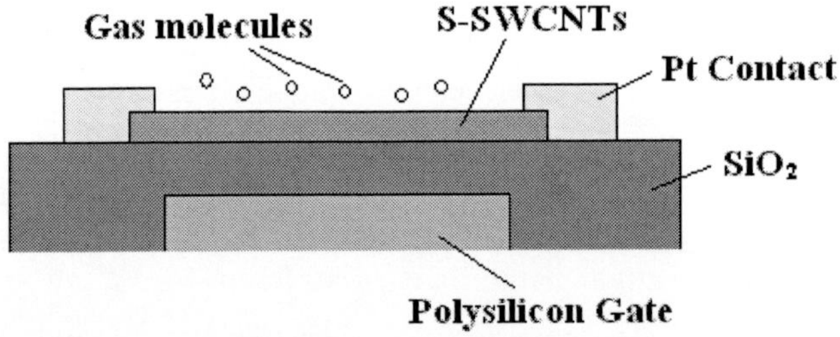

Figure 2. Schematic illustration of Single semiconducting SWCNT FET [14]. Reprinted with permission from AAAS.

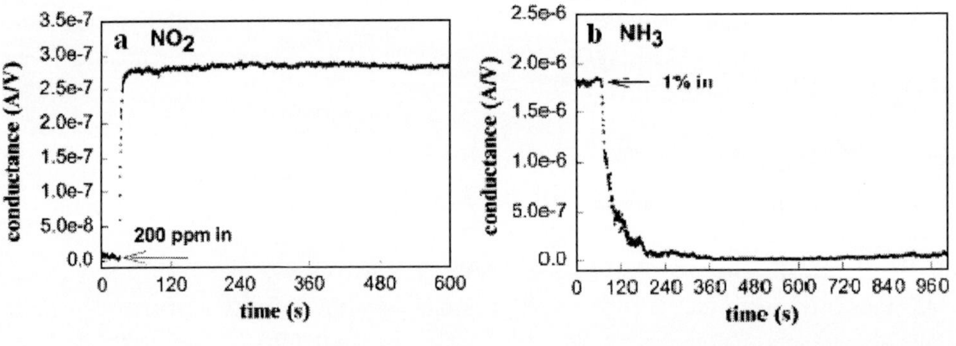

Figure 3. Change of sensor's conductance when exposed to (a) NO2 and (b) NH3 gases [14]. Reprinted with permission from AAAS.

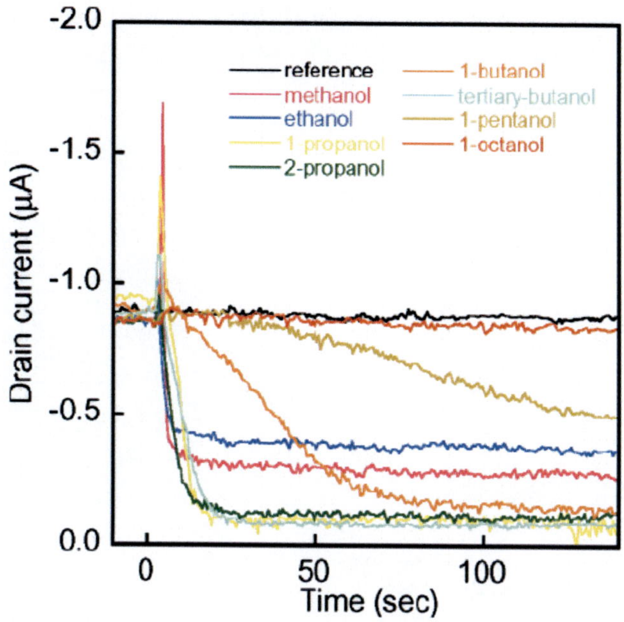

Figure 4. Drain currents a function of time upon exposure to saturated vaporof various kinds of alcohols: methanol, ethanol, 1-propanol, 2-propanol, 1-butanol, tertiary-butanol, 1-pentanol, and 1-octanol. Reprinted with permission from [22]. Copyright (2003) ACS

Although these individual semiconducting SWCNT transistors show pronounced response upon exposure to gas molecules at room temperature, it is generally very difficult to obtain pure and individual semiconducting SWCNTs. CNTs samples grown by all the three techniques mentioned earlier are mixture of both metallic and semiconducting nanotubes, and there is no efficient technique to realize the selective growth of semiconducting or metallic nanotubes up to now. This is one of the main issues related to S-SWCNT FET devices.

1.3.2. Resistance Gas Sensors Based on CNT Films

In contrast with individual nanotubes, preparation of CNTs mat or CNTs films are much less complex and more efficient. Therefore, they are able to provide much higher yield. Based on the change in electrical properties of CNTs by gas molecule adsorption, CNTs resistance gas sensors for different gases have been successfully developed by researchers with simple fabrication processes [23-26].

Jing Li et al. developed a resistance gas sensor by simply casting SWCNTs on an interdigitated electrode (IDE) [8], which provided a large sensing area and sufficient contact between the electrodes and coated CNTs. The electrodes were fabricated by photolithography and evaporation of Ti and Au (60 nm in thickness together) on silicon oxide (Figure 5(a)). As-grown SWCNTs were purified with acid first and then by air oxidation before integrated with IDE. As a result, the final SWCNTs had a relative high purity up to 99.6%, and the impact of impurities on the sensor's characteristic was minimized. The purified nanotubes were then dispersed in dimethylformamide (DMF) and drop-deposited onto the electrode area. A network of nanotubes subsequently formed after the evaporation of DMF, as shown in figure 4(b). Upon exposure to NO2, the conductance change of the sensor showed a linear response to the gas concentration with a detection limit of <4 ppb. The recovery time was about 10 hours but can be intensively shortened by UV light treatment. The sensor's response to organic vapor such as benzene, acetone, and nitrotoluene were also tested.

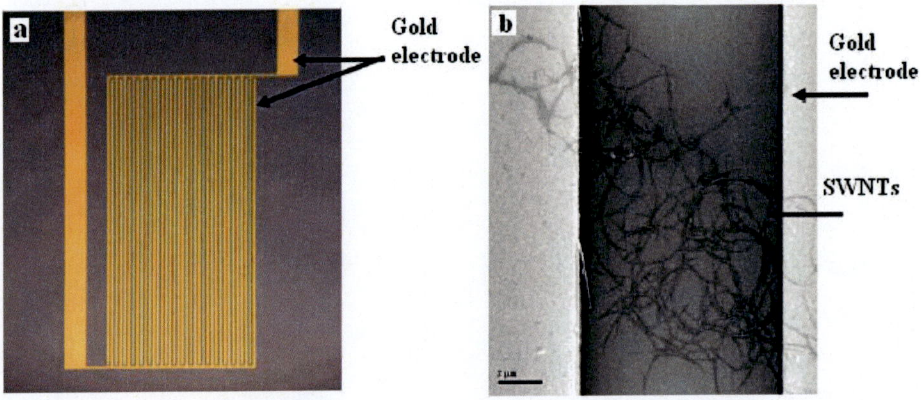

Figure 5. (a) interdigitated electrodes; (b) SEM image of SWCNTs across two gold electrodes. (Reprinted with permission from [8]. Copyright (2003) ACS)

Compared with SWCNTs, the mechanism of MWCNTs' response to gas adsorption is more complicated due to the multi-layer tube structure. However, they also show high sensitivity to specific gases experimentally. A MWCNTs resistance gas sensor was developed by laterally growth of nanotubes between two electrode arrays using CVD [23]. The growth of MWCNTs was guided by electrical field applied between the electrode arms to achieve the nano-wiring. Figure 6(a) shows the SEM image of aligned CNTs grown on silicon dioxide between electrodes, and the insert figure shows the image of the developed device. The resistance of the sensor increased when cycling the test chamber atmosphere from air to vacuum (shown in Figure 6b). This phenomenon is consistent with the fact that CNTs can be *p*-doped by oxygen in air because O2 acts as charge acceptor to the nanotubes. The sensor also showed good response to NH3 with almost linear relationship between the resistance and the gas concentration. By heating the sensor after exposure to ammonia, the recovery time can be intensively shortened from ~20h to 3h. This improvement in performance can be explained by the accelerated gas molecules desorption due to increased the ambient temperature.

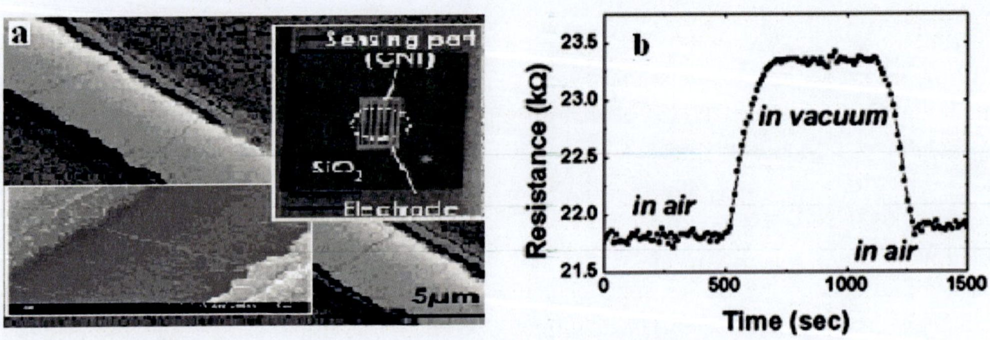

Figure 6. (a) SEM image of aligned CNTs resting on SiO_2 surface after directed growth; (b) electrical resistance of sensor by cycling chamber atmosphere from air to vacuum [23], Copyright Elsevier.

In general, slow recovery time is a common problem related to gas sensing based on adsorption and desorption of gas molecules. To overcome this issue, Woo-Sung Cho et al. integrated a micro heater under the MWCNTs sensing layer of a resistance gas sensor [24]. As shown in Figure 7(a), the sensor consists of a micro heater, an insulating layer, and a pair of electrodes with MWCNTs sensing film grown on top of them. The function of the heater was to accelerate the gas desorption rate during operation. Figure 7(b) shows the resistance change of the sensor in vacuum with and without heat treatment. The recovery time decreased significantly from approximate 6 hours to 10 minutes (the recovery time was defined as the time for 90% of total resistance change). When tested under cycles of NO2 exposure and evacuation (100 ppm), the sensor showed excellent reproducibility and stability with the resistance change being about 220% (Figure 7c).

The main advantage of resistance gas sensors is their simple fabrication process. By using IDE or electrode arm arrays, good and efficient contact between CNTs and the electrodes are obtained to ensure large signals. All the sensors mentioned above operated at room temperature and fast recovery time can be achieved by UV light or thermal treatment.

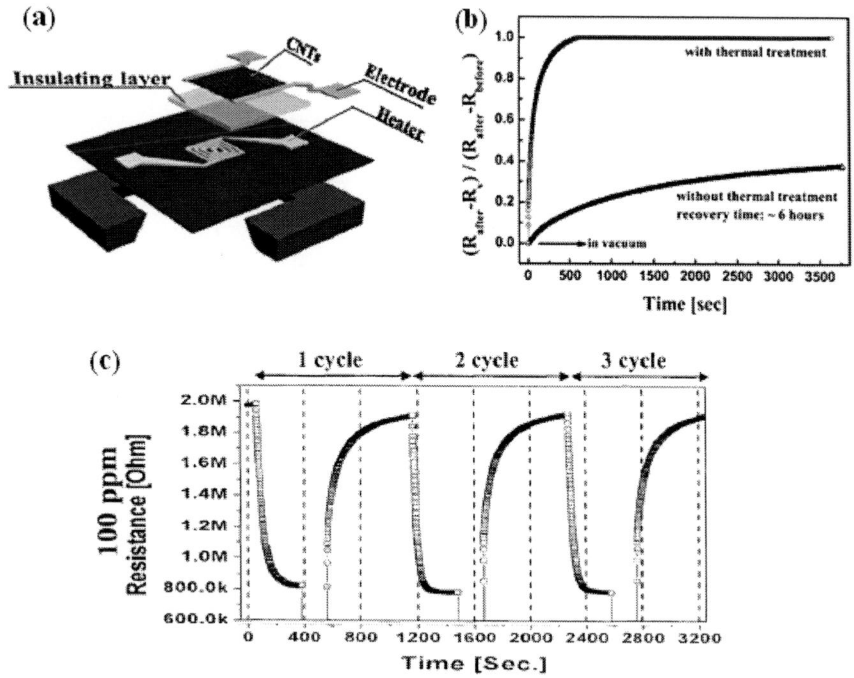

Figure 7. (a) Schematic of the sensor; (b) Recovery time of sensor with thermal treatment and without thermal treatment in vacuum; (c) Resistance changes of the sensor as a function of time when exposed to three cycles of a sequence of NO [24] (@ [2005] IEEE).

1.3.3. CNTs/Polymer Gas Sensors

Organic polymers are one of the principal materials applied in gas sensing systems. Some conducting polymers can behave like semiconductors due to their heterocyclic compounds which display physicochemical characteristics. As a result, reversible changes in the sensing layer's conductivity can be detected upon polar chemicals' adsorption on the surfaces at room temperature [27]. This effect is believed to be caused by the charge transfer between gas molecules and the polymer or the polymer film's swelling [28]. This sensing response has intensively driven the motivation to develop high sensitive and selective gas/chemical sensors by tailoring the compounds of different organic polymers.

Non-conductive polymers can also be used as gas sensors. For some polymers, the bulk dissolution of gas into the film can cause changes of their physical properties [28]. Since most physical property changes are not readily detectable; researchers have tried to integrate polymers with other functional sensing elements to achieve easier detection process. For example, a recent experimental and theoretical study on the chemical sensing characteristic of polymer coated SAW and STW resonators was reported by Ivan D. Avramov et al. [29,30].

Another promising sensing material based on polymers is conducting particles–insulating matrix composition. The polymer acts as the insulating matrix while dispersed conducting particles provides the conducting path for sensing [31]. Due to adsorption of interested analyte, there are volumetric changes of the matrix polymer. This can lead to a distinct change in percolation-type conductivity around a critical composition of the material, which is known as "percolation threshold". Generally, the percolation threshold is dependent on the shape of the conducting particle. Composite consisted of particles with higher aspect ratio shows lower threshold and higher sensitivity [32].

CNTs, with almost one dimensional thread like structure and good conductivity, are ideal as the dispersed particles in this conducting particles–insulating matrix composition for gas sensing systems. Seong M. Cho et al. investigated the chemical sensing characteristic of SWCNT-Ethyl cellulose (EC) composite [31]. To get uniform dispersion of CNTs and stable composite, the material was prepared carefully with several steps. Poly (p-phenylenevinylene-co-2,5-dioctoxy-m-phenylenevinylene) (PmPV) was used as the dispersant of the nanotubes and chloroform was used as the media. The nanotubes sol was sonicated for 4 hours before mixed with EC, and then the mixture was sonicated for another 6 hours. The mixed sol were sprayed on interdigitated electrodes and annealed at 80oC for 24 hours to form the thin sensing film. The sensor's response to benzene and ethanol was measured separately as a function of gas concentration and SWCNTs' content in the composite. The change of resistance showed almost linear relationship with concentration. The maximum sensitivity was 9% for the film with 5wt.% SWCNT contents when exposured to 5,000ppm of benzene. The sensor's response to ethanol was less obvious and with some noise, however it showed good linearity within the concerned range with short response and recovery time. Based on the same idea, Biju Philip et al. developed composite thin film of polymethylmethacrylate (PMMA) with both MWCNTs and oxidation modified MWCNTs dispersed in it for gas sensing. The resistance change of the composites was evaluated upon exposure to different gases including dichloromethane, chloroform and acetone et al. A 102-103 order of magnitude increase in resistance of the surface-modified-CNT/PMMA composite was obtained at room temperature with fast response and recovery time [32].

In addition of CNT/polymer composite, the sensing characteristics of polymer-coated CNTs film to specific gases were also investigated. In general, CNTs do not have sensing response to all gases and vapours but only the ones with high adsorption energy or that can interact with them. Therefore, coating or doping of an enhance element on CNTs may broaden the application range. Based on IDE SWCNTs resistance gas sensor that they developed, Jing Li et al. studied the sensing character of the same structure but with polymer coatings on CNTs [33]. Chlorosulfonated polyethylene was dissolved in tetrahydrofuran (THF) solvent as a coating solution for Cl2 sensing, and Hydroxypropyl cellulose was dissolved in chloroform as coating to detect HCl. The resistance of polymer coated SWCNT film changed upon exposure to Cl2 and HCl while uncoated pure SWCNTs showed no sensing signal. The sensing response demonstrates a huge potential of using modified CNTs gas sensing materials for a broad range of gases and chemical vapors.

1.3.4. CNTs Enhanced Ionization Chamber for Gas Sensing

All gas sensing systems introduced in the previous sections are based on chemical or physical adsorption of analytes on the sensing layers. Therefore, they are limited by the low adsorption energy and poor charge transfer of the gas species with CNTs. For ionization gas sensors, the sensing principle is based on the ionization characteristics of different gases, which is independent of the chemical interactions between gases and the sensing materials. Generally, ionization of neutral gas is carried out by collisions of molecules with electrons that are accelerated by the electrical field between the anode and cathode. The issue related to conventional ionization gas sensors is their considerable high breakdown voltage (approximately thousands of volts). This is quite inefficient and risky in operation. CNTs, due to their extremely sharp tip structure, can induce a large field enhancement factor and therefore intensively increase the electric field around the tips to initiate corona discharge at low voltage [34].

A miniaturized gas ionization sensor with aligned MWNT film as anode was developed by Ashish Modi et al. in 2003 [35]. Figure 8 (a) shows a diagram of the sensor's structure. It consistes of MWNT film anode, Al plate cathode and a 150μm thick glass insulator between them. The vertical aligned MWNT film was prepared by CVD on SiO2 substrate with tip diameter in the range of 25~30nm and 30μm in length (Figure8 (c)). As shown in Figure8 (b), the current between anode and cathode was tested under increasing DC voltage. By using CNTs film as anode, the breakdown voltage of air decreased dramatically from about 1000V to 346V. Different gases including NH3, CO2, N2, O2, He, Ar and air were tested and showed distinct and precise breakdown voltage (Figure 9).

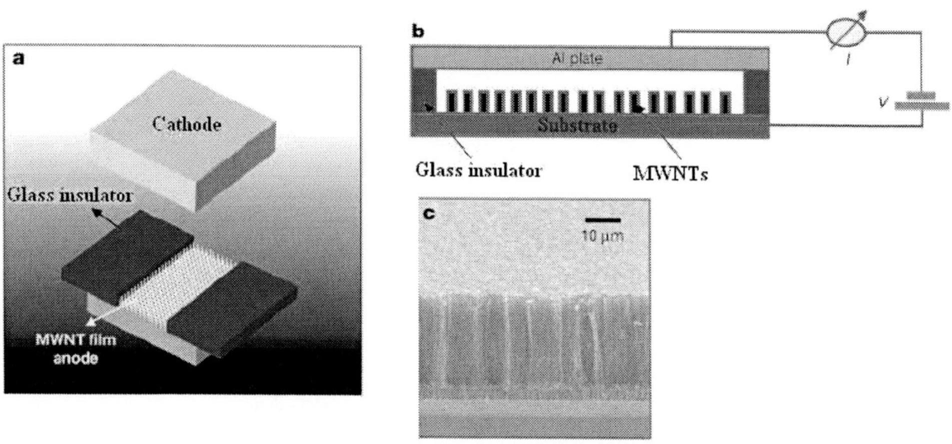

Figure 8. (a) Schematic diagram of the nanotube sensor device; (b) Diagram of actual test set-up; (c) SEM micrograph of vertically aligned MWCNT film used as the anode. Reprinted by permission from Macmillan Publishers Ltd: [23], copyright (2003).

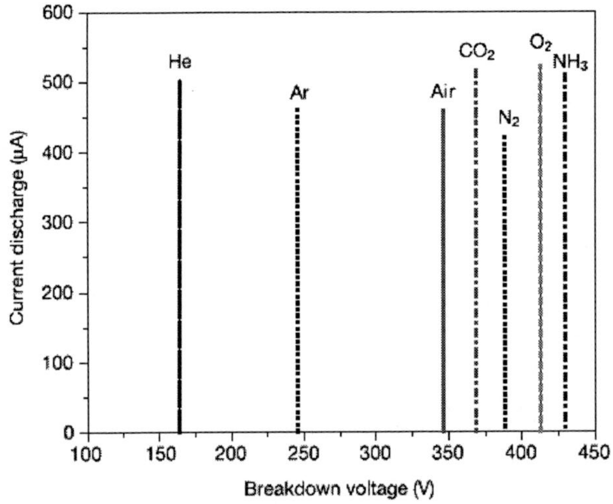

Figure 9. Current–voltage curves for electrical breakdown for NH3,CO2, N2, O2, He, Ar and air. Reprinted by permission from Macmillan Publishers Ltd: [23], copyright (2003).

The distinct breakdown voltage signatures indicated its potential for gas identification within a mixture of gases. The discharge current of each gas showed almost linear relationship with the gas concentration, but the breakdown voltage was rarely affected by the concentration. This can be explained by the fact that the discharge current is dominated by the number of molecules per unit volume and the breakdown voltage depends mainly on the intensity of electric field and the bonding energy of the gas molecules.

Nanotubes are good electron emitter due to their sharp tip curvature and low electron escaping work function. If CNT films are deposited as cathode of the device instead of anode, the breakdown process will be initiated by a negative corona. Field emission may occur under low pressure to enhance the ionization, and lower breakdown voltage can be obtained as a result [36]. Another important factor that dominates the ionization characteristic is the distance between anode and cathode. According to Paschen's curve, a certain gap distance is required for breakdown phenomenon to be induced, and the breakdown voltage decreases by bringing the electrodes closer. In the capacitor-architecture device, it is very difficult to control the electrode gap under 10μm to gain a safe operation voltage. Zhongyu Hou et al. [34] introduced a novel hollow slot electrode system with CNT sidewalls and electrode spacing of 6-12μm for gas ionization sensors, which dramatically decreased the threshold voltage to approximately 5V-40V.

2. Gas Sensors Based on Other Nanotechnologies

Besides carbon nanotubes, other nanomaterials such as nanowires, nanofibers, nanoparticles, and nano porous structures have also attracted significant interests due to their special morphology, property and the potential applications on various systems. With the development of synthesize methods, nano structures of metal, metal oxide and polymers have been obtained successfully and their properties have been intensively investigated.

In gas sensing systems that are based on the adsorption and desorption mechanism, the interface or interaction area between the sensing layer and analytes is a key factor to gain a large signal and to achieve a lowe detection limit. Compared to bulk or film materials, nanomaterials possess ultra high surface to volume ratio and porous structures. This can provide much larger sensing surface and easier diffusion of gas molecules through the sensing material. For sensing materials whose reaction with gas molecules mainly occurs on the grain boundaries, nanocrystalline structure can intensively increase their sensitivity. And in terms of functional modification of some gas sensing materials, the ultra small size of nanoparticals permits more efficient dispersion into other sensing elements. All of these advantages make nanomaterials excellent candidates in gas sensing systems. In this section, we will introuduce the application of nanomaterials other than CNTs on gas/chemical sensing systems.

2.1. Nanosized Semiconductor Metal Oxide Gas Sensors

Semiconductor metal oxide (SMO) gas sensors are one of the most commonly used chemical sensing systems due to their advantages of low cost, simple structure, fast response and recovery time, and insensitivity to moisture and temperature. Their sensing layer generally consists of polycrystalline metal oxide films, such as SnO2, ZnO, TiO2, and CeO2 et al. SMO gas sensing systems can be widely used for the detection of various gases such as O2, CO, NH3, NO, H2 et al. The detection principle of SMO gas sensors is based on the variations of energy barriers in presence of reducing or oxidizing gases, which is caused by the change of the depletion layer at the grain boundaries. The conductivity of the polycrystalline metal oxide, for example, increases when the magnitude of these barriers is reduced by adsorbed gases [37]. One of the main issues with SMO gas sensors is their high operation temperature (300~500℃) [38]. It is believed that by reducing the grain size of the metal-oxide semiconductor materials, the detection threshold can be lowered. As a result, a higher sensitivity and lower operation temperature can be achieved [40]. This is mainly due to the increase of surface area to volume ratio of the particles and the formation of grains with dimensions in the same order of the depletion layer depth [41]. Therefore, synthesize of nanosized metal oxide materials and their sensing characteristic has attracted extensive research interest.

Zinc oxide (ZnO) is a multifunctional semiconductor oxide material that exhibits piezoelectric and pyroelectric properties. It has wide direct band gap of 3.36 eV and large exciton binding energy (60 meV), which makes it a promising smart material for diverse optical and electronic devices. For SMO gas sensing devices, nanostructured ZnO has been synthesized and their sensing characteristics have been investigated recently. The properties of ZnO film depend strongly on the stoichiometry, the nature and quantity of impurities, and the growth parameters. By doing gowth modifications, ZnO with better sensing properties can be achieved. T.Shishiyanu et al. [41] developed Sn and Cu doped nanostructured ZnO thin film for the detecting of volatile organic compounds (VOC) gas (ethanol and methanol). Nanostructured thin films with various %Sn and %Cu were prepared on a glass substrate by chemical deposition, followed by rapid photothermal processing (RPP) at 350-550℃ for 20s to form homogeneous crystal. The synthesized particle size was in the range of 100~150nm. The sensors sensing characteristics were detected at different temperatures and gas concentrations. When the synthesized thin films were exposed to VOC gas, the conduction electrons were consumed. And the subsequent reactions resulted in the increase of the barrier height and decrease of the surface conductance. The highest sensitivity (defined as resistance change over original resistance) was approximately 28%/ppm when exposed to ethanol at room temperature. The Sn and Cu concentration of this film was 2% and 1% respectively. It was demonstrated that the Sn and Cu impurities improved the ZnO film's sensitivity, and shortened the response time as well. The temperature where the maximum sensitivity was shown was determined by the grain size of the nanostructure.

ZnO nanowires were also synthesized by researchers for gas sensing application based on their semiconducting property. Zhiyong Fan [39] built a field-effect transistor with single-crystalline ZnO nanowires that were grown via vapor trapping chemical vapor deposition method. The as-synthesized nanowires were suspended in alcohol and deposited onto Si substrates with a SiO2 insulating layer on the top. The metal contacts were fabricated using photolithography and ebeam evaporation. Figure 10(a) illustrates the AFM image of the device and the schematic of the circuit. The device's sensing characteristic to various gases including O2, NO2, NH3, and CO were studied. This nanowire FET sensor showed oxidizing property to O2, NO2, and NH3 with decrease conductance at room temperature. However, due to the shift in Fermi energy at elevated temperature, it showed reducing sensing property for NH3 at 500K. Figure 10(b) shows the sensor's response to different concentration of NO2. It was also demonstrated that by reducing the nanowires diameter, the material's sensitivity to O2 increased significantly.

Yin Hu and her coworkers prepared nanosized-SrTi1±xO3−δ series powders for oxygen sensing. The nanostructured powder was synthesized through high energy ball-milling technique. The powder were formed into pastes using organic binding vehicle and then screen printed onto Al2O3 substrates with precoated interdigital Au electrodes on the top. All the samples were annealed in air before tested. The sensors showed optimal operating temperature of 40℃. The deviating stoichiometry of the nanosized materials had obvious effects on their sensing properties [40].

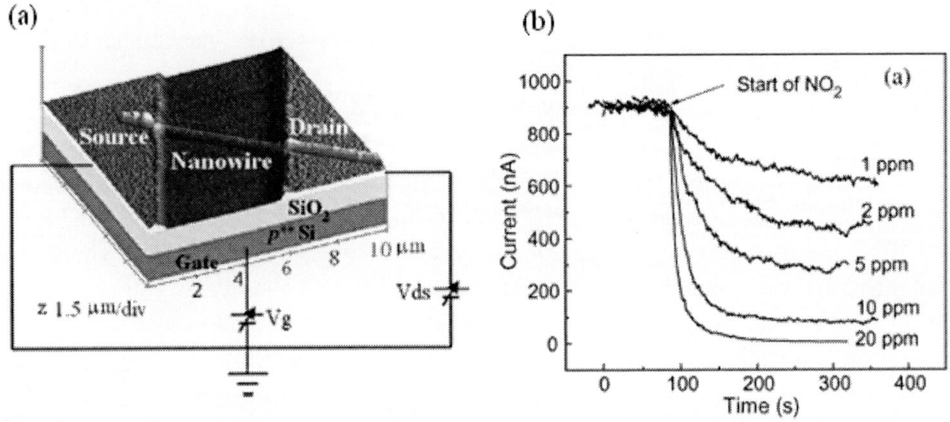

Figure 10. (a) AFM image of ZnO nanowire FET and the schematic of the measurement circuit; (b) ZnO nanowire sensing response to different concentration of NO_2 [39] (@ [2006] IEEE).

SnO_2 is another commonly used metal oxide material in SMO sensors. Liqin Shi et al. [42] developed a highly sensitive SnO_2 based gas sensor for indoor air quality monitoring. The nanostructured SnO_2 was grown on top of Au electrodes using RF induction plasma deposition (IPD) method. It showed a good sensitivity to indoor air pollutants (formaldehyde and nitrogen oxide) at extremely low gas concentration (approximately 20 ppb). Modified SnO_2 films also showed improved sensitivity for certain gases. Arijit Chowdhuri et al. [43] used CuO for H_2S sensing and obtained improved response by dispersing CuO nanoparticles onto SnO_2 film. Agrawal Rajnikant [44] developed nanoparticle layer that consists of indium oxide (In_2O_3)-doped tin oxide (SnO_2) semiconductor nanocrystalline particles with platinum (Pt) nanoclusters for hydrogen sensing. The nanoparticle layer was deposited on interdigitated electrodes and the device showed resistance increase by five orders of magnitude upon exposure to H_2 at room temperature.

2.2. Metallic Nanomaterials and Polyaniline Nanofiber for Gas Sensing

Palladium (Pd) metal film is widely utilized for hydrogen sensing due to its incremental change in resistance by exposure to hydrogen. The development of synthesis technology of nanomaterials motivated researchers interest to fabricate nanostructured Pd based sensors.

Pd nanowires or nanoporous Pd film have been fabricated and their sensing characteristic to hydrogen has been studied by different groups in past few years [45-47]. Figure 11 shows a bundle of vertical aligned Pd nanowires fabricated by electrodeposition into anodized aluminium oxide (AAO) films for hydrogen detecting [45]. The diameter of the nanowires was 80~90nm while the length of the bundle was about 1μm. Compared to Pd thick film hydrogen sensor, the resistance of the Pd nanowires decreased when H_2 is adsorbed into the sensing layer. This is due to the expansion of Pd grains caused by the adsorption of H_2. The swelling causes the gap reducing in grain boundaries. As a result, the intergranular resistance was lowered in a higher portion than the resistance increase factor due to the formation of Pd hydride [46].

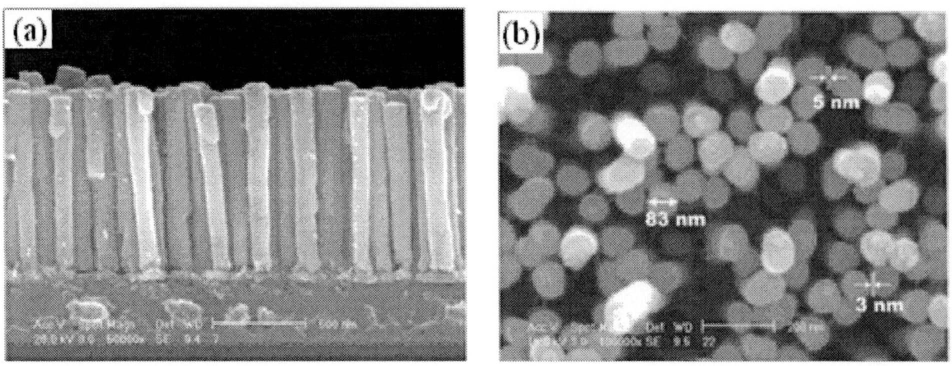

Figure 11. (a) SEM image of the freestanding Pd nanowires on Si substrate; (b) Top view image of Pd nanowires [45] (@ [2006] IEEE).

High sensitivity (approximate two orders decreasing in resistance), fast response time (700ms) and recovery time (2s), and room temperature operation were reported on this Pd nanowire sensors, which demonstrated significant improvement to conventional Pd thick film sensors.

As introduced in former section, conducting polymer is one of the commonly used materials in gas sensing systems. Polymer nanofibers, possessing all the advantages of nanostructures, is another promising material for gas sensing. Polyaniline nanofiber based SAW gas sensor, for example, was developed by Abu Z. Sadek et al [48]. The polyaniline nanofibers were synthesized using chemical oxidative polymerization of aniline by a template-free rapidly mixed reaction. The nanofibers were deposited onto layered ZnO/64o YX LiNbO3 surface-acoustic-wave transducers. The device showed enhanced sensitivity and linear response to hydrogen at room temperature. Fast response and recovery time with good repeatability were also obtained.

Summary

In this chapter, we introduced various gas/chemical sensing systems based on nanomaterials. The utilization of carbon nanotubes and other nanomaterials shows extensive enhancement on the performance of gas/chemical sensors. High sensitivity, room-temperature operation, and fast response and recovery time can be achieved simultaneously by nanotechnology based sensors. However, there are still some issues related with this technique. For example, the mechanisms of gas sensing with CNTs are not completely understood yet, especially for MWCNTs and nanotubes films. The effect of defects and impurities, the relative narrow sensing range of gases, and improvements on the stability and reversibility of the systems all demand further and in-depth research. The development of nanotechnology based gas sensing systems has a challenging but promising prospect.

References

[1] M. Meyyappan and Harry Partridge, *Nano and Biotechnology R*esearch at NASA Ames, http://www.ipt.arc.nasa.gov.

[2] M. S. Dresselhaus, G. Dresselhaus, and P. C. Eklund, *Science of Fullerenes and Carbon Nanotubes*, Academic Press, New York (1996).

[3] S. Ijima, *Nature*, 354, 56 (1991).

[4] T.W .Ebbesen and P.M. Ajayan, *Nature* 358, 220 (1992).

[5] D. S. Bethune, C. H. Kiang, M. S.de Vries, G. Gorman, R. Sa voy, J.Vazquez, and R. Beyers, *Nature* 363, 305 (1993).

[6] Hongjie Dai, Nanotube Growth and Characterization. In: M.S. Dresselhaus, G. Dresselhaus, *Carbon Nanotubes: Synthesis, structure,properties and applications*. New York: Springer, 2000, 29-54.

[7] Thomas W. Ebbesen, Production and Purification of Carbon Nanotubes. In: Thomas W. Ebbesen, *Carbon Nanotubes: Preparation and Properties*. CRC Press, 1997, 139-162.

[8] Jing Li, Yijiang Lu, and Qi Ye et al, *Nano Letters*, Vol.3, No.7 (2003): 929-933.

[9] Jun Li, Hou Tee Ng, *Encyclopedia of Nanoscience and Nanotechnology*, Volume 1: Pages (591–601).

[10] R. Saito, M. Fujita, G. Dresselhaus, and M. S. Dresselhaus, *Appl. Phys. Lett.* 60, 2204 (1992)
[11] Niraj Sinha, Jiazhi Ma, John T. W. Yeow, Journal of Nanoscience and Nanotechnology, Vol.6, 573-590, 2006.
[12] O. Lourie and H. D. Wagner, *J. Mater. Res.* 13, 2418 (1998).
[13] M. F. Yu, O. Lourie, M. J. Dyer, et al., *Science* 287, 637 (2000).
[14] J. Kong, Nathan R. Franklin, Chongwu Zhou, et al., *Science* 287, 622 (2000).
[15] A. Wadhawan, R. E. Stallcup, K. F. Stephens, and J. M. Pereza, *Appl. Phys. Lett.* 79, 1867 (2001).
[16] C. K. W. Adu, G. U. Sumanasekera, and B. K. Pradhan, et al., *Chem. Phys. Lett.* 337, 31 (2001).
[17] J. Zhao, A. Buldum, J. Han, and J. P. Lu, *Nanotechnology* 13, 195(2002).
[18] Shu Peng, James O'Keeffe, Chengyu Wei, and K. Cho, the 3rd International Workshop on Structural Health Monitoring, 2001.
[19] Seung-Hoon Jhi, Steven G. Louie, and Marvin L. Cohen, *Phys. Rev. Lett.* 85, 1710 (2000);
[20] G. U. Sumanasekera, C. K.W. Adu, S. Fang,2 and P. C. Eklund, *Phys. Rev. Lett.* 85, 1096 (2000);
[21] H E Romero, G U Sumanasekera, SKishore1 and P C Eklund, *J. Phys.: Condens. Matter* 16 (2004) 1939–1949.
[22] Takao Someya, Joshua Small, Philip Kim, Colin Nuckolls, and James T. Yardley, *Nano Letters*, Vol.3, No.7 (2003): 877-881.
[23] Yoon-Taek Jang, Seung-Il Moona, and Jin-Ho Ahnb, et al., *Sensors and Actuators* B 99 (2004) 118–122.
[24] Woo-Sung Cho, Seung-Il Moon, Yang-Doo Lee, et al., *IEEE Electron Device Letters*, Vol.26, No.7, JULY 2005.
[25] O.K. Varghese, P.D. Kichambre, D. Gong, et al., *Sensors and Actuators* B 81 (2001) 32-41.
[26] L. Valentini, C. Cantalini, I. Armentano, et al., *Diamond and Related Materials* 13 (2004) 1301–1305.
[27] Krishna C Persaud, *IEE Colloquium on Molecular Electronics*, Vol.19, 4/1 - 4/3 (1990).
[28] M.C. Petty and R. Casalini, *Eng. Sci. and Edu. J.* June (2001) 99-105.
[29] Ivan D. Avramov, Shigeru Kurosawa, and Michael Rapp, *IEEE Trans. on Microwave Theory and Tech.*, Vol. 49, No. 4, April 2001.
[30] Ventzislav M. Yantchev, Vesseline L. Strashilov, Michael Rapp, *IEEE Sensors J.*, Vol.2, No.4, August 2002.
[31] Seong M. Cho, Young Jun Kim, Yong Shin Kim, Sensors, 2004. *Proceedings of IEEE*, Vol.2, 701 -704 (2004).
[32] Biju Philip, Jose K Abraham, Anupama Chandrasekhar and Vijay K Varadan, *Smart Mater. Struct.* 12 (2003) 935–939.
[33] Jing Li, Yijiang Lu, and M. Meyyappan, *IEEE Sensors J.*, Vol.6, No.5, 2006.
[34] Zhongyu Hou, Dong Xu, and Bingchu Cai, *Appl. Phys. Lett.* 89, 213502(2006).
[35] Ashish Modi, Nikhil Koratkar, Eric Lass, Bingqing Wei, and Pulickel M. Ajayan, *Nature*, Vol 424, 171(2003).
[36] S J Kim, *J. Phys. D: Appl. Phys.* 39 (2006) 3026–3029.
[37] G. Bläser, Th. Rühl, C. Diehl, M. Ulrich, D. Kohl, *Physica* A 266 (1999) 218-223.

[38] Ying Hu, Ooi Kiang Tan, and Weiguang Zhu, *IEEE Sensors J.*, Vol.5, No.5, 2006.
[39] Zhiyong Fan and Jia G. Lu, *IEEE Trans. on Nanotechnology*, Vol. 5, No.4, July,2006.
[40] Ying Hu, Ooi Kiang Tan, and Weiguang Zhu, *IEEE Sensors J.*, Vol.6, No.6, 2006.
[41] T. Shishiyanu, S. Shishiyanu, 0. Lupan, V. Sontea, A. Bragorenco, *International Semiconductor Conference*, Vol.1, 201-204(2006).
[42] Liqin Shil, Y. Hasegawa, T. Katsubel, *The 13th international Conference on Solid-State Sensors, Actuators and Microsystems*, 1203-1206(2005).
[43] Chowdhuri, A., Gupta, V., Kumar, R., et al., Sensors, 2003. *Proceedings of IEEE*, Vol 1, 201 – 205(2003).
[44] Rajnikant, A., Shukla, S., Ludwig, L., et al. Sensors, 2004. *Proceedings of IEEE*, Vol 1, 395 – 398(2004).
[45] Kyun Tae Kim, Sang Jun Sim, and Sung Min Cho, *IEEE Sensors J.*, Vol.6, No.3, 2006.
[46] Massood Z. Atashbar, Deep Banerji, and Srikanth Singamaneni, *IEEE Sensors J.*, Vol.5, No.5, 2006.
[47] Dongyan Ding and Zhi Chen, *6th IEEE Conference on Nanotechnology*, 2006, Vol.2, 777- 780(2006).
[48] Abu Z. Sadek, Christina O. Baker, David A. Powell, *IEEE Sensors J.*, Vol.7, No.2, 2007.

In: Nanotechnology Research Developments
Editor: R. Jiménez-Contreras, pp. 295-314
ISBN: 978-1-60021-899-6

Chapter 8

SELF-ASSEMBLED HIERARCHICAL ARCHITECTURES BY LIQUID-PHASE PULSED LASER ABLATION

***Li Yang*[1*], *Paul W. May*[1] *and Lei Yin*[2]**
[1]School of Chemistry, University of Bristol, Cantock's Close, Bristol BS8 1TS, United Kingdom
[2]Department of Aerospace Engineering, University of Bristol, Queen's Building, University Walk, Bristol BS8 1TR, United Kingdom

ABSTRACT

Over the past decade, liquid phase pulsed laser ablation (LP-PLA) has been used as an effective method for preparation of many nanostructured materials. This technique involves the interaction of a laser with a solid target and surrounding liquid. The laser causes the target surface to vaporise in the form of an energetic ablation plume, containing species such as atoms, ions, and clusters, travelling with high kinetic energy. These can collide and react with molecules of the surrounding liquid phase, producing new materials, often on the nanometre scale. Due the intensity of the laser and the nanosecond timescales, high temperature, high pressure, and high density conditions within the reaction volume are produced which provide a 'brute force' method of synthesising novel materials that have hitherto been inaccessible using milder, more conventional techniques. In this chapter, we discuss more recent progress in the synthesis of hierarchical, complex, and ordered crystal architectures by this novel technique. It is believed that the ability of nanostructures to self-assemble in this manner provide potential building blocks for microscale or nanoscale devices. We show that for C3N4, a Group IV-V covalent compound, a range of self-assembled arrays of nanoparticles, nanorods, nanoleaf, and mesoflowers can be produced, and this also suggests that the route described here can be applicable to the synthesis of nanostructures of other similar materials. Therefore, we expect LP-PLA may play a key role in advancing complex nanomaterials into many applications in optics, electronics, biology, medicine and energy/chemical conversion.

* E-mail: li.yang@bristol.ac.uk

1. Introduction

Although self assembly has occurred in nature for thousands of years (for instance, the formation of salt crystals and the intricate structure of snowflakes), the term 'self assembly' as used in chemical synthesis is relative new. Self assembly refers to the creation by physical or chemical reaction, of small building units, which may be nanoparticles (NPs), nanorods (NRs), *etc*, which then aggregate together in specific arrangements to give one-dimensional (1D), two-dimensional (2D) or even three-dimensional (3D) superstructures [1]. The aggregation may be spontaneous without human intervention, or may be as a result of changing local conditions, *e.g.* temperature, concentration, drying, *etc.* Current understanding extends only to the most rudimentary of these stages for its use in materials, biology, chemistry, and condensed matter science [2].

Due to the emergence of a new generation of high-technology materials, 2D and 3D self-assembled, aligned superstructure materials have begun to be widely investigated over the last decade [3, 4]. It is believed that the ability of nanostructures to self-assemble with controlled crystalline orientation, size, complexity and crystal morphology provide potential applications in data storage, functional devices, communications and technology. As a result, a rapidly growing field of science has emerged to understand and control these self-organized architectures, involving the dedicated efforts of chemists, physicists, material scientists and biologists. In general, physical methods (including chemical vapor deposition [5], vapour phase transport [6], and pulsed laser ablation in vacuum [7]) and chemical methods (including hydrothermal methods [8], soft-template [9] and use of various surfactants [10, 11]) have been developed to fabricate nano- to microscale materials with a range of morphologies, such as compact hexagonal networks [12], rings [13] , dandelion-shaped hollow structures [14], strips [15], tubes [16], and flower-like structures [17]. However, the former techniques require relatively high temperature, vacuum conditions, expensive equipment and sometimes complicated processes, which limit them to smaller scale fabrication. Also, conventional chemical methods usually involve the use of catalysts, surfactants and possibly complex chemical reactions, which often produce a significant amount of unwanted byproducts requiring further purification. Therefore, a technique which combines the merits of both physical and chemical methods, while giving high yield at low cost, would be desirable, if novel self assembled materials are to be produced on an industrial scale.

One such technique, liquid-phase pulsed laser ablation (LP-PLA), has been applied to the problem of producing self-organized nanomaterials only relatively recently. LP-PLA involves focusing a high power laser beam onto the surface of a solid target, which is submerged beneath a liquid. The interaction of the laser with the target causes the surface to vaporise in the form of an ablation plume, which contains species such as atoms, ions, and clusters, travelling with high kinetic energy. The species in the plume collide and react with molecules of the surrounding liquid, producing new compounds containing atoms from both the original target and the liquid. Due the intensity of the laser and the nanosecond timescales, the instantaneous temperatures and pressures within the reaction volume can be extreme (many thousands of K at tens of GPa) [18]. Such high temperature, high pressure, and high density conditions provide a 'brute force' method of synthesising novel materials that have hitherto been inaccessible using milder, more conventional techniques.

In this chapter, we first briefly review the use LP-PLA and its applications in synthesis of nanocrystals. Furthermore, we reported recent progress in the synthesis of hierarchical, complex, and ordered crystal architectures by this novel technique. We shall show that the chemical and physical properties of these superstructures are intrinsic to the self-assembly induced by the close vicinity of the NPs or NRs. Finally, a summary and expectation are discussed with regard to the application of LP-PLA as a synthesis route for highly desirable complex architectures.

2. Laser Ablation of Solids in Liquid Environments

Pulsed laser deposition was first developed in the 1960s, shortly after the invention of the pulsed ruby laser. Since then, laser ablation in a vacuum or dilute gas has been studied by many researchers. By using different target materials and background gases, and varying parameters such as the laser wavelength, fluence, and pulse duration, it is possible to produce a wide variety of thin films. These include high temperature superconductors [19], metals, semiconductors, oxides, and other ceramics [20], and diamond-like carbon [21]. The thin films have a variety of applications, for example semiconductor devices, electrodes, and wear-resistant coatings [22].

The introduction of pulsed laser ablation at the solid-liquid interface was first reported by Patil and co-workers in 1987, who used a pulsed laser to ablate a pure iron target in water to form iron oxides with metastable phases [23]. Following their work, Ogale [24] extended the potential of LP-PLA for the surface modification of metals, such as metallic oxidation, nitriding, and carbiding. This pioneering work opened new routes for materials processing based on the PLA of solids in various liquids. Since then, the LP-PLA method has been used to produce a wide range of novel materials, such as nanodiamond and related nanocrystals, metallic nanocrystals, nanocrystal alloys, and metal oxides. A summary of the nanocrystals synthesized by LP-PLA is listed in Table 1.

These studies clearly indicate that LP-PLA has become a successful material fabrication technique, allowing versatile design through choosing suitable solid targets and confining liquids. Compared to the conventional methods mentioned above, the technique of LP-PLA has many distinct advantages: (i) a chemically "simple and clean" synthesis, the final product is usually obtained without byproducts and no need for further purification; (ii) low cost of experimental setup and easy controllable parameters; (iii) the extreme confined conditions and induced high temperature, high pressure region favour the formation of unusual metastable phases. A recent review by Yang [25] gives a comprehensive understanding of the nucleation thermodynamics, the phase transition, and the growth kinetics of nanocrystals by laser ablation of liquids. Therefore, in this chapter we will not discuss the effects of liquid confinement, thermodynamic nucleation and growth mechanism in detail. Instead we shall focus upon the recent progress in the synthesis of hierarchical, complex, and ordered crystal architectures by LP-PLA.

Table 1. Nanomaterials synthesis via LP-PLA in various liquids. NP = nanoparticles, NR = nanrods

Product	Solid Target	Liquid Solution	Laser parameters
Ag NP	Ag	H_2O [26, 27]	Nd: YAG laser (355, 532, 1064 nm) Laser energy: 90-340 mJ/pulse
		C_2H_5OH [28]	Cu vapour laser (510.5 nm) Laser fluence: 1-20 J/cm^2
Au NP	Au	H_2O [29]	Nd: YAG laser (532, 266 nm) Laser fluence: 5-250 J/cm^2
		Alkane liquid [30]	Nd: YAG laser (532 nm) Laser fluence: 1-200 J/cm^2
Co_3O_4 NP	Co	H_2O [31]	Nd: YAG laser (355 nm) Laser energy: 30 mJ/pulse
Cubic-BN NP	Hexagonal-BN	Acetone [32]	Nd: YAG laser (532 nm) Power density: 10^{10} W/cm^2
Cubic-C_3N_4 NR	Graphite	Ammonia [33]	Nd: YAG laser (532 nm) Power density: 10^{10} W/cm^2
Diamond NP	Graphite	Acetone [34]	Nd: YAG laser (532 nm) Power density: 10^{10} W/cm^2
		H_2O [35]	Nd: YAG laser (532 nm) Laser energy: max 125 mJ/pulse
Hf_2S NP	HfS_3	Tert-butyl disulfide [36]	Nd: YAG laser (532 nm) Laser energy: 30-50 mJ/pulse
$Mg(OH)_2$ Tubular structures	Mg	H_2O +SDS [37, 38]	Nd: YAG laser (355 nm) Laser energy: 100 mJ/pulse
Pt/TiO_2 NP	Pt/TiO_2	H_2O [39]	Nd: YAG laser (355 nm) Laser power: 160 mJ/pulse
SnO_2 NP	Sn	H_2O +SDS [40]	Nd: YAG laser (355 nm) Laser energy: 100 mJ/pulse
TiO_2 NP	Ti	H_2O H_2O +SDS [41, 42]	Nd: YAG laser (355 nm) Laser energy: max 150 mJ/pulse
ZnO NP core-shell	Zn	H_2O +SDS [43]	Nd: YAG laser (1064 nm) Laser energy: 70 mJ/pulse
ZnO NP	Zn	H_2O [44] H_2O +LDA [45] H_2O +CTAB [46]	Nd: YAG laser (355 nm) Laser energy: max 3.2 J/cm^2
$Zn(OH)_2$ layered nanocomposite	Zn	H_2O +SDS [47]	Nd: YAG laser (355 nm) Laser energy: 100 mJ/pulse

3. Highly Organized Growth of Carbon Nitride

To date, most nanomaterials made by LP-PLA have been either continuous film structures, zero-dimensional (0D) NPs, or 1D NRs. Since the 0D and 1D nanocrystals can serve as building blocks in forming 2D or 3D complex architectures with long term periodic structures, it should be expected that the LP-PLA approach would have great potential as a means to grow large arrays of hierarchical, complex, oriented and ordered superstructures. Indeed, our recent findings [48, 49] indicated that the instantaneous high temperature, high pressure and high density conditions that arise when a high-intensity focused laser beam impinges upon a graphite target confined by a thin layer of liquid ammonia can promote growth of crystalline carbon nitride nanoparticles.

Carbon nitride has been the subject of numerous publications since the prediction by Liu and Cohen [50] in 1989 that crystalline C3N4 should have superhard properties. However, successful synthesis of bulk amounts of this material still remains a challenge. Recently, we have developed two-step synthesis strategies for adding complexity to oriented nanocrystals, and shown the possibility of forming self-assembled nanostructures and microstructures of crystalline carbon nitride. This approach has involved initially using the LP-PLA process to prepare carbon nitride NP suspensions, and then sequential seeding these onto an appropriate substrate and carefully drying away the liquid. More complex nanostructures, such as observed in natural materials or biomineral 3D well-arranged architectures could also be generated [51, 52].

3.1. Nanocrystal Seed Preparation

The carbon nitride nanocrystal seed solution was prepared by LP-PLA, details of which are given elsewhere [53]. Briefly, a solid graphite target (Testbourne Ltd., 99.99%) was ablated at room temperature while submerged in a 5 ml solution of 25-35% ammonia solution (Fisher Scientific, used as received without further purification) inside a sealed stainless steel cell. In order to reduce the effect of target aging, the cell was rotated at 700 r.p.m. during ablation using a standard magnetic stirrer. A Q-switched Nd:YAG laser (532 nm, pulse duration 15 ns, frequency 10 Hz) was directed by a prism and then focused onto the graphite surface using a 25 mm-focal-length lens. The intense laser light passed through a quartz window in the top of the cell, then through ~5 mm layer of the liquid covering the graphite, to form a ~0.5 mm-diameter spot on the target surface. The ablation was typically carried out at laser fluences of 25-125mJ / pulse for reaction times $t = 0.5$-12 h. After ablation, a pale yellow colloidal suspension was obtained, which contained a mixture of unreacted graphite and ablation product, both in the form of NPs. The suspension was stable, with no precipitate being observed for months or even longer. The graphite sediments were filtered and removed as much as possible by boiling with 70% perchloric acid, before further characterization.

For analysis, a drop of the suspension was placed onto a transmission electron microscopy (TEM) copper grid. TEM revealed (Figure 1a) that pristine spherical particles appeared after only 10 min ablation (laser fluence at 75 mJ per pulse, 35% ammonia) with an average diameter 15.7 nm. The size distribution (Figure 1b) was obtained from statistical analysis of Figure1a by measuring the diameters of more than 900 particles, and fits well to a

Gaussian curve. The area highlighted by the rectangle in Figure 1a indicates that some of the particles may be starting to line up. After 3 h of ablation, one-dimensional rod-like structures are formed of length ~200 nm and width 30-50 nm (see Figure1c). Since both ends of the flat structures taper to a point, they resemble the shape of a leaf, and so henceforth have been termed 'leaf-like' structures. The inset in Figure 1c shows the selected area electron diffraction (SAED) ring pattern (obtained without any tilting) from the leaf-like nanostructures, which indicates that these structures are polycrystalline. The four inner rings, with interlayer *d*-spacings of 2.788, 2.256, 1.610 and 1.348 Å, respectively, were indexed as (200), (101), (211) and (221) with reference to the calculated structure for β-C3N4 ($a = 6.4017$ Å, $c = 2.4041$ Å, space group: P63/m (176)) [54, 55], which are identical to that reported in the literature for the same β-C3N4 crystal [56].

When the reaction time was increased to 5 h (Figure 1d) the size of nanoleaf structures increased to ~100 ×~500 nm. The high magnification TEM image (Figure 1e) from a single nanoleaf shows that these structures are, themselves, composed of a large number of smaller NPs that have close-packed together in an ordered arrangement to form the leaf-like shapes. In contrast, Figure 1f shows that these leaf-like nanostructures are mostly made up of component NRs, while NPs are the building blocks for the NRs. These tiny NRs are well-aligned and all point in the same direction, perpendicular to the propagating *c*-axis. Although individual leaf-like structures within a local region seem to be randomly oriented, the overall NR aggregates still exhibit long-range ordering [57] (Figure 1f).

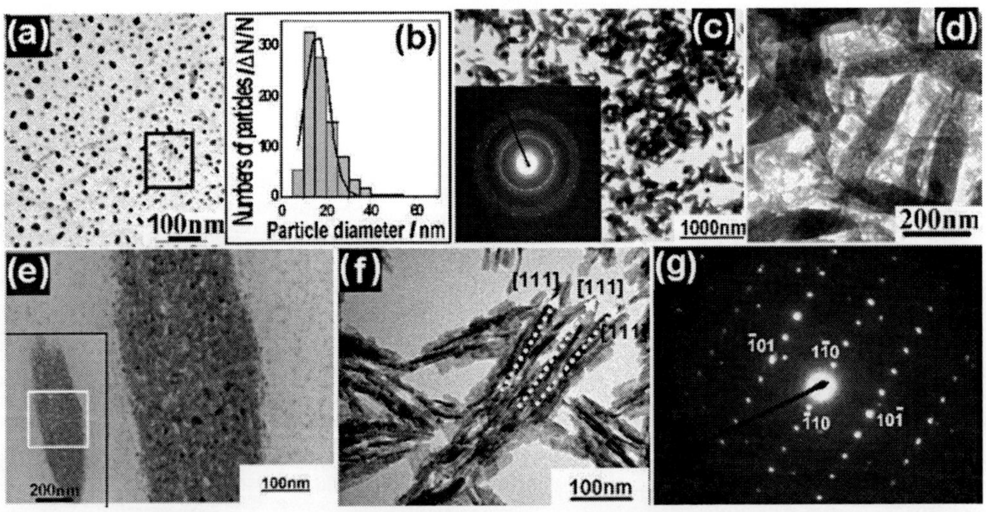

Figure 1. TEM images of C_3N_4 nanostructures synthesized with 5 ml 35% ammonia covering a graphite target, using 75 mJ / pulse laser fluence. (a) After 10 min ablation, spherical NPs appear, with an average size of 15.7 nm. Adjacent particles are aligned with one another along their main axes, highlighted by the rectangle. (b) Size distribution of the NPs, with a Gaussian fit to the data. (c) Overall appearance of the leaf-like nanostructures after 3 h ablation. Inset shows a typical SAED pattern taken from Figure1c. (d) Leaf-like structures of length ~500 nm produced following 5 h ablation time. (e) A single leaf-like unit formed from numerous small NPs (inset area framed with white box). (f) The NR aggregates exhibit a long-range order; arrows indicate the projecting direction along [111]. (g) A SAED pattern of a NR from the [111] zone-axis indicating the area is single crystal and consistent with crystalline β-carbon nitride. (Reproduced from [48] with the permission of American Chemical Society (ACS)).

A representative SAED pattern is given in Figure 1g, showing diffraction spots from the ($\bar{1}10$), ($1\bar{1}0$), ($\bar{1}01$) and ($10\bar{1}$) planes of the [111] zone-axis, which are aligned parallel each other.

3.2. Chemical Bonding Environment in Carbon Nitride

X-ray photoelectron spectroscopy (XPS) measurements estimated that the overall elemental composition of the nanoleaf structures (laser fluence at 100 mJ per pulse, 35% ammonia) was 86.3% C, 11.4% N, and 2.3 % O after 5 h, whereas for 7 h ablation values were 90.5% C, 7.8% N and 1.7% O. One possibility for the presence of the O signal is that the graphite target reacted directly with the water (or OH radicals generated in the water) during the high temperature conditions in the ablation plume. However, since the O signal decreased with longer ablation time, this idea has been discounted. An alternative explanation is that prolonged exposure of the samples to the laboratory atmosphere led to possible contamination by absorbed or adsorbed water vapor. For pure C3N4, we would expect a composition of 42.8% C and 57.2% N, which shows that our samples are considerably carbon rich. Since we know from TEM observations that regions of the samples contain pure C3N4, we must conclude that our samples are a composite of crystalline C3N4 and an amorphous carbon background. This observation is consistent with the conclusions by other workers [58, 59], who found that it was difficult to generate highly ordered structures containing large concentrations of N incorporated into the C network.

The breadth and asymmetry of the C (1*s*) and N (1*s*) core-level XPS spectra shown in Figure 2 indicate the presence of different bonding states within the crystallites. The core-level lines were fitted to the Gaussian function [60]. The N 1*s* feature is difficult to assign unambiguously, since it is known that N can give a broad XPS peak around 399 eV in some materials. Alternatively, a better fit can be achieved if the feature is deconvoluted into three peaks (as shown in Figure 2b and d), located at ~398.5 eV, ~399.6 eV and ~400.2 eV, respectively. The peaks situated at ~398.5 eV can be assigned to *sp*3 C-N bonds, and one at 400.2 eV assigned to *sp*2 C=N bonds [61]. A peak centred at a binding energy ~399.6 eV has been assigned by some groups [62, 63] to N atoms bonded to *sp*-hybridized C atoms (*i.e.* -C≡N). These assignments are also supported by *ab initio* binding-energy calculations [64] Nevertheless, within experimental accuracy it is impossible to distinguish between these two alternatives. Also, it should be noted that C≡N bonds can produce 3 peaks in this region [63], 398.2 eV [65], 398.1 eV [66] and 400.1 eV [67]. However, these assignments have been discounted since they are outside of the experimental accuracy of the XPS spectrometer. The presence of *sp*3 C-N in the ablated nanostructures is, again, in agreement with a C3N4 structure.

Similarly, the C 1*s* core-level spectra obtained by XPS (Figure 2) was deconvoluted into three peaks located at ~284.6 eV, ~285.5 eV and ~288.5 eV, respectively. However, unlike nitrogen, verification of the assignment by independent C (1*s*) spectra was relatively difficult due to small shifts in binding energy and multiple bonding environments of the carbon. Carbon atoms may have zero, one, two, or three bonds with nitrogen atoms, which complicate the situation, so that there is a wide range of interpretation in the literature for carbon spectral feature identification.

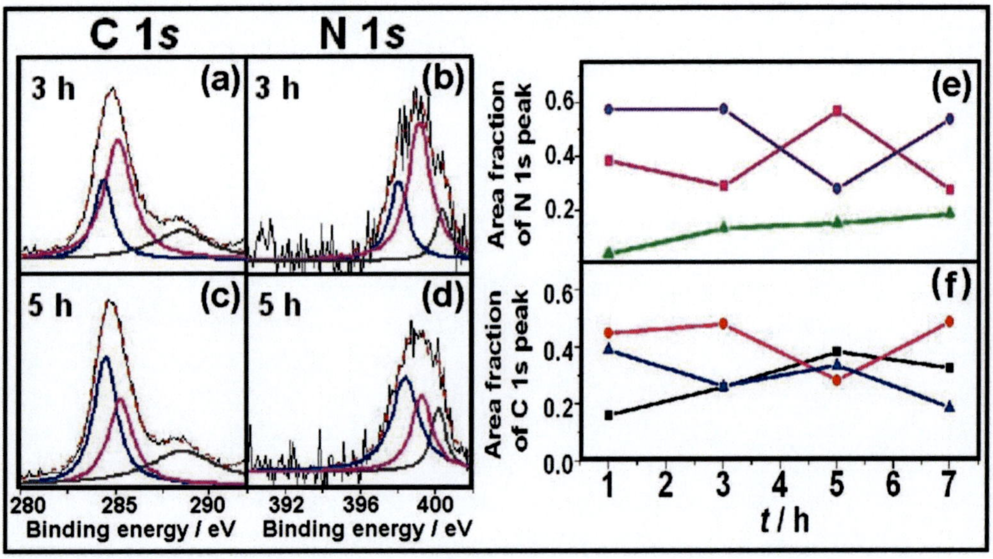

Figure 2. XPS spectra of C3N4 nanostructures (laser fluence at 100 mJ per pulse, 35% ammonia) for different ablation times *t* (a) C (1*s*) 3 h ; (b) N (1*s*) 3 h; (c) C (1*s*) 5 h; (d) N (1*s*) 3 h. Each spectrum has been deconvoluted by fitting to three Gaussian lineshape functions, with the total fitted line overlaid as a dashed line. Plots showing the time dependencies of the relative areas of the Gaussian functions used to fit (e) the N (1*s*) XPS peak (key: ■ 398.5 eV *sp*3 C-N bonds, ● 399.6 eV C≡N bonds, ▲ 400.2 eV *sp*2 C=N bonds), and (f) the C (1*s*) XPS peak (key: ■ 284.6 eV, ● 285.5 eV, ▲ 288.5 eV). (Reproduced from [53] with the permission of Institute of Physics (IOP)).

The peaks at ~284.6 eV and ~285.5 eV can be identified unambiguously as *sp*2 and *sp*3 hybridized carbon [68, 69]. But the assignment for the peak at ~288.5 eV has not yet reached a consensus. In our case, it is believed this peak corresponds to N≡C *sp* bonding [67] supported by the Raman data [53]. Le Normand *et al* [70] have claimed that it is inappropriate to assign components appearing at higher energies to specific C-C or C-N bonding environments. After searching a comprehensive range of organic polymers, Beamson and Briggs [71] also found that no significant difference in C 1*s* position can be appreciated between *sp*2 and *sp*3-hybridized carbon bonded to nitrogen.

The relationship between ablation time and relative peak area fraction is shown in Figure 2e. The peak area corresponding to C=N (*sp*2) bonding indicates a marked increase with prolonged ablation time, whereas the trends for C≡N (*sp*) and C-N (*sp*3) are relatively complex. Nitrogen inclusion into the CN network changes the shape and position of the C 1*s* peak, as seen as in Figure 2f. However, only the signal attributed to C=C (*sp*2) bonding shows smooth variation, *i.e.* increasing with ablation time, while all other signals show a more complicated trend.

3.3. Control of Self-Assembled Structures

To understand the formation mechanism of nanocrystalline carbon nitride, experiments were performed for different laser energies with constant ablation time (3 h) and ammonia concentration (35%). Figure 3 summarises the different C3N4 nanostructures synthesized with varying laser fluences. For low laser fluence (25 mJ/pulse), the leaf-like structures had a length and width of ~300 and ~120 nm, respectively, with smooth surfaces, and were composed of smaller-sized NPs (Figure 3a). There were also plenty of voids situated between some of the NRs (Figure 3b). The sharp SAED pattern for the leaf-like structures (inset Figure 3b) was consistent with crystalline hexagonal β-phase carbon nitride [55, 72]. When the laser power was increased to 50 mJ / pulse, the aligned leaf-like structures increased to a length and width of ~600 and ~250 nm, respectively (Figures 3c-d), and now consisted of a substantial number of connected NRs, although there was no apparent change in size of the component NRs. The surface of the leaf-like structure was no longer smooth, but contained numerous protruding NRs. For still higher fluences (75 mJ / pulse), there was no further significant change in the size or shape of the leaf-like structures and NR building units (Figure 3e-f), but there was an increase the number formed. With further increases in fluence to 100 mJ per pulse, micron-scale carbon nitride spheres (Figure 3g) were observed, which were formed following aggregation of a large number of the leaf-like structures. 'Flower-like' spiked crystallites (Figure 3h) were also seen, where the leaf-like structures coalesced at a common centre with multi-fold symmetry [14]. In general, the trend is that the size of the overall nanocrystallites and their basic building blocks increases with increasing laser power.

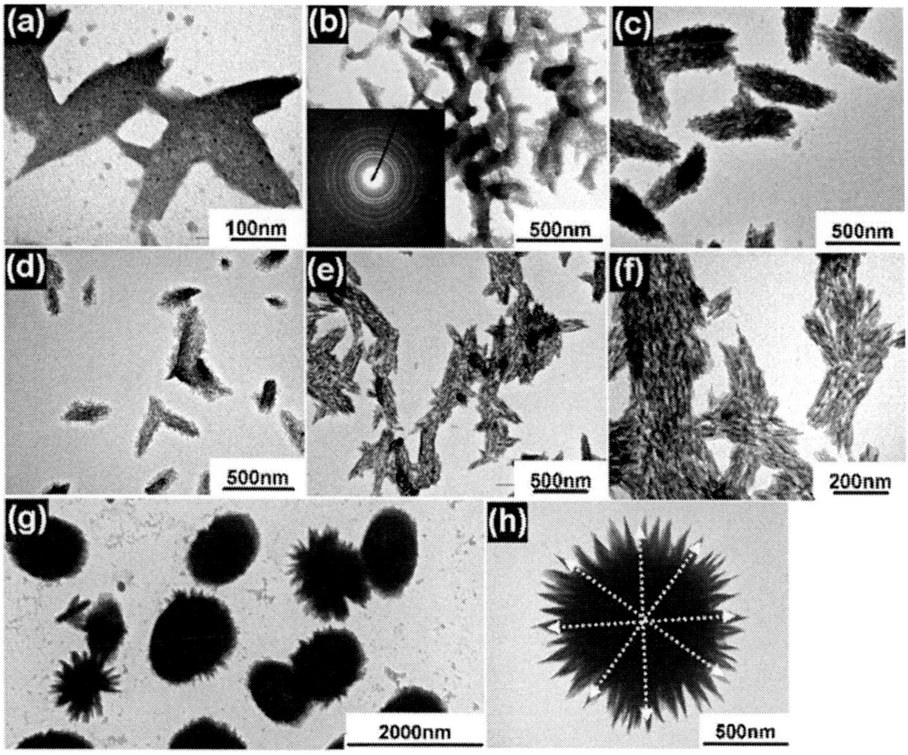

Figure 3. TEM images of C_3N_4 nanostructures synthesized for 3 h ablation time and 35% ammonia solution using different laser fluences. (a) and (b) 25 mJ/ pulse, leaf-like structures containing spherical NPs and short NRs. Inset in (b) is a sharp SAED pattern indicating the structures are crystalline. Note that there appear to be voids inside the leaf-like structures (the white areas in the grey area of image (b)). (c) and (d) 50 mJ/ pulse, denser leaf-like structures containing longer NRs which protrude from the surface. (e) and (f) 75 mJ/ pulse, leaf-like structures beginning to aggregate into larger clumps. (g) and (h) 100 mJ/ pulse, carbon nitride nanospheres and flower-like spiky crystallites. (Reproduced from [48] with the permission of ACS).

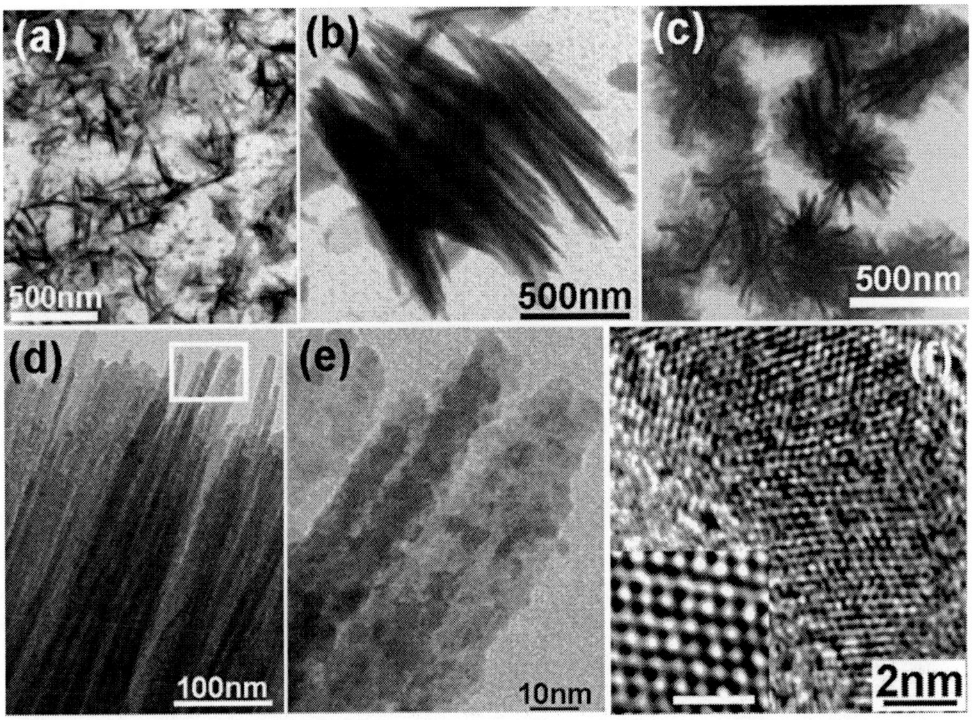

Figure 4. TEM images obtained by LP-PLA in 25% ammonia solution: (a) Isolated carbon nitride NRs (40 mJ/ pulse, t = 1 h) (b) Branched NRs (40 mJ/ pulse, t = 3 h) (c) Highly branched flowerlike architectures (100 mJ/ pulse, t = 12 h). (d) Rod-like structures showing straight, long and sharp tips. (e) Enlarged image of the region at the top of the NRs indicated by the open box in (d). (f) HRTEM image of a single NR, the inset shows the atomic arrangement, scale bar = 1 nm.

When 35% ammonia was replaced by 25% ammonia solution, carbon nitride was still formed on the copper grids; however, the morphology was different, as mentioned previously. For low laser power (40 mJ) and short ablation time (1 h), the product contained mostly a sparse collection of isolated NRs (Figure 4a). With longer ablation times (3 h), the NRs start to aggregate into branched structures (Figure 4b), serving as the starting points (or nucleation seeds) for the subsequent growth. However, with increasing laser power (100 mJ) and ablation time (12 h), the concentration of NRs increased, highly branched flower-like architectures completely made up of NRs were observed (Figure 4c). The NRs at the edge of the flower structures appeared to be protruding outward by ~10 nm. As shown in Figure 4d-4e, it should be pointed out that these NRs are themselves composed of a large number of smaller NPs that have packed together in an ordered arrangement to form the rod-like shapes. In particular, in the high-resolution (HR) TEM image in Figure 4f, taken from a single NR, the periodic lattices clearly show the atomic arrangement (Figure 4f inset) with very few defects, and reflect the relationship between the orientation of the NPs and the crystallography of the ordered NR array.

Based on above discussion, it was found that the size and morphology of the ablation products were a complicated function of laser power, ablation time, and ammonia concentration. Different experimental conditions can be used to optimize and control the nanocrystal nucleation and orientation independently.

4. SEQUENTIAL NUCLEATION AND GROWTH OF COMPLEX NANOSTRUCTURES

The fabrication of well-defined 2D and 3D nanostructures is usually difficult because control of the nucleation and growth is still a challenge. Complex architectures are normally related to multistep syntheses, which begin with substrate surface preparation for heterogeneous nucleation of oriented nanocrystals. Two basic approaches have been used: one is deposition on chemically modified interfaces using organic self-assembled monolayers (SAMs) [73]. Another, more straightforward method, is to seed the substrate with NPs of the desired film material [74]. The fundamental advantage of seeded growth is enhanced control, obtained by separating the NP film nucleation and oriented rod growth into two steps.

Extending this concept, we performed a two-step synthesis to produce oriented carbon nitride nanostructures. Many more complex nanostructures than simple NPs and NRs described in the previous section can be produced in a controlled fashion by simply altering the drying time and drying method of the suspension of ablated product. In this methodology, the carbon nitride materials produced in the initial LP-PLA step were physically deposited onto a silicon p-(100) substrate. The liquid was then removed by drying. Four different drying processes were used: (1) dry naturally in air; (2) dry in a sealed tube; (3) dry in an oven or on a hotplate; (4) dry in a critical point dryer (CPD). CPD is a widely used drying technique in which the surface tension can be reduced to zero during evaporation [75]. Those procedures allowed the time taken to evaporate the liquid to be controlled.

4.1. General Structural Features

By drying the suspension under different conditions, four main classes of structure have been identified in the ablation product, categorised based upon their shape and size (Figure 5a). The first class of structure had the shape of thin plates with rounded edges. Since they were the components of the larger 'flower-like' structures (described later), they have been termed 'nano-petals'. The quantity of these nanopetals and their location with respect to the larger structures (see later) were dependent upon the deposition and drying conditions. As shown in Figure 5b-5c, this indicated that these nano-petals were 2D aligned crystallites of carbon nitride which preferentially aligned themselves perpendicularly to the surface of the Si substrate. The number and length of these nano-petals increased with increasing laser ablation time from 0.5-2 h for the same laser fluence. It is also possible that these nano-petals began to aggregate and self-assemble (Figure 5d). When the concentration of nano-petals in the suspension increased, they tried to minimize their interfacial energy upon subsequent drying of the liquid by preferential tilting with respect to each other. This produced the second class of nanostructure, which has a 'grass-like' shape and exhibits several different morphologies (see later). However, all are produced in large quantities and cover the whole substrate.

By carefully controlling the drying process, 'flower-like' spiked, crystalline superstructures were formed (Figure 5e-5f). This third class of structure – now fully 3-dimensional – with sizes 1-20 μm, were created when many nano-petal structures coalesced at a common centre with multi-fold symmetry. One possible explanation is that the presence of the solid substrate physically hinders growth in that direction; so many branches are tilted away from the substrate, towards the solution.

When the evaporation speed of the liquid was rapid (for example, drying in an oven or hotplate), a fourth class of structure was observed (Figure 5g-5h). Instead of 3D flowers, the carbon nitride now formed 2D 'star-like' or flattened flower-like structures. New dendrites emanated from the core and acted as nucleation centres, eventually allowing the structure to expand into 2D horizontal flowers (Figure 5h). It is suggested that the higher water evaporation rate increases the interparticle capillary forces [76]. As the continuous flux of particles fill up the spaces on the substrate, successive layers are expected to be formed rather than 3D complex shapes.

Although these four kinds of structures have different densities and morphologies, they all exhibit high surface-to-volume ratio and so might have potential uses in semiconductor devices, anticorrosion protective coatings and new applications.

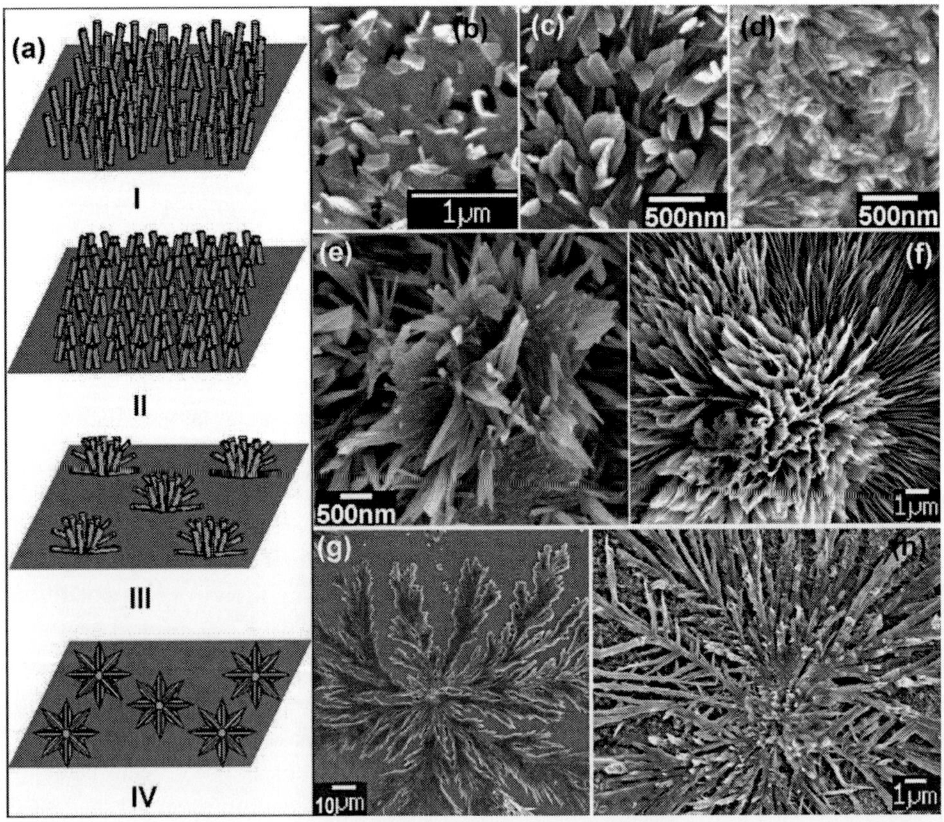

Figure 5. (a) Schematic illustration of the growth process leading to the observed four main classes (labelled as I-IV) of hierarchical structures, (b)-(e) SEM images of carbon nitride 'nano-petals' following ablation times of: (c) 0.5 h, (d) 2 h, and (e) 3 h. (e-f) Overall 'flower-like' structure following 5 h laser irradiation (synthesis conditions: laser power 125 mJ, 35% ammonia solution, drying in air). (g-h) 2D flattened flower, sample conditions identical to (e-f) except it was dried on a hotplate at 200°C (g) and an oven 80°C (h). (Reproduced from [51] with the permission of the Royal Society of Chemistry (RSC)).

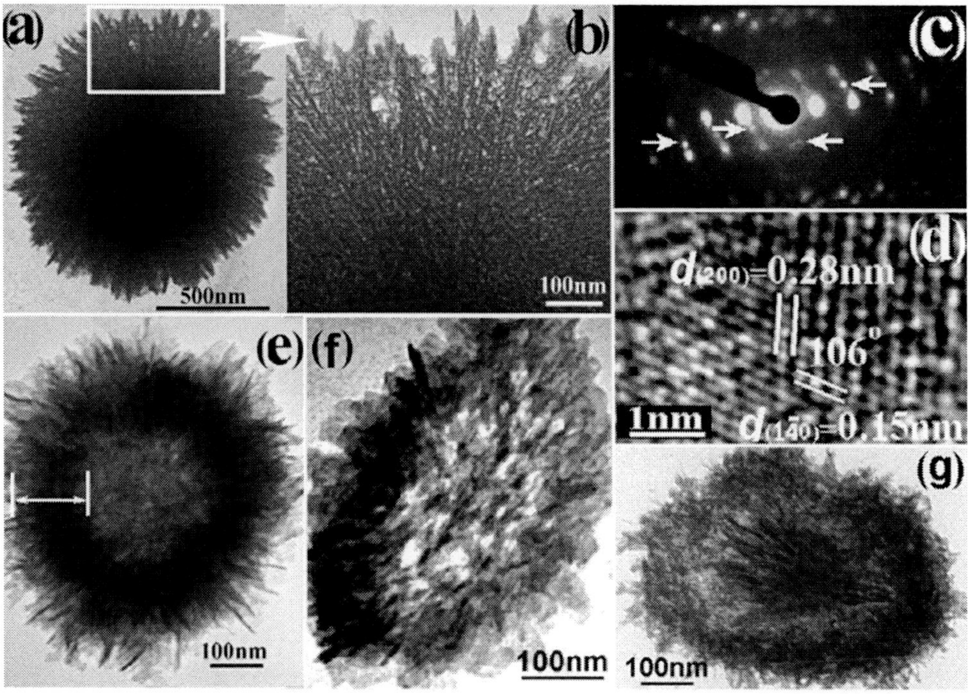

Figure 6. (a) TEM image of the flower-like structures produced by LP-PLA of a graphite target in 35% ammonia solution (laser fluence 125 mJ/pulse) for 5 h. (b) A higher magnification image of the framed region in (a), showing a high density of NPs surrounding the nano-petal framework. (c) [001] zone-axis MDA pattern from the tips of the nano-petals in (b), which corresponds closely to the calculated interlayer *d*-spacing of β-C3N4. Arrows point to different sets of [001] reflections (see text). (d) HR-TEM image recorded from the edge of the flower nanostructure that is oriented along [001]. (e) TEM image of a hollow flower formed after 8 h LP-PLA and prolonged drying. (f) Semi core-shell structure. (g) core-shell structure with a hollow tunnel. (Reproduced from [51] with the permission of the RSC).

Figure 6a shows a symmetric carbon nitride 'flower' together with its nano-petal building blocks (which look like flattened rods in the TEM). Figure 6b shows that these nano-petals appear fused together and 'interwoven' to form a lattice-like framework of the flower-like superstructure. The figure also shows the NPs that surround each nano-petal, and which fill in the holes within the framework to produce a dense, solid structure. Energy-dispersive X-ray (EDX) analysis confirms that carbon and nitrogen are present in all these structures, and micro diffraction analysis (MDA, Figure 6c) was also consistent with crystalline hexagonal β-phase carbon nitride oriented along the [001] zone-axis. Several [001] patterns in Figure 6c can be identified at the same time, indicating that the nano-petals consist of several domains, with different rotational orientations contributing to the diffraction pattern. The HR-TEM image in Figure 6d shows that the nano-petals at the very edge of the flower contain very few defects and are single crystalline, as was anticipated from the MDA pattern. Again, the lattice fringes ($d200$ = 0.28 nm, $d_{1\bar{4}0}$ = 0.15 nm) and their angles (106°) are in good agreement with the calculated values for hexagonal β-C3N4 [77].

The smaller NPs, which lie in and around the nano-petal-framework comprising the flowers, appeared to be mobile with respect to this framework, and diffused outward from the centre of the flower with longer drying times. The results of this diffusion can be seen in Figure 6e, where the solid carbon nitride flower (similar to that in Figure 6a) has become hollow. The NPs have diffused from the centre but remain loosely attached to the outside, making the outer shell of the flower appear fuzzy. The thickness of the fuzzy shell was ~ 140 nm and that of the hollow core was ~ 200 nm (shown as a lighter colour in the image). When the suspension was placed onto a hot-plate at 200°C for 0.5 h, the hollowing process was accelerated to form a semi-core-shell structure. The radial distribution of the NRs formed channels leading from the centre to the shell (Figure 6f). Another type of hollow structure was observed when the core-shell structures were essentially separated by a hollow tunnel, without linkage to the sphere (Figure 6g).

4.2. Arrangement of Components among the Architectures

The assembly process of hierarchical nanocrystals by repeated nucleation and growth is illustrated in Figure 7. By careful observation, it was found that the mesophase flower has a symmetrical structure (Figure 7a), consisting of many 2D nanopetals and radial growth from a central nucleus [78]. In addition, each nanopetal is constructed from many tiny nanocrystals or NR subunits (Figure 7b), as seen at the edge of the nanopetals (also see the later discussion). Surprisingly, just as for the nano-petals mentioned earlier in Figure 6b, these NRs also appeared to be 'interwoven' together into a similar lattice-like framework (Figure 7c). Interestingly, the NPs or NRs can self-assemble into a petal-like architecture along the *c*-axis direction. The tiny NPs attached side-by-side and fused into the wall of the nanopetal. This has the effect of making the nanopetal surfaces rough and curved, and can even make the two ends join together to form tubes (Figure 7d). Some nanopetals also exhibited structures inside the tubes, see the highlighted arrows in Figure 7d-f. The inset of Figure 7f shows a primary subunit with size of about 20 nm. It is worth pointing out that the length of the nanopetals is in the range of 1-20 μm, while the sizes of the subunit NPs or NRs have barely changed.

In principle, assembly is energetically favored because the formation of larger crystals can greatly reduce the interfacial energy of isolated NPs or NRs. Figure 7g summarizes a possible process for the carbon nitride self-assembly, mentioned above. By the aid of the solid substrate, hierarchically ordered nanocrystals assemble towards the surface of the substrate. New nanocrystals nucleate on the existing crystals and share the same edges. These structures are close-packed but still have spaces between them. It is still unclear why the NPs or NRs form such arrangements rather than random clumps. Further investigation of this mechanism is required.

The morphology of the self-assembled nanostructures is tuneable with the reaction conditions. By appling different building blocks, different final complex architectures on various scales can be achieved, as discussed in section 3.3. For lower ammonia concentration (25%), the carbon nitride formed 1D elongated nanoneedles or NRs. For low laser power (50 mJ) and short ablation time (2 h), the product contained mostly a sparse collection of these isolated NRs (Figure 8a). However, with increasing laser power (125 mJ) and ablation time (10 h), the concentration of NRs increased, and in some regions they began to coalesce into incomplete flowers (Figure 8b). The radial distribution of the NRs formed channels leading from the centre of the flower to the edge. These channels may form the conduits along which the NPs diffuse outwards. When a droplet of the suspension, identical to that shown in Figure 8a, was deposited on the substrate and dried for 12 h in a sealed tube, the carbon nitride produced flower-like structures constructed from pronounced tapered NRs (Figure 8c), instead of flowers made by the nanopetals (see Figure 7a). Figure 8d shows typical NR building blocks with a rod diameter of 10-20 nm.

Figure 7. (a) An individual flower exhibiting multi-fold symmetry. (Synthesis conditions: 35% ammonia solution, laser power 100 mJ, 3 h ablation and dried in a sealed tube) (b-f) Different arrangements of components that construct the final hierarchical superstructures. Arrows point to the inside surface of the tubular nanopetals. Inset of Figure 7f shows a single nanocrystal as building block, scale bar = 15 nm. (g) Schematic illustration of a suggested mechanism for the assembly of higher-order structures by nucleating new crystals through an edge-sharing configuration.

A new morphology was also observed when the carbon nitride was created using a laser power of 75 mJ for 3 h ablation and placed onto a silicon substrate, and then dried in a CPD. Figure 8e suggests that the product has a grass-like shape consisting of well-defined, protruding, short NRs. These structures can be assigned to the second class of morphology, mentioned previously. In contrast, if the substrate was dried in air (water evaporation time ~8 h), the individual NRs vertically assembled into bundles (see the highlighted black box region in Figure 8f) and formed stem-like structures. When the evaporation time of the liquid was increased to 24 h, the NRs developed a more complicated 3D structure, as shown in Figure 8g.

A possible explanation is that the longer duration for evaporation of the liquid allows sufficient time for the diffusion of all the NPs out of the flowers, leaving some liquid trapped in the gaps between the NR framework. With prolonged time, the wet NR framework has the opportunity to recrystallise or restructure into the shapes shown in Figure 8f-g.

Thus, by combining LP-PLA techniques with bottom-up self-assembly seeded growth, 3D complex well-arranged carbon nitride architectures can be created, with their structures and spatial organisation of the substrate controlled by deposition and drying conditions. The synthetic approach described above can also be extended to a range of other solid materials. For example, we have also demonstrated that a novel hierarchical and self-similar growth process can happen in zinc oxide materials by LP-PLA of a zinc plate submerged in aqueous SDS solution. The growth mechanism here also appears to involve an increase of the structural complexity from 0-dimensional nanoparticles to 1-dimensional nanorods, and then broadening of these into 2-dimensional 'nano-leaf' structures. Details can be seen in our previous report [79].

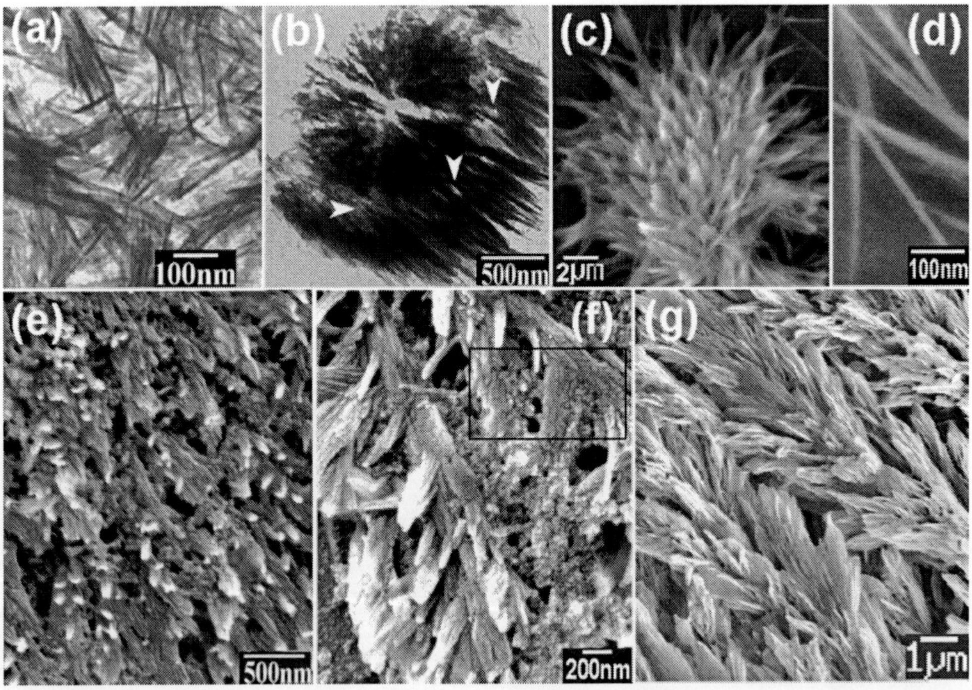

Figure 8. (a) TEM image of isolated carbon nitride NRs obtained by LP-PLA in 25% ammonia solution (laser fluence at 50 mJ per pulse, $t = 2$ h). (b) TEM image of an incomplete carbon nitride flower (25% ammonia solution, laser fluence at 125 mJ per pulse, $t = 10$ h) formed by the coalesced NRs. The arrows highlight the channels between the shell and the core. (c) After drying for 12 h (sample conditions identical to that in Figure 8a), numerous aligned NRs form the framework of the flower. (e) Grass-like structure constructed from many protruding NRs (25% ammonia solution, laser fluence at 75 mJ per pulse, $t = 3$ h, drying at CPD). (f) Stem-like structure consisting of vertical NR branches (region highlighted by black box), sample conditions identical to that in Figure 8e, but with drying in air. (g) After 24 h drying, the NRs have restructured to form 3D shapes. (Reproduced from [51] with the permission of the RSC).

5. Conclusion

With important discoveries continuing to be made in the field of nanomaterials, fulfilling the potential applications will require a greater control of material properties, along with a better understanding of the mechanisms for nanocrystal nucleation and growth. Extensive progress towards real-life applications have been made with ZnO, Cu2O, TiO2 and SiO2 nanostructures, because work on these materials is well established [74]. However, research on the understanding of the self-assembly processes of Group IV-V compounds, such as carbon nitride, is still in its infancy, and synthesis of complex nanostructures has only just begun. Our work has shown that it is possible to form self-assembled nanostructures and microstructures of crystalline carbon nitride. Through pulsed laser ablation of graphite in ammonia solutions, well-defined carbon nitride nanocrystal seeds were prepared. In the second step, 3D well-arranged architectures could be generated after these solutions were seeded onto a suitable substrate. The mechanism appears to involve an increase of the structural complexity from 0D NPs to 1D NRs, and then broadening of these into 2D nanopetals, which finally coalesce to 3D flowers. Factors such as the ammonia concentration, laser energy, the reaction time and the evaporation rate of the liquid are important in defining the structures observed. This unique process provides more insight into laser-induced chemical reactions in general, and of crystalline carbon nitride phases in particular, and may provide an alternative synthesis method for other composite nanostructures. A key advantage afforded by this aqueous multistage approach is the ability to systematically generate and control a diverse array of structures from a simple and inexpensive chemical route. With this improved understanding and concomitant seed control, LP-PLA may provide versatile and powerful industrial scale production processes for assembling complex architectures by specific design.

References

[1] Whitesides, G. M.; Grzybowski, B. *Science* 2002, *295*, 2418-2421.

[2] Whitesides, G. M.; Boncheva, M. *Proc. Natl. Acad. Sci. USA* 2002, *99*, 4769-4774.

[3] Tang, Z. Y.; Kotov, N. A.; Giersig, M. *Science* 2002, *297*, 237-240.

[4] Liu, B.;Zeng, H.C. *J. Am. Chem. Soc.* 2005, *127*, 18262-18268.

[5] Kempa, K.; Kimball, B.; Rybczynski, J.; Huang, Z. P.; Wu, P. F.; Steeves, D.; Sennett, M.; Giersig, M.; Rao, D. V. G. L. N.; Carnahan, D. L.; Wang, D. Z.; Lao, J. Y.; Li, W. Z.; Ren, Z. F. *Nano Lett.* 2003, *3*, 13-18.

[6] Duan, X. F.; Lieber, C. M. *Adv. Mater.* 2000, *12*, 298-302.

[7] Sun, Y.; Fuge, G. M.; Fox, N. A.; Riley, D. J.; Ashfold, M. N. R. *Adv. Mater.* 2005, *17* 2477-2480.

[8] Yang, H. G.; Zeng, H. C. *J. Phys. Chem. B* 2004, *108*, 3492-3495.

[9] Mirkin, C. A.; Letsinger, R. L.; Mucic, R. C.; Storhoff, J. J. *Nature* 1996, *382*, 607-609.

[10] Sun, Y. G.; Gates, B.; Mayers, B.; Xia, Y. N. *Nano lett.* 2002, *2*, 165-168.

[11] Li, Y.D.; Li, X. L.; Deng, Z. X.; Zhou, B. C.; Fan, S. S.; Wang, J. W.; Sun, X. M *Angew. Chem. Int. Ed.* 2002, *41*, 333-335.

[12] Motte, L; Billoudet, F.; Pileni, M. P. *J. Phys. Chem.* 1995, *99*, 16425-16429.

[13] Maillard, M.; Motte, L.; Pileni, M. P. *Adv. Mater.* 2001, *13*, 200-204.
[14] Liu, B.;Zeng, H.C. *J. Am. Chem. Soc.* 2004, *126*, 16744-16746.
[15] Petit, C.; Legrand, J.; Russier, V.; Pileni, M. P. *J. Appl. Phys.* 2002, *91*, 1502-1508.
[16] Ngo, A. T.; Pileni, M. P. *Adv. Mater.* 2000, *12*, 276-279.
[17] Liu, J. P.; Huang, X. T.; Li, Y. Y.; Sulieman, K. M.; He, X.; Sun, F. L. *J. Mater. Chem.* 2006, *16*, 4427-4434.
[18] Wang, J. B.; Zhang, C. Y.; Zhong, X. L.;Yang, G. W. *Chem. Phys. Lett.* 2002, *361*, 86-90.
[19] Dijkkamp, D.; Venkatesan, T.; Wu, X. D.; Shaheen, S. A.; Jisrawi, N.; Minlee, Y. H.; Mclean, W. L.; Croft, M. *Appl. Phys. Lett.* 1987, *51*, 619-621.
[20] Radhakrishnan, G.; Adams, P. M. *Appl. Phys. A* 1999, *69*, S33-S38.
[21] Pappas, D. L; Saenger, K. L.; Bruley, J.; Krakow, W.; Cuomo, J. J.; Gu, T.; Collins, R. W. *J. Appl. Phys.*, 1992, *71*, 5675-5864.
[22] Loir, A. S.; Garrelie, F.; Donnet, C.; Belin, M.; Forest, B.; Rogemond, F.; Laporte, P. *Thin Solid Films*, 2004, *453*, 531-536.
[23] Patil, P. P.; Phase, D. M.; Kulkarni, S. A.; Ghaisas, S. V.; Kulkarni, S. K.; Kanetkar, S. M.; Ogale, S. B.; Bhide, V. G. *Phys. Rev. Lett.*, 1987, *58*, 238-241.
[24] Ogale, S. B. *Thin Solid Films* 1988, *163*, 215-227.
[25] Yang, G. W. *J. Prog. Mater. Sci.* 2007, *52*, 648-698.
[26] Pyatenko, A.; Shimokawa, K.; Yamaguchi, M.; Nishimura, O.; Suzuki, M. *Appl. Phys. A.* 2004, *79*, 803-806.
[27] Tsuji, T.; Iryo, K.; Watanabe, N.; Tsuji, M.; *Appl. Surf. Sci.* 2002, *202*, 80-85.
[28] Dolgaev, S. I.; Simakin, A.V.; Voronov, V. V.; Shafeev, G. A.; Bozon-Verduraz, F. *Appl. Surf. Sci.* 2002, *186*, 546-551.
[29] Tarasenko, N. V.; Butsen, A. V.; Nevar, E. A.; Savastenko, N. A. *Appl. Surf. Sci.* 2006, *252*, 4439-4444.
[30] Compagnini, G.; Scalisi, A. A.; Puglisi, O. *J. Appl. Phys.* 2003, *94*, 7874-7877.
[31] Tsuji, T.; Hamagami, T.; Kawamura, T.; Yamaki, J.; Tsuji, M. *Appl. Surf. Sci.* 2005, *243*, 214-219.
[32] Wang, J. B.; Zhang, C. Y.; Zhong, X. L.; Yang, G. W. *Chem. Phys. Lett.* 2003, *367*, 10-14.
[33] Yang, G. W.; Wang, J. B.; *Appl. Phys. A* 2000, *71*, 343-344.
[34] Wang, J. B.; Zhang, C. Y.; Zhong, X. L.; Yang, G. W. *Chem. Phys. Lett.* 2002, *361*, 86-90.
[35] Yang, L.; May, P. W.; Yin, L.; J. A.; Rosser, K. N. *Diamond Relat. Mater.* 2007, *16*, 725-729.
[36] Nath, M.; Rao, C. N. R.; Popvitz-biro, R.; Albu-Yaron, A.; Tenne, R. *Chem. Mater.* 2004, *16*, 2238-2243.
[37] SDS is sodium dodecyl sulfate.
[38] Liang, C. H.; Sasaki, T.; Shimizu, Y.; Koshizaki, N. *Chem. Phys. Lett.* 2004, *389*, 58-63.
[39] Sasaki, T.; Liang, C. H.; Nichols, W. T.; Shimizu, Y.; Koshizaki, N. *Appl. Phys. A* 2004, *79*, 1489-1492.
[40] Liang, C. H.; Shimizu, Y.; Sasaki, T.; Koshizaki, N. *J. Phys. Chem. B* 2003, *107*, 9220-9225.
[41] Liang, C. H.; Shimizu, Y.; Sasaki, T.; Koshizaki, N. *Appl. Phys. A* 2005, *80*, 819-822.

[42] Liang, C. H.; Shimizu, Y.; Sasaki, T.; Koshizaki, N. *J. Mater. Res.* 2004, *19*, 1551-1557.
[43] Zeng, H. B.; Cai, W. P.; Li, Y.; Hu, J. L.; Liu, P. S. *J. Phys. Chem. B* 2005, *109*, 18260-18266.
[44] Ishikawa, Y.; Shimizu, Y.; Sasaki, T.; Koshizaki, N. *J. Colloid Interf. Sci.* 2006, *300*, 612-615.
[45] LDA is lauryl dimethylaminoacetic acid.
[46] CTBA is cetyltrimethylammonium bromide.
[47] Usui, H.; Sasaki, T.; Koshizaki, N. *Appl. Phys. Lett.* 2005, *87*, 063105-3.
[48] Yang, L.; May, P. W.; Yin, L.; Brown, R.; Scott, T. B. *Chem. Mater.* 2006, *18*, 5058-5064.
[49] Yang, L.; May, P. W.; Yin, L.; J. A.; Rosser, K. N. *J. Nanopart. Res.* 2007, DOI 10.1007/s11051-006-9192-4.
[50] Liu, A. Y.; Cohen, M. L. *Science* 1989, *245*, 841-842.
[51] Yang, L.; May, P. W.; Huang, Y. Z.; Yin, L. *J. Mater. Chem.* 2007, *17*, 1255-1257.
[52] Yang, L.; May, P. W.; Yin, L.; Huang, Y. Z.; Smith, J. A.; Scott, T. B. *Nanotechnology* 2007, *18*, 335605 (1-5).
[53] Yang, L.; May, P. W.; Yin, L.; Scott, T. B.; Smith, J. A.; Rosser, K. N. *Nanotechnology* 2006, *17*, 5798-5804.
[54] Liu, A.Y.; Wentzcovitch, R. M. *Phys. Rev. B* 1994, *50*, 10362-10365.
[55] Wang, J. B.; Lei, J. L; Wang, R. H. *Phys. Rev. B* 1998, *58*, 11890-11895.
[56] Yin, L. W.; Bando, Y.; Li, M. S.; Liu, Y. X.; Qi, Y. X. *Adv. Mater.* 2003, *15*, 1840-1844.
[57] Zeng, H. C. *J. Mater. Chem*, 2006, *16*, 649-662.
[58] Sjöström, H.; Staftröm, S.; Boman, M.; Sundgren, J. E. *Phys. Rev. Lett.* 1995, *75* 1336-1339.
[59] Terrones, M.; Redlich, P.; Grobert, N.; Trasobares, S.; Hsu, W. K.; Terrones, H.; Zhu, Y. Q.; Hare, J. P.; Reeves, C. L.; Cheetham, A. K.; Rühle, M.; Kroto, H. W.; Walton, D. R. M. *Adv. Mater.* 1999, *11*, 655-658.
[60] Rodil, S. E.; Muhl, S. *Diamond Relat. Mater.* 2004, *13*, 1521-1531.
[61] Hellgren, N.; Johansson, M. P.; Broitman, E.; Hultman, L.; Sundgren, J. E. *Phys. Rev. B* 1999, *59*, 5162-5169.
[62] Muhl, S.; Méndez, J. M. *Diamond Relat. Mater.* 1999, *8*, 1809-1830.
[63] Hellgren, N.; Guo, J.; Såthe, C.; Agui, A.; Nordgren, J.; Luo, Y.; Ågren, H.; Sundgren, J. *Appl. Phys. Lett.* 2001, *79*, 4368-4350.
[64] Souto, S.; Pickholz, M.; Dos Santos, M. C.; Alvarez, F. *Phys. Rev. B* 1998, *57*, 2536-2540.
[65] Ripalda, J. M.; Montero, I.; Galan, L. *Diamond Relat. Mater.* 1998, *7*, 402-406.
[66] Ogata, K.; Chubaci, J. F. D.; Fujimoto, F. *J. Appl. Phys.* 1994, *76*, 3791-3796.
[67] Khabashesku, V. N.; Zimmerman, J. L.; Margrave, J. L. *Chem. Mater.* 2000, *12*, 3264-3270.
[68] Diaz, J.; Paolicelli, G.; Ferrer, S.; Comin, F. *Phys. Rev. B* 1996, *54*, 8064-8069.
[69] Tabbal, M.; Merel, P.; Moisa, S.; Chaker, M.; Ricard, A.; Moisan, M. *Appl. Phys. Lett.* 1996, *69*, 1698-1700.
[70] Le Normand, F.; Hommet, J.; Szörényi, T.; Fuchs, C.; Fogarassy, E. *Phys. Rev. B* 2001, *64*, 235416.

[71] Beamson, G.; Briggs, D.; *High Resolution XPS of Organic Polymers*; 1992, John Wiley and Sons, Chichester, pp.182.
[72] Guo, Y. J; Goddard, W.A. *Chem. Phys. Lett.*, 1995, *237*, 72-76.
[73] Niesen, T. P; De Guire, M. R. *J. Electroceram*. 2001, *6*, 169-207.
[74] Sounart, T. L.; Liu, J.; Voigt, J. A.; Hsu, J. W. P.; Spoerke, E. D.; Tian, Z.; Jiang, Y. B. *Adv. Funct. Mater.* 2006, *16*, 334-344.
[75] Lu, S. N; Chung, J.; Ruoff, R.S. *Nanotechnology* 2005, *16*, 1765-1770.
[76] Brinker, C. J.; Lu, Y. F.; Sellinger, A.; Fan, H. Y. *Adv. Mater.* 1999, *11*, 579-585.
[77] Liu, A. Y.; Cohen, M. L. *Phys. Rev. B* 1990, *41*, 10727-10734.
[78] Li, Z. Q.; Xiong, Y. J.; Xie, Y. *Nanotechnology* 2005, *16*, 2303-2308.
[79] Yang, L.; May, P. W.; Yin, L.; Scott, T. B. *Nanotechnology* 2007, *18*, 215602 (1-5).

In: Nanotechnology Research Developments
Editor: R. Jiménez-Contreras, pp. 315-378
ISBN: 978-1-60021-899-6

Chapter 9

FORMATION AND ASSEMBLY OF HIERARCHICAL NANOARCHITECTURES BY BIOINSPIRED TECHNIQUES

Huilan Su, Di Zhang and Qun Dong
State Key Lab of Metal Matrix Composites,
Shanghai Jiaotong University, Shanghai 200030, China

ABSTRACT

The bioinspired technique has developed into an intersection of materials science, soft chemistry technique, and biotechnology. In this chapter, the bioinspired strategies integrating biotemplate, biomineralization, and biomimesis are adopted to fabricate functional nanomaterials with hierarchical architectures and interrelated unique properties.

Some natural biomaterials, such as eggshell membrane, silk fiber, etc., are introduced into different liquid processes to construct a series of functional hierarchical nano-architectures (containing oxides, chalcogenides, noble metals, and/or their composites). Under the direction of the biotemplates or biosubstrates, the nanocrystallites units of about 2-10 nm are formed, developed, and assembled into hierarchical architectures from the nanoscale to the macroscale. The nucleation, growth, and assembly of nanocrystallite units depend on the collaboration of special proteins structures and functional biomacromolecules of relevant biomaterials as well as the processing conditions. Moreover, the biomaterials usually provide both the physical substrates and the inductive substances to direct some complicated reactions. In addition, some amino acids of the biomaterials function as the capping agents to achieve the organized growth and assembly of small-sized nanocrystallites. As-synthesized hierarchical nanoarchitectures take on outstanding chemical or physical properties and have valuable applications in photocatalysis, gas sensing, ductile ceramics, and semiconductor technology.

This method and the relevant ideas provide a novel and versatile avenue to synthesize a new family of functional nanomaterials by integrating material science, chemistry, and biotechnology.

INTRODUCTION

Nanoscale functional materials have attracted much attention because of their potentially excellent and fascinating properties. In the progressive stage, the assembly of the nanoscale units into a hierarchical architecture would be required to obtain higher functionality and performance. Nevertheless, there are few feasible reports on the construction of hierarchical nanomaterials, especially practical and functional nanocomposites.

"Where nature finishes producing its own species, man begins, using natural things, with the help of this nature, and creates an infinite number of species". These words epitomizing the artist-scientist Leonardo da Vinci fully embody the competence of human beings to utilize natural substances for synthesizing new materials. Natural organisms or biological structures have evolved as a result of highly complex and elegant mechanism, the secrets gleaned from nature are a great inspiration for developing inorganic functional materials with hierarchical structures. Recently, mimicking vital functions in aqueous systems has been noteworthy as a promising nanofabrication technique for functional materials. Quite a few biological species have been utilized to construct inorganic materials, the list includes oyster shells, sea urchin spines, DNA, viruses, cuttlefish bone, nacre, and peptide nanotubes, diatoms and skeletal plates, butterfly wings, shell membrane, silk fiber, hair, wood, and other plants (leaves, paper, pollen grains), etc. These biogenic techniques enable the construction of hierarchical architectures including macro- and nanostructures by growing inorganic crystals with coexisting biospecies under ambient conditions.

Our bioinspired synthesis strategies integrating biotemplate, biomineralization, and biomimesis are based on the premise that amphiphilic and bioorganic macromolecules can be utilized for tailoring elaborate hierarchical biostructures through the self-assembly of nanoscale units to engender new functional properties.

1. SYNTHESIS OF HIERARCHICAL NANOSTRUCTURED METAL OXIDES WITH INTERWOVEN NETWORKS BY EGGSHELL MEMBRANE DIRECTED SOL–GEL TECHNIQUE

It has been discovered that in some biological systems such as diatoms and sponges, the formation of solid silica nanostructures with precisely controlled morphologies is the function of proteins and polysaccharides. These biomacromolecules can direct and bind the nucleation and the growth of inorganic particles. As a convenient and representative biomaterial, eggshell membrane (ESM) may be nominated to direct the fabrication of hierarchical nanostructured oxides.

The avian eggshell is composed of multilayered membranes and a calcified extracellular matrix, which are sequentially assembled during the motion of the egg along the oviduct. The examination of cross-fractured eggshell by FESEM, reveals that the calcified shell is composed of an inner mammillary layer (ML) and an outer palisade layer (PL) resting upon the shell membranes. The ESM is non-mineralized, collagen-based matrices interposed between the egg white and the mineralized shell, and consists of outer shell membrane (OSM), inner shell membrane (ISM), and limiting membrane (LM). The three layers are readily distinguished. Those deposited sparsely on the outer layer of the shell membranes are the mammillary knobs for penetrating into the tips of the mammillary cones.

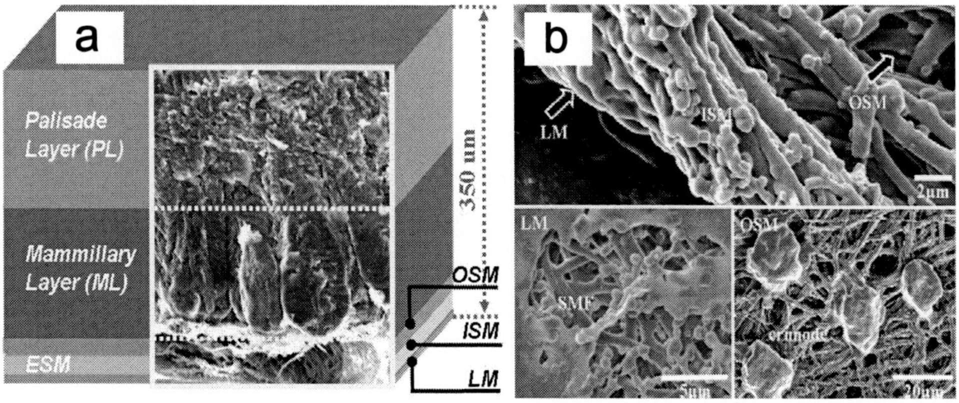

Figure 1. (a) FESEM images of eggshell membrane and the corresponding schematic illustration of eggshell structures. (b) FESEM images revealing the relative locations and the configurations of OSM, ISM, and LM.

In addition, the interwoven and coalescing nature of the shell membrane fibers (SMF) with an average width ranging from 500 nm to 2 μm is evident both in the OSM and the ISM. Furthermore, some of the broader fibers have a longitudinally striated appearance, where two or more narrower fibers join each other for part of their length. The amorphous LM shows an undulant appearance, which is due to a layer of spherical protein granules formed between the ISM and the LM. Within each layer of ESM, the fibers are randomly arranged.

Quoting from Cook et al. [1], "the synthetic replication of evidently useful biological structures by a simple casting process is expected to introduce some of the superb properties of biological structures into man-made materials." The most feasible method of attaining this goal at a high level seems to be the aqueous sol–gel technique. The colloid components can interact with some biomacromolecules and undergo surface-preferred gelation to form the coats on the biomaterials, even to veraciously replicate the biotemplate frameworks down to the nanoscale. Moreover, the sol–gel process offers several advantages over other methods, including simplicity of procedure, lower processing temperature, and better homogeneity. Commonly, metallic alkoxides are used as starting materials to prepare fine-scaled metallic oxides by the sol–gel processes. However, metallic alkoxides are both expensive and extremely sensitive to moisture, heat, and light, implying that the process is relatively difficult to control and not economical. Therefore, various metal salts are preferred as the precursors. Once biomaterials are introduced to the reformative sol–gel techniques, a series of intricate hydrolysis and condensation reactions happen in the sol–gel process. It seems to be a process that enable the easy preparation of hierarchical metal oxides, such as TiO2, SnO2, SiO2, ZnO, ZrO2, and so on. Therefore, biomimetic synthesis associated with the sol–gel technique may be an ideal approach for the design and fabrication of functional materials with predetermined physical and chemical properties. Our present study focuses on exploring the bottom-up assembly of hierarchical metal oxides by a biogenic synthesis technique introducing the ESM as the template.

1.1. Synthesis of Hierarchical Mesoporous Titania with Tunable Anatase and Rutile

Mesoporous titania has outstanding chemical and physical properties, and widespread applications in photovoltaics, photocatalysis, ductile ceramics, pigmentation, optics, and so on, which depend more on the phases and crystallite sizes of building particles and relevant hierarchical structures. Therefore, the design of hierarchical mesoporous titania with tailored structures and morphologies can provide promising properties.

Herein, ESM is chosen as the biotemplate, in view of the special proteins patterns and functional residues of the ESM biomacromolecules, to synthesis hierarchical interwoven titania through a sol–gel approach followed by thermal decomposition. In a typical procedure, titanium tetrachloride ($TiCl_4$) is added slowly into ice water under stirring to obtain 0.04 M $TiCl_4$ colloid medium, and the pH values of the media are adjusted to 1, 2, and 3 by adding 1 M NaOH solution, respectively. The membrane is then immersed in the above colloid media for 13 h at room temperature, subsequently rinsed with distilled water and dried naturally to obtain gel-ESM hybrids. The hybrids are calcined at 400, 550, 700, and 800 °C (heating rate 35, 18, and 1 °C/ min), respectively, to form the resulting white sheets.

TGA results show that the ESM template start pyrolyzing at 200 °C and absolutely pyrolyzed by 550 °C. The gel-ESM hybrids containing titanic and titanyl precursors also decompose around 200 °C, and 18.3% (by mass) inorganic ingredients remain when the template ESM is completely removed at 550 °C. XRD results indicate that natural ESM and the hybrids are both totally amorphous. After calcination at 400 °C, TiO_2 crystallites come into being. However, the calcination temperature about 550 °C is appropriate for TiO_2 formation attended by the thorough removal of the biotemplate. The crystallization of TiO_2 can be well established at 550 °C. An average crystallite size of about 6 nm is estimated according to the line width analysis of the TiO_2 (101) diffraction peaks based on the Scherrer formula. Higher calcination temperature makes for the increase in the size of titania crystallites. When the calcination temperature is increased to 800 °C, some diffraction peaks of rutile TiO_2 increasing abruptly, indicates that the phase transformation corresponding to rutile TiO_2 mostly occurs at this temperature.

Representative FESEM images of titania prepared at pH 2 and 550 °C compared with the original ESM templates including OSM, ISM, and LM, suggest that the refined reprography of the hierarchy is fully achieved. The obtained OSM-based TiO2 are composed of interwoven fibers and sparsely crunodes depositing on the ceramic fibers with the average interval about 11 μm, which is smaller than the original spacing 31 μm due to the heat shrinkage. It is obvious that the refined meshworks of natural OSM are also duplicated besides the larger intersectional blocks. The interwoven and coalescing fibers of TiO2 replica range from 0.2 to 1.2 μm in diameter in comparison with original SMF approximate 0.5 to 3.0 μm in diameter. Although considerable shrinkage is engendered during the calcination, it can be lowed by repeatedly immersing in the sol with no changes in the configuration. A typical image of the cross section shows that the fibers complecting the networks are tubular titania. Moreover, the thickness of obtained TiO2 networks is decided by the thickness of primeval ISM. As ISM is composed of multilayers of interlaced fibers, the thickness of the replica TiO2 can be adjusted according to the actual thickness of ISM layers retained up to 20 μm. When using LM as the template, a lumpy titania layer composed of contiguously hemispherical caps is obtained. These special silkworm-like structures come into being and array according to the spherical protein granules distributing under LM. These hemispherical prominences have the diameter of about 900 nm, which is in agreement with the size of slightly smaller protein granules. Therefore, it can be concluded that a precision replication of ESM hierarchical structures from OSM to innermost LM could be achieved by nanocrystalline titania.

To investigate the detailed morphologies and the structures of ESM-morphic titania, representative FESEM images of the samples calcined at different temperature (heating rate 35 °C/min) show the interwoven and coalescing TiO_2 fibers obtained at 550 °C ranging from 0.2 to 1.2 μm in diameter in comparison with that of the original membrane fibers. Although there exists considerable shrinkage, the overall refined reprography of the interwoven hierarchy is retained.

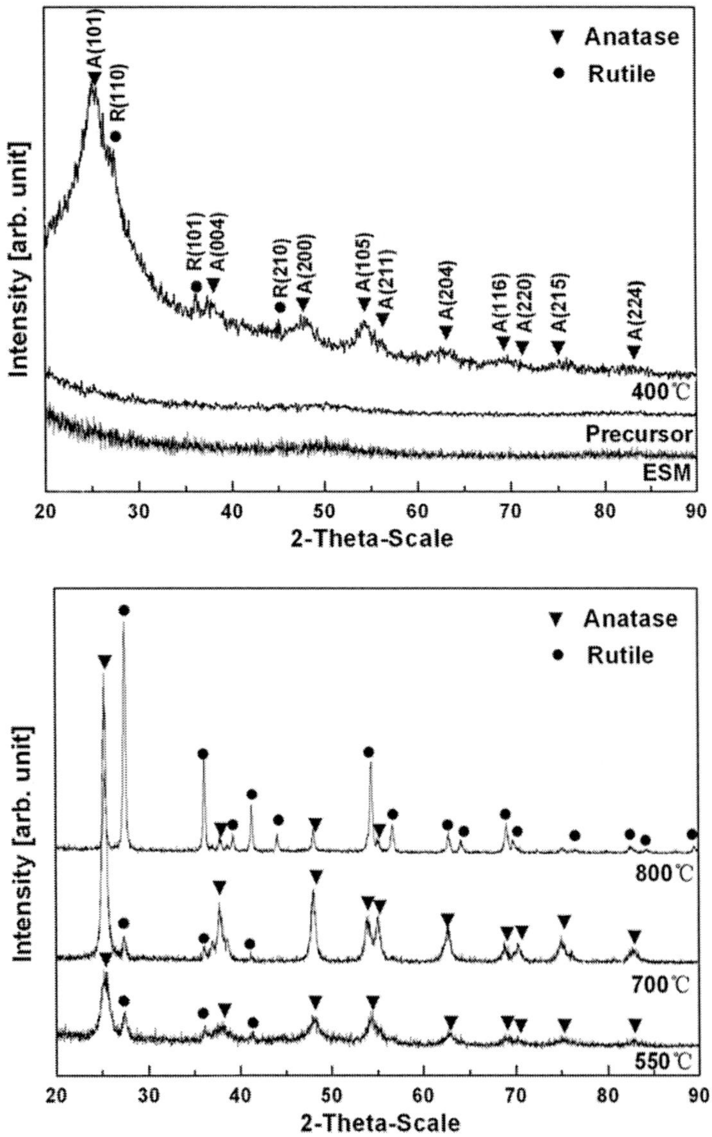

Figure 2. XRD patterns of ESM, gel-ESM precursor prepared at pH 2, and the sinters.

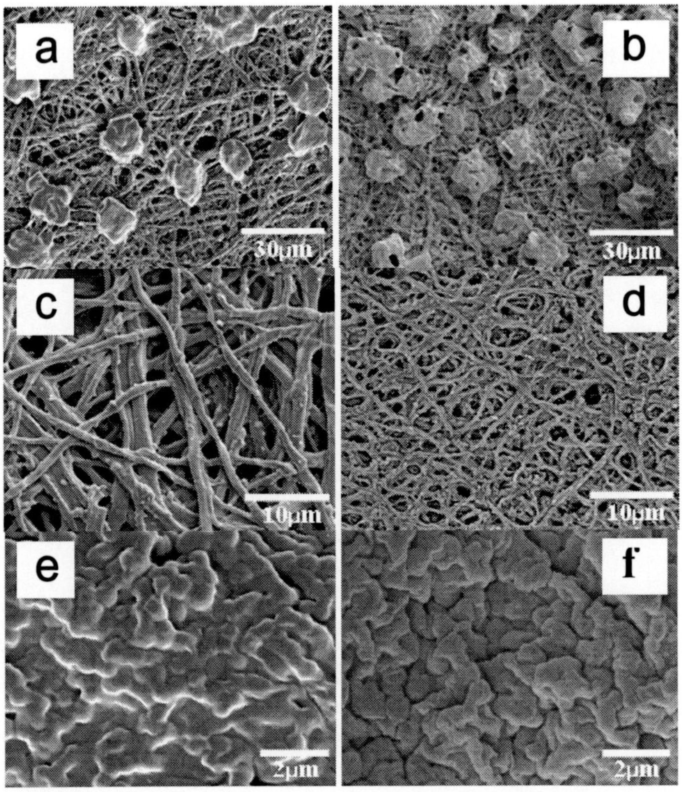

Figure 3. FESEM images showing the different layers of original ESM: (a) OSM, (b) ISM, and (c) LM; compared with the morphologies of as-prepared titania: (d) OSM-based, (e) ISM-based, and (f) LM-based TiO_2, respectively.

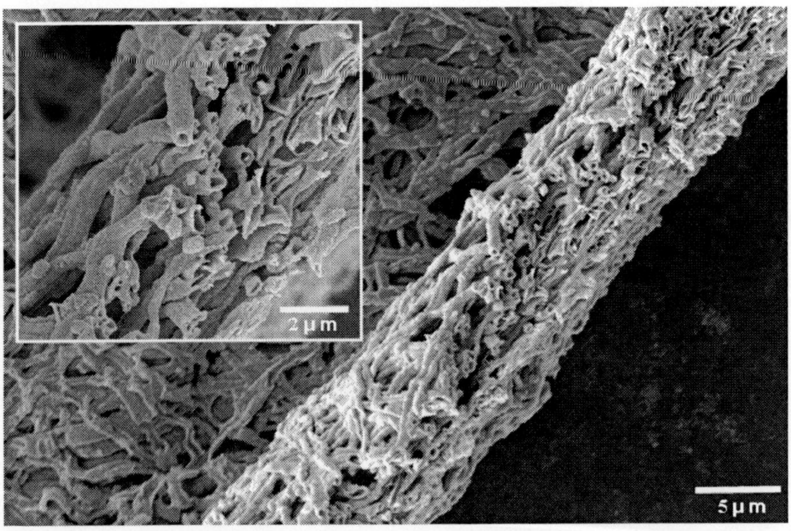

Figure 4. FESEM images of TiO_2 network by ESM using the sol-gel approach showing the overview of the cross section, and higher magnification of the cross section showing the broken hollow tubes.

When the calcination temperature is elevated to 700 °C, there are some coils due to higher temperature and presents interconnected networks with wavy figures. These characteristics are extended to an utmost as increasing calcination temperature to 800 °C. Obviously, titania networks have turned into unfolded mesh structures. With the higher magnification, it is observed that the fibers are vividly composed of short rods presenting porous figures. These fibers are inclined to stretch into belts, which are twice as large as ones at 550 °C in diameter. As a while, calcination temperature is crucial for the synthesis, which realizes the replication of ESM networks by titania below 700 °C while becoming laminated meshworks above 800 °C. At given calcination temperature, other parameters become more important. Heating rate of the whole calcination process may influence the morphologies and the crystal growth of target materials. Comparing with the higher heating speed 35 °C/ min, the heating rates about 18 and 1 °C/min are introduced to the corresponding process. With 18 °C/min at a medial level, ESM-morphic titania can also be formed, and some drapes along the fibers are found with careful observations.

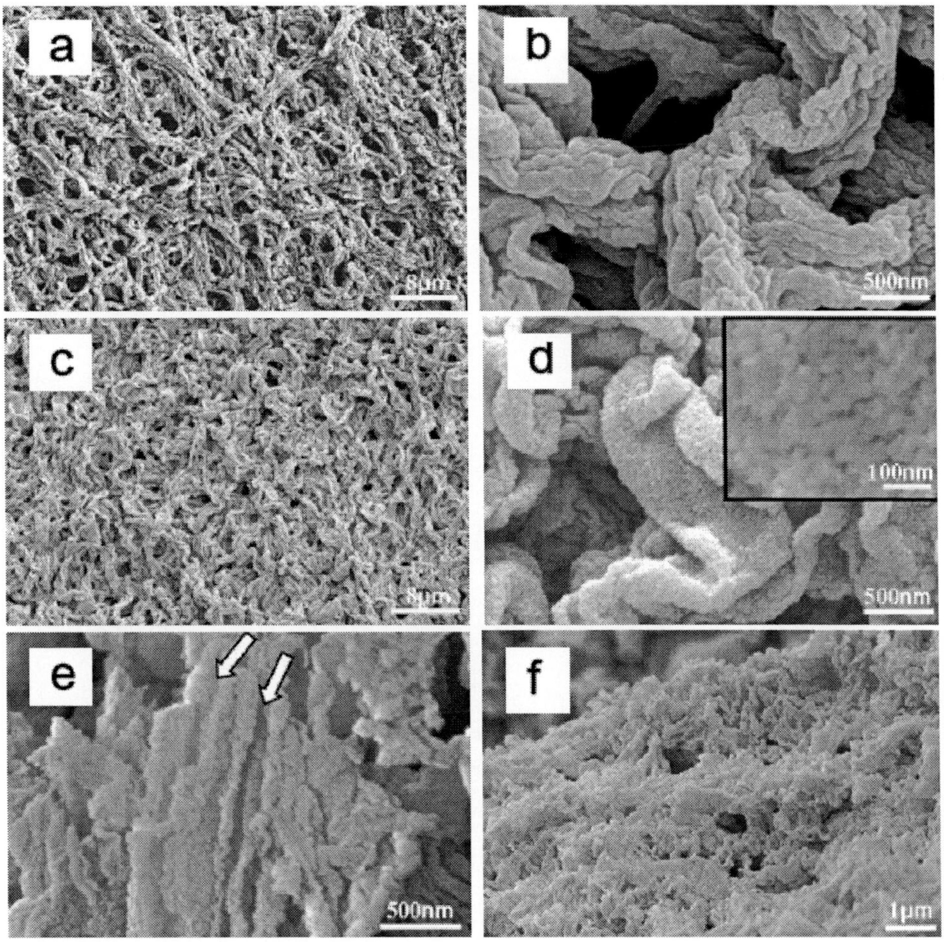

Figure 5. FESEM images of interwoven titania synthesized at 550 °C with different heating rates: (a) 18 °C/min and (c) 1 °C/min. Images (b) and (d) show relevant higher magnifications of (a) and (c). Inset of (d) shows porous structures of titania fibers. (e) The vertical section of TiO_2 sheets by the heating rate 18 °C/min. (f) The cross-section of TiO_2 sheets by the heating rate 1 °C/min.

Those subtle drapes may be formed during the sol–gel process, which contains the intricate interactions between the colloids and ESM biomacromolecules. When the heating rate reduced to 1 °C/min, the minutiae have vanished and more pores come into being. The low heating rate may result in particle agglomeration, with the crystallites grown at a larger scale to larger one, to blot out the minutiae and show porous structure. Moreover, interwoven fibers are arranged and fallen in loose folds. Probably based on the prominent dehydration of organic substances under the low heating rate, the framework of biotemplate ESM shrinks stepwisely before its completely pyrogenation. The TiO2 fibers constituting the porous networks might be tubular structures. The sample calcined at 550 °C with the heating rate 18 °C/min is consisted of channels about 200 nm in diameter, which are probably formed by the gelation and the crystallization of titania colloids along ESM fibers during the removal of the template. While the heating rate decreases to 1 °C/min, the sample calcined at 550 °C displays porous modality. Actually, these pores embody the channels as same as the above, and seems smaller as a result of overlay by the growth of building units with the lower heating rate.

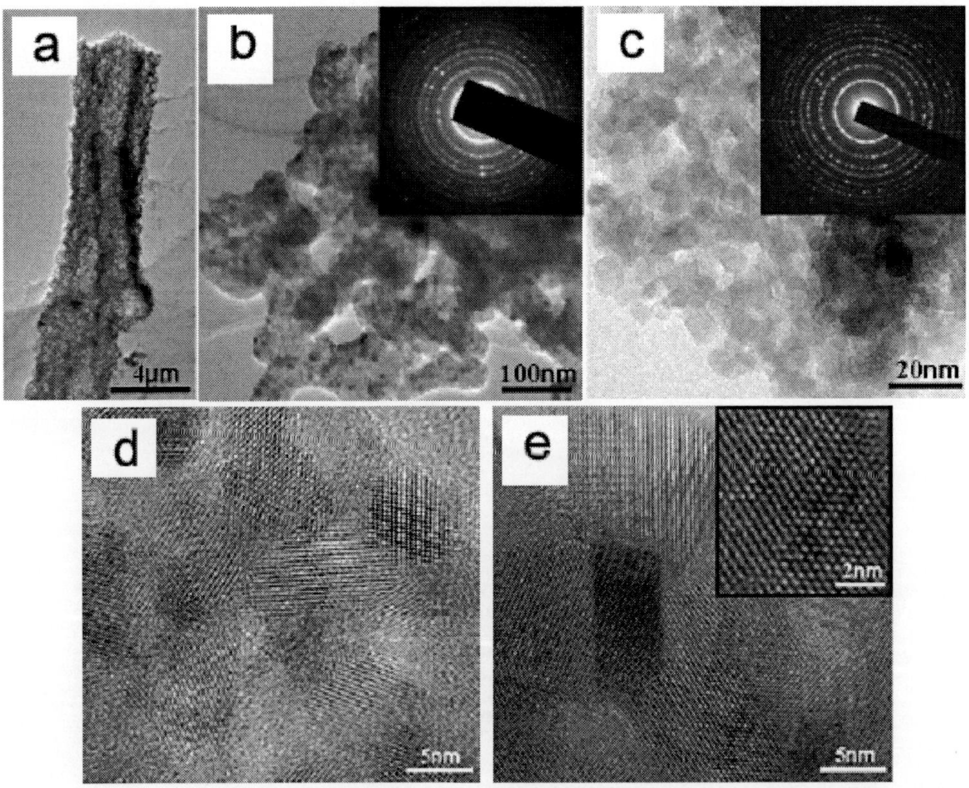

Figure 6. TEM images of (a) an individual TiO_2 fiber; the building particles of titania fibers synthesized at 550 °C with different heating rate: (b) 1 °C/min and (c) 35 °C/min. Insets in (b) and (c) show corresponding SAED patterns. HRTEM images of ESM-morphic TiO_2 obtained at 550 °C and pH 2 with the heating rate: (d) 35 °C /min; (e) 1 °C /min. Inset in (e) corresponds to anatase TiO_2.

The above analysis about the growth of the titania units is further proved by TEM results. The porous fibers synthesized at 550 °C with the heating rate 1 °C/ min are assembled by small particles. At higher magnification, these particles with an average size of about 50 nm, containing tiny nanocrystallites, are clearly seen. With the heating rate 35 °C/min, these particles show the serried and ordered assembly with the average size about 6 nm, which makes an allusion to the replication readily achieved by the small particles rather than large ones. The SAED patterns show that nanoscaled TiO_2 synthesized with different heating rate are polycrystalline on the basis of the uniform central and diffraction spots, and the diffraction rings are indexed to (101), (004), (200), (105), (204), and (116) diffractions of titania. HRTEM images show the difference of TiO_2 nanocrystallites prepared with distinct heating rate at the same calcination temperature 550 °C. The sizes of these nanocrystallites are close to 5–6 nm. However, there are more micropores of about 2 nm observed in the sample prepared at higher heating rate. A careful look also reveals that the present TiO_2 is characterized by ''nanostructured'' particles consisting of the nanocrystalline domains with structural defects such as twins.

The pH value of the impregnant has a significant effect on the phase of TiO_2 crystalline products. The high acidity is in favor of the formation of rutile phase. When the pH value of the precursor is about 1, the product is a mixture of rutile (majority) and anatase (minor). With increasing pH to 2, the product is formed by a remixture of rutile (minor) and anatase (majority). It is evident that the content of anatase phase in the products increases with the mount of the pH value. At pH = 4, the rutile phase disappears yet the anatase phase becomes dominant. As pH > 4, only anatase phase can be obtained. The concentration of $TiCl_4$ medium has direct affinity with the acidity of the impregnant. Decreasing the concentration of $TiCl_4$ medium favors the formation of anatase TiO_2. When the concentration is tuned from 0.04 to 0.03 M, the peak intensity of rutile titania is abruptly weakened. As the concentration is decreased to 0.02 M, further to 0.01 M, there are feeble diffraction peaks of rutile TiO_2. All the broader reflection peaks of different samples indicate that as-prepared TiO_2 nanocrystallites have small dimensions. The average crystallite size of about 5–8 nm is estimated according to the line width analysis of the TiO_2 (101) diffraction peaks based on the Scherrer formula.

FESEM image also indicates that the interwoven and coalescing TiO_2 fibers prepared at pH 1 and calcined at 550 °C range from 0.2 to 1.5 mm in diameter in comparison with the original membrane fibers. Although considerable shrinkage, there retains the overall refined reprography of the interwoven hierarchy. At pH 2, there are some drapes engendered on the surfaces of titania fibers while the replica of ESM hierarchy can also be obtained. However, when the pH value is elevated to 3, the morphology of obtained titania has a mutation, the titania fibers are distorted, and a sort of connected petalshaped structure comes into being. The two TiO_2 samples involving either 0.04 M or 0.01 M $TiCl_4$ impregnant display the analogical hierarchical interwoven conformations. However, there exists the evident distinction resulted from the effects of the impregnant concentration. The titania fibers synthesized at higher concentration are composed of lager particles while smaller titania nanocrystallites are obtained at lower concentration. The lumpy structures may form due to the agglomeration among TiO_2 particles at the higher concentration or higher acidity of the media precursor. A careful look at the surface of the biomorphic titania reveals that some mesopores are occurred, it is more clear for the sample involved in 0.01M $TiCl_4$ impregnant. So based on the alteration of the pH value and the concentration of $TiCl_4$ impregnant, different morphologies of TiO_2 samples can come forth.

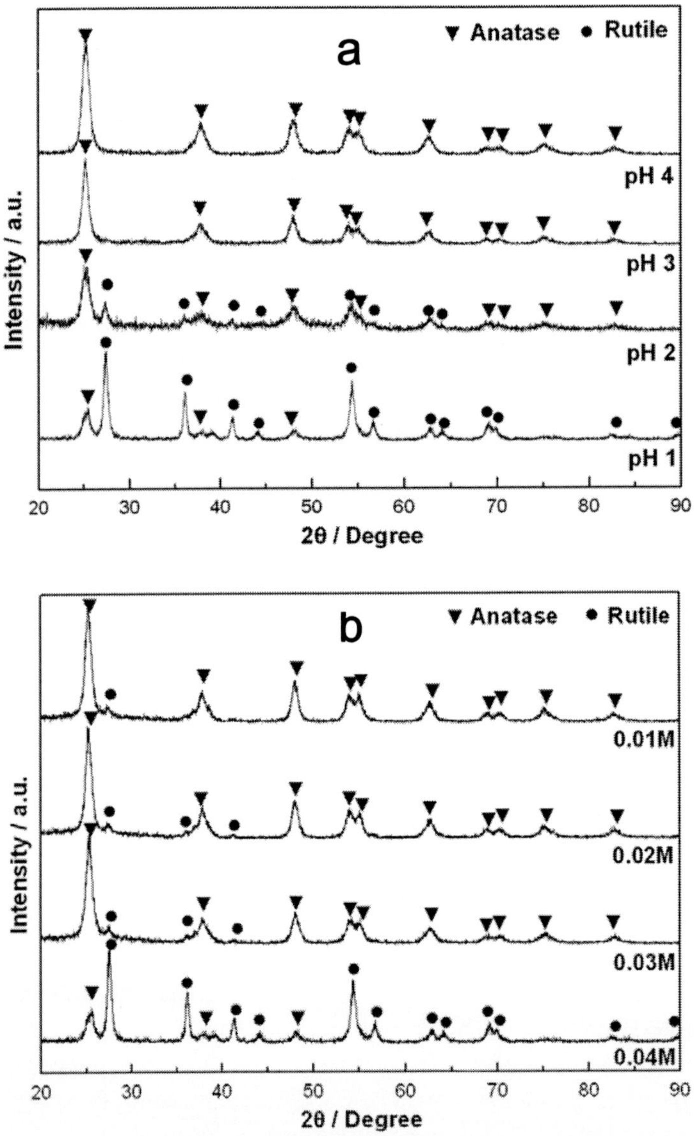

Figure 7. XRD patterns of Ti-ESM hybrids prepared under the different experimental conditions: (a) different pH value and (b) different concentration of the impregnant, and the sinter at 550 °C.

In balance, the synthesis conditions have momentous influences on the preparation of the interwoven titania. It observed that the higher acidity of the colloid impregnant is preferred to stabilize the ESM proteins, further to achieve the synthesis of interwoven titania. Dipping time is another concernful factor that may indirectly influence the morphology of resulting material. It has been affirmed that the proper dipping time can be tuned flexibly from 6 to 30 h. By decreasing the concentration of the colloid impregnant as well as shortening dipping time, an overall thinning of the wall and increased shrinkage are observed. The dilution of the nanoparticle dispersion leads to insufficient coating, thus the original structure is not maintained. Titania with ESM configuration is best obtained when the ESM immersed into pH 1 medium for 13 h, accompany a proper calcination treatment which is mainly responsible for the pyrogenation of the biotemplate and the crystallization of nanocrystalline TiO_2. The size of titania crystallites is mainly affected by the calcinations temperature, while the morphology is dominated not only by the calcination temperature, but also by the heating rate. However, the medium pH value and dipping time are crucial for the interfacial sol–gel process. The interactions between the colloids and ESM biomacromolecules present intricate and perplexing. It can be ascribed to a special sol–gel approach involving the biotemplate ESM, during which the biomacromolecules may direct the bottom-up assembly into hierarchical titania. To understand these investigations, a possible mechanism for this template-directed process is explored.

It is known that ESM are collagen-based matrices consisted of fibers of type I, V, and X collagen located in the core of each fiber. Each SMF is composed of a collagenous core, which contains smaller collateral channels, surrounded by a glycoprotein mantle. The core is predominantly composed of collagen and osteopontin proteins, while the mantle is rich in polyanions exhibiting a variety of keratan sulfate (KS) epitopes and Dermatan sulfate (DS). Lysozyme and clusterin are also present in ESM at high levels, and both have been detected within SMF, distributing discontinuously with no apparent partitioning between the core and the mantle. The cationic lysozyme can be bound to those anionic sites DS and particularly KS in the membrane. Elegant studies with clusterin have demonstrated that it exerts a powerful chaperone-like activity with lysozyme. In virtue of these special components and configurations, the inherent interactions of ESM and the colloids shall be daedal, and the correlations between the residues of SMF biomolecules and the colloid components are explored to understand the tamplate translation process. The FTIR spectrum of ISM shows the diagnostic peaks at 1641, 1532, and 1234 cm-1. These peaks are assigned to the amide I, amide II, and amide III bands of the proteins from natural ISM, respectively. The amide I peak, predominantly corresponding to the C=O stretch, shifts to 1632 cm-1 after ISM being immersed in the sol media. The red shift of the amide I peak indicates that the C=O bonds in the peptide chains are weakened because of the formation of new chelate bonds between C=O bonds and Ti ions involving in H-bonding with TiOH2+ or TiOH3+ surface groups. This indication implies that the carbonyl groups on the surface of glycoprotein biomolecules are the binding sites, and that the COO— symmetric stretching vibrations at 1446 shift to 1453 cm-1 of carboxylate anions complexed with surface Ti centers dominating in the spectra of the proteins with the sol precursor capped. The intensities of C—O—C and C—OH (at 1058 and 1400 cm-1, respectively) peaks are notably decreased, which indicates the interaction of hydroxy residues with cations and the transformation of glycoprotein configuration. It is evident not only the amide bands but also the characteristic COO— bands are not observable in the spectra of the sinters, thus confirming that the removal of the protein chains of the template. Above 2000 cm-1, both natural ISM and GMH exhibit the intense antisymmetric and symmetric —CH stretching vibrations (at 2926 and 2854 cm-1, respectively) of the —CH2— groups in the hydrocarbon moiety [2]. The shoulder at 2960 cm-1 can be associated with the asymmetric stretching of the terminal —CH3 group of the protein chains. In addition, the C—H stretching signals of the polypeptides at 3290 cm-1 are superimposed on an underlying surface —OH broad stretching band centered at 3400 cm-1. Below 900 cm-1, the characteristic vibrations of the Ti—O—Ti networks can be seen in the curves of GMH and the sinters according with the XRD results.

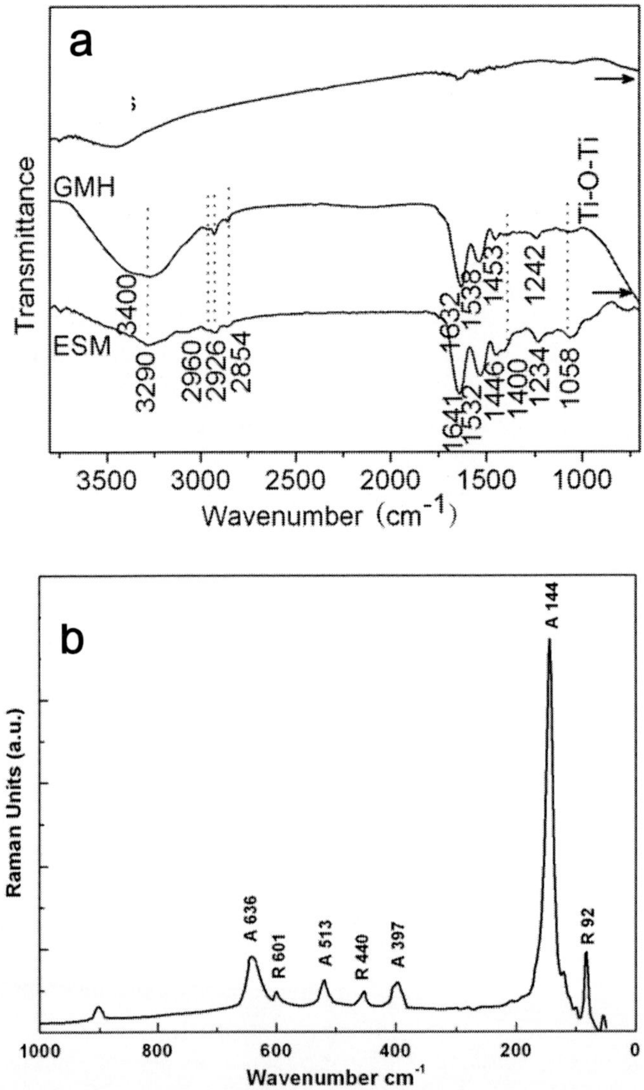

Figure 8. (a) FTIR spectra of original ISM, GMH prepared at pH 2, and the corresponding sinters at 550 °C, respectively. (b) Raman spectrum of ESM-templated titania prepared at pH 2 and 550 °C.

Herein, the interfacial sol-gel process is based on the adsorption of cationic matters from the sol media to the glycoprotein mantle of SMF containing abundant of carboxylic and hydroxy residues, subsequently a series of complicated reactions go along to drive the amorphous organogels for titanic and titanyl compositions, which further crystallize into titania nanocrystallites. The formation of nanocrystalline TiO2 can be denoted as: TiCl4 → TiOH3+ / TiOH2+ / TiO2+ → Ti (IV) oxo species → amorphous TiO2 → TiO2 nanocrystallites. The cationic TiOH3+ / TiOH2+ / TiO2+ in the sol system are prone to interact with carboxylic and hydroxy residues of glycoprotein, which has been proved by IR results. The interfacial sol-gel process would involve the strong affinity among some residues binding containing multiple hydrogen bonds, Van Der Waals interactions, and be enslaved to the ordering of surface polypeptide loops that bury a portion of the colloid particles in the protein interiors.

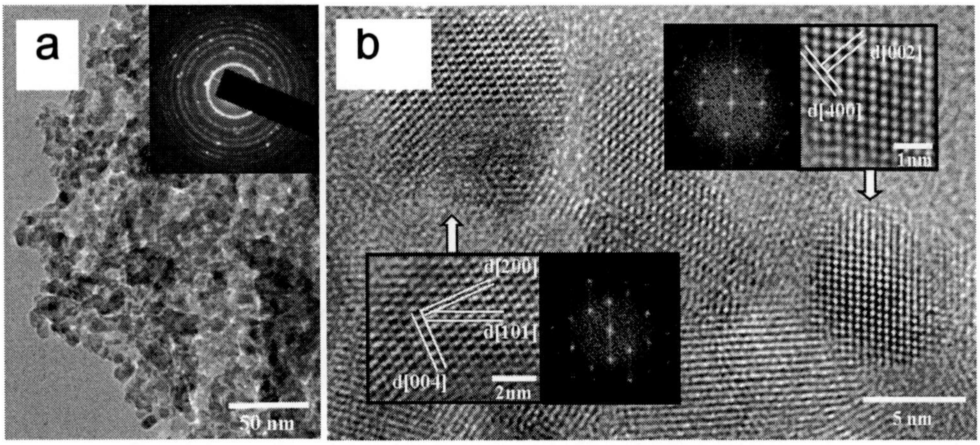

Figure 9. (a) TEM and (b) HRTEM Images of ISM-based TiO_2 obtained at 823K and pH 2. The inset in (a) showing the SAED pattern of as-prepared TiO_2; the insets in (b): FFT of the crystallites show verified the natase structure (left inset) and the rutile structure (right inset), which was established that the image was taken down the [100] zone axis.

The colloid particles can be bound to glycoprotein through the hydroxy groups of sugar chains linking to polypeptides, which are known to coordinate to titania gels [3]. Thereby the amorphous Ti-O networks can be derived from these reactions, and indeed an amorphous TiO2 film locating in the mantle is obtained with no change in the ESM configuration. The Ti (IV) oxo species is assumed to be an intermediate between TiO2+ and TiO2, consisting of partly dehydrated polymeric Ti (IV) hydroxide. Moreover, polyanions containing in the mantle tend to arrest the cationic H+. Therefore, in acidic media, the hydrolysis of Ti (IV) species is enhanced and the polycondensation of Ti–OH species is promoted, which further causes the homogeneous formation of a large number of seed nuclei, finally the crystallization of tiny titania nanocrystallites. Followed a calcination treatment, amorphous TiO2 further ripens into well-defined crystallites. Furthermore, high acidity is in favor of the formation of rutile TiO2 besides calcination temperature. Increasing calcination temperature usually accelerates phase transformation from thermodynamically metastable to stable phase and condenser phase. The heat treatment frequently brings on the agglomeration and large size. However, TiO2 in our work shows small size and good crystallization as well as a narrow size distribution. It probably ascribes to short-chained amino acides of glycoprotein, which functioned as the capping agents since the binding was strong enough for the control on the growth of titania nuclei. For comparison, the sample is synthesized through similar sol-gel approach and calcination treatment only without involving the template. Apparently, the above as-prepared TiO2 is distributed mussily with the average size of 28 nm, while TiO2 nanocrystallites with the average size of 5 nm in our work are arranged into orderly hierarchical structures, which further proves the significance of the template and supports the above analysis. HRTEM images show TiO2 nanocrystallites prepared at pH 2 and 550 °C are the admixture of rutile and anatase. Raman spectroscopy reveals that the four Raman bands at 636, 513, 397 and 144 cm-1 ranging between 100 and 800 cm-1 can be indexed as the anatase phase of TiO2. The Raman bands at 601, 440, and 92 cm-1 can also be indexed as the rutile phase of TiO2, which is in agreement with the XRD results.

Based on the above analysis, the interfacial sol-gel approach at different pH values are predominantly associated with the mantle proteins. The mantle of SMF containing lysozyme is prone to dissociate as the glucides in the sol or the solution other than the collagenous core, and lysozyme is a kind of protein with instability under alkaline medium. Therefore, the morphology of the mantle has the inherent affiliation with the pH value of the sol medium. The mantle can exist as a steady overcast at pH < 2.5 while the mantle protein seems flowing and reforming extremely at pH 3. Under that extreme circumstance, lysozyme is mostly activated, leading to the dissociation and disintegration of peptidoglycan-based mantle layer. Simultaneously, these proteins chains interact with the sol ingredients, and the clusterin exhibiting powerful chaperone-like activity with lysozyme can bind to and stabilize proteins aggregations. It prefers the irreversible precipitation by preferentially recognizing partially folded protein conformations that are responsible for this process, which conduces to moving and reforming of proteins molecules along the SMF intersections by the interfacial force. Thus, this special layer makes it possible to transform from interwoven to intermeshed structures. However, the interaction of strongly cationic lysozyme with collagenous matrix would be mediated by proteoglycans [4]. It is difficult for the colloid particles to interact with the core matrix under current concentration. Accordingly, hollow TiO_2 fibers form under higher acidity while the petal-shape titania is obtained with lower acidity after removing the templates.

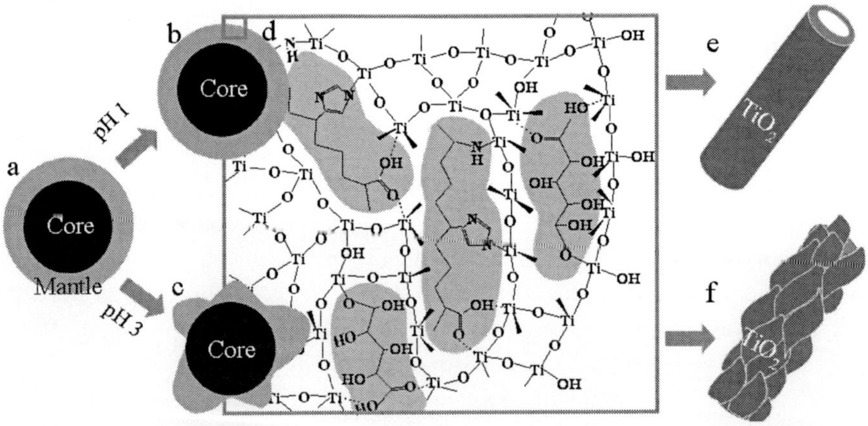

Figure 10. Schematic depiction of hierarchical TiO_2 synthesized through ESM directed interfacial sol-gel process: a) the cross-fractured SMF; b, c) the states of the mantle being immersed in the sol media at pH 1 and 3 for 13 h, respectively; d) the interaction between the sol ingredients and functional residues of ESM macromolecules; e, f) as-prepared tubular titania and petal-shaped titania by calcination treatment, respectively.

Based on N2 adsorption-desorption measurements, the isotherms and corresponding pore size distribution curves for interwoven TiO2 under different synthesis conditions are investigated in detail. Regardless of the increase of calcination temperature, the isotherms all exhibit IV-type with H3-type hysteresis loop in the Brunauer classification. As IV-type isotherm is caused by the capillary condensation of mesopores, and H3-type hysteresis loop is due to asymmetric channels and mesopores mentioned above in coincidence with the tubular characteristics pores during the synthesis process. The type of isotherms indicates an interconnected mesoporous system and very high pore connectivity according to the percolation theory, which goes well with the HRTEM results. The main pore sizes are about 5, 15, and 30 nm of the samples obtained at different calcinations temperature 550, 700 and 800 °C, while the corresponding pore volume decreases.

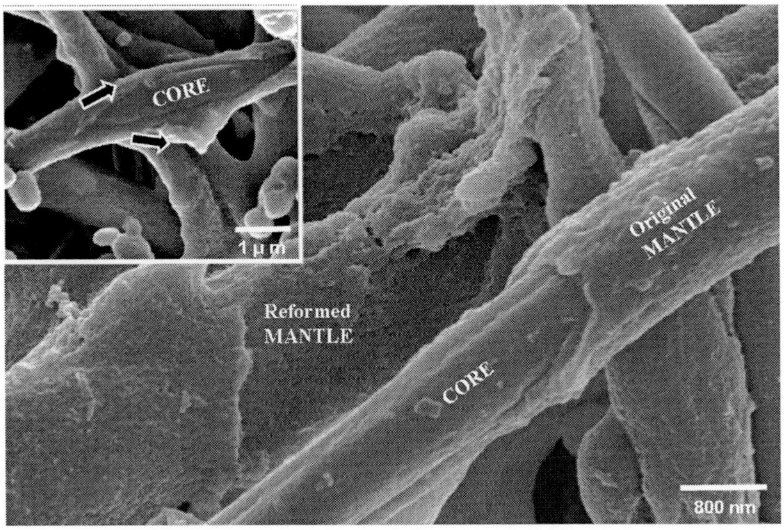

Figure 11. FESEM images with high magnification of SMF-based GMH in detail. The insetshows the mantle protein dissociating and disintegrating of the peptidoglycan-based mantle layer along the arrows direction.

Actually, pore size distribution is effected greatly by the size and morphology of building particles, which are directly dominated by the calcination temperature. Notably, the singular peak from IV-type curve of the sample at 550 °C is attributed to interwoven macropores compared with the other two mesoporous titania. Here, increasing temperature causes larger but less pores at the nanoscale.

By investigation on the isotherms and the pore size distribution curves for titania prepared at different heating rate, the trend variation of the isotherms is as same as that of the samples at different calcination temperature. By way of contrast, the size distribution is affected dominatingly by the calcination temperature rather than the heating rate. Pore sizes are less than 10 nm when reducing the heating rate below 18 °C/min. As the heating rate is further decreased to 1 °C/min, there is no evident difference in pore size distribution, due to the assembly of the target materials by the building particles with nearly same size.

The isotherms and corresponding pore size distribution curves for ESM-morphic titania at different media pH value show the presence of a pronounced hysteresis loop in each isotherm curve is associated with the filling and emptying of mesopores by capillary condensation. At relative low pressure (<0.5), the nitrogen adsorbing action of macropores is the same as the monolayer adsorption of mesopores. But above medium relative pressure (>0.5), there is no capillary condensation phenomenon in macropores. Hence, according to the types of isotherms, all these samples can be confirmed as a mixture of mesopores and macropores. The pores with larger size are formed whether medium pH value higher or lower than 2 is adapted to the system. It could be reasonably ascribed to the condensation of the hydrolytic products at low acidity and the contraction of frameworks at high acidity, which both bring on lager pores as demonstrated above. Herein the interwoven fibers are proved to be size-controllable porous, there exist several kinds of pores including micropore, mesopore and macropore. Prompted by the porous appearance, we suspect that as-prepared ESM-morphic hierachical titania must be useful in fabricating highly efficient catalyzer, photovoltaics, and so on. It makes the biotemplate-directed sol-gel synthesis to be a versatile and feasible route to the functional inorganic hierarchical materials.

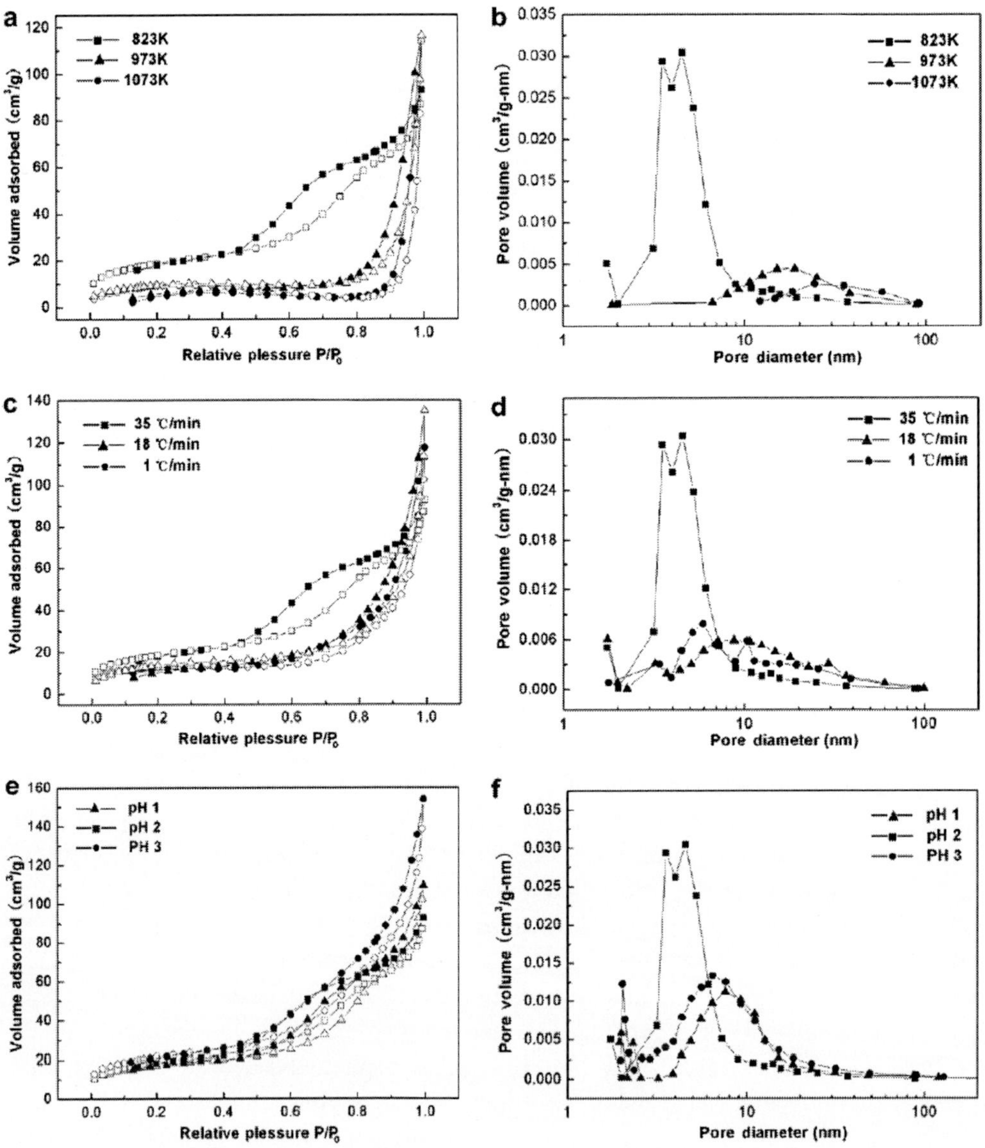

Figure 12. Nitrogen adsorption-desorption isotherms and corresponding pore size distribution plots of ESM-based TiO_2 prepared at (a, b) pH 2 and 35 °C/min, (c, d) pH 2 and 550 °C, and (e, f) 550 °C and 35 °C/min.

1.2. Fabrication and Gas-Sensing of ESM-Morphic SnO2

Having a large bandgap of 3.6 eV at 300 K, a high achievable carrier concentration of up to 6×1020 cm–3, and an expected exciton Bohr radius of about 2.4 nm, nanocrystalline tin oxide is of great technological and scientific interest as a key functional semiconductor that is used extensively for photovoltaic devices, transparent conductive electrodes, gas sensors, antistatic coatings, dye-sensitized solar cells, etc. In particular, as a gas sensor is one of its well-known applications. The involved mechanisms in the gas sensing processes are rather complex, which are related mainly with several steps such as gas adsorption, chemical reactions on the surface, and charge transport. All these processes mainly depend on the surface activity of gas-sensing materials, which is decided by the building unit and the assembly structure. It is usually claimed that nanometer-scaled materials as gas sensors have significant advantages because of their small grain size, which becomes important to enhance the adsorption capacity. Until now, few researches refer to the influence of surface activity derived from the structure and conformation on the gas sensing properties. Yet particular assembly of SnO2 nanocrystallites may provide promising gas-sensing performance. Attracted by bioinspiration, eggshell membrane in combination with a tin salt precursor is introduced to a sol-gel process, with the intent of synthesizing hierachical SnO2 semiconductors by suitable vehicles of special protein structures and functional residues of ESM biomacromolecules. In contrast to the metallic alkoxide precursors, this work employs a trilayer bio-scaffold in tin nitrate solution as the template for hierarchical nanostructured SnO2 with enhanced gas-sensing performance.

In a typical procedure, Sn powder is added stepwisely into ~2M HNO3 solution to obtain a clear solution incipiently, turn into a muddle subsequently, finally slight yellow Sn-colloid after being filtrated. The pH values of the colloid media are generally kept at 1 when used as the impregnants. For comparison, 1 M NaOH solution is added to modify the pH value of the media to 2 and 3. The prepared ESM immersed in the above-mentioned Sn-colloid systems for about 13 h, then being fished out, rinsed with distilled water, and dried, the soakage-treated ESM are subjected to calcination treatments in air at different temperatures to remove the organic ESM ingredients and achieve better crystallization of the inorganic sinters.

The calcination process was traced by conducting TGA and DTA surveys to follow the formation of hierarchical SnO_2. It is evident that the ESM starts pyrolyzing at 473 K and is absolutely decomposed by 823 K. Although GMH are also decomposed around 473 K, they can preserve up to 47.2% (by mass) inorganic matter at 823 K, at which temperature the organic ESM is completely removed. As seen from the differential curve, when GMH are heated from room temperature to 473 K, the weight loss of 9.8 wt-% is due to the removal of adsorbed water, and above 473 K they experience a two-step weight loss. Congruously, the DTA curve shows two peaks at 635 K and 731 K, which shall represent the decomposition of Sn–ligand hybrids and organic ESM components, respectively. After the calcination at 673 K, the amorphous hybrids are found to be crystallized, but the ESM is not thoroughly decomposed at this temperature. XRD results indicate the well-established crystallization process at 823 K. All the diffraction peaks can be indexed to tetragonal SnO_2 with a rutile structure (JCPDS 41–1445), which indicates that the specimens consist of pure single tetragonal SnO_2. The average crystallite sizes of about 2.8 nm at 673 K and 5.2 nm at 823 K, are estimated according to the line-width analysis of the diffraction peaks based on the Scherrer formula. The chemical states of tin and oxygen during the sol–gel process were investigated by XPS analysis. The corrected XPS spectra of the O1s emissions are different, where the peak with binding energy around 532.3 eV corresponds to the hydroxyl bonds of GMH, while the weaker peak around 531.1 eV corresponds to the Sn–O–Sn bonds of the sinters. Moreover, GMH appear as a spin-orbit doublet at 487.9 (3d5/2) and 496.4 eV (3d3/2) in the Sn3d emission spectra with an area ratio of 1.5:1. Thus, the chemical state of the element tin in as-prepared GMH corresponds to Sn^{IV}, but not to Sn^{II} or Sn^{III}. Representative FESEM images of as-prepared SnO_2 with various morphologies calcined at 823 K suggests that hierarchical SnO_2 accurately replicates the hierarchy of the biotemplate ESM from the OSM to the innermost LM, although the modest shrinkage. By repeating the immersing manipulation, heat constringency in size can be lowed to below 30%, and up to 47.2% of the mass of the sinters. Analysis with energy-dispersive X-ray spectroscopy also reveals that the atomic ratio Sn/O is close to 1:2, further confirming the stoichiometry of the replicas as SnO_2.

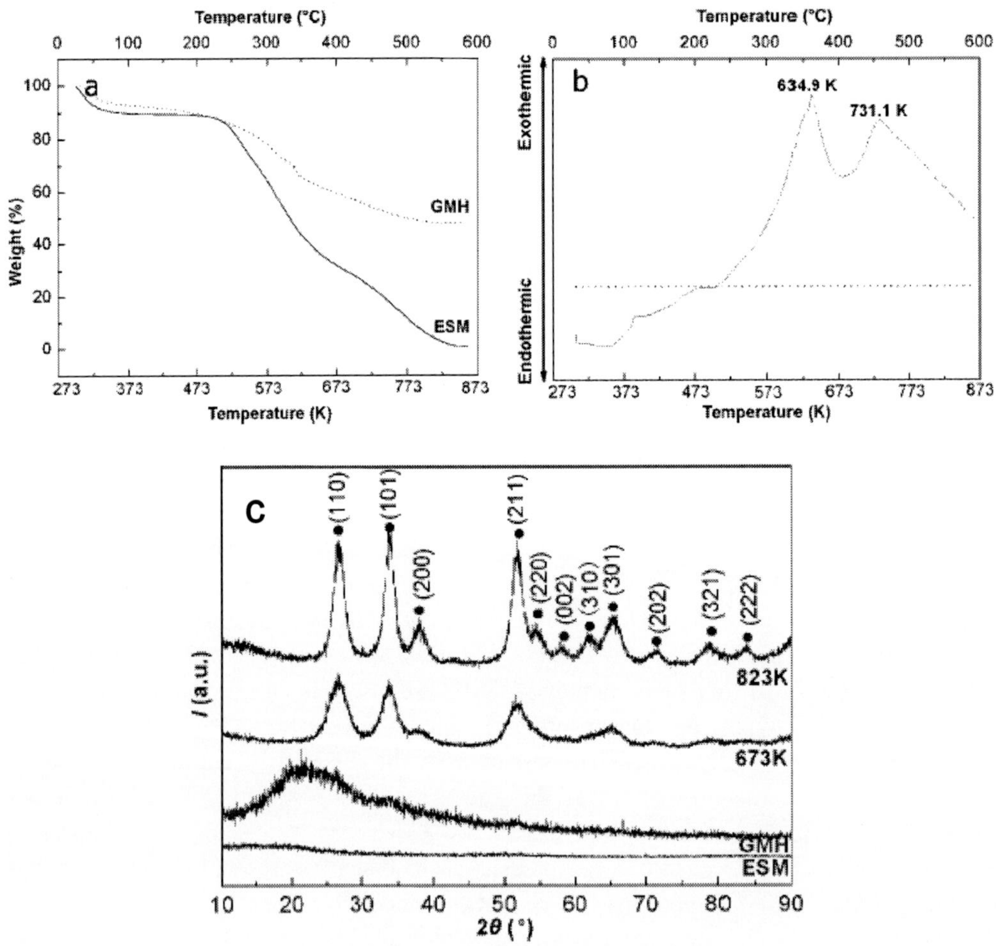

Figure 13. (a, b) TGA-DTA curves, (c) XRD patterns of the relevant samples.

FESEM images show the relative location of the hierarchical SnO_2 replicated from the three diverse membranes. It is clearly seen that the LM-based one has been sheltered by the upper layers composed of the ISM-based and OSM-based SnO_2 layers. However, the replica of the LM layer can be identified when the other side of this sheet is observed. The configuration of the intermeshed ISM consisting of interwoven fibers is intact, whereas the LM-based layer is curled up thoroughly. Moreover, the thickness of the obtained SnO_2 film can be tuned by removing the multilayers of the ISM. Thus, a SnO_2 film with a continuous thickness up to 20 μm can be achieved by varying the original thickness of remnant ISM layers. The tubular hierarchy can be observed clearly by the higher magnifications of the cross section and the vertical section of these SnO_2 interwoven fibrous films, respectively.

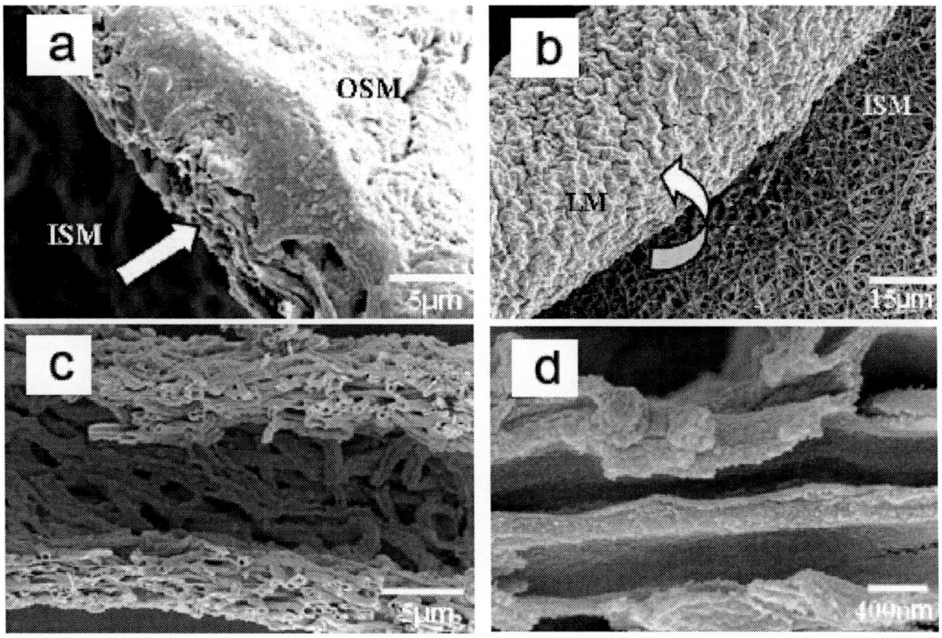

Figure 14. (a), (b) Typical FESEM images of as-prepared ISM-morphic SnO_2 revealing relative locations of the three distinct layers of the original ESM. (c), (d) FESEM images showing the cross-section and the vertical section of the ISM-based SnO_2, respectively.

The tubular structures have an inner-diameter range from 100 nm to 300 nm and an average wall thickness of less than 90 nm. These SnO2 tubes are rather uniform, and the hollow cores are glabrous and continuous throughout their length. Notably, these interwoven tubular structures are actually interconnected at junctions. A special kind of tube wrapping several smaller collateral cannulations is also observed, which likewise reproduces the original structures of the membrane fibers. To further investigate the replication process, the organic ISM is selected as the typical template to achieve the synthesis of hierarchical biomorphic SnO2 under different experimental conditions.

All the replicas at different calcination temperatures (823 to 1073 K) have similar microscopic features and preserve the interwoven tubular structures of the natural ISM, but their building blocks, SnO2 nanocrystallites, may grow with the increase in calcination temperature. The surface of SnO2 tubes synthesized below 973 K looks more glazed in virtue of the smaller particle units. Above 1073 K, the tubular structure is apparently composed of a network of interconnected SnO2 particles, and it may collapse down into zonary form at 1273 K. On the contrary, if the calcination temperature is lowered to 673 K, characteristic tubular structures appear partially, and some look puffy, which may be caused by the organic remnants. It is reasonable that some residues of organic ESM remain at 673 K even though SnO2 nanocrystallites have formed, which is in agreement with the TGA and XRD results.

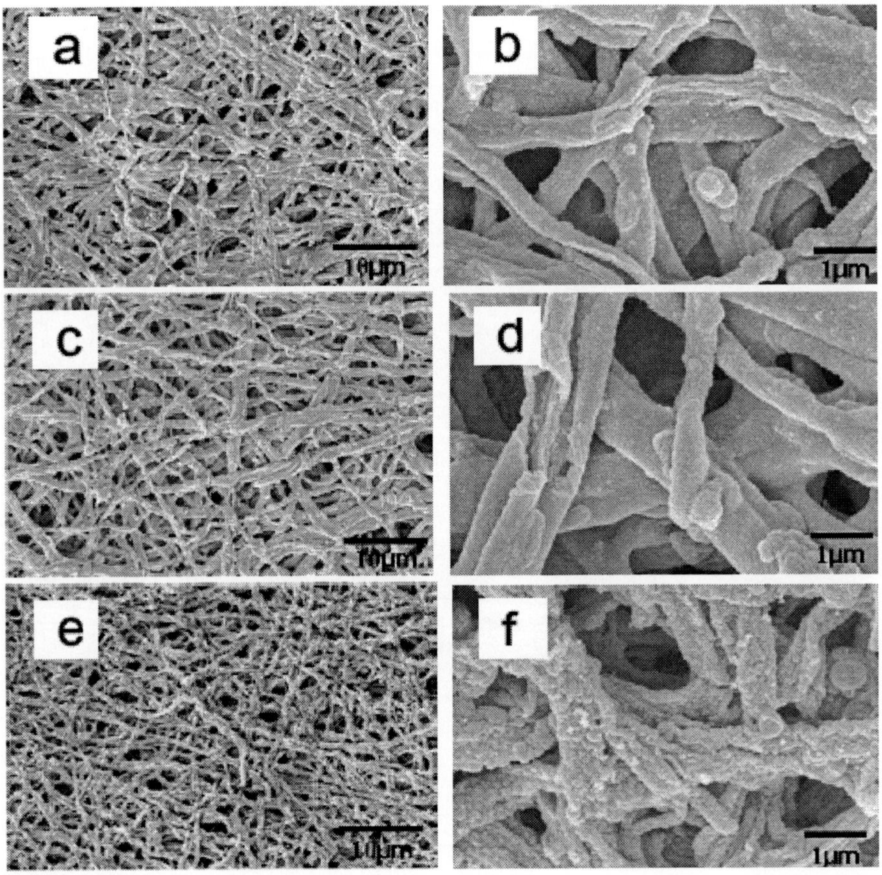

Figure 15. FESEM images of as-prepared ISM-morphic SnO_2 at different temperatures: (a) 823 K, (c) 973 K, and (e) 1073 K. Images (b), (d), and (f) show relevant higher magnifications of (a), (c), and (e), respectively.

All XRD diffraction peaks of the samples calcined at various temperatures between 823 and 1273 K, can be assigned to the tetragonal rutile structure of SnO2. The XRD peaks gradually become sharper with increasing temperature, indicating that the SnO2 nanocrystallites grow larger, and the crystallite sizes determined from the Scherrer equation are 5.2, 9.5, 10.3, and 20 nm, respectively. The TEM images further reveal the tubular interwoven character of ISM-based SnO2. The hollow tubular structure is clearly identified, and the wall thickness (50–60 nm) is found to be uniform along its entire length.

This indicates that at 1073 K each tube has been transformed from a dense structure with a smooth surface into a highly porous structure consisting of interconnected nanocrystallites of ca. 10 nm in size. In addition, the SAED patterns show that nanoscaled SnO2 at 1073 K has polycrystalline tetragonal rutile structure (cassiterite) due to the uniform central and diffraction spots, and the diffraction rings are indexed to the (110), (101), (200), (211), and (301) diffractions of rutile-phase SnO2. Although the sinters at 673 K contain some impurities from ESM residues, the characteristic crystalline structure of tetragonal SnO2 is observed in the electron diffraction image. This can be explained by the exothermic reactions occurring during the removal of the ESM: the oxidation of the organic biomaterial emits heat locally, causing an increase in the temperature around the adsorbed Sn colloids, thereby accelerating the gelation and bringing on nanocrystalline SnO2. These nanoparticles are ultrafine below 823 K while growing more quickly above 1073 K, and the average sizes of the nanoparticles are coincident with the crystallite sizes calculated from the XRD analyses. As the nanoparticles grow with the increase in calcination temperatures, the diffraction patterns gradually transform from ring (nanoscale) to spot (bulk scale). This indicates that accurately defined, size-controlled SnO2 nanoparticles in the range 2 to 20 nm are obtained to enable the construction of hierarchical nanomaterials at the corresponding calcinations temperatures. As observed the typical HRTEM images of SnO2 nanocrystallite units calcined at 823 K, the lattice image can be clearly seen as a result of phase contrast. The distances between the adjacent lattice fringes correspond to the interplanar distances of the tetragonal SnO2 (101) and (110), *d*101 = 0.264 nm and *d*110 = 0.335 nm, respectively. Considering that each domain having parallel lattice fringes is a single crystal of SnO2, the crystallite size is observed to be 5 nm. A careful look at the images also reveals that the present SnO2 is characterized by larger "nanostructured" particles consisting of nanocrystalline domains with structural defects such as twins.

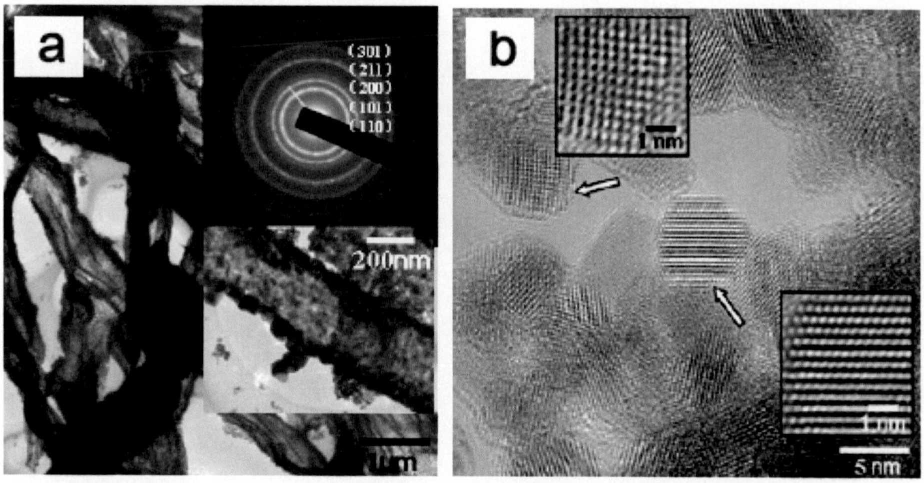

Figure 16. (a) TEM images of ISM-based SnO_2 synthesized at 1073 K and pH 1. Insets in (a) show an individual SnO_2 tube and corresponding SAED pattern. (b) HRTEM Image of ISM-based SnO_2 obtained at 823K and pH 1. Insets in (b) correspond to the tetragonal SnO_2 (101) plane (top left) and the (110) plane (bottom right), respectively.

To understand the multilevel interactions between the Sn colloids and the ESM biomacromolecules during the biomimetic sol–gel process, FTIR spectroscopic analysis was carried out on the original ESM, the GMH and the sinters. The spectrum of the original ISM shows the diagnostic peaks at 1641, 1532, and 1234 cm–1, which are assigned to the amide I, amide II, and amide III bands of the proteins from the natural ISM, respectively. After interaction with the sol precursor, the red shift to 1630 cm–1 of the peak in the spectrum of GMH may be partly attributed to the superposition of the bending mode of water with the broad band around 3450 cm–1 (not 3400 cm–1). The peak at 3430 to 3450 cm–1 is believed to be caused by adsorbed water. In general, the absorption band centered at 1630 cm–1, as well as the one at 3450 cm–1, is also ascribed to water adsorbed on the surface of the samples. Here, there may be no other corresponding peaks of amide I after immersion in the sol medium, if the peak at 1630 cm–1 (near 1620 cm–1) in the spectrum of GMH are only due to the bending mode of water. The spectrum shows an increase in intensity of the amide I band relative to that of the amide II band. It does confirm the presence of a water deformation mode. Below 700 cm–1, the intense bands appearing both in GMH and the sinters at lower frequencies around 600 cm–1 are typical of the Sn–O–Sn antisymmetric and symmetric vibrations. In addition, the bands appearing in the range between 850 and 1350 cm–1 are assigned to the bending mode of different types of surface hydroxy groups, which goes well with the previous research results. The surface of the GMH can be terminated with the –OH groups after the ESM is dipped in the Sn-sol medium. However, the calcinations treatment may bring on weak IR bands between 850 and 1350 cm–1, contributed by the decomposition of Sn–OH hydrate compounds into tin oxides.

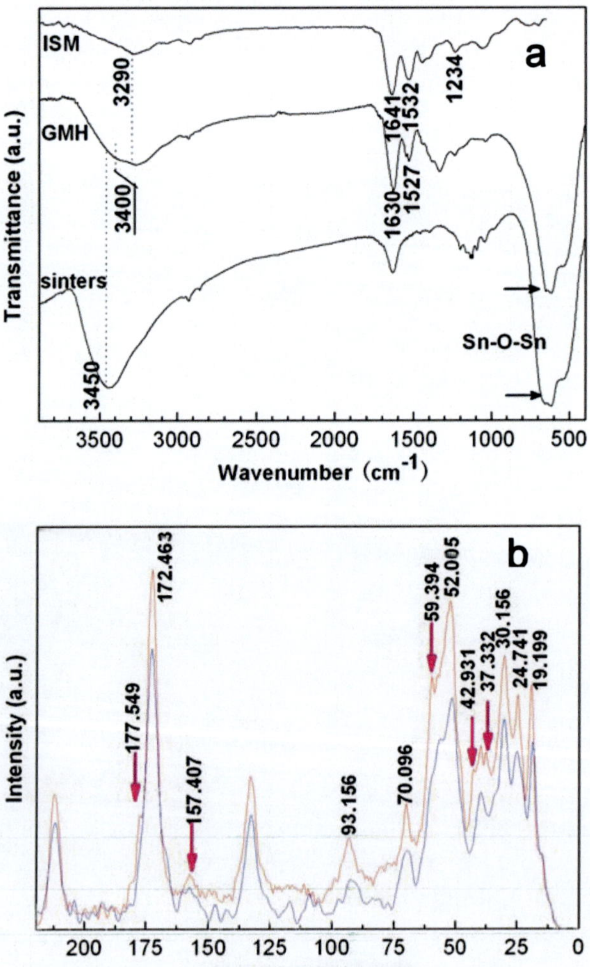

Figure 17. (a) FTIR spectra of original ISM, the corresponding GMH, and the sinters at 823 K. (b) MAS solid-state NMR spectra of original ESM (red line) and GMH (blue line).

Actually, the red shift of the amide I peak is probably overlapped by the above-mentioned water bending-mode peak. This red shift indicates that the C=O bonds in the peptide chains may be weakened because of the formation of new chelate bonds between C=O bonds and Sn colloids. This can be further corroborated by the NMR spectroscopic results. The chemical shift between 19 to 30 ppm is mainly due to the resonances of —CH3 and —(CH2)n— groups of Pro, Val, Arg and Metis.. (a)—COOH 177.549; (b) —COH 59.394; (c) —CH2NH2 157.407, 42.931, 37.332. It is also detected that the peak corresponds to the Sn–OH vibration band at around 3400 cm–1, which is superimposed on the underlying C–H stretching signals of the polypeptides at 3290 cm–1 and close to the hydroxy bands of the water. Accordingly, the interactions between Sn colloids and ESM macromolecules are mostly ascribed to the coordination of carboxyl, hydroxy, and other groups.

Herein, the interfacial sol–gel process is presumably based on the adsorption of cationic groups from the colloid medium into the glycoprotein mantle of the ESM fiber containing an abundance of carboxylic, hydroxy, and amino residues; subsequently a series of complicated reactions proceed to give hierarchical SnO2 films composed of ultrafine nanoparticles. The formation of SnO2 can be denoted as follows:

$$\mathbf{Sn^0 \rightarrow Sn^{IV} - ions\ (hydrate\ \ ions) \rightarrow Sn^{IV} - colloids \rightarrow Sn^{IV} oxo\ \ species\ (Sn - gel)}$$
$$\mathbf{\rightarrow amorphous\ \ SnO_2 \rightarrow nanocrystalline\ \ SnO_2}$$

The cationic SnIV species in the sol system are prone to interact with the residues of ESM glycoprotein proved by FTIR and NMR results. The formation of amorphous Sn-colloids/glycoprotein hybrids arises from strong affinity of some residues binding that involves multiple hydrogen bonds, Van Der Waals interactions, and the ordering of surface polypeptide loops. The Sn-colloids are bound on the biotemplate ESM by amino groups labeled glycoprotein through the -OH and -NH2 groups of sugar chains linked to polypeptides. Thereby the amorphous Sn-O networks can be derived from these reactions, and indeed an amorphous SnO2 film locating in the mantle is obtained without changing the original morphologies at room temperature. The Sn (IV) oxo species are assumed to be an intermediate, consisting of partly dehydrated polymeric Sn (IV) hydroxides.

Amorphous SnO2 ripens into well-defined nanocrystallites during the calcination treatment. Heat treatment frequently causes agglomeration and enlargement. However, SnO2 nanocrystallites in our work have small size and good crystallization as well as a narrow size distribution. This can probably be ascribed to some short-chained amino acids of glycoprotein, which function as capping agents, since the binding is strong enough for controlling the growth of tin dioxide nuclei. For comparison, a blank sample is synthesized through a similar sol–gel approach and calcination treatment at 823 K only without using the biomaterials. SnO2 nanoparticles in the blank sample are disorganized and have an average size of 21 nm, while the ESM-based SnO2 nanocrystallites with an average size of 5 nm are arranged into orderly hierarchical structures with higher surface area. This observation proves the significance of the biotemplate and supports the above analysis. It is also concluded that the ESM peptidoglycan functioning as the surfactant may play a key role in the aggregation and further crystallization of tin oxide nanoparticles. In addition, electrostatic forces assist the biomimetic sol–gel process. Because of the higher acidity of the system (with ample H+ ions), the tin oxide clusters in the colloid are likely to be positively charged at the surface by reference to the isoelectric point of cassiterite SnO2 (7.3) [5]. By electrostatic force, they will preferably attach onto the negatively charged DS– ions on the surface of the membrane. As surface coating is always dependent on surfactant concentration, it is difficult and complicated to determine the precise dosage.

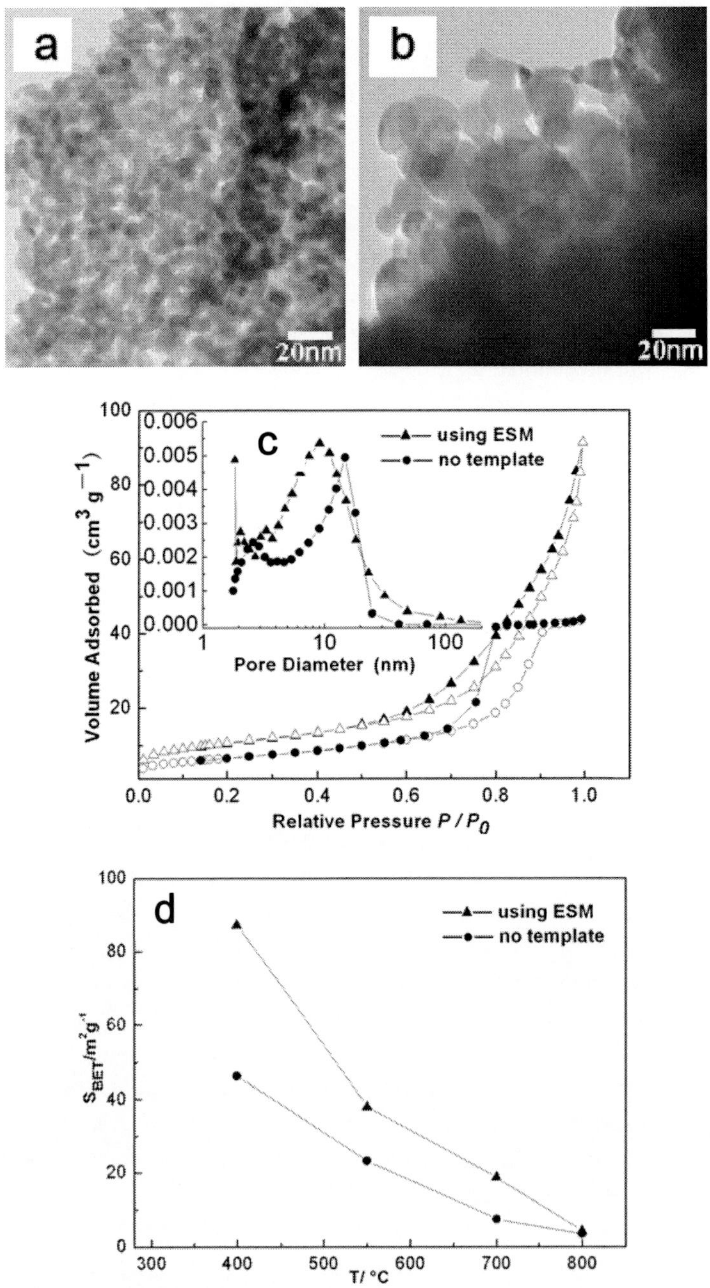

Figure 18. TEM images of (a) ISM-based SnO_2 nanocrystallites and the contrastive SnO_2 sample without introducing template through the same process at 823 K. (c) Nitrogen adsorption-desorption isotherms plots and corresponding pore size distribution plots of ESM-morphic SnO_2 and Blank sample, (d) Special surface area SBET vs calcination temperature.

However, in our work the "surfactant molecules" can be released systematically by the dissociation and reforming of ESM biomolecules. With the aid of "surfactant molecules", the construction of biostructure hierarchy can be achieved to synthesize subtle biomorphic functional nanomaterials, which is similar to biomoleculemicelle systems.

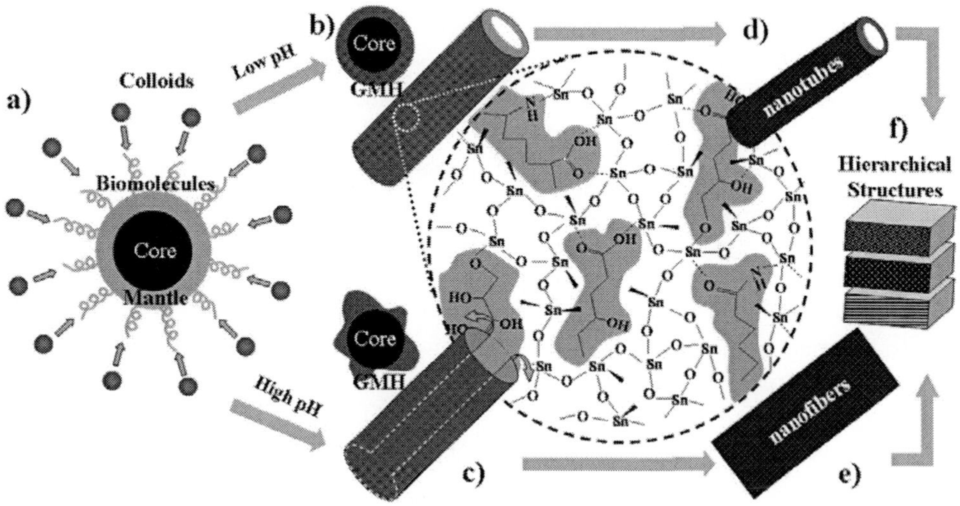

Figure 19. Schematic depiction of hierarchical SnO_2 synthesized through ESM directed interfacial sol-gel process: (a) the cross-fractured ESM fiber showing its biomolecules interacted with Sn-colloids; (b, c) the states of the mantle being immersed in the Sn-sol media at different pH; the inset in (b) illustrating the interaction between the sol ingredients and functional residues of ESM macromolecules; after the ESM removal, (d) tubular SnO_2 and (e) SnO_2 fibrous belts; (f) hierarchical SnO_2 achieved by nanotube or nanobelts arrays.

Notably, a completely morphologic conversion from SnO_2 tubes to SnO_2 belts occurred as a result of tuning the pH value of the sol media. As lysozyme is a kind of protein that is unstable in alkaline medium, it is clear that the mantle of ESM fiber containing lysozyme is prone to dissociation, its glucide contents not in the collagenous core end up in the colloid system. Therefore, the resultant morphologies of the SnO_2 products might have an inherent affiliation with the structures and constituents of the glycoprotein mantle. It is observed that the mantle can exist as a steady covering layer at pH < 2, while the distortions and dissociations of this protein layer occur as the pH value is elevated. The mantle protein seems to flow and reform evidently when pH > 3. Under these circumstances, lysozyme is most active, and it functions as a kind of enzyme leading to the dissociation and disintegration of the peptidoglycan-based mantle layer. These protein chains then interact with the Sn colloids. Simultaneously, the proteins formerly carrying tubular hybrids start to dissociate and stretch into planar fibers as a result of decreased acidity. Moreover, it is revealed that clusterin, exhibiting powerful chaperone-like action on lysozyme, binds to and stabilizes aggregating proteins, thus leading to irreversible precipitation by preferentially recognizing partially folded protein conformations. This process conduces to the stabilization of the final structures of SnO_2. The special layer thus makes it possible for SnO_2 fibers to transform from tubular to solid fibrous structures. However, the interaction of the strongly cationic lysozyme with the collagenous matrix may be mediated by proteoglycans. It is difficult for Sn colloids to interact with the core matrix under the current concentrations. Accordingly, hollow tubes form under higher acidity, while fibrous structures are obtained at lower acidity after the ESM is removed.

Prompted by their porous appearance and small-scale nanocrystalline units, as-prepared hierarchical ESM-based SnO_2 must be useful in fabricating semiconducting sensors. When the grain size is reduced to a scale comparable to the space-charge length (for SnO_2, this value is 6 nm), the sensitivity could be exponentially enhanced [6]. As an n-type semiconductor, SnO_2 has been extensively exploited in applications related to gas sensing. Most of these studies have focused on polycrystalline films made of small SnO_2 particles [7]. In contrast to some sensors made of individual single crystalline nanostructures or polycrystalline films, hierarchical nanostructured SnO_2 in our work, assembled by the small-scale, well-crystallized nanocrystallite units into tubular or zonary SnO_2 fibers with porous conformations and interconnected fibrous meshworks, would allow the sensors to be operated in the most sensitive, grain-controlled mode.

The sensor signal, S, is defined as the resistance ratio, Ra/Rg, where Ra and Rg stand for the electrical resistance in dry air and in the sample gas, respectively. The gas-sensing of as-prepared ESM-morphic SnO_2 films is measured continuously.

It is found that the dependence of the sensor signal (S) of the test gases on the synthesis temperature defining crystallite size (D). Here hierarchical SnO_2 film sensitivity to 50 ppm ethanol, 50 ppm H_2S, 500 ppm LPG, or 50 ppm gasoline 90#, is investigated at the working temperature 270 and 300 °C, respectively. At a lower working temperature, 270 °C, the sensitivity of as-prepared SnO_2 to ethanol is more prominent than that to H_2S, LPG and gasoline. However, when elevating the working temperature to 300 °C, the sensitivity of the sample SnO_2 calcined below 700 °C to LPG becomes remarkable among all these test gases. Regardless of the effects of the working temperature, the sensor signal to all other gases except H_2S shows downward trends with synthesis temperature. That is to say, the sensitivity to the test gas markedly increases with decreasing crystallite size of the sensor SnO_2 ranging from 26 to 3 nm.

There is the dependence of the sensitivity of hierarchical SnO_2 prepared at 550 °C to the test gases on the working temperature. It can clearly be seen that the sensor signal to LPG shows an upward trends, while that to ethanol shows a downward trends, but the sensor is insensitive to other gases at the working temperature. Combining the above analyses indicates that the target materials have good gassensing selectivity for LPG or ethanol by tuning the working temperature of the sensors.

As a powerful tool to reveal straightforwardly the surface characteristics of these nanocrystalline films, the evolution of Raman surface vibration modes is investigated to examine the surface reactivity with the target specimens. The relationships between the gas responses and the nanostructure surfaces derived from the hierarchy and small unit size characterized by the Raman band intensity are discussed, corroborating the significant influence of the assembly structures on the gas-sensing properties of SnO2 films.

The investigation on the evolution of Raman spectra versus the treatment temperature for the series, also confirms the rutile structure of the as-synthesized SnO2. The spectrum of well crystallized sample at 1000 °C, grain size 26 nm, is similar to that of microcrystalline SnO2. The three detectable Raman active modes have been identified: Eg (476 cm-1), A1g (629 cm-1), and B2g (772 cm-1), which usually appear in large single crystals or bulk polycrystalline SnO2 materials and in good agreement with previous reports.

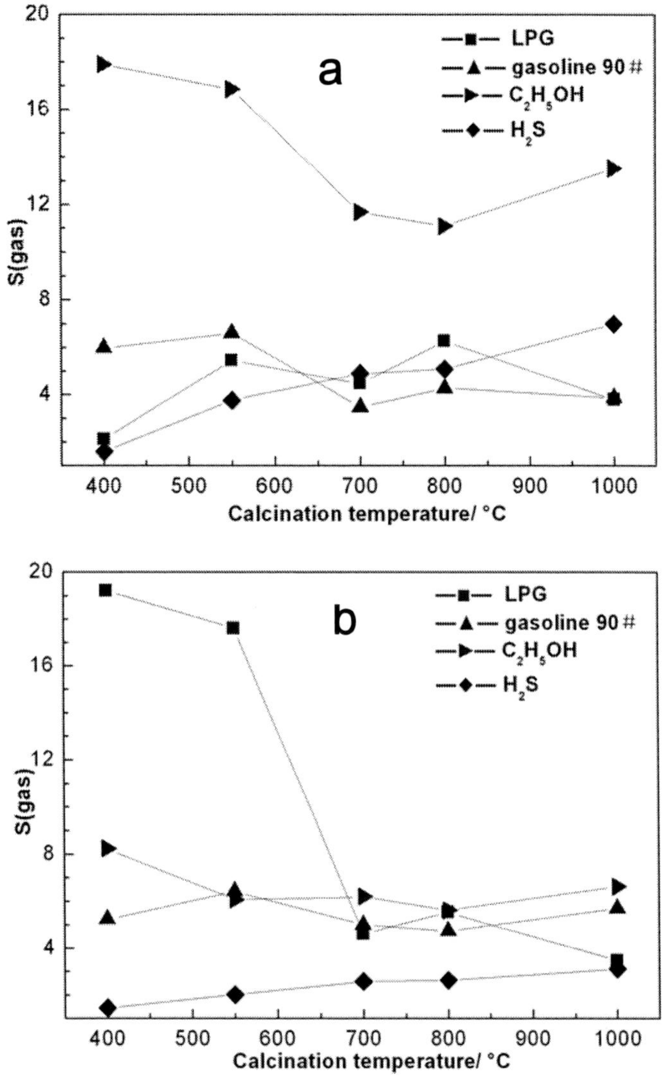

Figure 20. Dependence of the gas-sensitivity of SnO_2 films to 50 ppm ethanol, 50 ppm H_2S, 500 ppm LPG, and 50 ppm gasoline 90# on the synthesis temperature at the working temperature (a) 270 °C and (b) 300 °C.

As particle size decreased, the spectra of ESM700, ESM550, and ESM400 samples also exhibit the three bands and gradually descend around the A1g mode with a broad feature between 400 and 700 cm-1 attributed to surface modes. Additionally, three abnormal Raman lines are distinguished on the spectrum at 501.1, 541.9, and 693.9 cm-1, respectively, which have been detected before. Accordingly, these vibrational modes can be reasonably assigned to A2uTO (501.1), A2uLO (693.9) IR active, and B1u (541.9) Raman forbidden modes. The peaks are more sharpened with the smaller dimension of SnO2 particles. It is believed that the presence of IR modes and other forbidden Raman modes is attributed to the breaking down of the prevailing q0 = 0 selection rule as crystal size decreases to the nanoscale (limiting vibrations to the size of the crystal) or as the degree of disorder increases (reducing crystal symmetry). It is remarkable that high-frequency modes B2g, A2uLO, and A1g shifted to lower frequencies, whereas the low frequency modes, such as B1u, A2uTO, and Eg, shifted to higher wavenumbers accompanying the decrease in particle size. This is consistent with the findings reported by others, which can be explained using phonon dispersion curves.

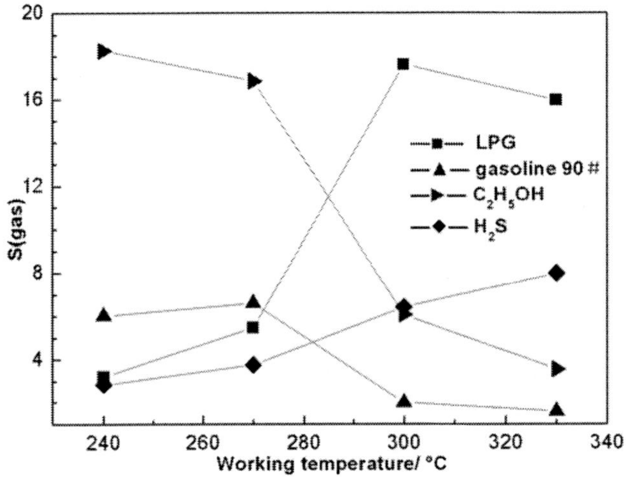

Figure 21. Dependence of the gas-sensitivity of SnO_2 films prepared at 550 °C to 50 ppm ethanol, 50 ppm H2S, 500 ppm LPG, and 50 ppm gasoline 90# on the working temperature.

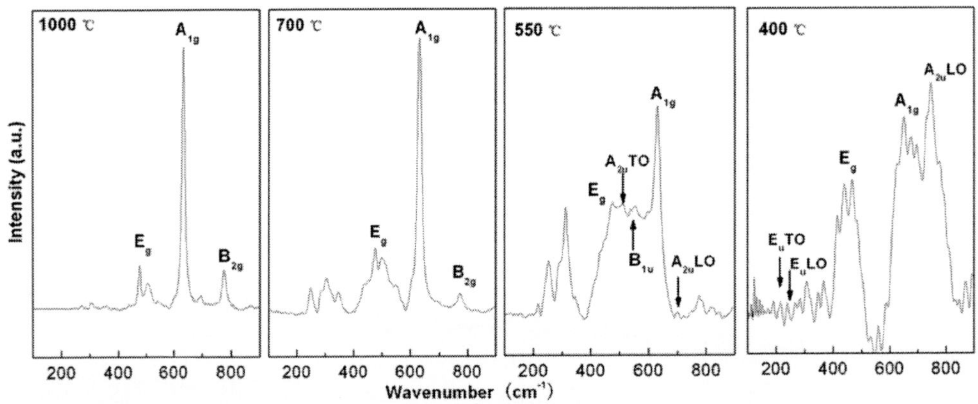

Figure 22. Raman spectrum evolution with the calcination temperature for ESM-templated SnO_2.

Furthermore, some new bands appear below the 400 cm-1 region and present a single broad feature composed of at least two components in the powders calcined at 400 °C. It reported that Raman bands located at 216.49 and 232.92 cm-1 in rutile SnO2 were attributed to EuTO and EuLO active modes. For SnO2 treated at higher temperatures, three bands are clearly observed at 245, 257, and 286 cm-1, together with the broad feature located at the region 310-350 cm-1, which has lost most of its intensity. These bands disappear for sample at 1000 °C. Other researchers associated the appearance of a 327 cm-1 centered band to the forced hydrolysis of Sn ions and the formation of SnO2 nanocluster, which might constitute a new kind of vibration mode. It is plausible to take into account that this band could be due to an active mode from the inactive Eu mode. Nevertheless, more work is still needed to clarify the origin of this band in very tiny SnO2 structures. It seems to strengthen the association of this band with the effects due to the nanostructure size against those due to the synthesis procedure. Those new bands, being related to the facet surface area of a crystal, arise from nanoscale SnO2 with small grain size and agree very well with the Matossi force constant model.

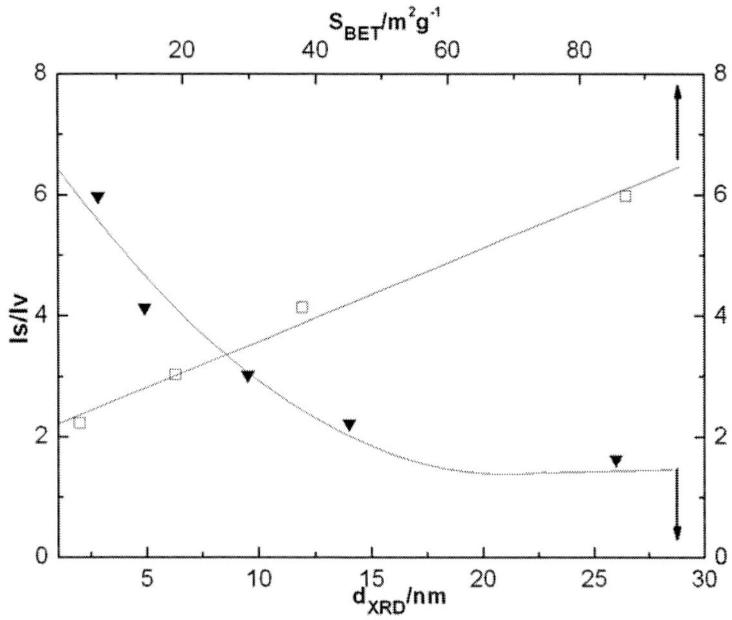

Figure 23. Raman relative intensity of surface modes I_S/I_V value vs SnO_2 crystallite size d_{XRD} and specific surface area S_{BET}.

The other peculiarity of Raman spectra of nanocrystalline SnO2 is the very broad feature between 400 and 700 cm-1. The origin of these bands has been connected to the low particle dimension of the materials. In a previous work, Diéguez et al. proposed that surface modes were due to the existence of a lattice softening taking place in the outer part of the grain, which is like a disordered lattice skin around each grain. Moreover, studies have shown that the relative intensity of surface bands depends strongly on the interactions with gases, which can modify the outer layers of the grain. So, to analyze these performances at the present cases, the relative intensity of surface modes have been taken as the ratio of the sum of their intensities *I*S to the intensity of A1g mode *I*V. These intensities are calculated by fitting A1g mode and the B2g mode when needed with Lorentzian-shaped peaks, and the surface modes with five large Gaussian peaks with Peakfit software. *I*V is taken as the area of the Lorentzian A1g peak, while *I*S is the sum of the five Gaussian peak areas. The variations of *I*S/*I*V value versus crystallite size *d*XRD and specific surface area *S*BET for SnO2 hierarchy are presented. There is a monotonic diminution of the surface modes as the grain size *d*XRD increases, which confirms a general trend. In addition, an excellent correlation between *I*S/*I*V and *S*BET suggests that surface modes are related to available surface sites to interact with target gas speciments. Therefore, it shall be corroborated from the behavior of the sensor response versus *I*S/*I*V.

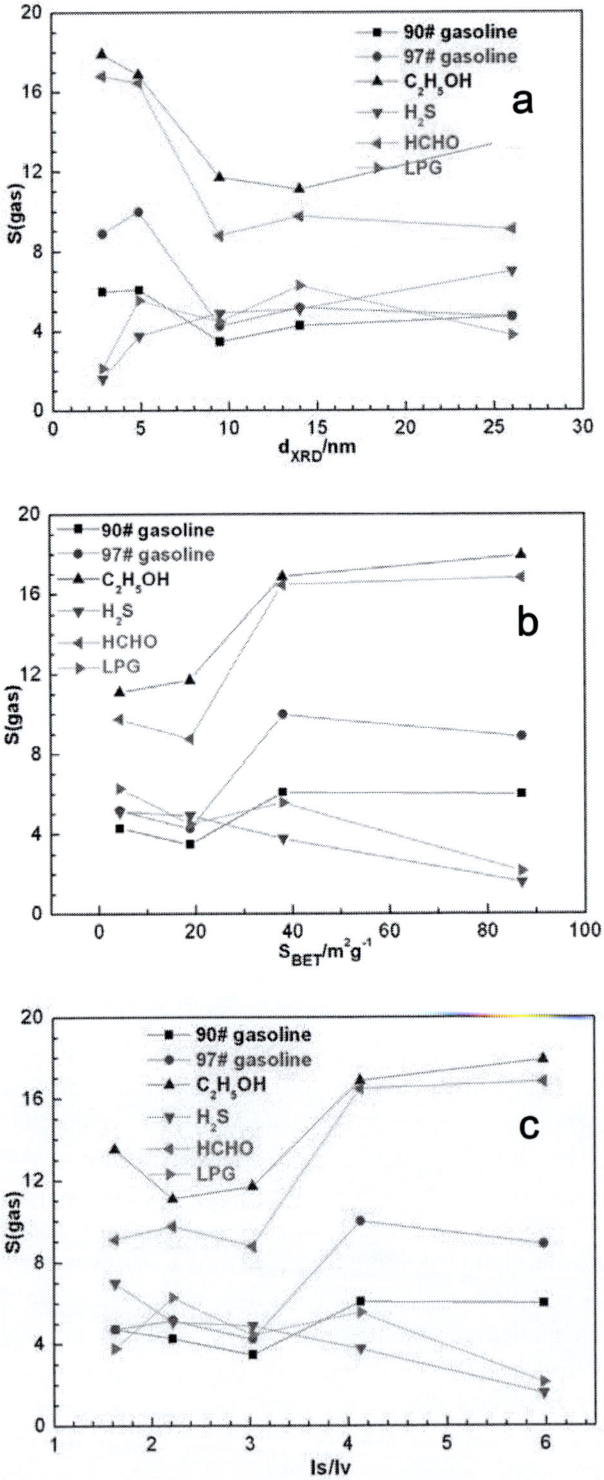

Figure 24. Sensing properties toward ethanol, formaldehyde, LPG, gasoline, and H_2S species S(gas) vs SnO_2 crystallite size d_{XRD} (a), specific surface area S_{BET} (b), and Raman parameter I_S/I_V (c).

With this aim, the values of sensor response of SnO2 hierarchy toward ethanol, formaldehyde, LPG, gasoline, and H2S species have been measured. The working temperatures are chosen at 270 °C for detection. Figure 19 presents the *S*(gas) value as a function of SnO2 crystallite size (a), specific surface area (b), and Raman parameter *I*S/*I*V (c) for the series. One can observe a general trend of the sensor response diminution as the SnO2 crystallite size increases. However, it is found that behavior of samples toward H2S molecules is different from that shown by the others. These series shows the inverse relationships between sensor response and microstructure parameters: *S* (H2S) increases as *d*XRD values rise too and *S*BET diminishes. The samples with *d*XRD 3-5 nm are characterized by an approximate sensor response value toward corresponding target gas. As for the larger crystallite size *d*XRD > 6 nm, the series display a lower *S* (gas) value. The dependency on *S*BET value appears to be more informative to characterize gas sensor properties of these tin oxide samples. All results can be approximatively presented as a linear function *S* (gas) – *S*BET taking even into account the accuracy of specific surface area and sensor response measurements. Additionally, an approximate linear correlation is also obtained plotting the sensor response against the Raman parameter *I*S/*I*V compared with the specific surface area. Herein, it confirms the usefulness of Raman spectroscopy for the characterization of nanocrystalline tin oxide interactions with gases as well as the specific surface area. Notably, the specific surface area is measured on powders with hierarchical structures demolished, while Raman spectra are taken on hierarchy and films. It corroborates that the special structures have some influence on valuable *S*BET of the materials, especially on the surface modes, further on gas-sensing response. Another important conclusion, which can be drawn from this strong correlation between *S*(gas) and *I*S/*I*V, is that the surface modes (*I*S) are straightforward connected with the outer layers of the grain, which, according to these results, depend on the hierarchical structures or introducing of the biotemplate ESM.

The examination of nanostructure influence on the sensor parameters of SnO_2 exhibits the opposite trends for detection of C_2H_5OH (or HCHO) and H_2S molecules. The *S* (C_2H_5OH) increases with the specific surface area growth, and vice versa, *S* (H_2S) increases as the specific surface area diminishes. These results can be explained from the mechanism analyzed in the literature from the point of view of the adsorption kinetics or from the temperature treatment influence.

The interaction of reducing or combustible gases with the surface chemisorbed oxygen, which exists in various forms, namely: O_2^-, O^-, O^{2-}, etc, can take place in different ways of physical or chemical reactions. Vancu et al. have suggested the following reaction to be favorable with the reducing gases: $R+O^- \Leftrightarrow RO + e^-$. It has also been pointed out that the gas-sensing phenomenon is intimately connected with the occurrence of surface catalyzed combustion. As we are aware from the catalytic chemistry, the surface acid/basic state could be advantageously utilized to favor some particular reactions. With the prior knowledge of the type of a target gas, the sensor surface could accordingly be modified in order to preferentially sense the reducing gas over the other interfering gases.

Based on the above theories, Gnanasekar et al.[8] have suggested that gasoline combustion on a basic surface where the following reactions would take place:

$$\text{Gasoline (or LPG)} + O_2 \rightarrow CO_2 + CO + H_2O$$

$$CO + O^-_{ads} \rightarrow CO_2 + e^-$$

Fu et al.[9] have suggested that aldehyde (CH_3CHO) was the intermediate product when ethanol was oxidated on the basic surface and the following reactions happened:

$$CH_3CH_2OH + O_2 \rightarrow CH_3CHO + H_2O$$

$$CH_3CHO + O^-_{ads} \rightarrow CO_2 + H_2O + e^-$$

H_2S and the chemisorbed oxygen O^- can take place the following reaction [10]:

$$H_2S + O^-_{ads} \rightarrow H_2O + S + e^-$$

Electrons released from the above reactions would annihilate the holes: h• + $e^- \rightarrow 0$. Hence, the resistance of gas-sensing materials increased and then realized the detecting to different gases. For gasoline or LPG, the population of negatively charged oxygen adsorbates effectively controls the conductance of the target SnO_2. The chemical reactions occurring at the oxide surface have been assumed to involve CO adsorption followed by CO_2 desorption and replenishment of the resulting surface oxygen vacancies by the adsorption of molecular oxygen. For the oxide SnO_2, the electron concentration depends on the stoichiometry deviation determined by oxygen vacancies, which are predominant atomic defects. The electrical properties of nanocrystalline SnO_2 depend strongly on the surface states, produced by oxygen and other gas molecules chemisorbed at the grain boundaries that result in the space charge appearance and band modulation. Therefore, the variation of the chemisorbed molecule density is supposed to be the mainly responsible for the electrical response.

It is well known that ethanol can act as the reducing atmosphere and the reducing ability is stronger than formaldehyde (HCHO). So the sensitivity of hierarchical nanostructured SnO_2 in our work to ethanol was higher than that to formaldehyde. In the similar conditions, materials with higher effective surface area offer for ethanol molecules a higher probability of interaction with adsorbed oxygen and hence they are able to have a higher electrical conductivity change. Moreover, the excellent correlation with the Raman surface mode intensity also reveals the existence of a relationship between these modes and the density of the available oxygen adsorption sites.

Contrarily, the situation concerning H_2S sensing mechanisms is different. As concerned as the sensing behavior to H_2S, a certain amount of H_2S was oxidated to condensed H_2O and S at such working temperature to reduce actual amount of the working H_2S, as a result there shows lower sensitivity although H_2S is a type of acid gas and easy absorbed on the basic surface of the target material SnO_2. Generally, the density of real adsorbed gas is influenced by the concentration of available electrons to form and absorb, and by the amount of available adsorption sites at the grain surface that can be altered by the competition with the oxygen. For example, a nanograin surface that is highly covered by adsorbed oxygen has a high density of negative charge that can prevent or render more difficult any action for increasing the surface negative charge such as is required by the H_2S molecule adsorption. Therefore, the existence of an inverse correlation is quite plausible between the sensor responses for C_2H_5OH and H_2S. Experimental results seem to corroborate that for high I_S/I_V values the sensor response to H_2S is low and it increases when the surfaces modes diminish. It is proved that there is a better correlation between the sensor response for H_2S and the active surface than if it is considered against the grain size, d_{XRD}.

At working temperature 270 °C, the sensitivity of as-prepared SnO_2 to C_2H_5OH and HCHO is more prominent than that to H_2S, LPG and gasoline. Moreover, the sensitivity to the test gas markedly increases with the crystallite size of the sensor SnO_2 ranged from 26 nm to 3 nm depending on synthesis temperature. However, H_2S shows downward trends with synthesis temperature, which gives rise to surface modification revealed by the diminution of the surface disorder or by the faceting of the grains. It facilitates the availability of electrons and specific sites for H_2S adsorption. Thus, after annealing at high temperature 400-1000 °C, the sensor response to H_2S is improved.

As-prepared hierarchical nanostructured films consist of interwoven interconnected tubular SnO_2 assembled by nanocrystallites with the grain size of 3–26 nm and specific surface area up to 87 m^2 g^{-1}. The influence of the calcination temperature, crystallite size, and specific surface area on the intensity of Raman surface vibration modes is studied, and an excellent correlation is found between the Raman surface vibration mode intensity and the active surface. It has been considered to corroborate that the Raman surface vibration modes reveal the availability of active surface sites for interaction with the target gas more valuably than specific surface areas, based on the fact that the Raman spectra are taken from intrinsic sensors keeping special nanostructure, while the specific surface area is measured on powders forfeiting the hierarchy.

1.3. Fabrication of ZnO and Other Hierarchical Nanostructured Oxides

In our optimized procedure, the membrane pieces are immersed into 1 M Zn(NO3)2 ethanol solution kept for 30 h at room temperature. This immersing step is repeated three to five times. After being fished out and rinsed with distilled water, these hybrids are dried naturally and calcined at 550 °C for heat treatment with different time (1h, 2h, 3h, 4h, and 6h) to remove the template and achieve better crystallization of the sinters.

The calcination process is traced to suggest the formation of ZnO nanomaterials by conducting TGA and DTA surveys. It is explicit that the natural ESM start pyrolyzing at 200 °C and absolutely pyrolyzed by 550 °C, while the templated hybrids are also decomposed around 200 °C and remained 22 % (by mass) inorganic matters at 550 °C with template completely removed. The weight loss of the hybrids may result from the charing of organic ESM and the following pyrogenation of the charing products. In addition, DTA curve shows two peaks at 357 °C and 531 °C with an interval about 100 °C, which shall represent the decompositions of Zn(NO3)2 and ESM with different organic components, respectively.

The Zn-ESM hybrids are mainly amorphous matters containing zinc precursors and organic bioingredients. Based on the TGA results, the biotemplate ESM is thoroughly decomposed at 550 °C, there present no other peaks of the impurities in the corresponding XRD pattern. All XRD diffraction peaks of the sinters with various heating time can be indexed to hexagonal ZnO with a wurtzite structure (SG: *P*63*mc*, JCPDS 36-1451). With the increase in treating time up to 4h, the XRD peaks intensify sharply. It is evident that the diffraction peaks of the sample calcined at 550 °C for 1h are broadened, assigned to ZnO nanoseeds with very small dimensions, while the diffraction peaks of the sample calcined at 550 °C for 4h are narrow. The average crystallite sizes are estimated to be about 6 nm (treated for 1h) and about 40 nm (treated for 4h), srespectively. All three samples calcined at 550 °C for about 1h, 2h, and 4h, respectively, have similar microscopic features and preserve interwoven and interpenetrated structures of the natural ESM.

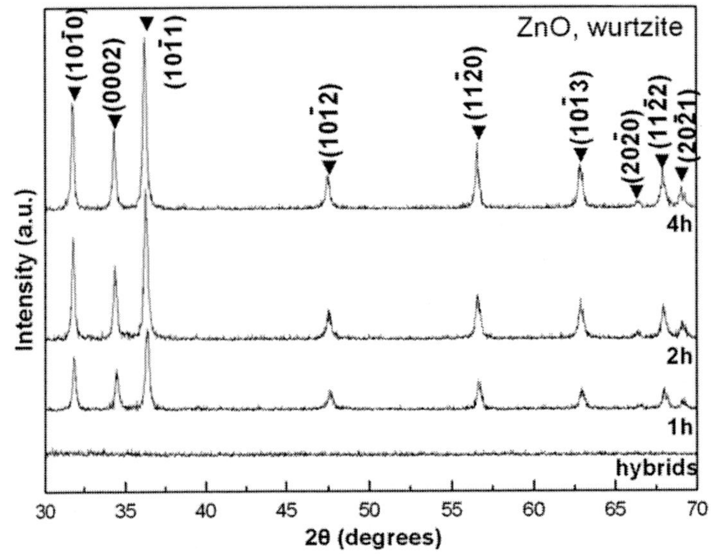

Figure 25. XRD patterns of the templated hybrids and the sinters at 550 °C for different time.

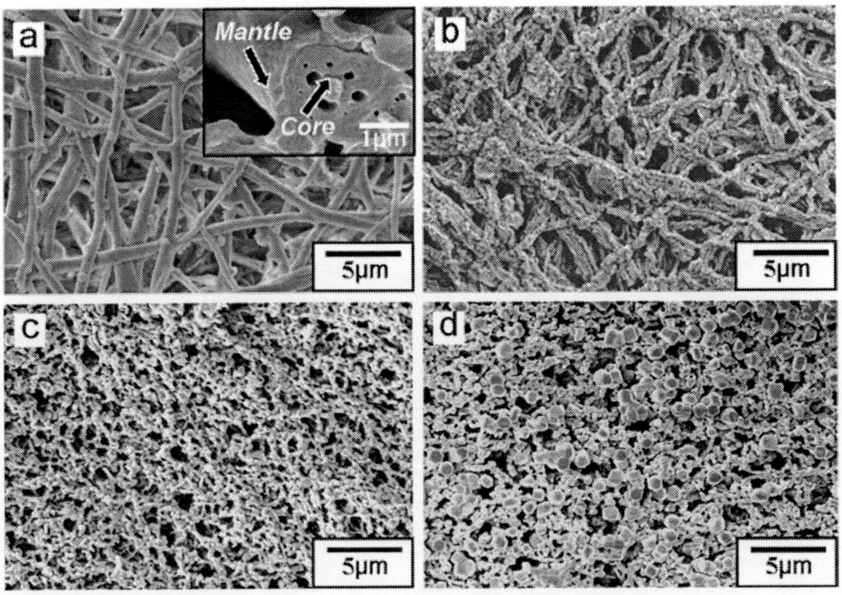

Figure 26. Typical FESEM images showing (a) the natural eggshell membrane and hierarchical interwoven ZnO calcined at 550 °C for different time: (b) 1h, (c) 2h, and (d) 4h. Inset in (a) reveals the conformations of one single membrane fiber.

A careful look at the FESEM images reveals the obvious morphologies evolution of the samples. The sample treated for 1h is characterized by repetitious networks consisting of the fibers ranging from 0.2 to 2 μm in diameter. The target materials are prone to exist as a film with the thickness of about 25 μm, and display hierarchical and subtle structures containing interconnected and penetrated meshworks at three dimensions. With the heat treating time prolonged to 2h, the ZnO fibers are slendered and exhibit porous structures. Notably, few particles with abmodality are appeared sticking on the networks. As heat treating time is further protracted to 4h, more ZnO polyhedron are *in situ* formed along the original ZnO fibers while the interwoven structures are weakened and nearly invisible. The density of the ZnO polyhedron is above 60% (volume).

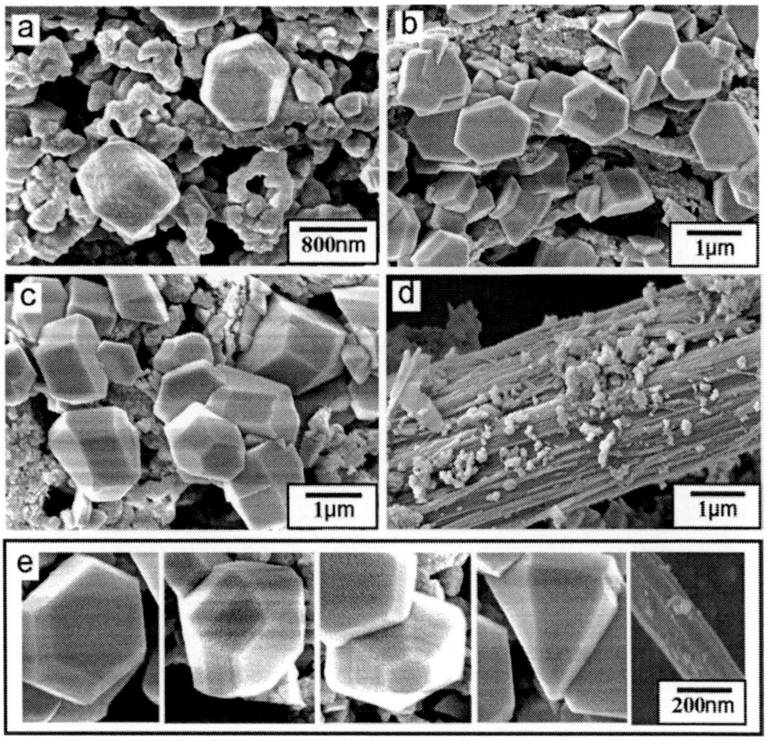

Figure 27. Typical FESEM images showing the extraordinary change in ESM-templated ZnO morphology by prolonging heat treating time: (a) 2h, (b) 3h, (c) 4h, and (d) 6h. Image (e) reveals the intuitionistic morphology evolution of one ZnO cell versus the heating time.

The higher magnification FESEM images show the extraordinary change in ESM-templated ZnO morphology by prolonging heat treating time. The ZnO crystallized along the interwoven matrix for about 2h consists of about 350 nm large polycrystalline agglomerates composed of about 20 nm large round-shaped nanoparticles. Notably, some ZnO crystals are formed like twinned platelets of hexagonal-shaped crystals with diameters of about 600 nm. It can be extended to a great amount of production as treated for about 3h. More hexagonal-shaped crystals with diameters of 400-500 nm come into being. Further treated for another hour, these hexagonal ZnO evolve into polyhedral crystals by the dominant growth with the whole thickness of about 1 μm. However, the use of a longer treating time of synthesis does not produce such regularly shaped ZnO crystals. For more than 6h heat treatment, a new kind of ZnO arrays are formed, which consists of nanorods with diameters of 100-200 nm and lengths of 10-20 μm.

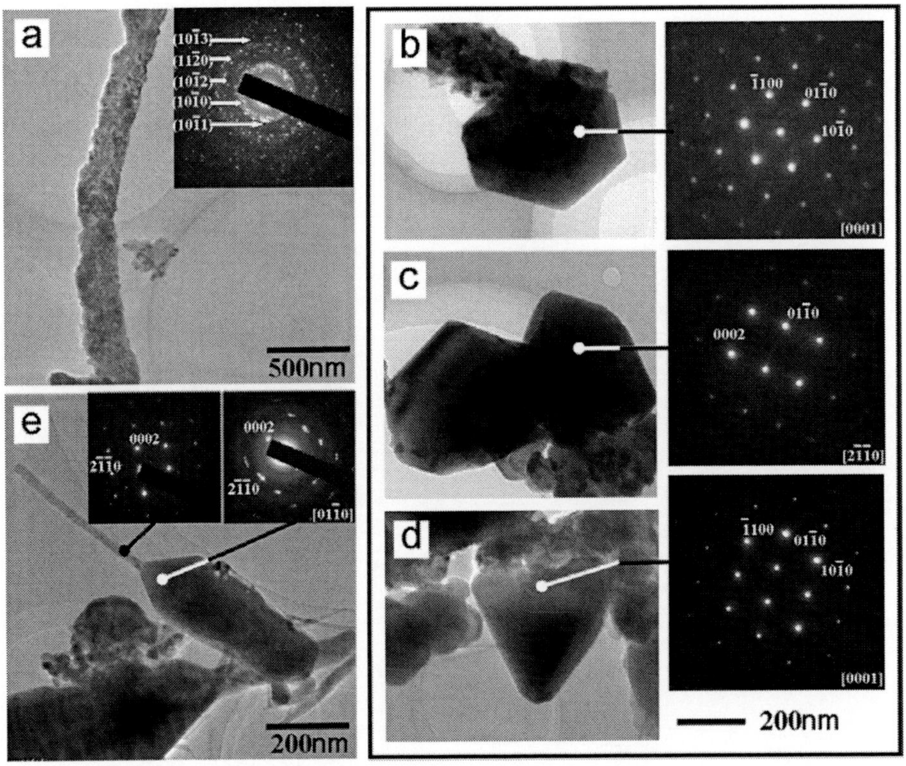

Figure 28. TEM images of the sinters at 550 °C for various heating time: (a) 1h, (b) 3h, (c,d) 4h, and (e) 6h. The insets show the corresponding SAED patterns.

Further structural characterization of the ZnO microcrystals is performed using TEM. One can that single ZnO fiber obtained at 550 °C for about 1h is assembled by nanocrystallites. The SAED patterns show that nanoscaled ZnO is polycrystalline nature, and the diffraction rings are indexed to (10-10), (10-11), (10-12), (11-20), and (10-13) diffraction of hexagonal ZnO, in agreement with the XRD results. The TEM images of the polyhedral ZnO microcrystals at different heat treated stage reveal the same size as determined from the FESEM images. The TEM image of hexagonal ZnO microcrystals as-treated for 3h shows disklike morphology with glazed surface and uniform size. The corresponding SAED pattern composed of bright spots in 6-fold symmetry indicates that the disklike ZnO microcrystal is a single crystal that is of hexagonal phase. The sample treated for 4h exhibits the TEM image of prism-like ZnO microcrystals, which actually grow along the [0001] orientation. The corresponding SAED pattern also indicates that the prism-like ZnO is of high crystallinity and hexagonal phase. There displays the TEM image of another kind of the ZnO crystals with hexagonal cone-shape viewed from the side of the hexagonal cone. The corresponding SAED pattern performed on the individual hexagonal cone-shaped crystal indicates that it is also single crystalline in nature and can be indexed as the hexagonal ZnO phase, which is analogous to the hexagonal disks treated for 3h. As continuously heat treated to 6h, morphology of ZnO crystals has changed. As treated for 6h, some ZnO nanorods are formed and grow along the [2-1-12] orientation basing on the former hexagonal cone structures. These nanorods incipiently have the diameters of 40-60 nm and the lengths of 500 nm, further grow up with the diameters of 200 nm and lengths of 10-20 μm.

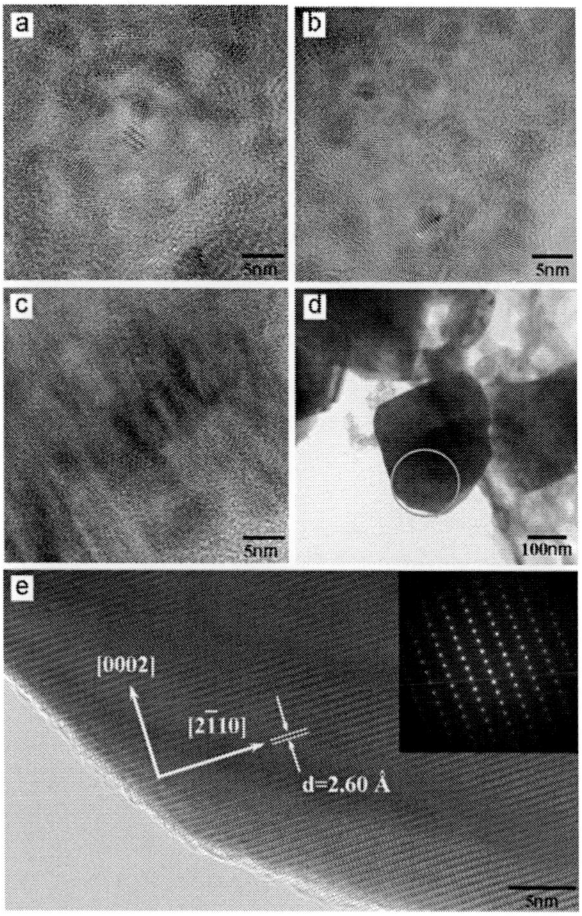

Figure 29. HRTEM images of the sinters at 550 °C and for various heating time: (a) 1h, (b) 2h, (c) 3h, and (d) 4h. Image (e) shows relative high magnification of circled area. The inset shows corresponding SAED patterns. The inset displays corresponding SAED pattern.

HRTEM image of the sinters at 550 °C for about 1h displays that the nanoseeds (ZnO nanocrystallites) are about 5-6.5 nm, and some nanocrystallites might be enwrapped by organic macromolecules of ESM remnants leading to the hazy lattice image. Herein the target material at 550 °C is assembled by such small nanocrystallites It probably ascribes to short-chained amino acides of ESM glycoprotein functioned as the capping agents to restrict the growth of zinc oxide nuclei. However, this limiting effect is weakened when the calcination is for longer time. As the heating time increased to 2h, the lattice image can be clearly seen due to the phase contrast. Considering that each domain having the parallel lattice fringes is a single crystal of ZnO, the average crystallite size is observed to be about 6.5 nm with better crystallization, which may result from the thorough pyrogenation of the organic matrix ESM. Notably, few grow up exceptionally. As-treated for 3h, obtained ZnO nanocrystallites have larger size up to 100 nm. TEM images reveal that the present ZnO crystals synthesized at 550 °C for about 4h is characterized by the polyhedral facets. The HRTEM and SAED images also indicate that the sample has larger crystal size and exhibits single-crystalline state. The (0002) planes are imaged of the hexagonal Wurtzite-type crystal with a lattice spacing of 2.60 Å.

The formation of the ZnO nanoseeds or the polyhedral ZnO structures is proposed to be a process comprised of solidification of biotemplated direction, surface oxidation, nucleation and growth. Herein, these interactions between dissolved zinc ions and specific surfaces affect not only the particle size and habit of nucleating crystals but also the stability of intermediate phases by dropping the activation energy required for the formation of specific crystal faces. Ascribing to the capping function of ESM glycoprotein short-chained amino acides, the growth of zinc oxide nanoseeds can be restricted. ZnO by itself tends to grow into rodlike structures like calcite, but ESM biomacromolecules specifically adsorb to the (0002) surface and force the crystal to grow into plates. The biomacromolecules of the bioorganic templates nucleate the mineral by controlling its crystallographic orientation and growth, by imitating the lattice of a two-dimensional face, or by the stereochemistry of the functional groups at the interface. Recent studies suggest that an epitaxial or geometrical match between macromolecules and crystalline faces does not occur. Instead, the charge density on the crystalline faces, and thus the electrostatic attraction, is alone sufficient for the heterogeneous nucleation. There has been much effort in trying to elucidate the mechanism of proteins and other biomolecules on biomineralization. As we know, in liquid medium, although the growth habit of ZnO microcrystals is mainly determined by its intrinsic structure, it is also affected by the external conditions such as temperature, pH value of solutions, and capping molecules. Among them, the capping molecules can tailor the surface energy according to capping the surface of nuclei and thus control the shapes of microcrystals. Many in vitro studies have demonstrated the importance of proteins and peptides in controlling the nucleation and/or growth of crystals.

Based on the directing and capping functions of ESM biomacromolecules for oxide nanocrystallites, natural ESM is chosen as the template to achieve 3D interwoven porous ceramic materials through similar bioinspired soakage processes with $Zn(NO_3)_2$, $Co(NO_3)_2$, and $PdCl_2$ solutions as the precursor impregnants, respectively. Thermogravimetric analysis of the impregnated ESM reveals that it begin to be pyrolyzed at 200 °C, completely decomposed at 500 °C, leaving behind 25.2%, 20.5%, and 36.8% (by mass) inorganic sinters of zinc, cobalt, and palladium oxides, respectively, along with the removal of organic ESM.

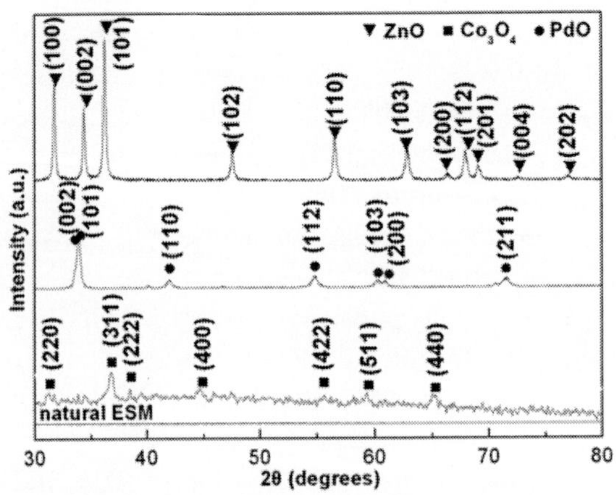

Figure 30. XRD patterns of the natural ESM template and the sinters at 550 °C.

All the XRD diffraction peaks of the calcined products at 550 °C can be assigned to the respective single phase of metal oxides without indication of other crystalline byproducts. All the peaks in the XRD pattern of zinc oxide can be attributed to the zincite structure of ZnO (JCPDS 36-1451), indicating well-established crystallization. For cobalt oxide, broadened peaks were found and are characteristic of smallsized crystallites with a cubic structure of Co_3O_4 (JCPDS 78-1970). The peaks for palladium oxide are fairly broadened and correspond to the tetragonal structure of PdO (JCPDS 75-0584). The average crystallite sizes of about 5.4 nm for ZnO, 9.5 nm for Co_3O_4, and 11 nm for PdO are estimated according to the line width analysis of the diffraction peaks based on the Scherrer formula, respectively. It indicates that by calcination at 550 °C, the ceramic products characterized as the target materials are composed of small-scaled nanocrystallites.

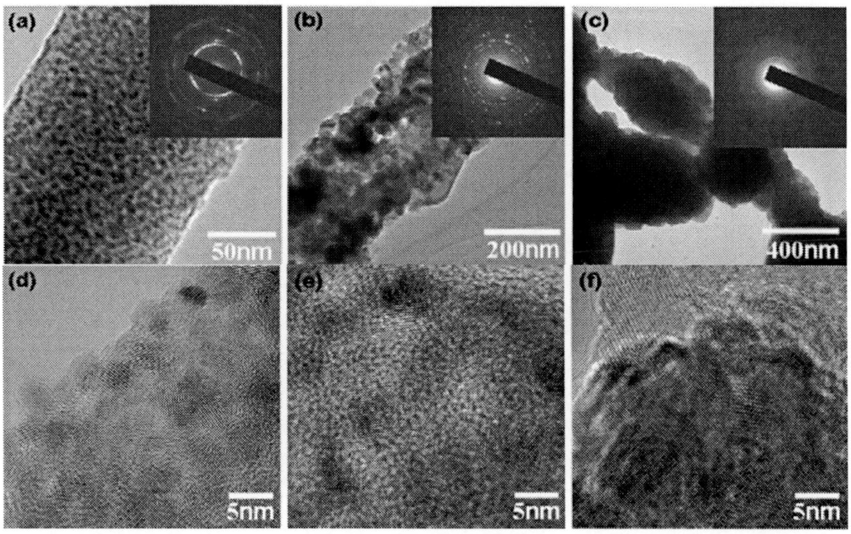

Figure 31. TEM overview images of (a) zinc oxide, (b) cobalt oxide, (c) palladium oxide fibers synthesized at 550 °C, the insets in (a,b,c) showing their respective SAED patterns. HRTEM images of above metal oxides are shown in (d), (e), and (f), respectively.

The overview TEM images of the series show that fibrous structures of about 200–400 nm in diameter are formed by the assembly of corresponding nanocrystallites. HRTEM images for the building unit nanocrystallites indicate that the lattice image can be clearly seen due to the phase contrast. Considering that each domain having the parallel lattice fringes is a single crystal, the average crystallite size is observed to be about 5, 9, and 11 nm for ZnO, Co_3O_4, and PdO, respectively. As expected from the XRD measurements, the zinc oxide nanocrystallites are particularly small, with the sizes ranging from 5 to 6 nm, and display more uniform morphology. It also reveals that the present ZnO hierarchy is characterized by the arrays of nanocrystallites with the structural defects such as micropores and mesopores. This indicates that accurately defined, size-controlled ZnO nanocrystallites of about 5 nm can be assembled to construct hierarchical and porous nanomaterials at the appropriate calcination temperature of 550 °C. However, cobalt oxide nanocrystallites exhibit a cubic shape with sizes of 9–10 nm and less pores compared with ZnO. For PdO, it is observed that the nanocrystallites of about 10 nm agglomerate compactly and the pores can hardly be found. In addition, the crystal structure of these nanoparticles is studied by selected area electron diffraction (SAED), in all cases, the SAED patterns coincide well with the XRD data.

The typical FESEM micrographs show that the size of coherent replicas depends on the original size of the natural ESM. Owing to the heat shrinkage of the biotemplated hybrids during the calcination process, the size of hierarchical replicas is up to about 70% of the original size of the natural ESM pieces. At low magnification, all the samples have similar microscopic features and preserve the interwoven structures of natural ESM fibers, but their detailed morphologies are quite different. At high magnification, fibrous ZnO and Co_3O_4 exhibited porous characteristics. On careful observation it is found that the pores in ZnO fibers are smaller than that for Co_3O_4, which may be ascribed to the lower agglomeration of ZnO building unit nanocrystallites, whereas the surfaces of PdO fibers appear more serried and glazed by virtue of the nonporous structures proved by TEM results.

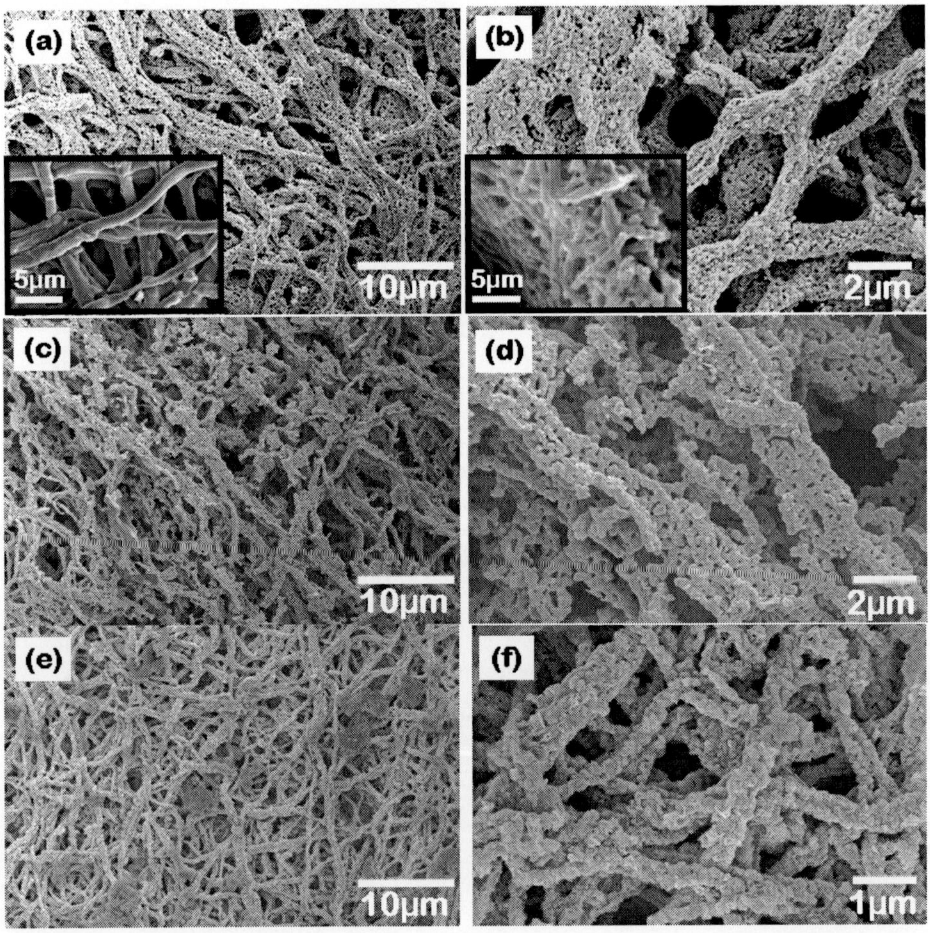

Figure 32. FESEM images showing (a) zinc oxide, (c) cobalt oxide, (e) palladium oxide with hierarchical interwoven structures prepared at 550 °C, respectively. Images (b), (d) and (f) show high magnifications of (a), (c) and (e), respectively. Insets in (a) and (b) displaying the feature of natural ESM and the cross-section of ESM-morphic ZnO, respectively.

From FTIR spectra of the samples calcined at 550 °C, it is found that the typical IR absorption peaks at about 460 cm^{-1}, 671 and 586 cm^{-1}, and 664 cm^{-1} and 603 cm^{-1} are characteristic vibrations of ZnO, Co_3O_4, and PdO, respectively, according to the above XRD results. For all the samples, the absorption bands centered at about 3440 and 1630 cm^{-1} are ascribed to adsorbed water on the surface of metal oxides. The bands appearing in the range between 850 and 1350 cm^{-1}, especially for Co_3O_4, are assigned to the bending mode of different types of surface hydroxyl groups [11], indicating that the surface of the nanoparticles is terminated with the —OH groups. Furthermore, an intense multiphonon scattering of ZnO nanoparticles is observed in the resonant Raman spectrum, where four major bands are observed to result mainly from the polar symmetry modes A_1(LO) and E_1(LO) at 582 cm^{-1} and their overtones. The first-order Raman mode is a superposition of the LO phonon mode in which Zn atoms and O atoms have the same vibration direction, respectively, to the neighbor lattices of the wurtzite ZnO [12]. The intense peak around 437 cm^{-1} is contributed to the E_2 (high) mode. Similarly, the Raman spectrum showing bands at 469, 512, 607, and 674 cm^{-1} with relative intensities corresponds to that of characteristic cubic Co_3O_4 [13], while the bands at 274, 418, 640, and 714 cm^{-1} correspond to that of characteristic tetragonal PdO. These results are consistent with the IR analyses.

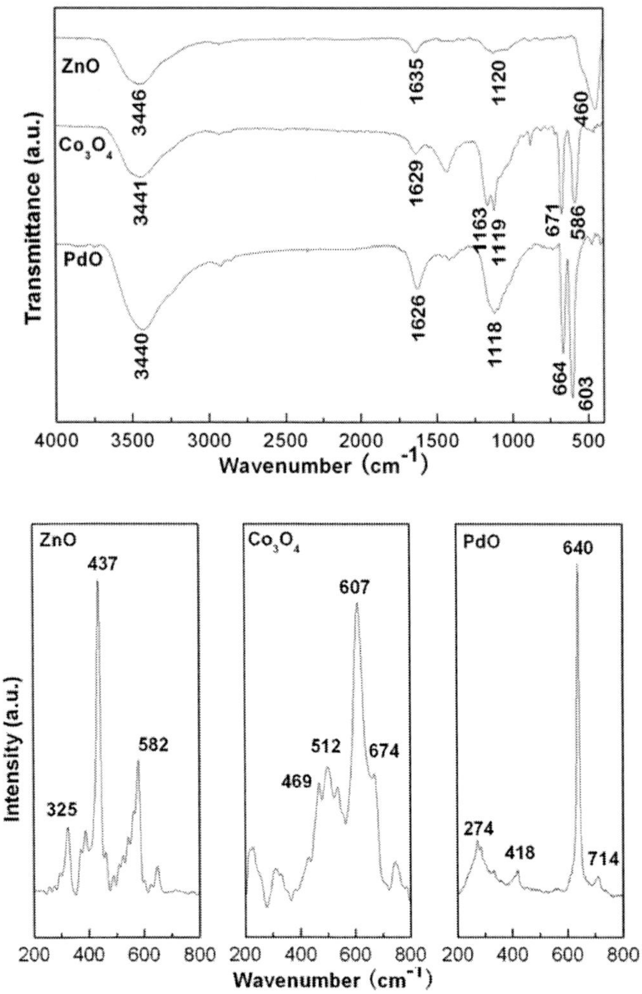

Figure 33. FTIR and Raman spectra of as-synthesized hierarchical metal oxides.

Typical nitrogen adsorption–desorption isotherms and corresponding pore-size distribution curves for ESM-templated metal oxides are examined. By the Brunauer classification, the isotherms change from IV-type for ZnO and Co_3O_4 to II-type for PdO where the hysteresis loop almost disappears completely. Different from IV-type curve based on mesoporous structures, the II-type isotherm is attributed to macroporous modes. At a low relative pressure (*o*0.5), the nitrogen-adsorbing action of macroporous materials is the same as the adsorption of mesoporous monolayer structures. But above the medium relative pressure (40.5), there is no capillary condensation phenomenon in macropores and then no hysteresis loop. Hence, according to the types of isotherms, three oxide samples can be confirmed as mesoporous ZnO and Co_3O_4, and macroporous PdO, individually. The average pore size increases from about 20–30 nm for ZnO, to 30 nm for Co_3O_4, to 80 nm for PdO, while the pore volume decreases. This means that the porous structures are quite different in nanometer scale depending on the different metal precursors and oxide materials. In addition, more than one pore diameter peaks in the pore-size distribution curves of ZnO and Co_3O_4 samples indicate several kinds of pore (micropore, mesopore, and macropore) characteristics. Actually, such mesopores and micropores have been observed in the above HRTEM images.

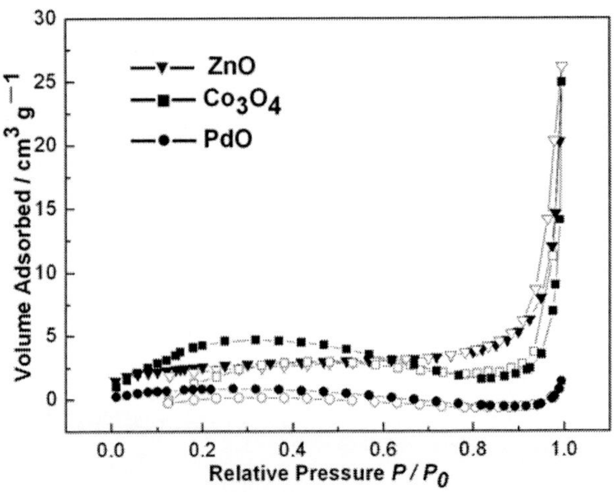

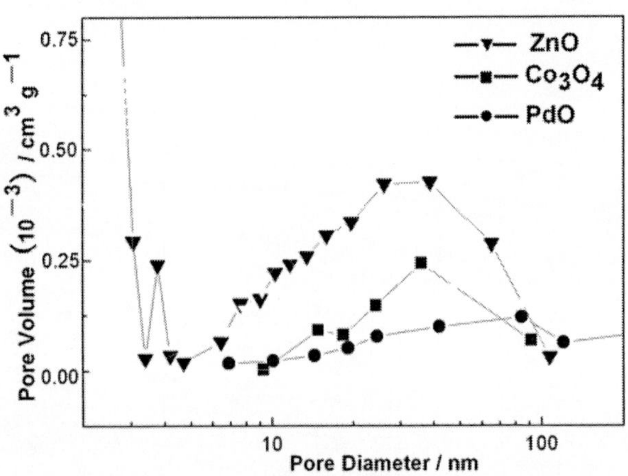

Figure 34. Nitrogen adsorption-desorption isotherms plots and corresponding pore size distribution plots of ESM-morphic metal oxides obtained at 550 °C.

The synthesis of biomorphic metal oxides can be separated into two steps: one contributes to the interplay of metal precursors and ESM constituents during the dipping pretreatment, and the other is the crystallization formation of metal oxides accompanied by the removal of the ESM template during the calcinations process. During the dipping treatment in the aqueous solution, due to the self-assembly behavior of the hydrophilic groups of glycoprotein, metal precursor ions are transported on the ESM surface; further, on calcination treatment at 550 °C, the final oxides can retain well the hierarchical structure of the biotemplate ESM when organic ESM is burned off. Moreover, it can probably be ascribed to short-chained amino acids of ESM glycoprotein, which functioned as capping agents to obtain well-distributed and small-scaled nanocrystallites as building blocks. However, there occur different reactions between corresponding metal precursor ions and ESM components; thus, final metal oxide unit nanocrystallites have divergent morphology and diameter. It has been found that the present ZnO or Co_3O_4 hierarchy is characterized by arrays of well-distributed and small-scaled nanocrystallites, and of micropores and mesopores. However, it is quite different for PdO that some Pd^{2+} ions could be reduced to Pd^0 by certain amino acids of ESM, which may further degrade the proteins by the catalysis. Thus, the conformations of ESM biomacromolecules may be altered and then the serried hybrids may be formed. The different porous conformation at the nanoscale can also be explained by the above mechanism. As a whole, ceramic oxides replicated well the hierarchical structure of the biotemplate ESM via a self-organized solution process, in which the biomaterial served both as a chemical revulsant and a physical substrate. Prompted by the special hierarchical buildup, we can speculate that as-prepared ESM-based metal oxides may be useful in fabricating optical materials (ZnO), magnetic storage devices (Co_3O_4), and catalysts or their carriers (PdO).

2. Formation and Assembly of Lead Chalcogenides on Eggshell Membrane Fibers

The synthesis control of metal chalcogenides and relevant nanocomposites has been the focus of intensive research due to their unique optical, electronic, and photovoltaic properties, which are chiefly determined by their shape, structure and distribution. Thereinto, nanostructured lead chalcogenides gain more and more research attention, due to their unique properties and valuable applications in highly efficient photovoltaic conversion devices, optical switching, biological labels, as well as electroluminescent devices and lasers.

A series of bioinspired methods involving some biospecies such as proteins, yeast, bacteria, virus, enzyme, eggshell membrane, rhodobacter sphaeroides, DNA, bean sprouts have succeeded in controlling small-scaled and well-dispersed chalcogenides nanocrystallites. In the progressive stage, the biogenic assembly of planned nanoarchitectures constitutes a main challenge to the "bottom-up" fabrication of nanocrystallites components for future nanodevices. Thus, research efforts are now invested in novel techniques for the deliberate organization and immobilization of metal chalcogenides nanoparticles on solid substrates to form hierarchical nanocomposites. However, it is still extremely difficult to find an appropriate biomaterial support on which the synthesis can undergo under ambient reaction conditions. In view of the special proteins patterns and functional residues of eggshell membrane (ESM) biomacromolecules, the convenient biological castoff can be utilized as an ideal biosubstrate for the preparation of metal chalcogenides and relevant nanocomposites.

Herein, a facile, moderate and green biosubstrate directed technique is designed to synthesize hybrid PbS/ESM nanocomposites at ambient conditions by immersing ESM into Pb(Ac)2-HAc and Na2S (or Na2SeSO3) solutions in turn. The functional residues such as amido and imido groups from glycoprotein mantle of ESM fibers, provide chelation and location functions to the formation and the assembly of nanocrystalline PbSe, and act as the surfactant and substrate to achieve fine nanocrystallites arraying well-distributed nanoclusters.

In a typical procedure for PbSe nanostructures, the substrate ESM is firstly immersed into the mixed solution of lead acetate and acetic acid at room temperature, taken out and rinsed with deionized water, subsequently the following two processes go along, respectively, namely that the membrane is dipped into fresh Na2SeSO3 solution (I); into ripen Na2SeSO3 solution laid still for 2 days (II). The final samples I and II as dark gray are obtained with correspondence to process I and II. XRD patterns indicate that the original ESM is totally amorphous, while all the reflection peaks of samples I and II can be indexed as cubic PbSe with the face-centered rock-salt structure (JCPDS No. 06-0354). The reflection peaks show markedly broadened, indicating that obtained products are small-sized nanocrystallites and sample I has smaller crystallite size than sample II.

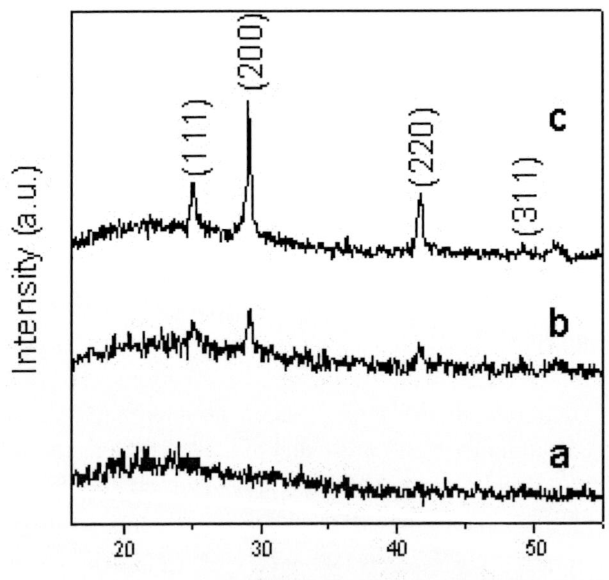

Figure 35. XRD patterns of the samples: (a) the original ESM; (b) sample I (PbSe/ESM); (c) sample II (PbSe/ESM), respectively.

The morphology and the configuration of the composite PbSe/ESM are investigated by SEM images. It can be clearly seen the ESM presents characteristic macroporous meshworks of diverse interwoven and coalescing fibers. PbSe nanocrystallites are stuck uniformly in ESM fibers while there is absent in the matrix among the membrane fibers. Fine PbSe nanocrystallites with the average size of 5 nm are assembled into 25–30 nm smaller nanoclusters distributing uniformly on the ESM support. And herein some smaller nanoclusters would like to arrange for bigger nanoclusters. While PbSe nanocrystallites are finally prone to ripen into nanocubes with lower thermodynamic energy. It is considered that these nanocrystallites can firstly be loosely assembled into smaller nanoclusters and subsequently compactly ripen into nanocubes with more glazed figures than the nanoclusters, as evidenced by the coexistence of PbSe nanoclusters and nanocubes when employing fresh Na2SeSO3 as the reactant with enough immersing time. All as-prepared nanocrystallites are well-dispersed and arrayed orderly on the surface of ESM fibers. The corresponding SAED pattern of PbSe nanocrystallites signifys the single-crystalline nature. The square spot array is indexed to (2 0 0) and (2 2 0) planes of the face-centered cubic (fcc) PbSe. The representative HRTEM images present well resolved lattice fringes revealing the high crystallinity, indexed to the (2 0 0) planes of the rock salt cubic phase with the interplanar spacing of 3.05 ° A, which are consistent with the XRD results of PbSe crystallites. On the basis of the TEM and HRTEM analyses, it can be considered that these nuclei bounded by (2 0 0) facets are assembled into nanoclusters and nanocubes on the ESM. In addition, as-prepared PbSe nanocrystallites with a characteristic feature of small size are displayed from the room temperature photoluminescence (PL) spectrum of sample I. PbSe nanocrystallites exhibit the emitting peak at 357 nm, with a 42 nm blue-shift in comparison with Guha's report, which reveals as-prepared PbSe nanocrystallites in Sample I have very small size and performed obvious 1Se~1Ph intraband transitions indicative of a short lifetime and obscuring the phonon bottleneck.

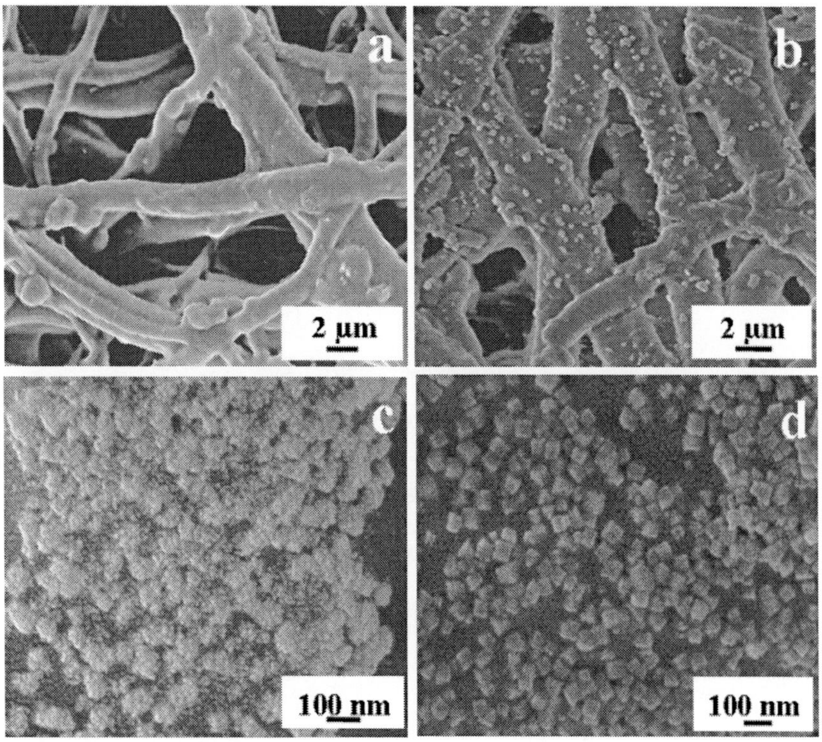

Figure 36. SEM images of (a) the overview of the surface of original ESM; (b) the overview of PbSe/ESM composites; (c) sample I and (d) sample II, respectively.

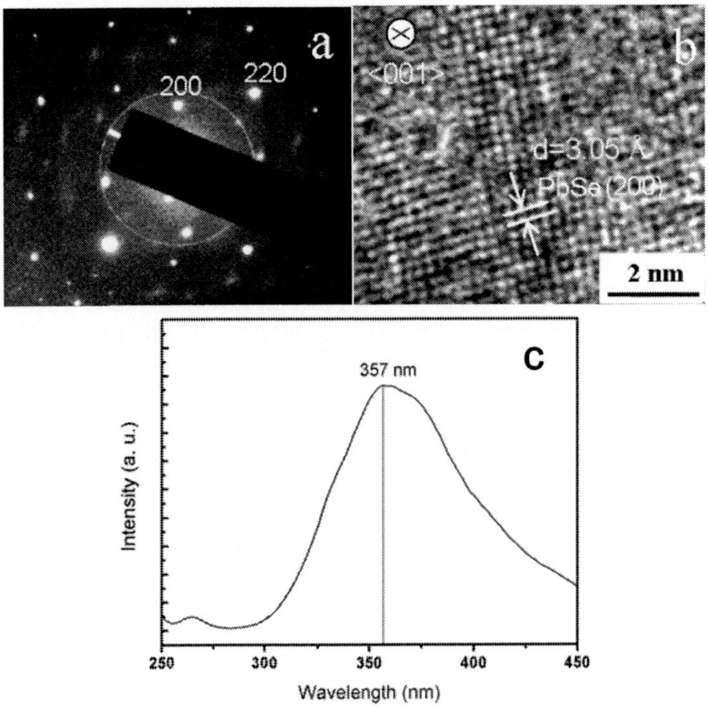

Figure 37. (a) Representative SAED pattern of PbSe nanocrystallites; (b) high-resolution TEM images showing PbSe nanocrystallites oriented attachment along (001) direction; (c) photoluminescence spectra of PbSe nanocrystallites, the excitation wavelength 250 nm.

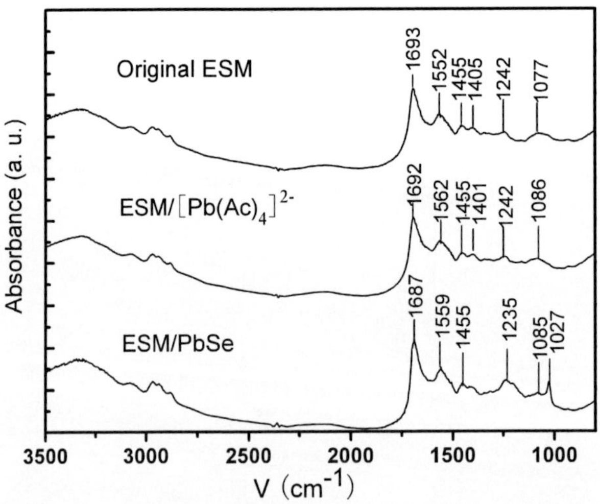

Figure 38. The typical FTIR spectra of (a) ESM; (b) [Pb(Ac)4]2-/ESM; (c) the composite ESM/PbSe, respectively.

FTIR spectra of [Pb(CH3COO)4]2⁻/ESM and PbSe/ESM are also be probed into. The absorption bands at 1693, 1552, and 1242 cm⁻1 are attributed to amide I, amide II, amide III bands of the glycoprotein mantle of original ESM, respectively. The peak appearing at 1455 cm⁻1 is due to the *v*s(COO) stretching vibration, and the band around 1077 cm⁻1 is assigned to the C–N stretching vibration in plane band. After the ESM being dipped into Pb-precursor solution, the shifts of amide II from 1552 to 1562 cm¯1 and *v*s(C–N) stretching vibration from 1077 to 1086 cm¯1 suggest that there occurred environmental changes around the imino groups and the C–N bond, which may be due to the condensation reaction of carboxyl groups from [Pb(CH3COO)4]2¯ with amine groups and imido residues from glycoprotein mantle of ESM [14]. Further soakage in the Na2SeSO3 solution, there clearly presents a new band at 1027 cm¯1 attended by the red shifts of amide I, amide II, amide III bands. These changes shall contribute to the formation of PbSe nanocrystallites on the ESM.

Herein, eggshell membrane as the biosubstrate, shall play crucial roles in the formation of PbSe nanocrystallites and the assembly into nanoclusters and nanocubes. The architecture of the ESM fiber can be depicted as the collagenous core which is mainly composed of collagen and osteopontin proteins, and the polyanionic glycoprotein mantle. When the ESM is dipped into Pb(Ac)2-HAc solution, the chelation prefers to occur on the surface of glycoprotein mantle of fibers while is hard to take place in the core.

Detailed amino acid composition (wt.%) of natural ESM fiber mantle is Asp 7.05, Thr 4.80, Ser 4.32, Glu 9.98, Pro 9.34, Gly 5.20, Ala 2.26, Cys 4.10, Val 5.30, Met 3.32, Ile 2.61, Leu 3.65, Tyr 1.87, Phe 1.35, His 2.97, Lys 2.98, Arg 5.93, Trp 1.80 [15]. It is well known that all amino acids are connected by peptide bonds. Especially, the lysine acid (2.98 wt.%), arginine acid (5.93 wt.%) and His (2.97 wt.%) containing two poorly alkaline amido groups, as functional moleculars, can beget the condensation reactions with the carboxyl residues from [Pb(CH3COO)4]2¯, thus the Pb-complexes in the solution may be caught on the surface of ESM fibers. Moreover, other acidic and neutral amino acids can also play some roles in attracting Pb-complexes but not evident. By sequential soakage treatments of the ESM into Na2SeSO3 solution, SeSO32¯ is gradually attracted towards lead sites on the ESM mantle and Se2¯ is prone to gently escape from SeSO32¯ under the electrostatic actions of lead cations and the inducement of imido residues and imino groups. Whereafter, in situ nucleus formation occurs and cubic PbSe comes into being at surface of ESM fiber mantle, since cubic PbSe has thermal stability which can effectively lower the system energy.

$$[Pb(CH_3COO)_4]^{2-} + 4R\text{-}NH_3^+ \rightarrow [Pb(CH_3CONH\text{-}R)_4]^{2+} + 4H_2O \quad (1)$$

$$SeSO_3^{2-} + H_2O \rightarrow SO_4^{2-} + Se^{2-} + 2H^+ \quad (2)$$

$$[Pb(CH_3CONH\text{-}R)_4]^{2+} + Se^{2-} \rightarrow (CH_3CONH\text{-}R)_4PbSe \quad (3)$$

$$\cdots(R = \text{alkyl residue})$$

ESM can also play an oriented role for the growth and the assembly of PbSe nanocrystallites. The shape of an fcc NC is determined by the ratio, *R*, of the growth rate in the <100> directions to that of the <111> directions [16]. During the growth stage, the facets of the crystallites tend to form on the low-index planes to minimize the surface energy, which is crucial for the growth rate of the crystallites. Herein, amido groups and imido residues contained in glycoprotein mantle may form weak binding ligands to the {100} faces while bind strongly to the {111} surfaces, thus compared with the {100} faces, the {111} faces can selectively have a low surface energy [17]. Accordingly, the growth rate in the {111} planes is significantly higher than that of the {100}, then {111} facets will vanish resulting in the formation of the nanocubes [18]. Moreover, the relative supersaturation of lead cations on glycoprotein mantle can in turn compensate for the negative electrostatic potential, which made seleniferous anions facile to contact with lead ions by getting over the spatial hurdle. As soon as seleniferous anions are captured on the ESM fiber mantle, the PbSe nucleus can come into being, which may also be an autocatalysis [19] for the growth of PbSe nuclei until well-distributed perfect nanocubes or uniform nanoclusters are achieved. In addition, the amino acid residues of the ESM macromolecules can be functioned as the surfactants to well-distributed and small-sized nanocrystallites. By capping roles, these fine nanocrystallites are feasible to evolve into nanoclusters and are finally self-assembled into thermodynamically stable perfect nanocubes in the absence of additional surfactants.

Moreover, the precursor Na2SeSO3 solution also plays an important role in the formation of PbSe nanoclusters and nanocubes. Upon the increase in the concentration of Na_2SeSO_3 solution and employing fresh Na2SeSO3, PbSe crystallites come into being at relatively fast speed corresponding to kinetic non-equilibrium of nucleus formation, thus large numbers of nuclei are formed, and the growth process is hard to control, it is less likely to form perfect crystal lattices and PbSe nanoclusters are obtained.

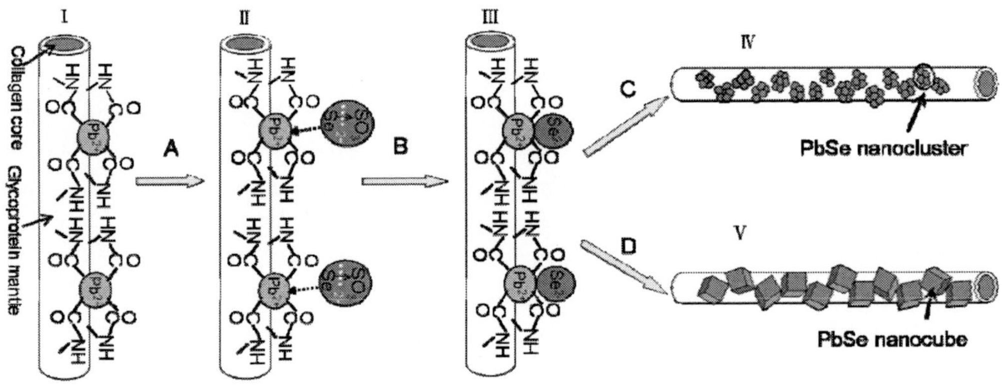

Figure 39. Formation illustration of lead selenide nanocubes and nanoclusters on ESM fiber. Models of (I) show the condensation reaction between the amido groups from glycoprotein mantle of a single ESM fiber and the carboxyls of $[Pb(CH_3COO)_4]^{2-}$ at the given locations, (II) $[(SeSO_3)^{2-}]$ ions in the solution approach to lead complexes on the mantle, based on the electrostatic actions, (III) in situ nucleation of PbSe nuclei, (IV) the formation of thermodynamic metastable PbSe nanoclusters, (V) tiny nanocrysatllites evolve into well-ordered nanocubes.

Since PbSe nanoclusters are thought to be thermodynamically metastable state, nanocubes are finally prepared, based on prolonging reaction time, using ripen precursor Na_2SeSO_3 and decreasing the concentration of Na_2SeSO_3 solution, which can meet all these conditions, such as the condense reactions between amidos and carboxyls, the electrostatic actions between lead cations and seleniferous anions, to the nucleation and the assembly of well-ordered PbSe nanocubes.

By similar procedure but Na2S solution as the corresponding impregnant, face-centered cubic PbS nanoparticles come into being on ESM fibers to obtain PbS/ESM hybrids. The formation scheme can be described as follows:

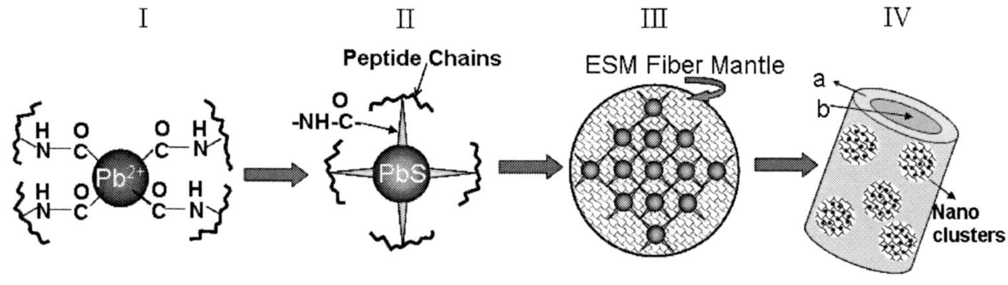

Figure 40. The illustration of the formation of well-distributed floriated lead sulfide nanoclusters on ESM fibers. Models of (I) the condensation reaction of the carboxyl of $[Pb(CH_3COO)_4]^{2-}$ with the amido groups from glycoprotein mantle of ESM fibers at the given locations; (II) in situ nucleation of PbS nuclei on the mantle by electrostatic actions; (III) the assembly of PbS nanocrystallites into the nanoclusters on the ESM fiber mantle; (IV) floriated PbS nanoclusters bound on the substrate ESM fibers to form the nanocomposites PbS/ESM. Single ESM fiber is structurally composed of glycoprotein mantle (marked as a) and collagen core (marked as b).

For further investigation on PbS nanoparticles, the Raman spectroscopy is thought to be effective and nondestructive. The strong peak at 155 cm^{-1} can be assigned to a combination of transverse acoustic and optic modes, TA+TO, over a large volume of the Brillouin zone. The strong band at 211 cm^{-1} can be indexed to fundamental longitudinal optical (LO) phonons near the Γ-point of the Brillouin zone [20]. It is noted that there exists a small shoulder at 204 cm^{-1} assigned to a surface phonon (SP) mode [21]. Based on Batonneau's reports [22], the bands observed at 155 and 204 cm^{-1} are presumably due to the luminescence of PbS semiconductor or caused by instrumental effects, and that the band at 450 cm^{-1} can be ascribed to the first overtone of the LO phonon (2LO) at the center (Γ) of the Brillouin zone, this overtone shall be induced by a Fröhlich-interaction and be Raman active. Meanwhile, the peak at 622 cm^{-1} can be labeled as the second overtone (3LO) [23]. Different to Krauss and Wise's report on the Raman spectrum of PbS quantum dots having three peaks at 215 (LO), 415 (2LO), and 630 cm^{-1} (3LO), respectively, the differences in the wavenumbers and the intensities to those of our Raman peaks can largely lie on the particle size, the excitation wavelength, and the operating temperature. In addition, it is demonstrated that the Raman frequency prefers to decrease with the increase in crystalline size and surface roughness. So it can be thought that in our work both PbS nanocrystallites and the nanoclusters have uniform small size and the nanoclusters are well-distributed on the ESM fibers. In addition, the observability of TO and SP modes proves that PbS nanocrystallites in the hybrids exhibit strong quantum size effects. Because of the Raman shifts of crystalline insulator or semiconductor, it is feasible to observe LO phonons while other modes such as TO and SP modes are usually not observable due to symmetry restrictions and their low intensities, respectively [24]. However, the SP mode by Raman scattering measurements can occur in Raman spectra of nanoscaled materials with quantum size effects [25]. Other peaks, such as located at 174, 511, 564 cm^{-1}, may attribute to some new bonds giving rise to the ESM intricate cooperative effects during the soakage treatment. Therefore, small-scaled PbS nanocrystallites on the ESM fibers are assembled into size-uniform nanoclusters distributing orderly on the ESM fibers.

Therefore, during the formation of lead chalcogenide nanocrystallites and nanoclusters on living ESM at room temperature, the ESM not only provides effectively revulsive sites to the formation of chalcogenides nuclei and the assembly into nanoclusters in virtue of the reciprocity of functional groups of ESM fiber mantle, precursors, but also acts as the surfactants bringing uniformly arranged and well-dispersed chalcogenides nanoclusters and nanocubes on the ESM. Other factors are also considered, such as reaction time, precursor concentration. The as-prepared chalcogenides nanoclusters behave good quantum size effect and dispersancy, and exhibit excellent biocompatibility when combined with the reactive biosubstrate ESM to be hybrid nanocomposites for applications in photoelectric, photonics, and electronics fields. This bio-inspired technique is a sort of green, moderate and effective strategy for the synthesis of other series of functional nanomaterials with pre-desired conformation and relevant valuable properties.

3. In Situ Synthesis of Noble Metal Nanoparticles on Biofibers at Room-Temperature

In a typical experiment for Pd nanoparticle, Ca. 15 mg ESM fiber is immerged into 200 ml 0.5 mM palladium chloride aqueous solution with the pH value tuned to about 8. The system is carried through in the dark for about 18 h. Then the as-immerged biofiber is fished out, subsequently rinsed with distilled water and directly dried with air flow to obtain final hybrids.

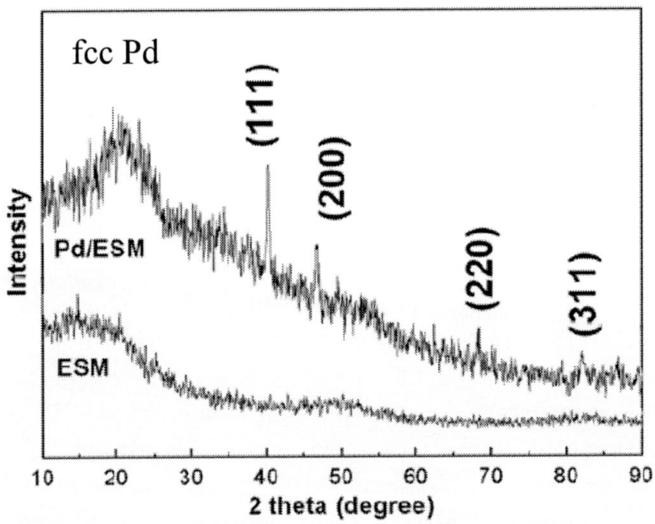

Figure 41. XRD patterns of natural eggshell membrane fibers (ESMF) and the composites Pd/ESMF.

One can be seen from the XRD patterns that the natural ESMF exhibits the diffractive peaks of crystalline domains around 20.64°, which are ascribed to the sequence and conformations of amino acids. As concerned as the hybrids samples being processed in dark for 18 h, it can be seen that the face-centered cubic (fcc) structure of palladium (JCPDS file No. 46-1043) comes into being, and the corresponding diffraction peaks can be indexed as (111), (200), (220), and (311) planes, respectively. In addition, the broadened reflection peaks indicate Pd nanocrystallites having small dimensions.

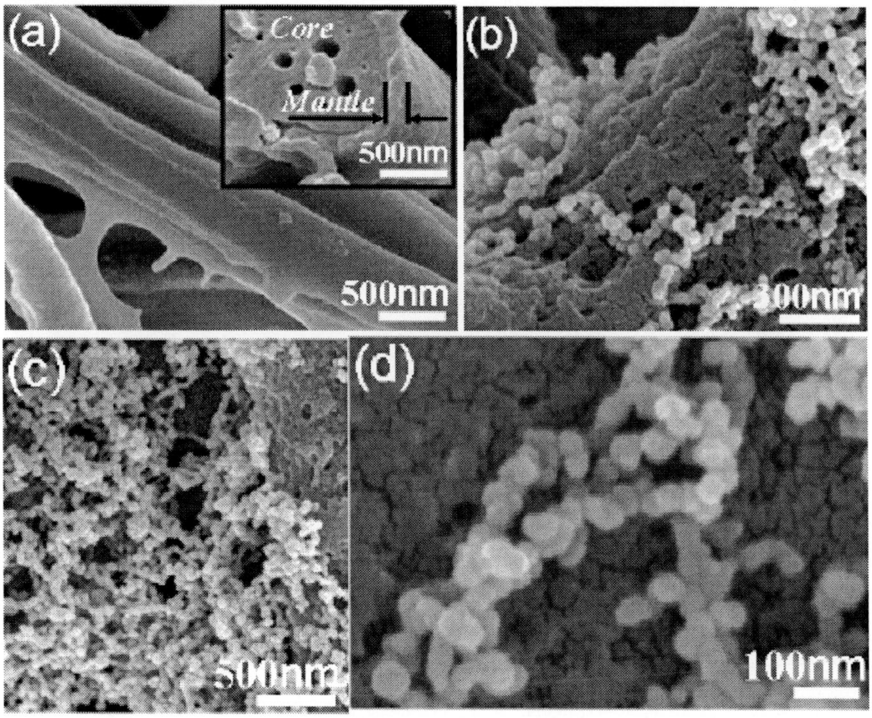

Figure 42. Typical FESEM images of (a) natural ESM fibers; as-prepared nanocomposites Pd/ESMF after reaction for (b) 5 h and (c) 18h; (d) high magnifications of image (c).

FESEM images reveal the morphologies of as-synthesized palladium nanoparticles on the biofibers with different reaction time about 5h and 18h. In the primeval stage the spherical Pd nanoparticles have formed and arrayed into the chained patterns, which is almost the same as the sample dealed with 18 h. One can see that there is no evident change in morphology by prolonging the reaction time besides a high yield of the target materials. A careful look at these fibers reveals that the previous glazed fibrous structures of the original ESMF have converted to the rugged forms, which may contribute to the nucleation and assembly of palladium nanocrystallites on ESMF. The high magnifications of the images, exhibit the pearl-necklace structures with the "pearls" about 40 nm in diameter and the "necklace" of 2-5 μm in the whole length. Figure 3 shows representative TEM and HRTEM pictures record from drop-coated films of the palladium synthesized by infusing ESM biofibers into $PdCl_2$ solutions for 18 h. At lower magnification, palladium nanoparticles show the pearl-necklace configuration with the "pearls" units of about 40 nm assembling into very open, quasilinear superstructures rather a dense, closely packed and chained form. The SAED pattern reveal that Pd particles are polycrystalline in nature and the rings could be indexed as the fcc structure of crystalline palladium, which are consistent with the XRD results. At higher magnification, it is striking that each "pearl" of the palladium necklaces virtually appear to be formed by the assembly of smaller nanoparticles containing a porous network, which can be described as "worm hole" variety. It is also observed that a second material coats the palladium nanoparticles.

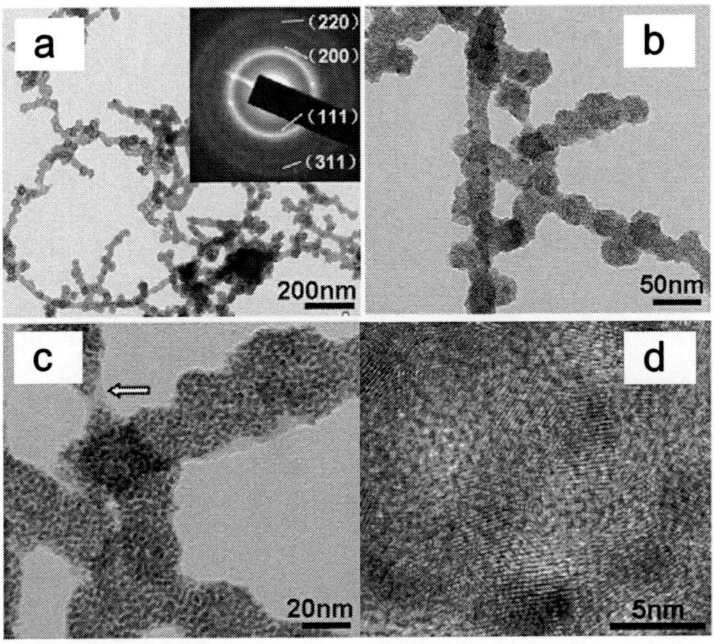

Figure 43. Representative TEM images of (a) necklace structured Pd nanoclusters; insets in (a) show the corresponding SAED patterns; (b, c) high magnifications of image (a); (d) HRTEM images of Pd nanoparticles.

It is believed that this coating material with an average thickness of 5 nm is made up of bioorganic components from ESMF glycoprotein and may form an amorphous shell around the palladium nanoparticles, giving rise to weak Bragg reflections. In addition, the lattice image can be clearly seen due to phase contrast. Considering that each domain having the parallel lattice fringes is a single crystal of Pd, the crystallite size is observed to be about 3 nm. It is noted that the Pd nanocrystallites are not in physical close contact but are monodispersed with a fairly uniform interparticle distance.

A recent review has highlighted on how intricate interplays during the crystallization and assembly of metal nanocrystallites under the directing of biomolecular template can give rise to cooperative transformation and reorganization of inorganic–organic hybrid building blocks to produce emergent nanostructures with complex form and hierarchy. An important point concerns the functionalization of inorganic nanoparticles by means of evolutionary optimized biological components. In this work, the aqueous synthesis of the hierarchical nanocomposites can be separated into two steps: one is the reduction execution of metal ions by reductive functional groups of the biofibers; and the other is the *in situ* crystallization and assembly of metallic nanoparticles under the direction of the biomacromolecules and arraying in special locations of the biofibers. As we known that shell membrane is a non-mineralized, collagen-based matrix consisted of type I, V, and X collagen located in the core of each fiber, which is surrounded by a soluble redox-active glycoprotein mantle. On the basis of the redox nature we expect that the glycoprotein mantle mainly containing amino and carboxyl groups can reduce Pd (II) to Pd (0) state. In above investigations, we have proved the assembly of hierarchical nanomaterials under the direction of the membrane macromolecules. Thus, we consider that the synthesis of hierarchical palladium nanostructures may rely on the reductive templating action of ESMF glycoprotein residues.

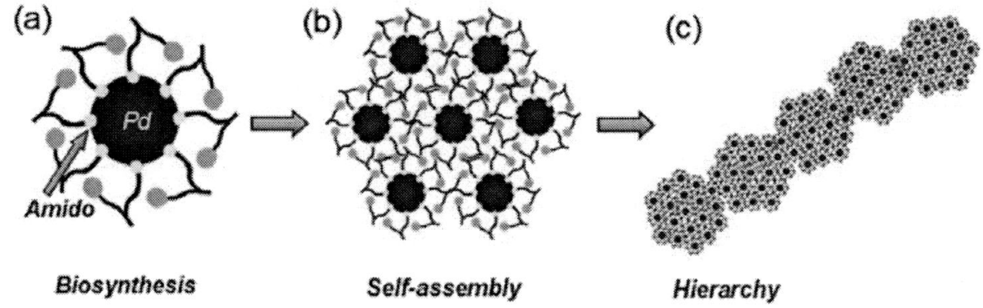

Figure 44. Schematic diagram of in situ synthesis of hierarchical nanocomposites involving redox-active biofibers at room temperature: (a) reductive mineralization of Pd nanocrystallites by amido residues of the eggshell membrane glycoprotein; (b) the assembly of Pd nanocrystallites into the nanoclusters under the direction of ESM biomacromolecules; (c) further assembly of Pd nanoclusters into Pd/ESMF nanocomposites.

The biomolecules are often coupled through noncovalent electrostatic interactions. At high pH values that are slightly above the isoelectric point of certain proteins, metal ions in aqueous solution are reduced by some biomolecule residues. However, as-formed palladium crystallites can be enwrapped by the soluble glycoprotein macromolecules in the hybrids composites. The palladium particles are overcastted by the cladding, which is maybe the organic glycoprotein remnants. It is also clearly displayed that the surface of formed hybrid composites exhibits rugged and rough pattern instead of the smooth appearance of the primeval ESMF.

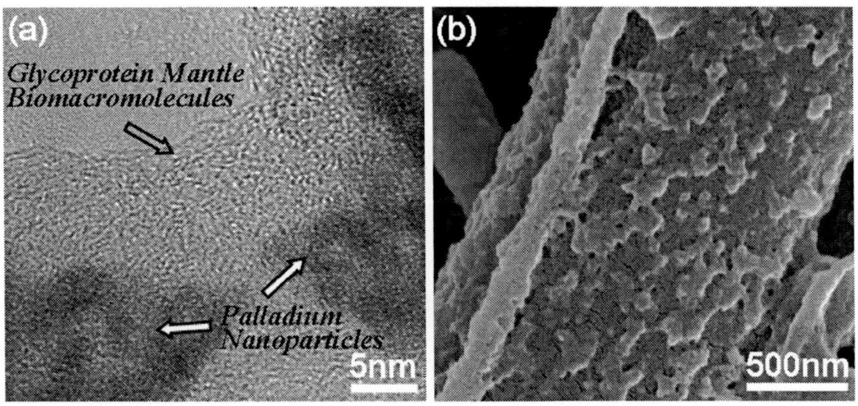

Figure 45. Representative (a) HRTEM and (b) FESEM images of Pd/ESMF hybrids prepared by infusing the eggshell membrane into PdCl2 solution for 5h, respectively.

It confirms that ESMF glycoprotein have interacted with the inorganic matters during the biogenic process. Further, these formed palladium particles may be functioned as the catalyst to efficiently coupling the highly charged peptides under alkaline conditions. It can be deduced that small parts of glycoprotein macromolecules may disengage from the ESMF into the solution and coat Pd crystallites on the membrane fibers to assemble ordered inorganic/biomolecules hybrids. In that way, as-prepared noble metal nanoparticles can be stabilized on the biofibers over long periods of time. Since the nanoparticles and biomolecules typically meet at the same nanometer length scale, this interdisciplinary approach will contribute to the establishment of a novel field, descriptively termed nanobiotechnology.

4. In Situ Synthesis of Noble Metal Nanoparticles on Silk Fibroin Fibers

More attentions have been focused upon building noble metals nanostructures due to their unusual optical properties as well as their novel chemical and catalytic properties combining characteristic hierarchical structures. Actually, the utilization of biological organisms has emerged as a novel method for the synthesis of noble metal nanoparticles. In a break from tradition, it has hitherto relied more on the use of prokaryotes such as bacteria, viruses and so on, as well as proteins, recurring to the intracellular synthesis of sliver nanoparticles. The biosynthesis of silver nanoparticles has also been achieved using peptides with the ability to bind sliver ions and particles [26]. Moreover, it has been published that some living plants, such as Alfalfa sprouts [27], Geranium leaf [28], have the capability to take up sliver ions from solid media and produce sliver nanoparticles resulting in the possibility of an environmentally friendly method of recovering precious metals from mining wastes and a new alternative to produce nanoparticles. All the processes can be easily controlled under mild and eco-friendly conditions. Here silver nanoparticles are chosen as the candidate linked to the biotemplate, with a view that silver particles with nanoscale building blocks have received a great deal of attentions recently owing to their promising applications in modern science and technology, for example, electron microscopy (contrast agents), analysis (chemical and biological sensors), electronics (single-electron transistors, electrical connects), materials (dyes, conductive coatings), and even catalysis. Whereas sliver nanoparticles usually exist as the colloids in the liquid system, which restricts the applications in some domains, for example, as the solid catalyzer that the catalysis reckons on the sizes and morpologies of both sliver particles and solid supports, as well as their reciprocities. It is extremely difficult to find an appropriate solid support on which Ag(I) ions can be reduced to Ag(0) at ambient temperature [29]. However, it may be achieved by selecting the proper biological template as the substrate, on which silver nanoparticles can be deoxidized and *in situ* grow up.

A convenient biomaterial silk extracted from the silkworm *Bombyx mori* cocoons, what with redox-active silk fibroin fibers (SFF) containing manifold amino acids and chemical residues, among which tyrosine residue (Tyr) component having strong electron donating complex property, shall been given priority to prepare noble metal nanomaterials through bio-template techniques under mild conditions. Elicited for some bioinspired processes, we did not redissolve the silk into silk fibroin (SF) proteins, but directly utilized SFF as the reductive template to synthesize silver nanostructures. The redox-active SFF is obtained by removing the sericin from original silks, the gummy binding protein coating the silk fibroin filaments, with the degumming process that the cocoons are boiled in 0.05 mol·L-1 Na2CO3 solution. In a typical experiment, silver nitrate solution with volumes 20, 5, 1, and 0.2 mL of 0.01 M are injected into conical flasks and diluted to 100 mL, respectively. A certain amount of SFF ca. 8.5 mg is immersed into the solutions in the dark. The bio-reduction of sliver ions on SFF substrates is monitored by periodic sampling of aliquots (2 mL) of the aqueous component and measuring the UV-vis spectrum of the solution.

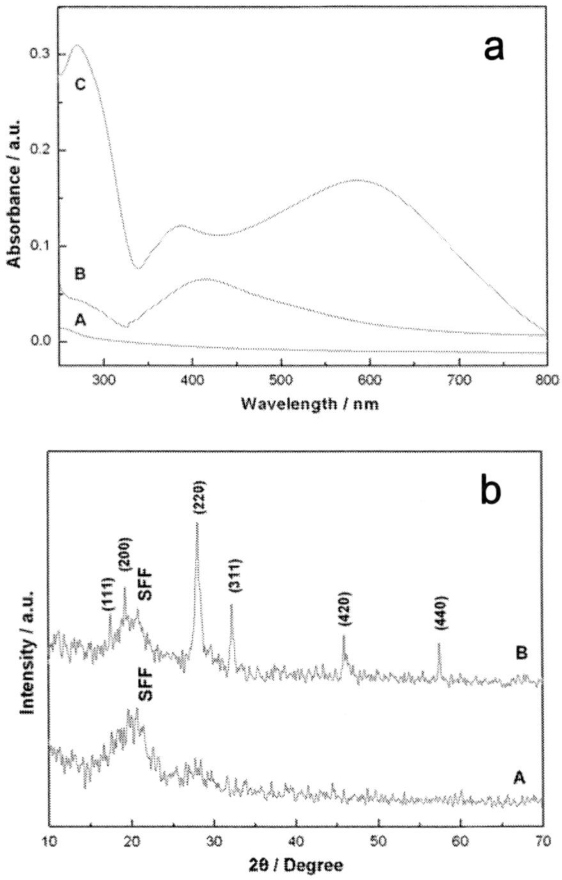

Figure 46. (a) UV-vis absorption spectra of the aqueous solutions of (A) SFF and (B, C) Ag/SFF with the additions of 20 ml and above 20 ml 0.01 mol·L-1 AgNO3 kept in the dark for 4 hours. (b) XRD patterns of (A) the original degummed silk and (B) silver nanoparticles on SFF supports.

UV-vis spectra clearly show that pure SFF does not absorb in the visible region, but the surface plasmon resonance band occurs at ca. 410 nm with addition of 20 mL AgNO3 solution kept 4 hours in the dark. Since silver nanoparticles are known to exhibit a size-dependent characteristic surface plasmon resonance band that can be measured, the characteristic absorption may be due to the redox of Tyr component in SFF in the case of the AgNO3 solution and the formation of silver nanoparticles, which is further proved by XRD results, and this absorption arising due to excitation of surface plasmons in the nanoparticles [30]. Actually it has been reported that the amino acid Tyr is an excellent reducing agent only under alkaline conditions [31]. However, it is not ambivalent with the current result owning to the pretreatment of SFF in Na2CO3 solution (a base medium) making for the ionization of the phenolic group of Tyr. Pretreated SFF thus has reduction activation and can *in situ* nucleate sliver nanocrystallites on SFF supports, regardless of rigorous pH value. Although the total content of Tyr only comes up to 5.17 mol% of silk fibroin, the Tyr is mainly located around SFF for its complex aromaticity, which may reduce silver ions around the membrane as the role of biotemplate. If the addition of AgNO3 solution is increased above 20 ml, the intensity of this plasmon peak changes very little with a blue shift, and a new peak appears at 570 nm. The two peaks, respectively, can be attributed to the transverse and longitudinal plasmon resonance of rod-shaped silver nanostructures [32]. However, no evident peaks can be tracked with addition of AgNO3 solution less than 1 mL due to the formation of few silver nanoparticles. It is considered that 4 hour is enough and the volume of 0.01 mol·L-1 AgNO3 solution shall range from 1 mL to 20 mL for the synthesis of silver nanocrystallites on SFF supports.

It can be seen from the XRD patterns that pure SFF containing plenty of amines, amides and carboxylic surface functional groups are nearly amorphous whereas Glycine–X repeats covering 94% of the silk sequence (X is Alanine in 65%, Serine in 23%, and Tyr in 9% of the repeats) [33] mainly form crystalline domains in silk fibers resulting in the peak around 20.64 degree [34]. A number of prominent Bragg reflections can be indexed as the face-centered cubic (fcc) structure of sliver with the corresponding diffraction peaks of (111), (200), (220), (311), (420) and (440) planes. The reflection peaks appear broader indicating as-prepared sliver nanoparticles with small dimensions, which is in agreement with UV-vis spectra results.

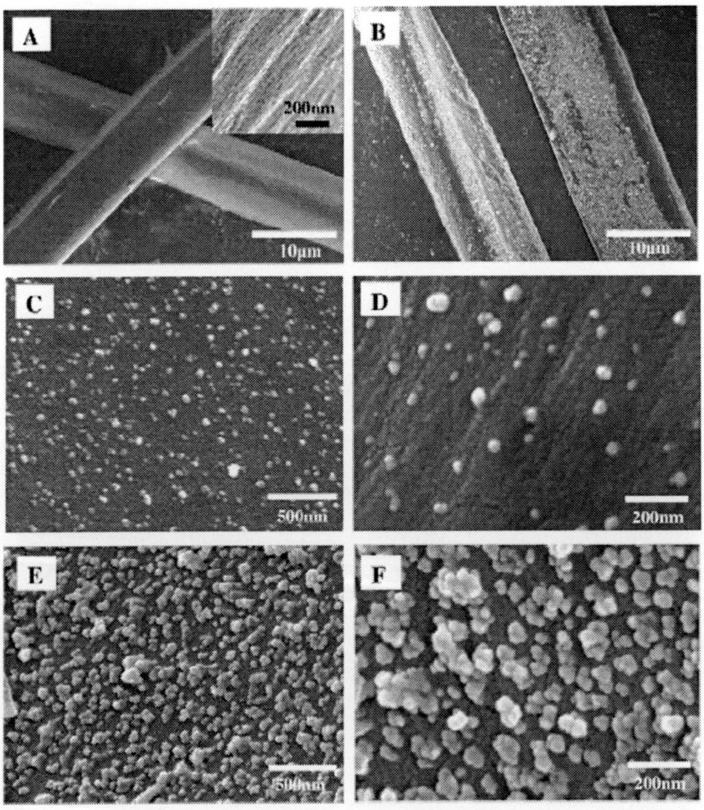

Figure 47. FESEM images of (A) the original SFF template, with the inset showing high magnification of the surface skin coat of aligned protein compared with the panorama of Ag/SFF composites, sliver nanoparticles on SFF supports with different dosages of 0.01 mol·L-1 AgNO3 as the reagents, (C) and (D) 1 ml, (B), (E) and (F) 20 ml. Images in Figure D and Figure F show high magnifications of Figure C and Figure E, respectively.

The characteristic FESEM images of original SFF supports seems as clean as a whistle under low magnification, as the surface skin coat of aligned protein obviously appears in the magnified images. It reveals that the surface has a regular rugged framework containing proteins chains with certain chemical residues. The typical images of silver nanoparticles depositing on SFF supports for 4 h, contrasting against original SFF supports, can be observed from the panorama that sliver nanoparticles distribute uniformly on SFF substrates. As a general rule, it is necessary to precisely control the reaction conditions, for example, the concentration of the silver ions, besides the base system during degumming process using Na2CO3 solution. As concerned as the images of Ag/SFF with 1 mL and 20 mL 0.01 mol·L-1 AgNO3 solution as the reactants, respectively, under low concentration (with 1mL AgNO3 added), as-prepared sliver nanoparticles distribute uniformly and present spherical shape with a mean size ca. 21 nm, few grow up exceptionally. As sliver ions concentration elevated (with 20 mL AgNO3 added), sliver nanocrystallites come into being ca. 30-50 nm clusters with the floriated figuration conjoined by about 5 nm small grains. It indicates that an increase in silver ions concentration leads to an obvious morphologic evolution of sliver nanostructures from dispersed grains to nanoclusters, which would be likely caused by the interactions of dissociative sliver ions and the chemical residues of SFF supports.

To further explore the *in situ* depositing process, ATR-FTIR measurements are carried out to identify possible biochemical radicals among biomolecules, which can be responsible for the reduction of Ag ions and capping silver nanoparticles. The degummed SFF supports are extracted from the solution, subsequently rinsed with distilled water and absolute alcohol, respectively, finally the fibers are collected for FTIR analysis. The FTIR spectrum of degummed silks and shows the diagnostic peaks at 1620, 1511, and 1224 (connected 1260) cm-1. These peaks are assigned to the amide I, amide II and amide III (crystal region, 1224 cm-1; amorphous zone, 1260 cm-1) bands of proteins released by SFF, respectively, which predominantly indicates β-sheet structure. The peaks at 1443 and 1372 cm-1 may be assigned to symmetric stretching vibrations of carboxylate groups of amino acid residues with free carboxylate groups and methylic groups of alanine in the proteins. The peaks at 1620 and 1160 cm-1 arises from carbonyl stretching vibration and phenolic groups of Tyr, respectively, which shows the carbonyl stretching vibration from the carboxylate ions and the hydroxyl stretching vibration from the phenolic ions in Tyr of the pure SFF. As for treated SFF, the peak of SFF at 1160 cm-1 disappears which indicates the phenolic residue exhausted, simultaneity showing the carbonyl stretching vibration peak at 1725 cm-1, which may be attributed to the formation of a quinone structure due to oxidation of phenolic group in Tyr. It has been reported that proteins can bind to metal nanoparticles through either free amine groups or cysteine residues in the proteins [35]. In the circumstances, oxidized tyrosine molecules bound Ag crystallites through amine groups so that Ag nanoparticles can stabilitate, which is also supported by the shift of the characteristic peaks. The amide I band at 1620 cm-1 has broadened, which also indicates the capping manner of silver nanoparticles by SFF macromolecules.

It has been further confirmed by MAS solid-state NMR analysis of pure SFF comparing with Ag/SFF. The resonances of phenyl group of Tyr is usually measured to be at 130-150 ppm, assigned to sp2 hybridized carbons of phenyl replaced by hydroxy. Due to the oxidation of Tyr components, phenyl group is converted into a semi-quinone; the chemical shift of quinone 13C is measured to be in the region of 180-190 ppm while the resonances at 160-170 ppm is just arose by the C of carbonyl in the carboxyl residues of SFF. In the case of Ag/SFF, a peak at 189 ppm shall be corresponding to semi-quinone hailing from the phenyl group during the reduction of Ag(I) to Ag(0), this peak is clearly missing in the NMR spectrum of pure SFF.

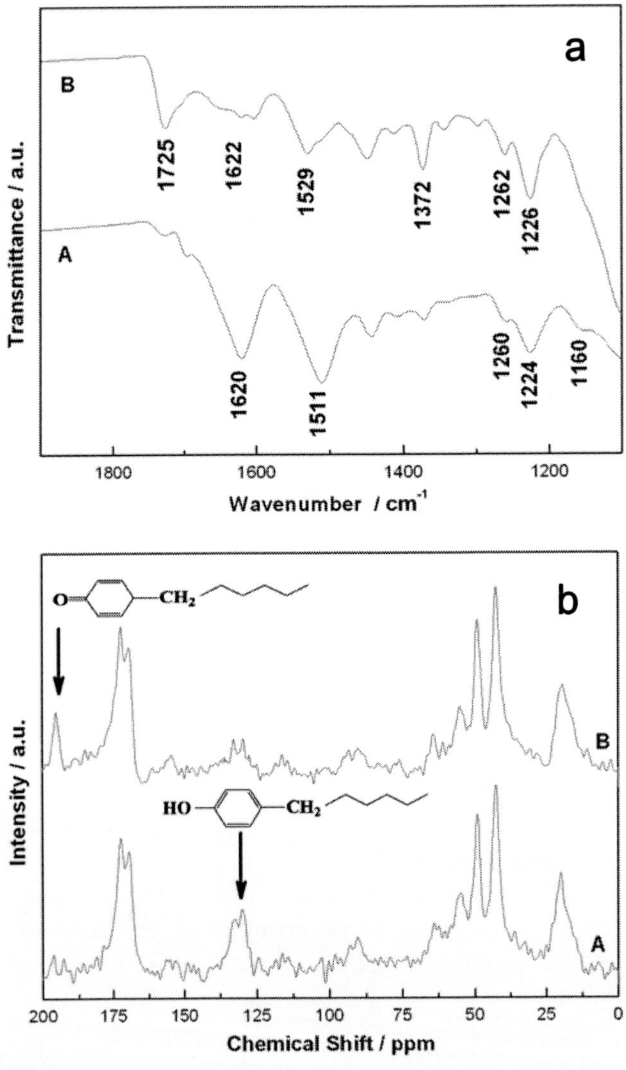

Figure 48. (a) ATR-FTIR spectra of (A) original degummed silk and (B) sliver nanoparticles on SFF supports. (b) MAS solid-state NMR spectra of (A) original degummed silk and (B) sliver nanoparticles on SFF supports. The inset shows a transformation of functional structure of Tyr component, which is a consequent on the reduction of sliver ions and formation of silver nanoparticles.

As proved above, it is ensured that the redox-active Tyr residues, mainly locating around the surface of SFF, can farthest reduce sliver ions on account the ionization of SFF during the pretreatment under alkaline conditions. A tentative degummed process substituting distilled water for 0.05 mol·L-1 Na2CO3 solution can not bring off ionization and there present no evident sliver grains, which indirectly proves that the reducing ability of Tyr achieved predominantly in alkaline system. Thus the reducing capability of Tyr is enhanced for phenolic group of Tyr at high pH. Consequently, the silver ions can be reduced by amines and amides, and interacted with carboxylic surface functional groups containing in those proteins chains around SFF substrates. The redox process shall occur in the present case between phenolic groups of Tyr exposed from SFF and captured Ag(I) ions, ultimately Ag nanoclusters can stabilize on SFF.

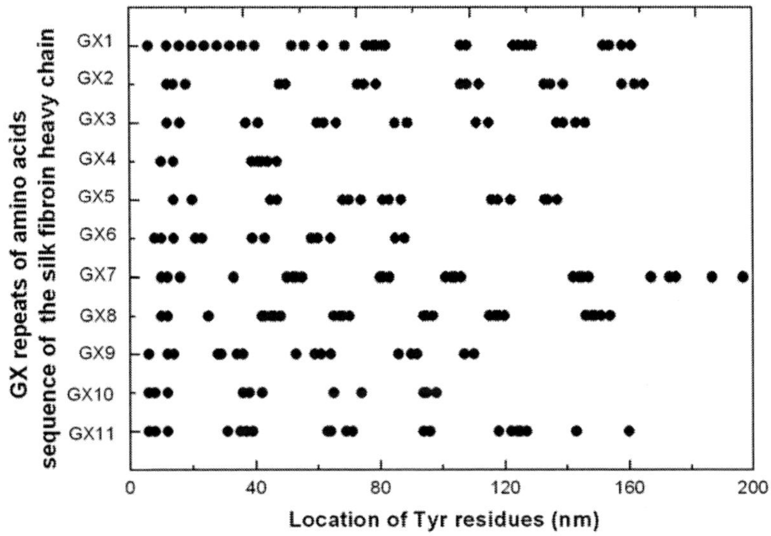

Figure 49. Distributing locations of Tyr residues according to the GX repeats of the amino acid sequence of the silk fibroin heavy chain.

The morphologies of silver nanoclusters reckon on special macromolecules structures of the biotemplate SFF as well as silver ions concentration. The fibroin is an insoluble protein containing up to 90% of the amino acids glycine, alanine, serine and tyrosine leading to significant content of antiparallel β-pleated sheet formation in the fibers. On examination of the primary sequence for the fibroin heavy chain, a pattern of hydrophobic and hydrophilic blocks has been identified that includes larger hydrophilic blocks at the chain ends, smaller hydrophilic blocks and hydrophobic blocks interval jointed. This pattern suggests the possibility to form micellar structures in liquid systems [36]. Fibroin molecules thus act as the hydrophilic–hydrophobic–hydrophilic polymers, during the formation of irregular sized micelles depending on chain folding and hydrophobic interactions. The smaller hydrophilic blocks can remain hydrate, and the larger terminal hydrophilic blocks at both the amino and carboxy termini define the outer edges of the micelles. Tyr residues are distributed according to the silk sequence of the amino acids and present all the way through the molecule, both in the hydrophilic and hydrophobic regions. Hereby, the organization does not give significant spatial ordering of the tyrosine. Distributing locations of Tyr residues are calculated according as the GX repeats formed the bulk of the polypeptide chain. Clearly, Tyr residues don't distributed well-proportioned but behave separated tufts at ca. 20~50 nm level. Through Coulombic interactions, the Ag(I) ions can be captured by phenolic groups of chemical residues. With this premise, the concentration of silver ions will affect the morphologies of as-prepared silver nanoparticles. Under higher concentration as-formed sliver crystallites present shapely spheres with the size of ca. 5 nm, due to homogeneous nucleation of ample silver ions. It is worth noting that sliver nanoparticles grow into nanoclusters. Tyr residues on SFF can contribute to the formation of silver nanoparticles in both reduction and location. Hence peculiar structures of Tyr residues and SFF shall be responsible for directing some morphological controls over the placement of silver nanocrystals.

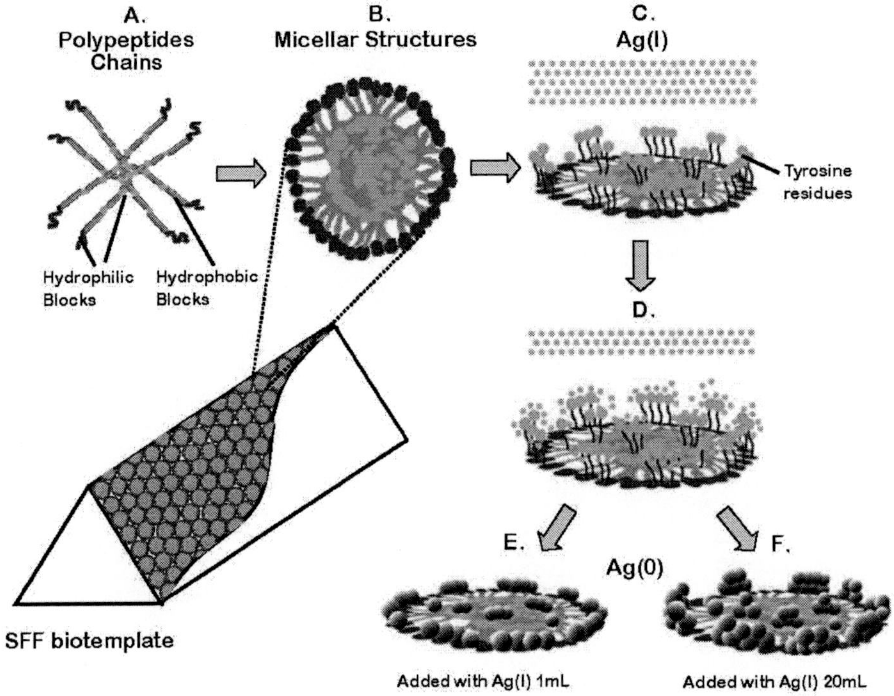

Figure 50. Illustration of in situ depositing Ag nanoparticles on SFF biotemplate for different morphologies and arrangements reckoning on the concentration of silver ions and macromolecules structures of SFF. It shows the model of (A) chain folding, (B) micelle formation, and (C, D) the reaction of sliver ions with chemical residues brings out two morphologies formed under different dosages of 0.01 $mol{\cdot}L^{-1}$ $AgNO_3$ as the reagents (E) 1 mL and (F) 20 mL, respectively.

The interaction of Tyr residues with silver nuclei provides a chemically reducing environment, thereby allowing further accelerated reduction of silver ions with sufficient metal ions at the interface between Tyr residues and sliver nanoparticles. It exists continuous growth by the diffusion of species from the solution with a great deal of silver ions towards the surfaces of primeval nuclei which build up into combined series leading to floriated clusters morphology.There are intervals between every two Tyr residues aggregations, which are consistent with the distributing locations of Tyr residues. Although the reductive reaction is the same as the former, lower concentration of silver ions generates single, long burst of nuclei, these nuclei can be bound by Tyr and slowly grow up. In some areas, a broad range of size distribution indicates the occurrence of multiple nucleation events [37]. Some sliver nanoparticles with the size of ca. 45 nm are formed anisomerously due to the transitory undulation of silver ions concentration. Other results also convincingly support the model. Sliver nanoparticles can present in interrupted arrays, in respect that amino acid residues such as arginine, cysteine, lysine and methionine are known to interact with silver ions, which leads to the faulty arrangement differing from the ideal state. In any case, the interactions among protein blocks, as well as distributing and spacing of these residues, play a critical and complicated role in the nucleation and growth of sliver nanocrystallites. Thus sliver nanoparticles present in arrays unconspicuously or in array out-of-order.

Representative TEM images of as-prepared silver nanoparticles separated from SFF and ones sticking on SFF proteins are shown in Figure 51A-B (with 20 mL silver nitrate solution added), respectively. It reveals that silver crystallites show itself nanoclusters. The SAED pattern reveals that the particles are polycrystalline in nature and the rings can be indexed as the fcc structure of crystalline silver which is consistent with the XRD results. As shown in Figure 51B, there are few spherical nanoparticles with the size of 10 nm on SFF proteins, and these special particles are indexed to single-crystalline silver. The uniformity in size for as-formed nanocrystallites can have been achieved through a self-sharpening growth process during which small particles grow more rapidly than larger ones.

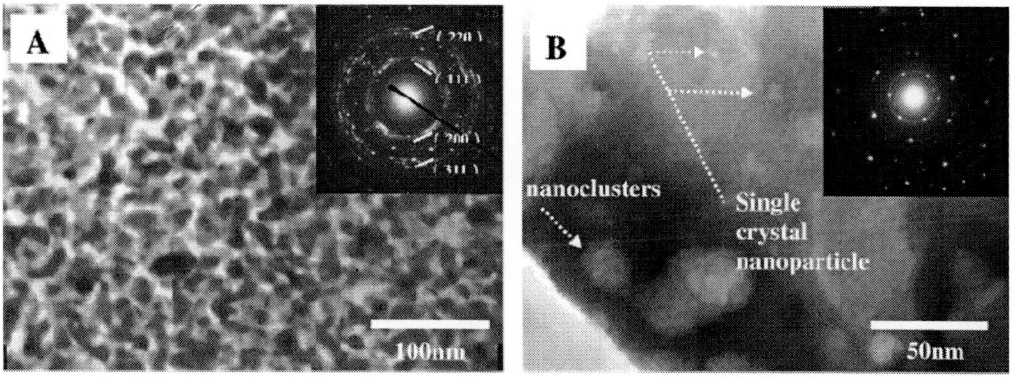

Figure 51. Representative TEM images of as-prepared Ag nanoclusters (added with 20 mL 0.01 mol·L^{-1} $AgNO_3$). (A) shows Ag nanoclusters separated from SFF substrates, the inset shows the selected area electron diffraction (SAED) pattern of the nanoclusters. (B) shows Ag particles sticking on SFF proteins, the inset shows the convergent beam electron diffraction patterns recorded by directing the electron beam perpendicular to the (111) facet.

In addition, it is found that sliver nanoparticles combine into larger swarms with ultraviolet radiation. A number of recent studies suggest that it shall rather be the conglomeration of some smaller subunits (or primary crystallites) than continuous growth by the diffusion of species from the solution towards the surfaces of nuclei. The as-prepared silver nanoclusters are rather stable on SFF supports over a period of several months, even under knocking on SFF substrates.

By the convenient biochemical technique with silk fibroin fibers as the biotemplate, support and reducer, silver nanoparticles/SFF composites has been successfully synthesized. The loading of silver nanoparticles is as high as 60% of the whole surface area, it is clear that the "biomaterial frameworks" provided highly effective sites for deoxidization and loading of noble metal nanoparticles, composing structures of macroscopic to nanoscopic dimensions. Our discovery that Tyr of SFF provides a redox-active framework to construct nanostructured silver crystallites is an important advance in combining biotemplate and chemical synthesis in the direction of organic-inorganic advanced materials synthesis.

CONCLUSION

Soft chemistry technique combined with materials science and biotechnology is presently developed to construct a variety of well-ordered functional nanostructures through appropriate bioinspired approaches.

A series of facile, green and versatile aqueous routes have been developed for the fabrication of hierarchical oxide films via different sol-gel soakage techniques combined relevant calcination treatment and involved the biomaterial eggshell membrane (ESM) as the template. The biomorphic replication of three diverse layers of ESM is systematically investigated by the control over synthesis conditions such as media pH value, dipping time, and calcination temperature. Herein, the biomaterial ESM serves both as the physical substrate and the functional macromolecule template to realize the precise biomorphic replication, by the rights of the interactions between ESM macromolecules (containing carboxyl, hydroxyl, amino groups, etc.) and relevant colloid ingredients. Moreover, some biomacromolecules of ESM components also act as the surfactants to obtain small-scaled and well-distributed oxide nanocrystallites, based on the strong bondage of short-chained amino acids within ESM glycoprotein with oxide nuclei.

Through in situ biotemplate soakage approaches at room temperature, a series of nanoclusters of chalcogenides and noble metal nanocrystallites have been successfully prepared and encrusted on degummed silk fibroin fibers (SFF) or eggshell membrane fibers (ESMF). Some amino acids of SFF or ESMF mainly provide both reactant and location functions under different solution conditions. The morphologies of these nanoclusters and the hierarchy of hybrid nanocomposites are mostly attributed to the concentration of relevant impregnant solutions as well as particular configurations and special functional groups of SFF or ESMF biomacromolecules. The biofibers not only serve as the substrates and templates, but also function as the surfactants, thus small-sized, well-distributed, and well-crystallized nanocrystallites can be organized into subtle hierarchical nanostructures combining organic biomaterials to form well-organized inorganic-organic hybrids.

This method and the relevant ideas offer an avenue to fabricate other functional nanomaterials with novel properties and valuable applications, on the basis of the integration among chemistry, materials sciences and biotechnologies.

REFERENCES

[1] Cook, G.; Timms, P. L.; Göltner-Spickermann, C. *Angew. Chem. Int. Ed.* 2003, *42*, 557.

[2] Cozzoli, P. D.; Kornowski, A.; Weller, H. *J. Am. Chem. Soc.* 2003, *125*, 14539.

[3] Lee, S. W.; Ichinose, I.; Kunitake, T. *Langmuir* 1998, *14*, 2857.

[4] Moss, J. M.; VanDamme, M. P. I.; Murphy, W. H.; Stanton, P. G.; Thomas, P.; Preston, B. N. *Arch. Biochem. Biophys.* 1997, *339*, 172.

[5] Liang, C. H.; Shimizu, Y.; Sasaki, T.; Koshizaki, N. *J. Phys.Chem. B* 2003, *107*, 9220.

[6] Dieguez, A.; Romano-Rodriguez, A.; Morante, J. R.; Kappler, J.; Barsan, N.; Gopel, W. *Sens. Actuators B* 1999, *60*, 125.

[7] a) Zhang, D. F.; Sun, L. D.; Jia, C. J.; Yan, Z. G.; You, L. P.; Yan, C. H. *J. Am. Chem. Soc.* 2005, *127*, 13492; b) Wang, Y.; Lee, J. Y.; Zeng, H. C. *Chem. Mater.* 2005, *17*, 3899–3903.
[8] Gnanasekar, K. I.; Jayaraman, V.; Prabhu, E.; Gnanasekaran, T.; Periaswami, G. *Sens. Actuators B* 1999, *55*, 170.
[9] Fu, M. G.; Wang, X. J.; Chen, J.; Wang, H. Y. *J. Zhengzhou Inst. Light Ind.* 1994, *28*, 160.
[10] Fang, G. J.; Liu, Z. L.; Zhang, Z. C. *J. Chin. Rare Earth Soc.* 1997, *15*, 10.
[11] Srivastava, D. N.; Chappel, S.; Palchik, O.; Zaban, A.; Gedanken, A. *Langmuir* 2002, *18*, 4160.
[12] Damen, T. C.; Porto, S. P. S.; Tell, B. *Phys. Rev.* 1966, *142*, 570.
[13] Hadjiev, V. G.; Iliev, M. N.; Vergilov, I. V. *J. Phys. C: Solid State Phys.* 1988, *21*, L199.
[14] van den Beucken, J. J. J. P.; Vos, M. R. J.; Thune, P. C.; Hayakawa, T.; Fukushima, T.; Okahata, Y.; Walboomers, X. F.; Sommerdijk, N. A. J. M.; Nolte, R. J. M.; Jansen, J. A. *Biomaterials* 2006, *27*, 691.
[15] Yi, F.; Guo, Z. X.; Zhang, L. X.; Yu, H.; Li, Q. *Biomaterials* 2004, *25*, 4591.
[16] Lu, W. G.; Fang, J.Y.; Stokes, K.L.; J. Lin, *J. Am. Chem. Soc.* 2004, *126*, 11798.
[17] Lee, S. M.; Cho, S.N.; Cheon, J. *Adv. Mater.* 2003, *15*, 441.
[18] Petroski, J. M.; Wang, Z. L.; Green, T. C.; El-sayed, M. A. *J. Phys. Chem. B* 1998, *102*, 3316.
[19] Iwahori, K.; Yoshizawa, K.; Muraoka, M.; Yamashita, I. *Inorg. Chem.* 2005, *44*, 6393.
[20] Smith, G. D.; Firth, S.; R. J.; Clark, H.; Cardona, M. *J. Appl. Phys.* 2002, *92*, 4375.
[21] Nanda, K. K.; Sahu, S. N.; Soni, R. K.; Tripathy, S. *Phys. Rev. B* 1998, *58*, 15405.
[22] Batonneau, Y.; Bremard, C.; Laureyns, J.; Merlin, J. C.; *J. Raman Spectrosc.* 2000, *31*, 1113.
[23] Krauss, T. D.; Wise, F. W. *Phys. Rev. B* 1997, *55*, 9860.
[24] Hayashi, S.; Sanda, H.; Agata, M.; Yamamoto, K. *Phys. Rev. B* 1989, *40*, 5544.
[25] Mlayah, A.; Brugman, A. M.; Carles, R.; Renucci, J. B.; Valakh, M. Ya.; Pogorelov, A. V. *Solid State Commun.* 1994, *90*, 567.
[26] Naik, R. R.; Stringer, S. J.; Agarwal, G.; Jones, S. E.; Stone, M. O. *Nat. Mater.* 2002, *1*, 169.
[27] Gardea-Torresdey, J. L.; Gomez, E.; Peralta-Videa, J. R.; Parsons, J. G.; Troiani, H.; Jose-Yacaman M. *Langmuir* 2003, *19*, 1357.
[28] Shankar, S. S.; Ahmad, A.; Sastry, M. *Biotechnol. Prog.* 2003, *19*, 1627.
[29] Hong, B. H.; Bae, S. C.; Lee, C. W.; Jeong, S.; Kim, K. S. *Science* 2001, *294*, 348.
[30] Kerker, M. *J. Colloid Interf. Sci.* 1985, *105*, 297.
[31] Selvakannan, P. R.; Swami, A.; Srisathiyanarayanan, D.; Shirude, P. S.; Pasricha, R.; Mandale, A. B.; Sastry, M. *Langmuir* 2004, *20*, 7825.
[32] Sun, Y. G.; Gates, B.; Mayers, B.; Xia, Y. N. *Nano Lett.* 2002, *2*, 165.
[33] Zhou, C. Z.; Confalonieri, F.; Jacquet, M.; Perasso, R.; Li, Z. G.; Janin, J. *Proteins* 2001, *44*, 119.
[34] Rathore, O.; Sogah, D. Y. *Macromolecules* 2001, *34*, 1477.
[35] Gole, A.; Dash, C.; Ramakrishnan, V.; Sainkar, S. R.; Mandale, A. B.; Rao, M.; Sastry, M. *Langmuir* 2001, *17*, 1674.
[36] Jin, H. J.; Kaplan, D. L. *Nature* 2003, *424*, 1057.

[37] Xia, Y. N.; Gates, B.; Yin, Y. D.; Lu, Y. *Adv. Mater.* 2000, *12*, 693.

In: Nanotechnology Research Developments
Editor: R. Jiménez-Contreras, pp. 379-387

ISBN: 978-1-60021-899-6

Chapter 10

WEAR DEPENDENT FRICTIONAL PROPERTIES OF ALKANETHIOLS SELF-ASSEMBLED FILMS

Mario D'Acunto[*]
Department of Physics, University of Pisa, Largo B. Pontecorvo 3, I-56127, Pisa, Italy
Department of Chemical Engineering and Materials Science, University of Pisa, Italy

ABSTRACT

Self-assembled monolayers (SAMs) present a 2D regular lattice, and the lateral force measured with an atomic force microscopy (AFM) scanning probe tip displays the typical stick-slip behavior. In this paper, we study the possible connection between an AFM induced damaged SAM surface and irregular behavior of the sticking phases.

In fact, the repeated passages of an AFM probe tip on an SAM surface can produce wear stripping off radicals from the carbon chain backbones, and irregularities of the SAM lattice take place. As a consequence, the frictional properties are changed because new extra space is made available to the final alkanethiol chain groups and the stick and slip phases are not time invariant and hysteresis between the forward and backward scanning direction is altered. The paper discusses if and how such friction irregularities can monitor the progressive modifications of the SAM worn surface.

Keywords: *Self-assembled monolayers, atomic force microscopy, friction, wear, stick-slip.*

INTRODUCTION

Self-assembled monolayers (SAMs) are very useful for the systematic modification of the physical, chemical and structural properties of a surface by varying the chain length, tail group and composition [1]. Many of these properties can be studied making use of the atomic

[*] E-mail: m.dacunto@ing.unipi.it

force microscopy (AFM), and the interaction between the AFM probe tip and the SAMs can also be considered an excellent reference to study the fundamentals properties of dissipation phenomena and onset wear for viscoelastic materials on the nanoscale [2-7], figure 1. It is well established that complete alkanethiol monolayers formed by [-SH(CH2)n-1CH3] molecules on Au(111) adopt a commensurate (√3×√3)R30° structure, with a nearest neighbour spacing of 0.497nm, which corresponds to a surface area of 0.2nm2 per chain [8]. AFM sliding friction experiments on Cn-alkanethiols SAMs showed a friction behaviour dependent on the chain length [9]. The coefficient of friction decreased as the chain length was increased. The low friction of methylic group terminated SAMs is due largely to the low surface energy coming from their hydrophobic tail groups and the high load bearing capacity is due to their compact structure minimizing the van der Waals interaction energy.

The first observation of an atomic stick-slip process was made by Mate *et al.* [10]. Making use of a tungsten tip sliding over graphite, they found a sawtooth-like modulation of the lateral force with the periodicity of the graphite's honeycomb structure. Tománek et al. gave an accurate description based on Tomlinson model for molecular friction [11]. Such description can be summarized as follows: the AFM tip is dragged over the periodic potential defined by the atomic structure of the surface and the contact. The lateral force is measured by means of a cantilever spring holding the tip. The tip sticks to a certain surface position until the force exerted by a moving support of the spring is high enough to activate a slip of the tip to the next atomic position. The stick-slip behavior shows up as the sawtooth-like modulation of the lateral force. Stick-slip movements in sliding friction are generally characterized by alternating periods of energy storage and period of energy dissipation in the driving cantilever spring.

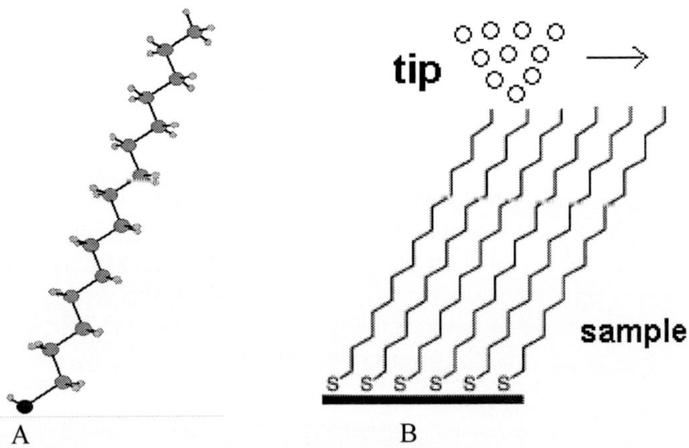

A B

Figure 1. Single C_{15}-alkanethiol chain, [-SH(CH2)$_{14}$CH$_3$], the sulfur atom is drawn in black, figure 1A. Schematic sketch of the AFM probe tip-SAM system, figure 1B.

At any time, the total power can be considered the summation of the storage power and the dissipated power, where dissipated power refers to all power that is not stored in the driving spring and that can be converted to heat or to an increase of the free energy of the substrate. Measuring slip time is rather complicated for a number of reasons. First of all, drift and noise limit the smallest value of velocity that can reliably be measured. Uncertainties in cantilever stiffness as well as in substrate compliance add complications in ruling out that the tip is moving at extremely low speed with respect to the surface.

In this paper, we focus the attention on the frictional response at the contact between probe tip and damaged SAM surface when single AFM induced defects correspondent to methyl group stripped off are progressively produced. Stick-slip dynamics of such damaged surface present irregularities for both stick and slip scale times. Moreover, reduction of hysteresis between left-to-right (forward) and right-to-left (backward) is observed, making possible in some case, the overlap between the forward and backward scanning direction. We propose an explanation of such overlap of lateral force hysteresis loop based on the spatial re-orientation of methyl groups confining to defects. Such re-orientation is meanly increased in proximity of defects, because the van der Waals energy cohesion between chains is reduced and the free space available to the methyl group is larger respect to primitive undamaged profiles.

Results and Discussions

The motion of a probe tip scanning a SAM lattice is characterized by two different timescales. The resonant frequency of a cantilever is around 105Hz, while the atomic vibration frequency is of the order of 1013Hz. As a consequence, in the contact area we have two kinds of dynamics, the slow dynamics of the cantilever and the fast vibrations of atoms. At the time scale of picoseconds, we encounter thermal vibrations of atoms, while the cantilever is stationary. On contrary, at the timescale of a microsecond, we observe the scanning vibrations of the cantilever in contact with a cluster of atoms. The energy modes between the cantilever and the atomic substrate can be decoupled, because the relaxation time of atomic cluster in contact with the probe tip is smaller than the cantilever timescale.

We focus the attention on the changes of the frictional properties produced by a damaged SAM lattice when end methyl groups can be stripped off by the progressive passage of an AFM probe tip at low applied loads. To do this, we use a one-dimensional Langevin equation describing an AFM probe tip sliding on a periodic potential, eq. (1):

$$M\ddot{x}(t) + M\gamma\dot{x}(t) + \frac{\partial U(x, x_{tip})}{\partial x} = \eta(t) \tag{1}$$

where M=3.0x10-11Kg, the damping factor is γ=8.9×105Hz, $\eta(t)$ is a random noise satisfying the fluctuation-dissipation relation described by $<\eta(t)\eta(t')>=2M\gamma k_B T\delta(t-t')$.

The eq. (1) describes the motion of the center of mass of the probe tip moving along the sample potential profile. It does not include parameters such as the dimension of the probe tip, and for a complete description of the influence of the AFM induced stresses also an equation along the z-dimension should be addressed in order to describe the balance of the applied load and deformation of the SAM surface [5-6]. The effective potential $U(x, x_{tip})$ can be written in absence of lattice defects as:

$$U(x, x_{tip}) = -U_0 \cos\left(2\pi \frac{x}{\lambda}\right) + \frac{k_x}{2}\left(x_{tip} - x\right)^2 \tag{2}$$

where the lattice potential barrier is U_0=0.6632×10-19 J, at room temperature the thermal energy is k_BT=0.041×10-19 J, λ is the lattice period =0.497nm, k_x =2.5N/m corresponds to an effective spring constant given by the combination of the stiffness of the cantilever and the lateral stiffness of the contact, *x*tip=*vt* is the instantaneous position of the probe tip, where the tip velocity is *v*=100nm/s. The parameters have been considered as effective in the load range of nearly 1nN, in the sense that their precise value may depend on the load applied and they can have a weak dependence on temperature [12]. The sinusoidal term in eq. (2) describes the regular lattice potential acting on the tip due to the interaction with the sample and the second term describes the elastic energy stored in the cantilever.

The actual interest is to reproduce the instantaneous lateral friction force that is given by the force $F(t)=k_x(vt-x)$. The stick-slip motion and continuous sliding are the two main regimes observed. Transitions between the two regimes as a function of the effective spring constant can be observed. The movement of the tip from one minimum to the next can be continuous or jumping, depending on the relation between corrugation U_0 and elastic energy which can be described by $\approx U_0/k_x\lambda^2$. When this ratio is greater than 1, then the stick-slip regime is observed [13]. The stick-slip dynamical behavior and the continuous regime for the probe tip sliding on a rigid substrate as described by eq. (1) are summarized in figure 2 for a forward and a backward scan. Hard-spring cantilever constants imply continuous regime.

Figure 2 demonstrates that to observe irregular stick-slip transitions, we need to work with soft spring cantilever constant. For regular periodicity, the saw-tooth modulation has the periodicity of the SAM lattice. The maximum lateral force is linearly related to the potential height corrugation of the SAM surface. The area enclosed in this hysteresis loop is the energy dissipated in one cycle, composed by the forward and the backward tip motion on a length of 2.5nm, including five methyl groups. For a pattern including defects and deformations, the potential as described by eq. (2) has to be modified locally accounting such defects and enclosed in the Langevin equation (1). As a consequence, when the SAM profile is progressively damaged by the probe tip, we should re-calculate a potential $U_{eff}(x)$ describing the SAM lattice profile to be inserted in eq.(1). This means that the effective potential can now be written as

$$U_{eff}\left(x, x_{tip}\right) = -U_w(x) + \frac{k_x}{2}\left(x_{tip} - x\right)^2 \tag{3}$$

where the subscript is used to distinguish such damaged profile potential respect to the potential used in eq. (2). *U*w(*x*) represents the local damaged resultant potential describing the wear induced irregular SAM one-dimensional profile. The paper deals out the numerical results obtained by the Langevin, eq. (1), when an effective potential, as given by eq. (3) is used instead of the undamaged potential, eq. (2).

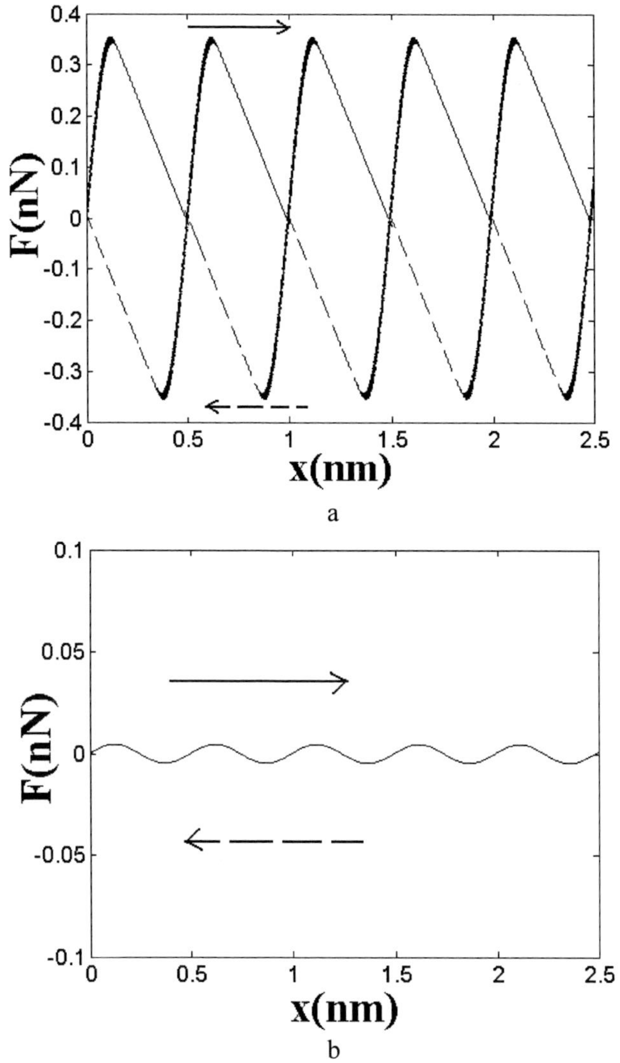

Figure 2. Numerical results for the lateral force F acting on a tip described by eq. (1) sliding on a C15 SAM surface for two different spring constants. Figure 2A shows the hysteresis between the forward (continue line) and backward (dashed line) scanning direction for a tip sliding on a regular period lattice for a soft effective spring constant k_x=2.5N/m. Figure 2B shows the continuous sliding regime for the lateral force as described by eq. (1) for a hard effective spring constant k_x=118N/m. The applied load, L, is =1nN and the tip velocity equals to 100nm/s, in both the figures.

Figure 3 shows a typical 2D lattice C18 SAM surface progressively damaged by the repeated passage of an AFM probe tip. The description of such wearing process is performed making two general assumptions, the first one is that the wear process takes place only stripping off the chain end methyl groups and the second assumption is that the n-1 backbone chain C carbons are locked, so that only the final chain carbon atom can be subject to be stripped off. It has been proposed that the progressive damaging starts from initial surface defects following a nucleation process [14]. The nucleation of wear domains starting by the initials SAM surface defects can be explained by the fact that defect involves a lowering of the van der Waals energy cohesion between neighboring chains and a consistent local change in the lateral force in correspondence of such defects. The AFM probe tip sliding on the SAM surface shows a typical stick-slip dynamics, where timescales for stick and slip phases are invariant and hysteresis between the forward and backward scanning direction is maintained. When the SAM surface is progressively damaged, as showed in figure 3, the probe tip moves on an irregular path, and the irregularity of the path reflects on the friction force.

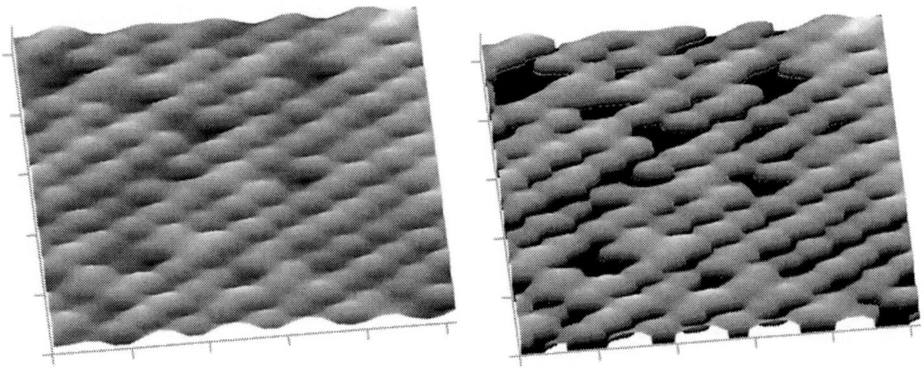

Figure 3. C18-chain progressive damage of simulated methyl group end surface. The initial image (left) is based on an AFM experimental acquired data on a C18 sample, the typical commensurate c(4×2) superlattice of (√3×√3)R30° arrangement is slightly altered by the applied load [11]. The final image (right) shows that nearly 5% of methyl groups have been stripped off by the AFM tip after 50 scans with an applied load, L, of 1nN. The black portions represent the wear damaged domains.

The effective potential $U_{eff}(x)$, as given by eq. (3), describes the upper level of the alkanethiol chains. When an end methyl group is stripped off, the potential profile should take in account that the upper level is lowered to the subsequent chain carbon of a distance of nearly 0.153nm, that is the C-C distance in the alkanethiol chain, figure 1. When stressed by the probe tip, the carbon backbone flexibility is generally modeled by rigid bonds connecting the C-C atoms. The total energy of a given chain can be written as a superposition of a short-range valence or internal terms such as bond stretch, bond-angle bending and angle torsion terms, and a long-range non bonded interaction led by van der Waals intermolecular interaction [7], details of the hybrid molecular approach based on a Montecarlo code are included in [14]. The one-dimensional trajectory for the probe tip described by the Langevin eq. (1) implies that when we calculate the effective potential $U_{eff}(x)$ we have to project the effects of the lattice surrounding carbon atoms along the scanning direction of the probe tip. The lateral force calculated making use of the Langevin eq. (1) with the effective potential $U_{eff}(x)$ presents the results summarized by the figure 4. Changes of lattice periodicity accomplished by vacancy involve irregular paths for the probe tip and consequently, irregular stick-slip motions, figure 4A. Again, the variation of the potential walls involves a growth in the lateral friction force. An unexpected result is the drastic reduction of the hysteresis loop of the lateral force, figure 4B. The reduction allows an overlap of the lateral force between the forward and backward scanning directions. Moreover, the symmetry of the lateral force between the forward and backward scanning directions is lost, and the two directions present different stick phases. The reduction of hysteresis loop and the alteration of the lateral force loop symmetry should be a consequence of re-orientation of chains during forward and the backward scanning directions, i.e., the effective potential changes due to made easier re-orientations of the end C atoms. In fact, the van der Waals energy cohesion ε_{vdW} responsible for the chain packaging, generally falls around 1×10-20J, but is can be lowered of about 50% at the edge of a defect, because the extra space available due to the presence of larger defects.

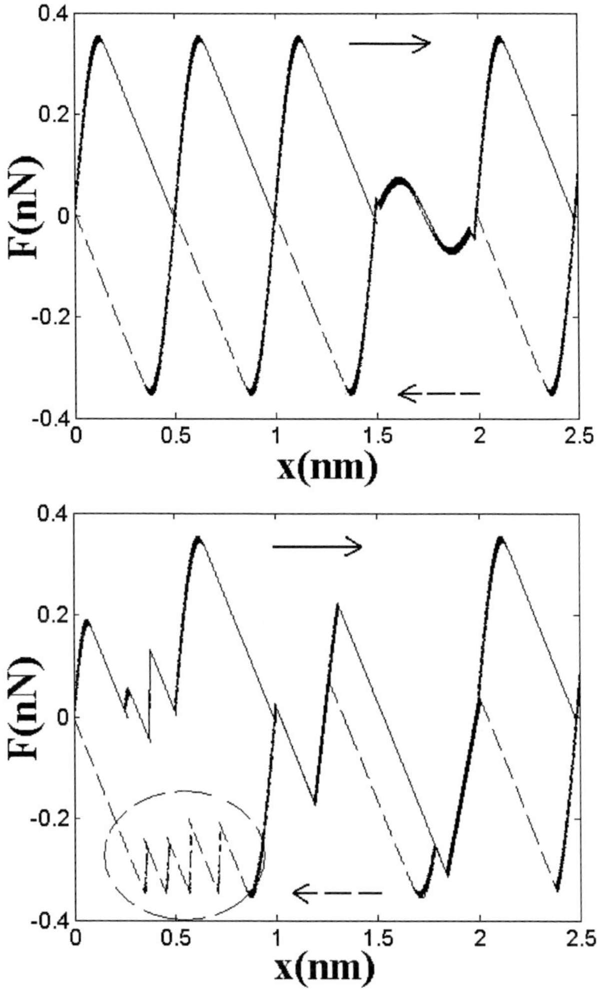

Figure 4. The figures represent the lateral force F for worn lattice profiles for a soft spring k_x=2.5N/m. The lateral force is increased with respect to the regular SAM periodicity and the directional symmetry is lost. As in figure 3, the applied load, L, is =1nN, the tip velocity equal to 100nm/s. The circle in figure 4B shows a regular stick-slip dynamics with a short periodicity respect to undamaged lattice period.

Figure 4 displays the typical stick-slip profiles for SAM surface where onset damages involving end chain groups taking place. In the left figure, the stick-slip profile corresponds to a 5 final methyl group where the 4-th methyl group has been previously stripped off by the AFM probe tip. It should be noted the corresponding frictional force has not been reduced, in fact while the other end methyl carbon atoms produce a stick phase starting by zero to nearly 0.35nN, the stick phase correspondent to the vacancy site of approximately 0.153nm starts with a zero force that has its initial value located below respect to the previous zero force. The maximum stick force must be calculated in this case considering that the stick value starts from a hole and the energy barrier meets by the scanning probe tip is $U_w > U_0$, where Uw is the energy barrier correspondent to a defect. Generally, for hole defects limited to a single stripped off methyl carbon atoms, we have a sticking force that should be twice respect to regular pattern stick force as confirmed by the numerical results reproduced in figure 4.

In the right figure, the potential represents a strong damaged surface and it is displayed in figure 5. The extra available space is now enhanced. When the probe tip meets a carbon atom, the extra space available to the carbon atoms permits movements of the carbon atoms that can change the stick force. In combination with the effects of the surrounding carbon atoms, limited domains of sawtooth-like modulated stick-slip dynamics can be observed, as highlighted by the circle in figure 4B.

The time scale for the atoms movements is rather fast falling in the range of picoseconds, while the collective probe movements fall in the milliseconds range. This implies that the two types of motion can be decoupled and the probe tip slides on a freeze potential configuration, which morphology is determined by the equilibrium positions of the atoms. Such equilibrium positions can be modified by internal rearrangements of end chain carbons atoms, or by the AFM probe tip induced stress causing surface damaging and friction.

It remains difficult to observe a direct connection between the effective potential and the different stick resultant phases. In our simulation, the stripping off process is limited to end chain methyl group carbon atoms and this is still an unrealistic approach. The physical situation is more difficult, nevertheless, we hope that such preliminary results can help to highlight the challenges presented in this paper, i.e., the possible monitoring of damaged surfaces making use of the sticking phases of lateral force generated at the contact between an AFM probe tip and a viscoelastic surface.

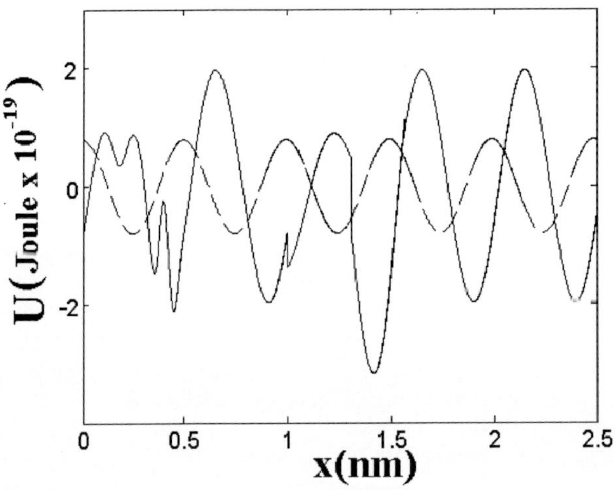

Figure 5. Damaged resultant potential $U_{eff}(x)$ correspondent to the stick-slip loop displayed in figure 4B. Such potential is the effective potential *seen* by sliding the probe tip. The dashed line represents the undamaged potential expressed by the sinusoidal term in the potential as in eq. (2).

CONCLUSIONS

Viscoelastic materials can exhibit irregular stick-slip dynamics when the regular lattice is progressively damaged by wear mechanisms. This irregularity could be used to monitor the damaged surface of a self-assembled monolayers (SAM). In order to demonstrate such behaviour, we consider an atomic force microscopy (AFM) probe tip sliding on a regular lattice of a SAM and progressively damaging the substrate. The collective coordinate of probe tip is described by a Langevin equation, where the potential describing the underlying surface is continuously recalculated as the damages take place. If we focus the attention on the not invariant time stick phase, it should be possible, almost in principle, to create a correspondence between the different stick phases and the corrugated potential describing the positions of the final SAM carbon atoms. In turn, worn SAM profiles can be also associated to anomalous hysteresis loops of frictional force between the forward and the backward sticking dynamics.

REFERENCES

[1] A. Ulman, *Introduction to Ultrathin Organic Films: From Langmuir-Blodgett to Self-Assembly*, Academic Press, San Diego, CA, 1991.

[2] V.N. Bliznyuk, M.P. Everson and V.V. Tsukruk, *J. of Tribology*, 120, 489 (1998).

[3] E. Barrena, S. Kopta, D.F. Ogletree, D.H. Charych, and M. Salmeron, *Phys. Rev. Lett.*, 82, 2880 (1999).

[4] B. Bhushan and H. Liu, *Physical Review* B, 63, 245412 (2001).

[5] Y.S. Leng and S Jiang, *J. Chem. Phys.*, 113, 8800 (2000).

[6] Y.S. Leng and S Jiang, *Phys. Rev. B*, 63, 193406 (2001).

[7] S. Jiang, *Molecular Physics*, 100, 2261 (2002).

[8] J. Hautman and M.L. Klein, *J. Chem. Phys*. 91, 4994 (1989).

[9] A. Lio, D.H. Charych and M. Salmeron, *J. Phys Chem*. B, 101, 3800 (1997).

[10] C.M. Mate, G.M. McClelland, R. Erlandsson, and S. Chang, *Phys. Rev. Lett.*, 59, 1942 (1987).

[11] D. Tomanek, W. Zhong, and H. Thomas, *Eurphys. Lett.* 15, 887 (1991).

[12] Y. Sang, M. Dubè, and M. Grant, *Phys. Rev. Lett.* 87, 174301 (2001).

[13] A. Sokoliuc, R. Bennewitz, E. Gnecco and E. Meyer, *Phys. Rev. Lett.*, 92, 134301 (2004).

[14] M. D'Acunto, *Nanotechnology*, 17, (2006).

In: Nanotechnology Research Developments
Editor: R. Jiménez-Contreras, pp. 389-397
ISBN: 978-1-60021-899-6

Chapter 11

FABRICATION OF MICRO/NANO-COMPOSITE STRUCTURES AND THEIR PHOTOVOLTAIC BEHAVIORS

Jin Zhai, Xiangliang Sheng, Yong Zhao and Liping Heng
Institute of Chemistry, Chinese Academy of Sciences,
Beijing 100080

ABSTRACT

Photovoltaic behavior in both organic thin films and organic/inorganic hybrid systems is an important subject not only in basic research but also in application fields such as organic solar cells, dyes sensitized solar cells etc. In order to improve the light collection efficiency, tuning film morphology is an efficient way to improve light harvest, then conversion efficiency. In this paper, some fabrication methods for different photoanodes with controlled morphology and their photovoltaic behaviors will be reviewed. While using electro-hydrodynamics (EHD) technique to prepare lotus leaf-like photoanodes will be discussed in detail to illustrate the effect of micro-/nano- composite structure on the enhancement of photo-electric conversion behavior in dye-sensitized solar cells (DSSCs). Further more, there will some discussion on the morphology control for the organic thin film system to increase photocurrent generation.

1. INTRODUCTION

In the past century, the consumption of non-renewable sources of energy has caused more environmental damage than any other human activity. Electricity generated from fossil fuels such as coal and crude oil has led to high concentrations of harmful gases in the atmosphere. This has in turn led to many problems being faced today such as ozone depletion and global warming. That do harm to the human life. Therefore, alternative and clean sources of energy have become very important and relevant to today's world. These sources, such as the sun and wind, can never be exhausted and therefore are called renewable. They cause less emission and are available locally. Their use can, to a large extent, reduce chemical, radioactive, and thermal pollution. Among all the renewable energy sources, photovoltaic technology utilizing solar energy is considered as the most promising one. Now, silicon-based and other inorganic semiconductor solar cells have made much progress not only in basic research but also in industrialization with the total conversion efficiencies between 15 and 20% [1]. But, the relatively high cost and complicated manufacturing have prevented their widespread application. Photovoltaic behavior in organic thin films and organic/inorganic hybrid systems is an important subject in organic solar cells and dye-sensitized solar cells (DSSCs). In order to improve the cells' conversion efficiencies, tuning film morphology, such as coupling the photonic crystal layer to electrode as dielectric mirror [2, 3], hollow sphere voids [4] and large size solid particles as scattering center [5-7], is an efficient way to improve light harvest. Therefore, porous structured photo-anodes with predominant penetration effect of viscous and solid electrolytes, fast charge transport and effective light collection efficiency are studious orientation for researchers.

Here some fabrication methods for different photo-anode with controlled morphology and their photovoltaic behaviors will be reviewed. At the same time, using electro-hydrodynamics (EHD) technique to prepare lotus leaf-like photo-anodes will be discussed in detail to illustrate the effect of micro-/nano- composite structure on the enhancement of photo-electric conversion behavior in DSSCs. Furthermore, the morphology control for the organic thin film system to increase photocurrent generation will be shown in this review.

2. Photo-Anodes with Different Morphologies in DSSCs

In 2002, A. P. Alivisatos et al [13] fabricated polymer solar cell by blending CdSe nanorods and P3HT (polythioene derivative). Their results illustrated that semiconductor nanorods showed better electron transport behavior than nanoparticles. Thereafter, P. D. Yang et al [14] successfully fabricated a dense array of oriented crystalline ZnO nanowires as photo-anode by mild aqueous chemistry and featured a surface area up to one-fifth as large as that of nanoparticle cell. The direct electrical pathways provided by the nanowires ensured the rapid collection of carriers generated throughout the device. As an ordered topology that increases the rate of electron transport, a nanowire electrode may provide a means to improve the quantum efficiency of DSSCs in the red region of the spectrum. Important differences in transport, internal electric field distribution and light scattering should make comparative studies of wire and particle devices fruitfully.

Due to the advantages of nanowires in electron transport, oriented one-dimensional nanomaterials were more and more attractive. In the same year, S. Uchida et al [15] synthesize TiO_2-based nanotubes through TiO_2 nanoparticles and fabricated photo-anodes. M. Adachi et al [16] formed single-crystal-like anatase TiO_2 nanowires in a network structure by surfactant-assisted self-assembling processes at low temperature. The crystal lattice planes of the nanowires and networks of such wires composed of many nanoparticles were almost perfectly aligned with each other due to the "oriented attachment" mechanism, resulting in the high rate of electron transfer through the TiO_2 nano-network with single-crystal-like anatase nanowires. A single-crystalline anatase TiO_2 exposing mainly the {101} plane has been prepared, which adsorbed ruthenium dye over 4 times higher than P25, and a high light-to-electricity conversion yield of 9.3% was achieved. C. A. Grimes et al [17] described the use of highly ordered transparent TiO_2 nanotube arrays in dye-sensitized solar cells (DSSCs) with

2.9% of overall conversion efficiency. Highly ordered nanotube arrays were grown perpendicular to a fluorine-doped tin oxide-coated glass substrate by anodic oxidation of a titanium thin film. Voltage-decay measurements indicated that the highly ordered TiO_2 nanotube arrays, in comparison to nanoparticulate systems, have superior electron lifetimes and provide excellent pathways for electron percolation. Some other ordered nanomaterials used in DSSCs include nanorods [18], nanosheet [19], nanotrees [20]. In addition, spinning TiO_2 fibers was also successful prepared and fabricated cells [21, 22]. All those results showed that one-dimensional nanomaterials have excellent electron transport and separation rate, which can boost the development of DSSCs.

Effective utilization of sunlight (especially above 600 nm light) can improve the performance of DSSCs, and multiple scattering structures are effective. A. Usami [23, 24] once pointed out that submicron multiple structures were in favor of light scattering and collecting through theory simulation. Theoretical calculation showed that the introduction of scattering layer or scattering center can effectively achieve useful photoelectric conversion efficiency [4-7, 25, 26]. T. E. Mallouk et al [2, 3] reported that the light harvesting efficiency of dye-sensitized photo-anodes was enhanced by coupling a TiO2 photonic crystal layer to a conventional film composed of TiO2 nanoparticles. In addition to act as a dielectric mirror, the inverse opal photonic crystal caused a significant change in dye absorbance that depended on the position of the stop band. Absorbance was suppressed at wavelengths shorter than the maximum stop band and enhanced at longer wavelengths. This effect arises from the slow group velocity of light in the vicinity of the stop band, and the consequent localization of light intensity in the voids (to the blue) or in the dye-sensitized TiO2 (to the red) portions of the photonic crystal. By coupling photonic crystals, the short circuit photocurrent efficiency across the visible spectrum (400-750 nm) could be increased by about 26%, comparing to an ordinary dye-sensitized nanocrystalline TiO2 photo-anode.

Spherical voids as light scattering centers in nanocrystalline TiO2 films were realized by S. Hore et al [4] with polystyrene particles of diameter 400 nm through sintering. As reported, spherical voids enhanced the scattering properties of the films and hence the light capture inside the device. Thus, it increased the overall conversion efficiency of the devices by 25%. Additionally, light scattering layers [25, 26] were also used to enhance light utilization and raised the performance of cells.

In all, there are great achievements of the photo-anode as the pivotal part in DSSCs during the past a few years. As a perfect photo-anode, it must have the following features such as fast electron injection and separation, slow electron recombination, excellent electron transport, high surface area, outstanding light collecting and so on. Composite structure, ordered one-dimensional nanomaterials and multiple structures provide the opportunities to further development in DSSCs.

3. Electro-Hydrodynamics (EHD) Technique in DSSCs

Since traditional preparation methods of photo-anodes such as doctor blade method, layer-by-layer deposition method and screening print method have the disadvantages of complex operation and processing, development of novel methods for fabricating photo-anodes in DSSCs with simple, efficient, large-scale traits is pursued by many researchers. At the same time, although DSSCs based on liquid electrolytes have reached efficiency as high as 10% under AM 1.5 (1000W/m2) [27], the devices sealing and long-term stability with organic liquid electrolytes (OLE) remain an important issue in view of the practical application. Consequently viscous molten-salt/polymer electrolytes without volatility, quasi-solid-state and solid-state electrolytes [28] have been introduced to substitute the organic liquid electrolytes. However superior penetration of those electrolytes into the nanocrystalline network structure and the quick charge transport in pure nano-structured films can't be realized easily. Therefore, porous structured photo-anodes with predominant penetration effect of viscous and solid electrolytes, fast charge transport and effective light collection efficiency are studious orientation for researchers.

The electro-hydrodynamic (EHD) technique has been confirmed to be effective and simple for fabricating micro to nano-scale fibers or particles from a variety of materials, such as organic [29–31], inorganic [32–34] and hybrid organic/inorganic [35, 36] compounds. It is widely applied in membrane technology [37, 38], drug delivery [39], tissue engineering [40], optical sensors [41] and biosensors [42]. Figure 1 shows the schematic diagram of the experimental setup used for EHD processing.

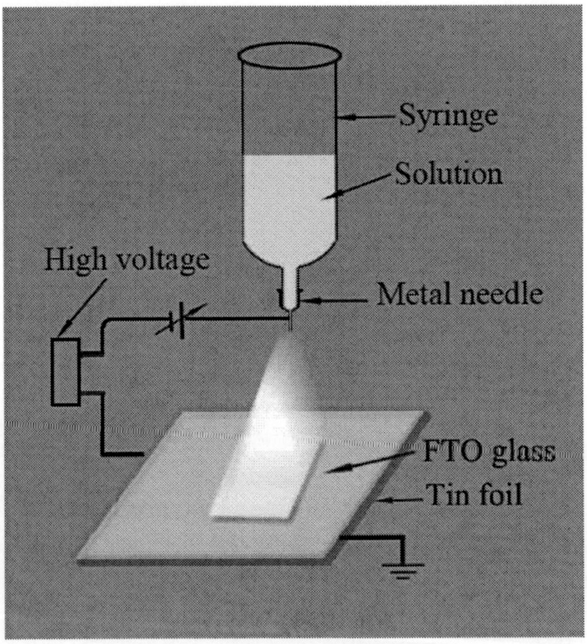

Figure 1. The schematic diagram of the experimental setup used for EHD processing.

In a typical process, the mixed solution containing the colloidal solution of nano-particles and polymer is sprayed from a nozzle under an applied electric field as strong as 1.5-3 kV/cm. The build-up of electrostatic charges on the surface of the liquid droplet induces the formation of jet, which is subsequently separated to form the mist. The solvent was removed from the solution under the influence of high voltage before depositing on the substrate. After the film was dried in the air, the calcinations were carried out to remove polymers. Hence, EHD technique can be used to realize not only large area but also continuous fabrication, which may be an alternative technique for future cell industrialization.

A one-dimensional TiO_2 nano-fibre structure prepared by electrospun and calcined was applied as an electrode in a DSSC and it facilitated the penetration of the viscous polymer gel electrolytes into the electrode. The photocurrent generation of the DSSC using the nano-fibre film as a photo-anode with a polymer gel electrolyte was over 90% of the value for an equivalent liquid DSSC. A η of 6.2% and an ff of 60% were obtained [21, 22]. As reported recently, TiO_2 photo-anode prepared by electro- spraying method had higher photocurrent than conventional photo-anode because of better ionic paths in the TiO_2 layer [43].

Very recently, EHD technique was used in our group and the photo-electric conversion behavior was studied in detail [44, 45].

Figure 2(a) and (b) show TiO2 composite electrodes prepared by EHD method, which has been confirmed to be effective and simple for fabricating micro to nano-scale fibers or particles from a variety of materials. From the SEM image, it can be seen clearly that the films was composed of the micro-scale and sub-micro-scale near spherical clusters with the diameter of 0.5-1 μm. Except for the spherical clusters, lots of nano-pores and micropores with the size of 2 nm -1 μm exist in this film. From the magnified image (Figure 2(b)), it can be seen that the macro-sphere was assembled by TiO2 nanoparticles and the size of nano-pores among the particles centralizes at ca. 10-20 nm. Micro-pores (micro-spheres) and nano-pores (nanoparticles) form a micro/nano composite structured film. Figure 2(c) illustrates the ideal sketch map of the composite electrode to show the structural specialties. Lots of submicro-pores are formed among the interspace of TiO2 submicro-spheres and hence the continuous micro-channels (red line). The inset is the magnified image. Micro-spheres formed by process of EHD method are composed of nanoparticles. Nano-pores exist among the nano-particles to form continuous nano-channels (blue line). The hierarchical branched inner channels composed of continuous micropores (red line) and nano-pores (blue line) in electrode, just like the blood circulation system of human body constructed by main vessel and capillary vessel, maybe hold nice capability of charge transport in viscous electrolytes.

Figure 2. (a) SEM images of film prepared by EHD method. (b) The magnified image of (a). (c) Simulative pore structure of electrodes with micro-channels and nano-channels.

Based on the composite hierarchical porous structure, we used the micro-nano electrode to assemble the DSSCs with viscous ionic liquid electrolytes, and also a nano-structured film was selected as a reference electrode to compare different performance of micro-nano structure and nano-structure. The analyzed data indicated that the photoelectrical parameters (fill factor and total conversion efficiency) were enhanced largely after incorporating composite hierarchical porous structure into the TiO_2 photo-anode, about increasing 10% for fill factor, 7% for open-circuit photovoltage, 6% for short-circuit photocurrent, and resulting enhancement of 20% for total conversion efficiency with viscous ionic liquid electrolytes.

The improved conversion efficiency was ascribed to enhancement of light collection efficiency and ionic diffusion rate in viscous electrolyte, which was induced by hierarchical porous channels in the micro-nano composite thin film.

The light-to-electric conversion efficiency of DSSCs is determined by light collection efficiency, capability of charge separation and charge transport. So the kinetics of ionic transportation and light scattering effect, which can be simultaneously tuned by morphology structure of the electrodes, are important parts in the system of DSSCs. They influence each other and there must be a balance for the best photovoltaic performance. If the pores are too large, despite of ions transporting easily for viscous electrolytes and possessing light-scattering effect, the thin films can't absorb enough dye molecules for light harvesting; if the pores are too small, the transportation kinetics of redox species limits the charge transfer rate and absence of light scattering effect though the amount of dye absorption is increased. Composite structure with hierarchical channels balances these two effects of ionic transportation and light scattering, and then obtains the optimal photoelectrical action. Except for TiO2 composite porous electrode with P25 and hydrothermal anatase nano-particles, we also extend the EHD technique to fabricate ZnO porous electrode, the improvement of photoelectrical performance was also observed compared to that of nano-structured ZnO electrode [46]. Based on the research of micro/nano electrodes, it can be deduced that the composite hierarchical porous structure is benefit for the high efficiency device with viscous electrolytes, such as ionic liquid and quasi-solid-state electrolyte. The idea of film structure modulation will provide a novel space for high efficiency devices.

4. Photocurrent Generation of Honeycomb Structures in Organic Thin Films

Many studies on organic solar cells have been conducted in the donor-acceptor linked molecules or hetero-structured films prepared by various methods. However, the surface structure effect has never been taken into account. Considering that light scattering can be achieved by adding pores or photonic crystals in the inorganic semiconductor films of TiO2 [2-4], which plays an important role in the efficiency of light harvesting in dye sensitized solar cells, fabrication of ordered honeycomb structure films with photo-electric functional polymers may benefit the photocurrent generation in organic thin film system accordingly. Breath figure formation is a simple and useful technique to prepare porous polymer films [47-51]. Under humid conditions, water vapors are condensed into water droplets at the solution surface during solvent evaporation, and then the water droplets are self-organized and closely packed hexagonally by the convectional flow or the capillary force generated at the solution front. After solvent and water evaporate completely, traces of water droplets remain in the polymer film to form honeycomb structures. The copolymer with the rod-coil block-poly (distyrylbenzene - co - triethylene glycol) (PDTG) was chosen and ordered honeycomb structures were introduced in the film by breath figure formation method [52].

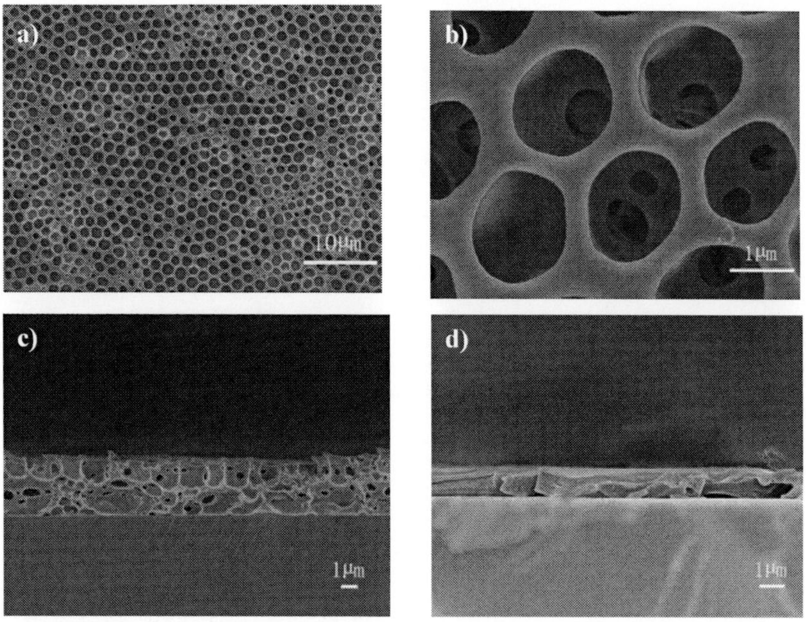

Figure 3. FE-SEM images of the as-prepared honeycomb film (a) low-magnification (b) high magnification. (c) cross-section and (d) cross-sectional FE-SEM image of the smooth honeycomb film.

Figure 3(a-c) shows the typical FE-SEM images of the as-prepared honeycomb structure films of PDTG under a humid environment (RH~70%) and (d) cross-sectional FE-SEM image of the smooth honeycomb film. It can be clearly seen that the porous films can be prepared over a large area, and uniform pores with hexagonal arrangement are formed with an average diameter of ca. 1.4 μm (Fig.3a). From the magnified image (Fig. 3b), the under layer can be observed through the pores of the top layer. The cross-sectional image (Fig. 3c) confirms that this porous film is made of double layers of empty spheres voids with the total thickness of 3.2 μm.

A steady and rapid cathodic photocurrent of 293 nA/cm2 was produced when the irradiation of the honeycomb structure film was switched on and off. The cathodic photocurrent response of its smooth film was 103 nA/cm2, which was about one third of the value obtained by the ordered porous film. It can be deduced that in the photo-electric conversion systems, the porous structures not only significantly improve light capture ability by enhance the light absorption, but also benefit the electron/hole separation and charge transfer by introducing PDTG/electrolyte interface. Therefore through the molecular design and the introducing of micro-/nano- structures on the photo-electric active surface, there will be a great prospect for improving the photo-electric conversion performance in the organic thin film system.

5. CONCLUSION

Photo-electric conversion is one of the most important issues in the field of energy conversion field. The era of nanotechnology, especially the bridging of the disparate fields, offers opportunities to challenge the conventional wisdom regarding the efficiency of alternative energy sources. According to the research achievements reviewed above, we are sure that by introducing the micro-/nano- structures on various photo-electric active surfaces, there will be a great prospect for improving the photo-electric conversion performance in the photovoltaic systems.

REFERENCES

[1] Green, M. A.; Emery, K.; King, D. L., Hishikawa, Y.; Warta, W. *Prog. Photovolt: Res. Appl. 2006*, 14, 455–461.

[2] Nishimura, S.; Abrams, N.; Lewis, B. A.; Halaoui, L. I.; Mallouk, T. E.; Benkstein, K. D.; Lagemaat, J.; Frank, A. J. *J. Am. Chem. Soc. 2003*, 125, 6306-6310.

[3] Halaoui, L. I.; Abrams, N. M.; Mallouk, T. E. *J. Phys. Chem. B 2005*, 109, 6334-6342.

[4] Hore, S.; Nitz, P.; Vetter, C.; Prahl, C.; Niggemannb, M.; Kern, R. *Chem. Commun. 2005*, 15, 2011-2013.

[5] Ferber, J.; Luther, J. Sol. Energy Mater. *Sol. Cells 1998*, 54, 265-275.

[6] Rothenberger, G.; Comte, P.; Grätzel, M. Sol. Energy Mater. *Sol. Cells 1999*, 58, 321-336.

[7] Wang, Z.-S.; Kawauchi, H.; Kashima, T.; Arakawa, H. *Coordinat. Chem. Rev. 2004*, 248, 1381-1389.

[8] Mobius, D. *Acc. Chem. Res. 1981*, 14, 63– 68.

[9] Souza, F. D.; Deviprasad, G. R.; Rahman, M. S.; Choi, J. P. *Inorg. Chem. 1999*, 38, 2157-2160.

[10] Tkachenko, N. V.; Rantala, L.; Tauber, A. Y.; Helaja, J. P.; Hynninen, H.; Lemmetyinen, H. *J. Am. Chem. Soc. 1999*, 121, 9378-9387.

[11] Ikeda, A.; Hatano, T.; Shinkai, S.; Akiyama, T.; Yamada, S. *J. Am. Chem. Soc. 2001*, 123, 4855-4856.

[12] Fushimi, T.; Oda, A.; Ohkita, H.; Ito, S. *Langmuir 2005*, 21, 1584-1589.

[13] Huynh, W. U.; Dittmer J. J.; Alivisatos, A. P. *Science 2002*, 295, 2425-2427.

[14] Law, M.; Greene, L. E.; Johnson, J. C.; Saykally, R.; Yang, P. D. *Nat. Mater. 2005*, 4, 455-459.

[15] Uchida, S.; Chiba, R.; Tomiha, M.; Masaki, N.; Shirai, M. *Electrochemistry 2002*, 70, 418-420.

[16] Adachi, M.; Murata, Y.; Okada, I.; Yoshikawa, S. *J. Electrochem.Soc. 2003*, 150, G488-G493.

[17] Mor, G. K.; Shankar, K.; Paulose, M.; Varghese, O. K.; Grimes, C. A. *Nano. Lett. 2006*, 6, 215-218.

[18] Guo, M.; Diao, P.; Wang, X.; D. Cai, S. M. *J. Solid State Chem. 2005*, 178, 3210 -3215.

[19] Hosono, E.; Fujihara, S.; Honma, I.; Zhou, H. S. *Adv. Mater. 2005*, 17, 2091 -2094.

[20] Baxter, J. B.; Aydil, E. S. *Appl. Phys. Lett. 2005*, 86, 053114

[21] Song, M. Y.; Kim, D. K.; Ihn, K. J.; Jo, S. M.; Kim, D. Y. *Nanotechnology 2004*, 15,1861-1865.

[22] Song, M. Y.; Ahn, Y. R.; Jo, S. M.; Kim, D. Y. *Appl. Phys. Lett. 2005*, 87, 113113.

[23] Usami, A. Sol. Energy Mater. *Sol. Cells 1999*, 59, 163-166.

[24] Usami, A. Sol. Energy Mater. *Sol. Cells 2000*, 62, 239-246.

[25] Nazeeruddin, M. K.; Kay, A; Rodicio, I; Humphry-Baker, R.; Müller, E.; Liska, P.; Vlachopoulos, N.; Grätzel, M. *J. Am. Chem. Soc. 1993*, 115, 6328-6390.

[26] Hore, S.; Vetter, C.; Kern, R.; Smit, H.; Hinsch, A. Sol. Energy Mater. *Sol. Cells 2006*, 90, 1176-1188.

[27] Grätzel, M. J. Photochem. *Photobiol. A-Chem. 2004*, 164, 3-14.

[28] Li, B.; Wang, L. D.; Kang, B. N.; Wang, P.; Qiu, Y. Sol. Energy Mater. *Sol. Cells 2006*, 90, 549-573.

[29] Madhugiri, S.; Dalton, A.; Gutierrez, J.; Ferraris, J. P.; Balkus, K. J. *J. Am. Chem. Soc. 2003*, 125, 14531-14538.

[30] Bognitzki, M.; Czado, W.; Frese, T.; Schaper, A.; Hellwig, M.; Steinhart, M.; Greiner, A.; Wendorff, J. H. *Adv. Mater. 2001*, 13, 70-72.

[31] Jiang, L.; Zhao, Y.; Zhai, J. *Angew. Chem. Int. Ed. 2004*, 43, 4338-4341.

[32] Ko, F.; Gogotsi, Y.; Ali, A.; Naguib, N.; Ye, H.; Yang, G.; Li, C.; Wills, P. *Adv. Mater. 2003*, 15, 1161-1165.

[33] Li, D.; Xia, Y. *Nano Lett. 2003*, 3, 555-560.

[34] Viswanathamurthi, P.; Bhattarai, N.; Kim, C. K.; Kim, H. Y.; Lee, D. R. *Inorg. Chem.* Commun. *2004*, 7, 679-682.

[35] Caruso, R. A.; Schattka, J. H.; Greiner, A. *Adv. Mater. 2001*, 13, 1577-1579.

[36] Demir, M. M.; Gulgun, M. A.; Menceloglu, Y. Z.; Erman, B.; Abramchuk, S. S.; Makhaeva, E. E.; Khokhlov, A. R.; Matveeva, V. G.; Sulman, M. G. *Macromolecules 2004*, 37, 1787-1792.

[37] Chen, C. H.; Emond, M. H. J.; Kelder, E. M.; Meester, B. Schoonman, J. *J. Aerosol Sci. 1999*, 30, 959-967.

[38] Benítez, R.; Soler, J.; Daza, L. J. *Power Sources 2005*, 151, 108-113.

[39] Kenawy, E. R.; Bowlin, G. L.; Mansfield, K.; Layman, J.; Simpson, D. G.; Sanders, E. H.; Wnek, G. E. *J. Control. Release. 2002*, 81, 57-64.

[40] Wnek, G. E.; Carr, M. E.; Simpson, D. G.; Bowlin, G. L. *Nano Lett. 2003*, 3, 213-216.

[41] Wang, X. Y.; Drew, C.; Lee, S. H.; Senecal, K. J.; Kumar, J.; Sarnuelson, L. A. *Nano Lett. 2002*, 2, 1273-1275.

[42] Wang, X. Y.; Drew, C.; Lee, S. H.; Senecal, K. J.; Kumar, J.; Sarnuelson, L. A. Nano Lett. *2004*, 4, 331-334.

[43] Fujimoto, M.; Kado, T.; Takashima, W.; Kaneto, K.; Hayase, S. *J. Electrochem. Soc. 2006*, 153, A826-A829.

[44] Zhao, Y.; Zhai, J.; Tan, S. X.; Wang, L.F.; Jiang, L.; Zhu, D. B. *Nanotechnology 2006*, 17, 2090-2097.

[45] Zhao, Y.; Sheng, X. L.; Zhai, J. Jiang, L. Yang, C. H.; Sun, Z. W.; Li, Y. F.; Zhu, D. B. *ChemPhysChem 2007*, 8, 856-861.

[46] Sheng, X. L.; Zhao, Y.; Zhai, J.; Jiang, L.; Zhu, D. B. *Appl. Phys.* A *2007*, 87(4), 715-719

[47] Widawski, G.; Rawiso, M.; Franois, B. *Nature 1994*, 369, 387 –389.

[48] Francois,B.; Pitois, O.; Francois, *J. Adv. Mater. 1995*, 7, 1041 –1044.

[49] Jenekhe, S.; Chen, X. *Science 1999*, 283, 372–375.

[50] Stenzel, M. H. *Aust. J. Chem. 2002*, 55, 239– 243.

[51] Karthaus, O.; Maruyama, N.; Cieren, X.; Shimomura, M.; Hasegawa, H.; Hashimoto, T. *Langmuir 2000*, 16, 6071 –6076.

[52] Heng, L.; Zhai, J.; Zhao, Y.; Xu, J.; Sheng, X.; Jiang, L. *ChemPhysChem 2006*, 7, 2520-2525.

INDEX

A

B

C

D

E

F

G

H

I

J

K

L

M

N

O

P

Q

R

S

T

U

V

W

X

Y

Z